Collected Papers

Volume II

Springer
*New York
Berlin
Heidelberg
Barcelona
Hong Kong
London
Milan
Paris
Singapore
Tokyo*

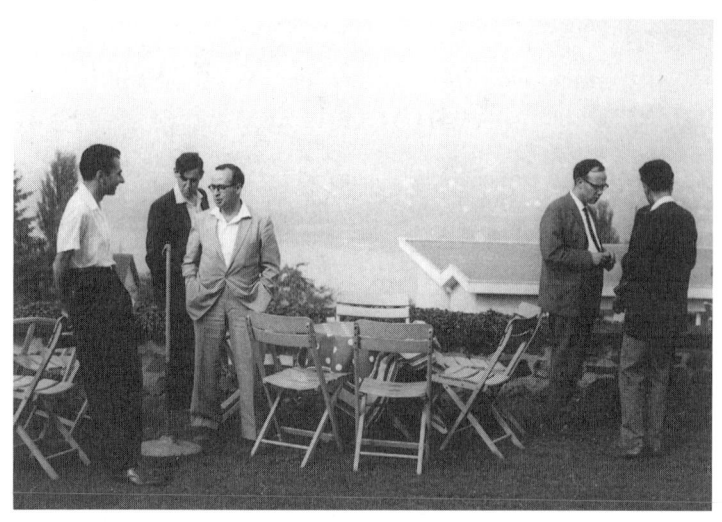

Arbeitstagung 1962

Arbeitstagung 1963

Serge Lang

Collected Papers

Volume II

1971–1977

Springer

Serge Lang
Department of Mathematics
Yale University
10 Hillhouse Avenue
PO Box 208283
New Haven, CT 06520-8283
USA

Mathematics Subject Classification (1991): 11Gxx, 14Kxx, 11Fxx, 11Jxx. 14H

Library of Congress Cataloging-in-Publication Data
Lang, Serge, 1927–
    Collected papers / Serge Lang.
      p.   cm.
    Includes bibliographical references.
    Contents: v. 1. 1952–1970 — v. 2. 1971–1977.
    ISBN 0-387-98802-5 (v. 1 : hardcover : alk. paper) — ISBN 0-387-98803-3
(v. 2 : hardcover : alk. paper)
    1. Mathematics. I. Title.
QA7.L288   2000
510—dc21   99-017359

Printed on acid-free paper.

© 2000 Springer-Verlag New York, Inc.
All rights reserved. This work may not be translated or copied in whole or in part without the written permission of the publisher (Springer-Verlag New York, Inc., 175 Fifth Avenue, New York, NY 10010, USA), except for brief excerpts in connection with reviews or scholarly analysis. Use in connection with any form of information storage and retrieval, electronic adaptation, computer software, or by similar or dissimilar methodology now known or hereafter developed is forbidden. The use of general descriptive names, trade names, trademarks, etc., in this publication, even if the former are not especially identified, is not to be taken as a sign that such names, as understood by the Trade Marks and Merchandise Marks Act, may accordingly be used freely by anyone.

Production managed by Michael Koy; manufacturing supervised by Joe Quatela.
Photocomposed copy prepared from the author's original manuscripts.
Printed and bound by Sheridan Books, Inc., Ann Arbor, MI.
Printed in the United States of America.

9 8 7 6 5 4 3 2 1

ISBN 0-387-98803-3 Springer-Verlag New York Berlin Heidelberg   SPIN 10715932
ISBN 0-387-91591-5 (Volumes I–V)   SPIN 10738493

Contents

Bibliography (through 1999)   vii

[1971a] Transcendental Numbers and Diophantine Approximations 1

[1971b] On the Zeta Function of Number Fields 44

[1971c] The Group of Automorphisms of the Modular Function Field 53

[1972a] Isogenous Generic Elliptic Curves 55

[1972b] Continued Fractions for Some Algebraic Numbers
(with H. Trotter) 69

[1973] Frobenius Automorphisms of Modular Function Fields 78

[1974a] Higher Dimensional Diophantine Problems 102

[1974b] Addendum to "Continued Fractions of Some Algebraic Numbers" (with H. Trotter) 111

[1975a] Diophantine Approximations on Abelian Varieties with Complex Multiplication 120

[1975b] Division Points of Elliptic Curves and Abelian Functions over Number Fields 122

[1975c] Units in the Modular Functions Field I (with D. Kubert) 178

[1975d] Units in the Modular Functions Field II (with D. Kubert) 208

[1975e] Units in the Modular Functions Field III (with D. Kubert) 223

[1976a] Diophantine Approximations on Abelian Varieties with
Complex Multiplication (with J. Coates) 236

[1976b] Distributions on Toroidal Groups (with D. Kubert) 241

[1976c] Units in the Modular Function Field (with D. Kubert) 260

[1976d] *Frobenius Distributions in GL_2-Extensions* (with H. Trotter) 289

[1977a] Units in the Modular Field IV (with D. Kubert) 567

[1977b] Primitive Points on Elliptic Curves (with H. Trotter) 587

Bibliography (through 1999)

(Boldface items are books or Lecture Notes.)

[1952a] On quasi algebraic closure, *Ann. of Math.* **55** No. 2 (1952) pp. 373–390.

[1952b] Hilbert's nullstellensatz in infinite dimensional space, *Proc. AMS* **3** No. 3 (1952) pp. 407–410.

[1952c] (with J. TATE) On Chevalley's proof of Luroth's theorem, *Proc. AMS* **3** No. 4 (1952) pp. 621–624.

[1953] The theory of real places, *Ann. of Math.* **57** No. 2 (1953) pp. 378–391.

[1954a] Some applications of the local uniformization theorem, *Am. J. Math.* **76** No. 2 (1954) pp. 362–374.

[1954b] (with A. WEIL) Number of points of varieties in finite fields, *Am. J. Math.* **76** No. 4 (1954) pp. 819–827.

[1955] Abelian varieties over finite fields, *Proc. NAS* **41** No. 3 (1955) pp. 174–176.

[1956a] Unramified class field theory over function fields in several variables, *Ann. of Math.* **64** No. 2 (1956) pp. 285–325.

[1956b] On the Lefschetz principle, *Ann. of Math.* **64** No. 2 (1956) pp. 326–327.

[1956c] L-series of a covering, *Proc. NAS* **42** No. 7 (1956) pp. 422–424.

[1956d] Sur les séries L d'une variété algébrique, *Bull. Soc. Math. France* **84** (1956) pp. 385–407.

[1956e] Algebraic groups over finite fields, *Am. J. Math.* **78** (1956) pp. 555–563.

[1957a] (with J.-P. SERRE) Sur les revêtements non ramifiés des variétés algébriques, *Am. J. Math.* **79**, No. 2 (1957) pp. 319–330.

[1957b] (with W.-L. CHOW) On the birational equivalence of curves under specialization, *Am. J. Math.* **79**, No. 3 (1957) pp. 649–652.

[1957c] Divisors and endomorphisms on an abelian variety, *Am. J. Math.* **79** No. 4 (1957) pp. 761–777.

[1957d] Families algébriques de Jacobiennes (d'après IGUSA), *Séminaire Bourbaki* No. 155, 1957/1958.

[1958a] Reciprocity and correspondences, *Am. J. Math.* **80** No. 2 (1957) pp. 431–440.

[1958b] (with J. TATE) Principal homogeneous spaces over abelian varieties, *Am. J. Math.* **80** No. 3 (1958) pp. 659–684.

[1958c] (with E. KOLCHIN), Algebraic groups and the Galois theory of differential fields, *Am. J. Math.* **80** No. 1 (1958) pp. 103–110.

[1958d] *Introduction to algebraic geometry*, Wiley-Interscience, 1958.

[1959a] (with A. NÉRON) Rational points of abelian varieties over function fields, *Am. J. Math.* **81** No. 1 (1959) pp. 95–118.

[1959b] Le théorème d'irreductibilité de Hilbert, *Séminaire Bourbaki* No. 201, 1959/1960.

[1959c] *Abelian varieties*, Wiley-Interscience, 1959; Springer-Verlag, 1983.

[1960a] (with E. KOLCHIN) Existence of invariant bases, *Proc. AMS* **11** No. 1 (1960) pp. 140–148.

[1960b] Integral points on curves, *Pub. IHES* No. 6 (1960) pp. 27–43.

[1960c] Some theorems and conjectures in diophantine equations, *Bull. AMS* **66** No. 4 (1960) pp. 240–249.

[1960d] On a theorem of Mahler, *Mathematika* **7** (1960) pp. 139–140.

[1960e] L'équivalence homotopique tangentielle (d'apres MAZUR), *Séminaire Bourbaki* No. 222, 1960/1961.

[1961] Review: Elements de géométrie algébrique (A. Grothendieck). *Bull. AMS* **67** No. 3 (1961) pp. 239–246.

[1962a] A transcendence measure for E-functions, *Mathematika* **9** (1962) pp. 157–161.

[1962b] Transcendental points on group varieties, *Topology* **1** (1962) pp. 313–318.

[1962c] Fonctions implicities et plongements Riemanniens, *Séminaire Bourbaki* 1961/1962, No. 237, May 1962.

[1962d] *Introduction to Differential Manifolds*, Addison Wesley, 1962.

[1962e] *Diophantine Geometry*, Wiley-Interscience, 1962.

[1963] *Transzendente Zahlen*, Bonn Math. Schr. No. 21 (1963).

[1964a] Diophantine approximations on toruses, *Am. J. Math.* **86** No. 3 (1964) pp. 521–533.

[1964b] Les formes bilinéaires de Néron et Tate, *Séminaire Bourbaki* 1963/64 Fasc. 3 Exposé 274, Paris 1964.

[1964c] *First Course in Calculus*, Addison Wesley 1964; Fifth edition by Springer-Verlag, 1986.

[1964d] *Algebraic and Abelian Functions*, W.A. Benjamin Lecture Notes, 1964. See also [1982c].

[1964e] *Algebraic Numbers*, Addison Wesley, 1964; superceded by [1970d].

[1965a] Report on diophantine approximations, *Bulletin Soc. Math. France* **93** (1965) pp. 177–192.

[1965b] Division points on curves, *Annali Mat. pura ed applicata*, Serie IV **70** (1965) pp. 229–234.

[1965c] Algebraic values of meromorphic functions, *Topology* **3** (1965) pp. 183–191.

[1965d] Asymptotic approximations to quadratic irrationalities I, *Am. J. Math.* **87** No. 2 (1965) pp. 481–487.

[1965e] Asymptotic approximations to quadratic irrationalities II, *Am. J. Math.* **87** No. 2 (1965) pp. 488–496.

[1965f] (with W. ADAMS) Some computations in diophantine approximations, *J. reine angew. Math.* Band **220** Heft 3/4 (1965) pp. 163–173.

[1965g] Corps de fonctions méromorphes sur une surface de Riemann (d'après ISS'SA), *Séminaire Bourbaki* No. 292, 1964/65.

[1965h] *Algebra*, Addison Wesley, 1965; second edition 1984; third edition 1993.

[1966a] Algebraic values of meromorphic functions II, *Topology* **5** (1960) pp. 363–370.

[1966b] Asymptotic diophantine approximations, *Proc. NAS* **55** No. 1 (1966) pp. 31–34.

[1966c] *Introduction to transcendental numbers*, Addison Wesley, 1966.

[1966d] *Introduction to Diophantine Approximations*, Addison Wesley 1966; see [1995d].

[1966e] *Rapport sur la cohomologie des groupes*. Benjamin, 1966.

[1967] *Algebraic structures*, Addison Wesley 1967.

[1968] *Analysis I*, Addison Wesley, 1968; (superceded by [1983c]).

[1969] *Analysis II*, Addison Wesley, 1969; (superceded by [1993c]).

[1970a] (with E. BOMBIERI) Analytic subgroups of group varieties, *Inventiones Math.* **11** (1970) pp. 1–14.

[1970b] Review: L.J. Mordell's *Diophantine Equations*, *Bull. AMS* **76** (1970) pp. 1230–1234.

[1970c] *Introduction to Linear Algebra*, Addison Wesley 1970; see also [1986b].

[1970d] *Algebraic Number Theory*, Addison Wesley 1970; see also [1994c].

[1971a] Transcendental numbers and diophantine approximations, *Bull. AMS* **77** No. 5 (1971) pp. 635–677.

[1971b] On the zeta function of number fields, *Invent. Math.* **12** (1971) pp. 337–345.

[1971c] The group of automorphisms of the modular function field, *Invent. Math.* **14** (1971) pp. 253–254.

[1971d] *Linear Algebra*, Addison Wesley 1971; see also [1987b].

[1971e] *Basic Mathematics*, Addison Wesley 1971; Springer-Verlag 1988.

[1972a] Isogenous generic elliptic curves, *Amer. J. Math.* **94** (1972) pp. 661–674.

[1972b]   (with H. TROTTER) Continued fractions for some algebraic numbers, *J. reine angew. Math.* **255** (1972) pp. 112–134.

[1972c]   *Differential manifolds*, Addison Wesley, 1972.

[1972d]   *Introduction to Algebraic and Abelian Functions*, Benjamin-Addison Wesley, 1972; second edition see [1982c].

[1973a]   Frobenius automorphisms of modular function fields, *Amer. J. Math.* **95** (1973) pp. 165–173.

[1973b]   *Calculus of Several Variables*, Addison Wesley 1973; Third edition see [1987d].

[1973c]   *Elliptic functions*, Addison Wesley 1973; second edition see [1987d].

[1974a]   Higher dimensional diophantine problems, *Bull. AMS* **80** No. 5 (1974) pp. 779–787.

[1974b]   (with H. TROTTER) Addendum to "Continued fractions of some algebraic numbers," *J. reine angew. Math.* **267** (1974) pp. 219–220.

[1975a]   Diophantine approximations on abelian varieties with complex multiplication, *Advances Math.* **17** (1975) pp. 281–336.

[1975b]   Division points of elliptic curves and abelian functions over number fields, *Amer. J. Math.* **97** No. 1 (1975) pp. 124–132.

[1975c]   (with D. KUBERT) Units in the modular function field I, Diophantine Applications, *Math. Ann.* **218** (1975) pp. 67–96.

[1975d]   (with D. KUBERT) Units in the modular function field II, A full set of units, *Math. Ann.* **218** (1975) pp. 175–189.

[1975e]   (with D. KUBERT) Units in the modular function field III, Distribution relations, *Math. Ann.* **218** (1975) pp. 273–285.

[1975f]   La conjecture de Catalan d'après Tijdeman, *Séminaire Bourbaki* 1975/76 No. 29.

[1975g/85] $SL_2(\mathbf{R})$, Addison Wesley, 1975; Springer-Verlag corrected second printing, 1985.

[1976a]   (with J. COATES) Diophantine approximation on abelian varieties with complex multiplication, *Invent. Math.* **34** (1976) pp. 129–133.

[1976b]   (with D. KUBERT) Distribution on toroidal groups, *Math. Z.* **148** (1976) pp. 33–51.

[1976c]   (with D. KUBERT) Units in the modular function field, in *Modular Functions in One Variable* V, Springer Lecture Notes **601** (Bonn Conference) 1976, pp. 247–275.

[1976d]   (with H. TROTTER) *Frobenius distributions in GL_2-extensions*, Springer Lecture Notes **504**, Springer-Verlag 1976.

[1976e]   *Introduction to Modular Forms*, Springer-Verlag, 1976.

[1977a]   (with D. KUBERT) Units in the modular function field IV, The Siegel functions are generators, *Math. Ann.* **227** (1997) pp. 223–242.

[1977b]   (with H. TROTTER) Primitive points on elliptic curves, *Bull. AMS* **83** No. 2 (1977) pp. 289–292.

[1977c]   *Complex Analysis*, Addison Wesley; second Edition Springer-Verlag 1985; fourth edition Springer-Verlag 1999.

[1978a]   (with D. KUBERT) The p-primary component of the cuspidal divisor class group of the modular curve $X(p)$, *Math. Ann.* **234** (1978) pp. 25–44.

[1978b]   (with D. KUBERT) Units in the modular function field V, Iwasawa theory in the modular tower, *Math. Ann.* **237** (1978) pp. 97–104.

[1978c]   (with D. KUBERT) Strickelberger ideals, *Math. Ann.* **237** (1978) pp. 203–212.

[1978d]   (with D. KUBERT) The index of Strickelberger ideals of order 2 and cuspidal class numbers, *Math. Ann.* **237** (1978) pp. 213–232.

[1978e]   Relations de distributions et exemples classiques, *Séminaire Delange-Pisot-Poitou (Théorie des Nombres)*, 1978 No. 40 (6 pages).

[1978f]   *Elliptic curves: Diophantine Analysis*, Springer-Verlag 1978.

[1978g]   *Cyclotomic Fields* I, Springer-Verlag, 1978.

[1979a]   (with D. KUBERT) Cartan-Bernoulli numbers as values of L-series, *Math. Ann.* **240** (1979) pp. 21–26.

[1979b]   (with D. KUBERT) Independence of modular units on Tate curves, *Math. Ann.* **240** (1979) pp. 191–201.

[1979c]   (with D. KUBERT) Modular units inside cyclotomic units, *Bull. Soc. Math. France* **107** (1979) pp. 161–178.

[1980]   *Cyclotomic Fields* II, Springer-Verlag 1980.

[1981a] (with N. KATZ) Finiteness theorems in geometric classfield theory, *Enseignement mathématique* **27** (3–4) (1981) pp. 285–314.

[1981b] (with DAN KUBERT) *Modular Units*, Springer-Verlag 1981.

[1982a] Représentations localement algébriques dans les corps cyclotomiques, Séminaire de Théorie des Nombres 1982, Birkhauser, pp. 125–136.

[1982b] Units and class groups in number theory and algebraic geometry, *Bull. AMS* **6** No. 3 (1982) pp. 253–316.

[1982c] *Introduction to algebraic and abelian functions, Second Edition*, Springer-Verlag, 1982.

[1983a] Conjectured diophantine estimates on elliptic curves, in Volume I of *Arithmetic and Geometry*, dedicated to Shafarevich, M. Artin and J. Tate editors, Birkhauser (1983) pp. 155–171.

[1983b] *Fundamentals of Diophantine Geometry*, Springer-Verlag 1983.

[1983c] *Undergraduate Analysis*, Springer-Verlag, 1983.

[1983d] (with GENE MURROW) *Geometry: A High School Course*, Springer Verlag, 1983 Second edition 1988.

[1983e] *Complex Multiplication*, Springer-Verlag 1983.

[1984a] Vojta's conjecture, *Arbeitstagung Bonn 1984*, Springer Lecture Notes **1111** 1985, pp. 407–419.

[1984b] Variétés hyperboliques et analyse diophantienne, *Séminaire de théorie des nombres*, 1984/85, pp. 177–186.

[1985a] (with W. FULTON) *Riemann-Roch Algebra*, Springer-Verlag, 1985.

[1985b] *The Beauty of Doing Mathematics*, Springer-Verlag, 1985 originally published as articles in the *Revue du Palais de la Découverte*, Paris, 1982–1984, specifically:
   Une activité vivante: faire des mathématiques, *Rev. P.D.* Vol. **11** No. **104** (1983) pp. 27–62
   Que fait un mathématicien pure et pourquoi?, *Rev. P.D.* Vol. **10** No. **94** (1982) pp. 19–44
   Faire des Maths: grands problèmes de géométrie et de l'espace, *Rev. P.D.* Vol. **12** No. **114** (1984) pp. 21–72.

[1985c] *Math! Encounters with High School Students*, Springer-Verlag, 1985 (French edition *Serge Lang, des Jeunes et des Maths*, Berlin, 1984).

[1986a] Hyperbolic and diophantine analysis, *Bulletin AMS* **14** No. 2 (1986) pp. 159–205.

[1986b] *Introduction to Linear Algebra*, Second Edition, Springer-Verlag 1986.

[1987a] Diophantine problems in complex hyperbolic analysis, *Contemporary Mathematics AMS* **67** (1987) pp. 229–246.

[1987b] *Linear Algebra*, Third Edition, Springer-Verlag 1987.

[1987c] *Undergraduate Algebra*, Springer-Verlag 1987.

[1987d] *Elliptic functions*, Second Edition, Springer-Verlag 1987.

[1987e] *Introduction to complex hyperbolic spaces*, Springer-Verlag 1987.

[1988a] The error term in Nevanlinna theory, *Duke Math. J.* **56** No. 1 (1988) pp. 193–218.

[1988b] *Introduction to Arakelov Theory*, Springer-Verlag 1988.

[1990a] The error term in Nevanlinna Theory II, *Bull. AMS* **22** No. 1 (1990) pp. 115–125.

[1990b] Old and new conjectured diophantine inequalities. *Bull. AMS* **23** No. 1 (1990) pp. 37–75.

[1990c] *Lectures on Nevanlinna theory*, in *Topics in Nevanlinna Theory*, Springer Lecture Notes **1433** (1990) pp. 1–107.

[1990d] *Cyclotomic Fields I and II*, combined edition with an appendix by Karl Rubin, Springer-Verlag, 1990.

[1991] *Number Theory III*, Survey of Diophantine Geometry, Encyclopedia of Mathematical Sciences, Springer-Verlag 1991.

[1993a] (with J. JORGENSON) On Cramér's theorem for general Euler products with functional equations, *Math. Ann.* **297** (1993) pp. 383–416.

[1993b] (with J. JORGENSON) *Basic analysis of regularized series and products*, Springer Lecture Notes **1564** (1993).

[1993c] *Real and Functional Analysis*, Springer-Verlag, 1993.

[1993d] *Algebra, Third Edition*, Addison Wesley 1993.

[1994a] (with J. JORGENSON) Artin formalism and heat kernels, *J. reine angew. Math.* **447** (1994) pp. 165–200.

Bibliography

[1994b] (with J. JORGENSON) *Explicit Formulas for regularized products and series*, in Springer Lecture Notes **1593** pp. 1–134.

[1994c] *Algebraic Number Theory, Second Edition*, Springer-Verlag 1994.

[1995a] Mordell's review, Siegel's letter to Mordell, diophantine geometry, and 20th century mathematics, *Notices AMS* March 1995 pp. 339–350.

[1995b] Some history of the Shimura-Taniyama conjecture, *Notices AMS* November 1995 pp. 1301–1307.

[1995c] *Differential and Riemannian Manifolds*, Springer-Verlag 1995.

[1995d] *Introduction to Diophantine Approximations, new expanded edition*, Springer-Verlag 1995.

[1996a] La conjecture de Bateman-Horn, *Gazette des mathématiciens* January 1996 No. 67 pp. 82–84.

[1996b] Comments on Chow's works, *Notices AMS* **43** (1996) No. 10 pp. 1117–1124.

[1996c] (with J. JORGENSON) Extension of analytic number theory and the theory of regularized harmonic series from Dirichlet series to Bessel series, *Math. Ann.* **306** (1996) pp. 75–124.

[1996d] *Topics in Cohomology of groups*, Springer-Verlag, Springer Lecture Notes **1625**, 1996 (English translation and expansion of *Rapport sur la Cohomologie des Groupes*, Benjamin, 1966).

[1997] *Survey of Diophantine Geometry*, Springer-Verlag 1997 (same as *Number Theory III*, with corrections and additions).

[1998] The Kirschner article and HIV: Scientific and journalistic (ir)responsibilities.

[1999a] (with J. JORGENSON) Hilbert-Asai Eisenstein series, regularized products and heat kernels, *Nagoya Math. J.* **153** (1999) pp. 155–188.

[1999b] Response to the Steele Prize, *Notices AMS* **46** No. 4, April 1999 p. 458.

[1999c] *Fundamentals of Differential Geometry*, Springer-Verlag 1999.

[1999d] *Complex analysis*, fourth edition, Springer-Verlag 1999.

[1999e] *Math Talks for Undergraduates*, Springer-Verlag 1999.

The Zurich Lectures

I was invited by Wustholz for a decade to give talks to students in Zurich. I express here my appreciation, also to Urs Stammbach for his translations, and for his efforts in producing and publishing the articles in *Elemente der Mathematik*.

Primzahlen, *Elem. Math.* **47** (1992) pp. 49–61
Die abc-vermutung, *Elem. Math.* **48** (1993) pp. 89–99
Approximationssätze der Analysis, *Elem. Math.* **49** (1994) pp. 92–103
Die Wärmeleitung auf dem Kreis und Thetafunktionen, *Elem. Math.* **51** (1996) pp. 17–27
Globaler Integration lokal integrierbarer Vektorfelder, *Elem. Math.* **52** (1997) pp. 1–11
Bruhat-Tits-Raüme, *Elem. Math.* **54** (1999) pp. 45–63

Articles on Scientific Responsibility

Circular A-21. A history of bureaucratic encroachment, J. Society of Research Administrators, 1984
Questions de responsabilité dans le journalisme scientifique, *Revue du Palais de la Découverte* Paris February 1991 pp. 17–46
Questions of scientific responsibility: The Baltimore case, *J. Ethics and Behavior* **3(1)** (1993) pp. 3–72
The Kirschner article and HIV: Scientific and journalistic (ir)responsibilities, Refused publication by the *Notices AMS*, dated 5 January 1998

Books on Scientific Responsibility

The Scheer Campaign, W.A. Benjamin, 1966
The File, Springer-Verlag, 1981
Challenges, Springer-Verlag, 1998

TRANSCENDENTAL NUMBERS AND DIOPHANTINE APPROXIMATIONS[1]

BY SERGE LANG

ABSTRACT. This is a survey article on the state of knowledge concerning the transcendence and algebraic independence of various numbers, obtained as values of certain classical functions, mostly of exponential and logarithmic type. The diophantine approximation considerations are taken from the point of view of quantitative results concerning the above numbers, i.e. give an explicit lower bound for values $P|\alpha_1, \cdots, \alpha_n|$, where P is a polynomial with integer coefficients, and $\alpha_1, \cdots, \alpha_n$ are the numbers under consideration. The lower bound should depend on the degree of P, the size of its coefficients, and the numbers α_i. Some discussion is given as to what "best possible" such lower bounds have been or could be obtained.

Let f be a classical function (for instance exponential, elliptic, zeta, etc.). Starting with the rational numbers, one can construct a field inductively by adjoining values of f with arguments in the field already obtained, taking algebraic closure, and iterating these operations (as already suggested in [48]). We may call the numbers so obtained the "classical" numbers. Our point of view is that the theory of transcendental numbers determines which of the numbers so obtained are transcendental (over the rational numbers Q). This is the qualitative theory. Given numbers $w_1, \cdots, w_n$ in this field, and a nonzero polynomial $F(X_1, \cdots, X_m)$ with integer coefficients, one then wants to give a lower bound for the absolute value

$$|F(w_1, \cdots, w_n)|,$$

as a function of the degree of F, the absolute value of its coefficients and of course the w_i. This is the quantitative theory, and we view diophantine approximations from this point of view in the present survey. (If $w_1, \cdots, w_n$ have already been proved to be algebraically independent then $F(w_1, \cdots, w_n) \neq 0$. Otherwise, one has to assume this latter condition.) Because of the present point of view, I do not discuss the full range of the theory of diophantine approximations

---

AMS 1970 *subject classifications.* Primary 10F35, 10F40; Secondary 10F45, 33A10, 33A35, 32A20, 14L10.

Key words and phrases. Transcendental numbers, diophantine approximations, exponentials, logarithms, several complex variables.

[1] A survey article printed by invitation of the editors; received by the editors March 12, 1971.

Copyright © American Mathematical Society 1971

and a large body of results are omitted which would otherwise find their place.

The number of monographs on these subjects is still small. We refer the reader generally to Siegel [87], Gelfond [44], Schneider [80], and Lang [49] for transcendental numbers; and to Cassels [24], Khintchine [46], Schmidt [77] and Lang [50], for diophantine approximations. More precise references are given in the course of the report.

The theory of transcendental numbers offers ground for research over a very broad spectrum of tastes. One can work on the ground floor of mathematics, with very little knowledge, and still prove very deep results, or one may wish to work in the general context of the parametrization of algebraic varieties by uniformizing maps, and formulate or prove transcendence results for such objects. I have included a report of both types of results. The reader can always disregard those aspects which he may dislike. On the other hand, solutions of problems which have simple statements (e.g. Siegel's theorem on the finiteness of integral points on curves of genus ≥ 1) require machinery for their proofs which is fairly elaborate, so that such proofs are out of the reach of those who dislike, say, abelian or automorphic functions. Furthermore, the present relations with the theory of several complex variables are clearly becoming very fruitful, and may draw the analysts into number theory, or vice versa.

The present report attempts to cast results in fairly comprehensive terms. Especially in diophantine approximations, where results are less extensive at present than comparable results in transcendental numbers, I have attempted to tie together what I regard as very partial results by making conjectures. We are dealing with a branch of mathematics where practically any example is already a theorem.

TRANSCENDENTAL NUMBERS

1. The ordinary exponential function. The theory of transcendental numbers started with Hermite's proof [45] that

(1.1) *e is transcendental.*

A few years afterwards, this result was extended by Lindemann [56] who showed that:

(1.2) *If α is algebraic $\neq 0$, then e^α is transcendental. In particular, π is transcendental (because $e^{2\pi i} = 1$).*

Lindemann proved much more, namely:

(1.3) *If $\alpha_1, \cdots, \alpha_n$ are algebraic numbers linearly independent over the rationals, then*

$$e^{\alpha_1}, \cdots, e^{\alpha_n}$$

are algebraically independent.

(By algebraically independent, we always mean over the rationals, unless otherwise specified.)

These results were to be extended in two directions, stemming from the differential equation satisfied by e^z, or from its addition theorem. They fit in the general context of determining conditions under which classical functions, suitably normalized, take on transcendental values at algebraic points. More generally, under suitable conditions one expects that algebraically independent functions take on algebraically independent values at a certain point, unless there is a "structural" reason for it being otherwise. We shall see later concrete examples of this. In the case of the Lindemann theorem, the functions are $e^{\alpha_i t}$, $i=1, \cdots, n$, with algebraic α_i, linearly independent over the rational numbers Q. However, proofs of algebraic independence results are considerably more difficult at present than proofs for transcendence results.

In the same line as above, we also have the classical theorem of Gelfond-Schneider [44], [80], [87], [49]:

(1.4) *If α, β are algebraic, $\alpha \neq 1$, and β irrational, then α^β is transcendental.*

This was extended in a significant way recently by Baker [11], who shows:

(1.5) *If $\alpha_1, \cdots, \alpha_n$ are non zero algebraic numbers, multiplicatively independent (or equivalently, whose logarithms, together with $2\pi i$, are linearly independent over the rationals), and $\beta_1, \cdots, \beta_n$ are algebraic, and such that $1, \beta_1, \cdots, \beta_n$ are linearly independent over the rationals, then*

$$\alpha^{\beta_1} \cdots \alpha^{\beta_n}$$

is transcendental. Furthermore, the numbers

$$\log \alpha_1, \cdots, \log \alpha_n$$

are linearly independent over the algebraic numbers.

We shall return to this later in connection with diophantine approximations.

The preceding results use the differential equation, as well as the addition theorem. Using only the addition theorem, one obtains the following statement.

(1.6) *Let β_1, β_2 be complex numbers, linearly independent over $\mathbf{Q}$, and let z_j ($j = 1, 2, 3$) be complex numbers, also linearly independent over $\mathbf{Q}$. Then at least one of the numbers*

$$e^{\beta_1 z_j}, \quad e^{\beta_2 z_j} \qquad (j = 1, 2, 3)$$

is transcendental.

Thus for instance, if y is real and x^y is algebraic for all positive rational $x \neq 0$, then y is rational. Although this statement was known to Siegel [9], I rediscovered it independently and Siegel wrote me once that the proof I gave in [49] apparently was the first in the literature. Since this proof is the simplest in the theory of transcendental numbers, but also exhibits some basic features from all proofs, we shall summarize it below.

In investigating values of α^β when β is transcendental and α is algebraic, there may be one algebraic α such that α^β is algebraic. For instance,

$$2^{\frac{\log 3}{\log 2}} = 3.$$

By the Gelfond-Schneider theorem, $\log 3/\log 2$ is transcendental, and constitutes such a number β for which 2^β is algebraic. The previous theorem shows that there are at most two multiplicatively independent possibilities, and conjecturally there is only one, i.e. one can shrink 3 to 2 in Theorem (1.6).

The most general conjecture concerning transcendence and algebraic independence of values of the exponential function is due to Schanuel, and runs as follows.

(1.7) *Let $\alpha_1, \cdots, \alpha_n$ be complex numbers, linearly independent over the rationals. Then the transcendence degree of the field*

$$\mathbf{Q}(\alpha_1, \cdots, \alpha_n, e^{\alpha_1}, \cdots, e^{\alpha_n})$$

is at least n.

For instance, this implies Lindemann's Theorem (1.3), and also implies the old conjecture that the logarithms of multiplicatively independent algebraic numbers are algebraically independent. (Baker's theorem 1.5 concerns their linear independence.) Note that it is unknown even if $\log 2$ and $\log 3$ are algebraically independent, or even if $(\log 2)(\log 3)$ is algebraic. It is unknown if $e+\pi$ is algebraic. The algebraic independence of e, π would follow from Schanuel's conjecture by considering $1, 2\pi i, e, e^{2\pi i}$. In fact, Schanuel's conjecture implies at once all other known conjectures concerning values of the exponential function. For instance, I had conjectured that π cannot lie in the field obtained by starting with the algebraic numbers, adjoining values of the exponential function, taking algebraic closure, and iterating these two operations. This follows from Schanuel's conjecture as follows. Let $\alpha = (\alpha_1, \cdots, \alpha_n)$ be linearly independent algebraic numbers. Use vector notation, so that $e^\alpha = (e^{\alpha_1}, \cdots, e^{\alpha_n})$. By (1.7) it follows from the linear independence of $(\alpha, 2\pi i)$ that

$$\alpha, 2\pi i, e^\alpha, 1$$

has transcendence degree $\geq n+1$, whence π is transcendental over the field F_1 obtained by adjoining all values e^α to Q. Now take n large, let $u = (u_1, \cdots, u_n)$ be algebraic over F_1 and linearly independent over F_1. Consider

$$\alpha, e^\alpha, u, 2\pi i; \qquad e^\alpha, e^{e^\alpha}, e^u, 1.$$

Selecting u to have sufficiently many elements, one sees again from (1.7) that $2\pi i$ is transcendental over $Q(e^\alpha, u)$. One then proceeds by induction.

Schanuel also formulated his conjecture for formal power series in lieu of complex numbers. In this context the conclusion amounts to the algebraic independence of formal power series. This was proved by Ax [10].

The theory of transcendental numbers can also be developed p-adically. Let C_p be the completion of the algebraic closure of the p-adic numbers Q_p. Then C_p plays for the p-adic absolute value the same role as C for the ordinary absolute value. Mahler first extended the transcendence proof of e^α and α^β to the p-adic case [57]. For further results, see also Adams [7]. Brumer [23] proved the p-adic analogue of Baker's theorem, and thereby showed that Leopoldt's p-adic regulator does not vanish [54], for abelian extensions of the rationals. The p-adic analogue of (1.6) given in [49] has been used by Serre in his theory of p-adic representations [82]. The methods of complex variables in the standard case can be replaced by the p-adic Schnirelman integral and Cauchy formula. For a convenient exposition, see Adams [7].

2. **Sketch of proofs.** Gelfond was the first to realize explicitly the connection between transcendence problems and algebraic values of

entire, or meromorphic functions. Many years before he proved the α^β theorem, investigating special cases in 1929 (cf. [43]), he saw that such a problem was related with a question (I believe raised by Polya) concerning the possibility of an entire function taking integral values at integers, and the interpolation problem arising from it. If f, g are, say, entire functions taking algebraic values in a set S, then one reduces the study of S to an analytic problem concerning *zeros* of an auxiliary function, a polynomial in f, g, namely

$$F = \sum a_{ij} f^i g^j,$$

with coefficients in a number field (finite extension of the rationals). From the assumption that f, g take on values in the number field in S, one can construct such a function F having many zeros by a simple lemma of Siegel concerning linear equations with integral coefficients, [87] or [49]. For simplicity we state Siegel's lemma over the ordinary integers Z.

(2.1) *Let*

$$u_{11}x_1 + \cdots + u_{1n}x_n = 0,$$
$$\cdots$$
$$u_{r1}x_1 + \cdots + u_{rn}x_n = 0,$$

be a system of linear equations with integer coefficients u_{ij}. Let A be a bound for the absolute values of all u_{ij}, and assume $n > r$. Then this system has a solution in integers x_j not all 0, satisfying

$$\max |x_j| \leq (nA)^{r/(n-r)}.$$

Suppose that f, g take on values in the integers Z, and are algebraically independent. On the one hand, using the lemma, one can construct a function F having enough zeros in the set S so that at some point w of S, the value $F(w)$ is very small. On the other hand, if S has sufficiently many elements, one can also pick w so that $F(w) \neq 0$, and since $F(w)$ is an integer, one gets the contradiction.

A function like F was used first by Siegel when he proved transcendence results for the Bessel function (stated precisely later). A similar function led to other results like the transcendence of α^β. As an example, we shall now sketch a proof of (1.6), in a special case, with the functions

$$f(t) = e^{\beta_1 t} \quad \text{and} \quad g(t) = e^{\beta_2 t},$$

showing that they cannot take on values in Z at three points z_1, z_2, z_3

linearly independent over $\mathbf{Q}$. Suppose that they did. We let n be a large integer, assumed square free for convenience. Let $r=(4n)^{3/2}$. By Siegel's lemma we can find integers a_{ij} not all 0 such that the function $F=\sum_{i,j=1}^{r} a_{ij}f^i g^j$ has a zero at every point

$$k \cdot z = k_1 z_1 + k_2 z_2 + k_3 z_3, \quad 1 \leq k_\nu \leq n.$$

This amounts to solving linear equations in r^2 unknowns, with $r^2=(4n)^3$ and the number of equations is equal to n^3. The coefficients of these equations are bounded by $C^{n^{5/2}}$ for some constant C. Hence by Siegel's lemma, we can find integers a_{ij} not all 0, satisfying a similar bound. Since f, g are algebraically independent, it follows that F is not identically zero, and takes on values in $\mathbf{Z}$ for all arguments $k \cdot z$. On the other hand, F cannot vanish at all such linear combinations, with all triples of integers k, because the linear combinations $k \cdot z$ are not discrete, or alternatively because F is entire of order 1, and in a circle of large radius R, there are more such linear combinations than the bound $O(R)$ for the number of possible zeros of F. Let s be the largest integer such that $F(k \cdot z)=0$, for all k with $1 \leq k_\nu \leq s$. Then $s \geq n$. Let

$$w = k_1 z_1 + k_2 z_2 + k_3 z_3,$$

with some $k_\nu = s+1$, and $1 \leq k_\nu \leq s+1$ for all ν, and $F(w) \neq 0$. Then

$$|F(w)| \leq C^{s^{5/2}},$$

with a suitable constant C. We now estimate $|F(w)|$ by considering the expression

$$F(w) = \frac{F(t)}{\prod(t - k \cdot z)} \prod(w - k \cdot z)\bigg|_{t=w},$$

the products being taken over all k_ν with $1 \leq k_\nu \leq s$. There are s^3 terms in the product. The function on the right of this last equality is an entire function, and we apply the maximum modulus principle on a circle of radius $R = s^{3/2}$. Note that for $|t| = R$, we have $|t - k \cdot z| \geq R/2$ (for s large), and also

$$\frac{|w - k \cdot z|}{|t - k \cdot z|} \leq \frac{C_1 s}{R} \leq \frac{C_1}{s^{1/2}}$$

for some constant C_1 and s large. Hence

$$\log|F(w)| \ll \log|F|_R + s^3 - \tfrac{1}{2} s^3 \log s,$$

where the sign $\ll$ means that the left-hand side is smaller or equal to a constant times the right-hand side (Vinogradov's notation), and $|F|_R$ is the maximum of F on the circle of radius R. A trivial estimate shows that

$$\log |F|_R \ll s^3,$$

whence

$$\log |F(w)| \ll s^3 - s^3 \log s.$$

For n (and hence s) large, this contradicts the fact that $|F(w)| \geq 1$, and completes the proof.

The extension to number fields and algebraic values involves only minor technique in algebraic numbers. If K is a number field (finite extension of Q), let $\{\sigma\}$ range over the embeddings of K in C. For any element α in K, we call $\{\sigma\alpha\}$ the conjugates of α. A positive integer d such that $d\alpha$ is an algebraic integer is called a denominator for α. We let

$$\text{size}(\alpha) = \max(\log d, \log|\sigma\alpha|),$$

where d is the smallest denominator for α, and $\sigma\alpha$ range over the conjugates of α. Taking the norm of $d\alpha$, which is an ordinary integer $\neq 0$ if $\alpha \neq 0$, one gets the fundamental inequality

(2.2) $$-2[K:Q]\,\text{size}(\alpha) \leq \log|\sigma\alpha|,$$

for any σ. This inequality then can be used to replace the last part of the argument in the preceding proof. Generally speaking, working with algebraic numbers in this context is no harder than working with ordinary integers.

As an example, we shall give one of Baker's theorems [11], concerning an effective lower bound for linear combinations of logarithms of algebraic numbers, with algebraic coefficients. Baker's proof used previous ideas of Gelfond [44] and Feldman (cf. his list of papers). Although Gelfond's method was effective, it applied only to the case of two logarithms, and Baker saw how to extend it to linear combinations of n logarithms. The importance of doing this had already been observed by Gelfond, who remarked that such an extension would lead to an effective improvement of Liouville's inequality (see below). Baker's theorem is as follows.

(2.3) Let $\alpha_1, \cdots, \alpha_{n+1}$ be algebraic numbers, whose logarithms are linearly independent over the rational numbers. Let $\tau > n+2$. Then there exists an effectively computable constant $c(\alpha) = c$ such that for any algebraic numbers $\beta_1, \cdots, \beta_n$ of size $\leq h$ we have

$$\log|\beta_1 \log \alpha_1 + \cdots + \beta_n \log \alpha_n - \log \alpha_{n+1}| > -ch^\tau.$$

We now sketch Baker's proof, and assume that $\beta_1, \cdots, \beta_n$ are such that

$$\log|\beta_1 \log \alpha_1 + \cdots + \beta_n \log \alpha_n - \log \alpha_{n+1}| \ll -h^\tau,$$

with $h \geq c(\alpha)$ size β. We construct the auxiliary function

$$F(z_1, \cdots, z_{n+1}) = \sum a_{(j)} \alpha_1^{j_1 z_1} \cdots \alpha_n^{j_n z_n} (\alpha_1^{\beta_1} \cdots \alpha_n^{\beta_n})^{j_{n+1} z_{n+1}}$$

to have zeros of high order at certain points. We shall use vector notation, so that we write

$$F(z) = \sum a_{(j)} \alpha^{jz} (\alpha^\beta)^{j_{n+1} z_{n+1}}.$$

The sum is taken for $0 \leq j_\nu < J$, with a suitable integer J. Using vector notation, we abbreviate this in the form $j \leq J$. Let $\lambda = (\lambda_1, \cdots, \lambda_n)$. Let

$$D^\lambda = D_1^{\lambda_1} \cdots D_n^{\lambda_n}$$

be the standard differential operator, and for an integer $k \geq 0$ consider

$$D^\lambda F(k, \cdots, k) = \sum a_{(j)} \alpha^{jk} \alpha^{\beta j_{n+1} k} (j + j_{n+1} \beta)^\lambda.$$

We shall want to make this derivative essentially equal to 0 for certain values of k and λ. This amounts to solving linear equations. Indeed, α_{n+1} is very close to $\alpha_1^{\beta_1} \cdots \alpha_n^{\beta_n}$, and one solves the corresponding linear equations with α^β replaced by α_{n+1}. Let

$$G_\lambda(k) = \sum_{j<J} a_{(j)} \alpha^{jk} \alpha_{n+1}^{j_{n+1} k} (j + j_{n+1} \beta)^\lambda.$$

Select δ such that $1 < \delta < \tau/(n+2)$. Let

$$J \doteq h^\delta \quad \text{and} \quad L \doteq h^{\delta + (\delta-1)/n}.$$

(Since for instance h is not necessarily an integer, by these equalities $\doteq$ we mean that J or L are equal to the largest integers less than or equal to the right-hand sides.) Solve the linear equations

$$G_\lambda(k) = 0$$

for $\lambda \leq L$ and $1 \leq k \leq h$. Then the number of variables is approximately equal to J^{n+1}, the number of equations is approximately equal to $L^n h$, and the size of the coefficients is $\ll Lh$. Note that the number of variables is approximately equal to the number of equations. By using the existence of zeros of high order, and the usual estimate with the maximum modulus principle, one then finds that

(*) $$\log\left| D^\lambda F(k, \cdots, k) \right| \ll - h^\tau$$

for $1 \leq k \leq h$. Proceeding inductively, we wish to achieve the inequality (*) for

$$k \leq h_\nu = h^{1+\nu(\delta-1)/n} \quad \text{and} \quad \lambda \leq L/2^\nu,$$

with ν ranging from 1 to approximately n^3. The inductive step is done by taking $l \leq h_{\nu+1}$ and $\lambda \leq L/2^{\nu+1}$. Let

$$f(t) = D^\lambda F(t, \cdots, t).$$

Then

$$\left. \frac{f(l)}{\prod_{k \leq h_\nu}(l-k)^{L/2^{\nu+1}}} = \frac{f(t)}{\prod_{k \leq h}(t-k)^{L/2^{\nu+1}}} \right|_{t=l}.$$

We use the maximum modulus principle with a circle of radius

$$R = h^{1 + \frac{(\delta-1)}{n} + \nu \frac{(\delta-1)}{n}}$$

and estimate $\log|f(l)|$. From the fundamental inequality (2.2) we conclude that $G_\lambda(l) = 0$ by induction. (It is this inductive step which represents the improvement by Baker over the earlier Gelfond method.)

Thus our inductive procedure ultimately gets us to

$$G_0(k) = \sum_{j \leq J} a_{(j)} \alpha^{jk} \alpha_{n+1}^{j_{n+1}k} = 0$$

for $k \leq h^{\delta(n+1)}$. This is a system of linear equations, whose matrix of coefficients has a Vandermonde determinant, whence the contradiction which proves Baker's theorem.

Baker's theorem has gone through successive improvements, lowering the exponent τ, and recent papers of his adjust the arguments to avoid the Vandermonde determinant at the end. This is advantageous because in more complicated situations requiring an extension of the proof, it is not obvious to show that the corresponding determinant does not vanish. One of these deals with elliptic functions. Cf. the papers of Baker and Coates on this subject.

3. Algebraic values of meromorphic functions.
The importance of solutions to differential equations in connection with transcendence

results first appeared in Siegel's results on the Bessel function [86]. We state these in §5. Schneider [79] gave a general criterion under which an entire, or meromorphic function satisfying a certain type of algebraic differential equation with algebraic coefficients can take on values in a number field at only a finite number of points. His type of differential equation was sufficiently general to include the ordinary exponential function, the Weierstrass elliptic function, and the elliptic modular function. Using Schneider's ideas properly, and proving the right estimates, I then extended the result to the most general algebraic differential equation (needed for other applications e.g. abelian functions). We recall that an entire function f on C is said to be of strict order $\leq \rho$ if

$$\log |f|_R \ll R^\rho.$$

A meromorphic function is said to be of strict order $\leq \rho$ if it can be expressed as a quotient of entire functions of order $\leq \rho$. We then have [49]:

(3.1) *Let K be a number field. Let $f_1, \cdots, f_N$ be meromorphic functions of strict order $\leq \rho$. Assume that the field $K(f_1, \cdots, f_N)$ has transcendence degree ≥ 2 over K, and that the derivative $D = d/dt$ maps the ring $K[f_1, \cdots, f_N]$ into itself. Let $w_1, \cdots, w_m$ be distinct complex numbers not lying among the poles of the f_i such that*

$$f_i(w_\nu) \in K$$

for all $i = 1, \cdots, N$ and $\nu = 1, \cdots, m$. Then $m \leq 20\rho [K:Q]$.

The transcendence of e^α then follows by considering the ring $K[t, e^t]$. Assuming that e^α is algebraic, one takes K to contain both α and e^α. Then the infinite number of points $\alpha, 2\alpha, 3\alpha, \cdots$ provides the contradiction. Similarly, for α^β one considers $K[e^t, e^{\beta t}]$. Still following Schneider, one gets the transcendence of $\wp(\alpha)$ for a Weierstrass $\wp$-function, with algebraic α, and algebraic g_2, g_3 by considering the ring $K[\wp, \wp']$.

Schneider had proved the transcendence of the periods of abelian functions by dealing with several complex variables [78]. I formulated a theorem analogous to (3.1) in several variables [49]. To get a similar bound on the set of points where the functions take on values in K, I had to assume that the set S of such points was a product of sets on the coordinate axes, to be able to use Cauchy's formula in several variables as an iteration of the formula in one variable (as Schneider had done). The product condition was unnatural, as one sees from the example of functions

$$z_1, \cdots, z_n, e^{P(z_1,\cdots,z_n)}$$

where P is a polynomial with integer coefficients. These functions are algebraically independent and take on algebraic values at the algebraic zeros of P. By using deep techniques from the theory of several complex variables (potential theory and Hörmander L^2-estimates) Bombieri was able to remove the product condition, and proved the following theorem [21].

(3.2) *Let K be a number field and let $f = (f_1, \cdots, f_N)$ be meromorphic functions in $\mathbf{C}^d$ of strict order $\leq \rho$. Assume that the transcendence degree of $K(f)$ is $\geq d+1$, and that the partial derivatives $D_i = \partial/\partial z_i$ map the ring $K[f]$ into itself. Then the set of points $w \in \mathbf{C}^d$ where $f(w)$ is defined and $f(w) \in K^N$ is contained in an algebraic hypersurface of degree*

$$\leq d(d+1)\rho[K:\mathbf{Q}] + 2d.$$

In the simple case when the above set of points S is of type

$$S = S_1 \times \cdots \times S_d,$$

where $S_j \subset \mathbf{C}$, then Bombieri's theorem implies the bound which I had obtained for the cardinality of S, of the form $b\rho[K:\mathbf{Q}]$, for some easily computable constant b depending on d. This special case is actually sufficient for a number of applications (see below, and [49]).

When no differential equation is available, then one has to rely on other properties of the set of points where the functions take on values in K. Usually such a set S is expressed as a union of subsets $\{S_n\}$, where each S_n is contained in a ball of radius $\ll n$, and where the functions have a specified arithmetic order of growth, e.g. there exists a constant C such that for all n and all $z \in S_n$ we have

$$\text{size } f(z) \leq Cn^\rho.$$

Schneider first formulated such a theorem [79] in one variable. For useful variants and applications, see [49]. We shall state the applications on group varieties in §4, including the several variables version.

We note that instead of taking values in a number field, one may take values in a finitely generated extension, provided that one assumes a condition to replace (2.2). See [49] for the appropriate statements, raising possibilities for an inductive argument. In dimension 1, Waldschmidt [92] eliminated the extra condition. We discuss this again in connection with diophantine approximations.

4. General exponential functions. A group variety is a group in affine or projective space, which is also an algebraic variety (connected), i.e. its points are the set of solutions of algebraic equations, and such that the law of composition and inverse have graphs which are also algebraic varieties. An important example is the linear group $GL(m)$ of invertible $m \times m$ matrices. If K is a subfield of the complex numbers, we say that the group variety is defined over K if all the above mentioned algebraic equations can be chosen to have coefficients in K. If that is the case, then we denote by G_K the set of points of G having coordinates in K, and it follows that G_K is a group. When K is a number field, we view G_K as a discrete group. When $K = \mathbf{C}$, we view $G_\mathbf{C}$ as a complex analytic manifold, i.e. a complex analytic group.

Let G be a group variety. By a 1-parameter subgroup of G we mean a complex analytic homomorphism $\phi : \mathbf{C} \to G_\mathbf{C}$ of the complex line into $G_\mathbf{C}$ whose derivative at the origin is injective. Thus ϕ is an analytic curve in $G_\mathbf{C}$. We define a d-parameter subgroup $\phi : \mathbf{C}^d \to G_\mathbf{C}$ in a similar way. For instance, when G is the linear group, a 1-parameter subgroup is given by the exponential series

$$t \mapsto \sum \frac{t^k M^k}{k!}$$

where M is a matrix. We are interested both in conditions under which a point $\phi(z)$ is algebraic, and also under which the image of ϕ is an algebraic subgroup (i.e. is closed). Most of the time, of course, the image just winds around (in the compact case). Indeed, the compact case is obtained as follows. We take a lattice L of real dimension $2n$ in complex n-space. Let us assume that the factor group $\mathbf{C}^n/L$ can be embedded complex analytically in projective space $P_\mathbf{C}^N$, so that its image in projective space is necessarily an algebraic group (Chow's Theorem, for instance). Let us also assume that this group is defined over a number field. A 1-parameter subgroup may be viewed as winding around the torus $\mathbf{C}^n/L$, which is called an abelian manifold. The algebraic group corresponding to it in projective space is called an abelian variety. In dimension $n = 1$, the parametrization of the abelian variety is done by means of the Weierstrass elliptic function,

$$z \mapsto (1, \wp(z), \wp'(z)),$$

the coordinates on the right being projective coordinates, on the elliptic curve

$$y^2 = 4x^3 - g_2 x - g_3.$$

An analytic subgroup of an abelian variety is closed if and only if it is an abelian subvariety (i.e. an algebraic subgroup). If this analytic subgroup is algebraic and has dimension 1, it is an elliptic curve.

For 1-parameter subgroups, we have the following theorem [49].

(4.1) *Let G be a linear group variety or an abelian variety, defined over the field of algebraic numbers. Let $\phi: \mathbf{C} \to G_\mathbf{C}$ be a 1-parameter subgroup. Let Γ be a subgroup of $\mathbf{C}$ having at least three linearly independent elements over $\mathbf{Z}$ in the linear case, and seven in the abelian case. If $\phi(\Gamma)$ is contained in the group of algebraic points of G, then the image of ϕ is closed, i.e. it is an algebraic subgroup of $G_\mathbf{C}$.*

(In the mixed case of a product, say, one has to take the maximum of three and seven to get the same conclusion.)

The proof follows the same lines as the proof given in §2, but to make the required estimates in the abelian case, one must use the quadratic form of Néron-Tate [64]. Conjecturally, the numbers 3 and 7 can be shrunk to 2, as for (1.6).

The extension of (4.1) presented difficulties of two types. The first concerns pure analysis, namely the need for a Schwarz lemma in several variables. It was surmounted by Bombieri-Lang [22], who prove the following result. Let λ be a positive number. Let $\{S_n\}$ be a sequence of subsets of $\mathbf{C}^d$. Let B_r be the ball of radius $r > 0$, centered at the origin in $\mathbf{C}^d$. We say that the family $\{S_n\}$ is λ-distributed in B_r if the following condition is satisfied. There exists N_0 such that given $w \in B_r$ and $N \geq N_0$ there exists a point $u \in S_N$ such that

$$|u - w| \leq 1/2N^\lambda.$$

If Γ is a finitely generated subgroup of $\mathbf{C}^d$, with generators $\{u_1, \cdots, u_m\}$ then we let S_n be the set of all linear combinations

$$k_1 u_1 + \cdots + k_m u_m, \quad |k_j| \leq n.$$

We have:

(4.2) *Let G be a linear or abelian group variety. Let $\phi: \mathbf{C}^d \to G_\mathbf{C}$ be a d-parameter subgroup, and let Γ be a finitely generated subgroup of $\mathbf{C}^d$, which is λ-distributed in a ball B_r. Assume that $\lambda > (d+1)/2$ in the linear case, and $> d+1$ in the abelian case. If $\phi(\Gamma)$ is contained in the group of algebraic points of G, then $\phi(\mathbf{C}^d)$ is an algebraic subgroup of $G_\mathbf{C}$, i.e. it is closed.*

The condition of λ-distribution is a condition of diophantine approximation. For further comments on it, cf. [22]. The p-adic analogue of (4.2) was done by Serre [83]. It is a deep problem to reduce the value of λ, and to prove that it has the "expected" value arising from the theory of diophantine approximations.

If V is a variety (algebraic, irreducible) defined over a number field K, and $(x_1, \cdots, x_n)$ are affine coordinates for a point on V, then we say that this point is rational over K if $K(x)=K$, algebraic over K if $K(x)$ is algebraic over K, and transcendental over K if $K(x)$ is not algebraic over K, i.e. if at least one coordinate is transcendental over K. The other type of theorem which one has on group varieties can then be stated as follows [49].

(4.3) *Let G be a group variety defined over the field of algebraic numbers. Let $\phi: C \to G_C$ be a 1-parameter subgroup, whose differential at the origin is algebraic. If $\alpha \in C$, $\alpha \neq 0$ is algebraic and $\phi(t)$ is not an algebraic function of t, then $\phi(\alpha)$ is a transcendental point on G_C. On the other hand, if there exists a point $u \neq 0$ in C such that $\phi(u)$ is algebraic, then $\phi(C)$ is an algebraic subgroup of G_C.*

The first assertion of (4.3) had been conjectured by Cartier. The two formulations generalize the Lindemann theorem (1.2) and the Gelfond-Schneider theorem (1.4). In the first case, the 1-parameter group is

$$t \mapsto e^t,$$

and in the second case, it is

$$t \mapsto (e^t, e^{\beta t}).$$

The only case when $\phi(t)$ is an algebraic function of t occurs in the linear case, and when ϕ is formed with the exponential series of a nilpotent matrix, in which case ϕ is even a rational function of t.

Theorem (4.3) applied to abelian varieties yields the transcendence of the period vectors, originally proved by Schneider [78]. It also shows that in the representation

$$\Theta: C^n \to A_C$$

of an abelian variety as a quotient of C^n, normalized to have algebraic derivative at the origin, the image of an algebraic point in C^n is a transcendental point on A_C, by considering the line passing through the point in C^n, and its image under Θ.

Under the normalization at the origin by means of the differential equation (algebraic derivative), Theorem (4.3) extends to the higher dimensional case of a d-parameter subgroup

$$\phi: C^d \to G_C.$$

Instead of one point u, one must then assume the existence of d points linearly independent over $\mathbf{C}$, whose values under ϕ are algebraic [49]. Note the similarity with (4.2), where we do not normalize the map, but then assume the existence of more points linearly independent over $\mathbf{Z}$ (not $\mathbf{C}$).

In the 1-dimensional case, Schneider had proved:

(4.4) *Let j be the elliptic modular function. If τ is algebraic and not imaginary quadratic, then $j(\tau)$ is transcendental.*

By means of the higher dimensional results, one then gets a higher dimensional analogue as follows [49].

(4.5) *Let A be an abelian variety parametrized as above, $\Theta: \mathbf{C}^n \to A_\mathbf{C}$, normalized to have algebraic derivative at the origin. Assume that the period matrix is normalized so that the principal matrix has the usual canonical form, and let the period matrix be $\Omega = (W_1, W_2)$. Let $T = W_2 W_1^{-1}$. If T is algebraic, then T (viewed as a linear transformation) maps the period lattice tensored with $\mathbf{Q}$ into itself.*

When $n=1$, then $T=\tau$, $W_1=\omega_1$ and $W_2=\omega_2$. Note that $W_1^{-1}W_2$ is the moduli point associated with the abelian variety in the Siegel upper half space H_n. The theorem concerns $W_2 W_1^{-1}$, and the relation between these points is not clear.

One may attempt to formulate the Schanuel conjecture in the present context. One then has to consider a point in the product space

$$(\alpha, \Theta(\alpha)) \in \mathbf{C}^n \times A_\mathbf{C}.$$

The general expectation is that the transcendence degree is $\geq n$ under obvious conditions on the curve passing through α. The Riemann relations provide a counterexample to the analogue of Schanuel's conjecture in this context, but hopefully, if one stays away from the period relations, i.e. if α is not a period point, then the expected lower bound for the transcendence degree will hold up. For a general discussion, cf. [49]. Grothendieck has also formulated a conjecture concerning the possible polynomial relations for the elements of the period matrix, to the effect that they should all be due to algebraic cycles on the product of the variety with itself. The situation is as follows. Let W be a variety defined, say, over the rational numbers. Let $\{\gamma_i\}$ be a basis for its homology $H_*(W, \mathbf{Q})$, and let $\{\omega_j\}$ be a basis for $H^*(W, \mathbf{Q})$, where H^* is the cohomology defined by the algebraic complex of differential forms of De Rham. We assume that the basis $\{\omega_j\}$ is obtained by taking together bases for the Hodge spaces

$H^{p,q}$. Let Z be a subvariety of W, of dimension k. Then $Z \sim \sum n_i \gamma_i$. If ω is a differential form not of type (k, k), then

$$\sum n_i \int_{\gamma_i} \omega = \int_Z \omega = 0.$$

In this way we obtain a linear relation among the periods, with rational coefficients. Consider this when $W = V^r$ is the product of a variety V with itself taken r times. Then $H^*(W, \mathbf{Q})$ is the tensor product of $H^*(V, \mathbf{Q})$ with itself r times, and applying the above shows that an algebraic cycle on V^r gives rise to a polynomial relation of degree k among the periods.

The graph of a nontrivial endomorphism of an elliptic curve (complex multiplication) gives an example, which would explain the algebraic (linear) dependence between two fundamental periods over the algebraic numbers.

The simplest "Riemann" relation is the Legendre relation (it involves the multiplicative group), arising from the parametrization of the group variety associated with integrals of the second kind by the map

$$C \times C \to G_C$$

given by

$$(t, u) \mapsto (1, \wp(t), \wp'(t), u - \zeta(t)),$$

where ζ is the Weierstrass zeta function. This map has periods (ω_1, η_1) and (ω_2, η_2) (classical notation), and the Legendre relation

$$\eta_1 \omega_2 - \eta_2 \omega_1 = 2\pi i$$

is nothing but the Riemann relation for this case. It is of degree 2. On the other hand:

(4.6) *Any nonvanishing linear combination of η_1, η_2, ω_1, ω_2, $2\pi i$ with algebraic coefficients is transcendental.*

Without the $2\pi i$, this was proved by Baker [13], and the more general statement is due to Coates [28]. The technique of proof extends Baker's technique.

If G is a commutative group variety, defined over a number field K, then its tangent space at the origin can be identified with C^n ($n = \dim G$). Then the general exponential map

$$\phi: C^n \approx T_C \to G_C$$

provides a geometric setting generalizing the ordinary function e^z. Its inverse mapping is a generalized log, and the points in C^n which map on 0 are the (vector) periods.

One can also look at the parametrization of curves and varieties of higher genus. Thus already in [48], I conjectured that if

$$\phi: D \to V_C$$

is the uniformizing map of a curve V of genus ≥ 2, from the disc D centered at the origin, normalized so as to have algebraic derivative at the origin, and if α is an algebraic point of the disc $\neq 0$, then $\phi(\alpha)$ is transcendental on V. (We assume that the curve is defined over a number field.) One can take the upper half plane instead of the disc. Curves can also be parametrized by the "noncompact" case of modular functions, so that I was thus led to the conjecture that if τ is a point of the upper half plane, not equivalent to i or $\rho = e^{2\pi i/3}$ under the modular group (i.e. such that $j'(\tau) \neq 0$) and such that $j(\tau)$ is algebraic, then $j'(\tau)$ is transcendental. Siegel [88] also makes remarks concerning the possible transcendence of $j'(\tau)$ at quadratic imaginary irrationalities other than i or ρ. If one normalizes g_2, g_3 to be algebraic, and if one takes into account the formula relating the derivative j' with j (cf. Siegel [88] or the seminar on complex multiplication [84]), then one sees that the conjecture concerning the transcendence of $j'(\tau)$ is equivalent to the following concrete statement: *Let g_2, g_3 be algebraic, and let ω be a nonzero period of the corresponding $\wp$-function. Then ω^2/π is transcendental.* Indeed, one has the relation

$$j'(z) = \frac{9\omega^2 g_3}{\pi i g_2} j(z).$$

(The formula in Siegel [88] is not yet properly normalized so that the ω^2 does not appear in it. The point is that g_2, g_3 can be viewed as functions of lattices, and one has to take a lattice in the given class which makes g_2, g_3 algebraic. Then ω^2 appears as the quotient of ω^6 by ω^4.) The problem is analogous here to determining the transcendence of $(\log 2)^2/\log 3$, say.

Observe that the parametrizations of the curves by j (or the parametrization of curves of higher genus corresponding to modular functions of level N), and the parametrizations of the curves by the universal covering map normalized to send the origin of the disc to an algebraic point, and to have algebraic derivative at the origin, correspond to two different normalizations of the uniformizing map. (The fact that the fundamental domain of j is not compact is not essential here, the same phenomenon occurs in the compact case, as in the work of Shimura.)

As an incidental question for the parametrization of V_C as above, one can ask if the radius of the disc D is transcendental when ϕ is normalized as stated in [48].

For some results concerning the field generated by values of the j-function and its derivatives over the algebraic numbers, cf. Ramachandra [70], who makes use directly of the differential equation satisfied by the j-function. For instance, he proves:

(4.7) *Let f be a modular form having algebraic q-expansion coefficients (classical terminology). Then the singular values $f(\tau)$, $f'(\tau), f''(\tau), \cdots$ for τ taking values in an imaginary quadratic field of discriminant d lie in the field*

$$\bar{Q}(\pi, \Delta^{1/m}(\sqrt{d}))$$

for a suitable positive integer m.

(The integer m is made explicit, but this gets too technical here.) As usual, Δ is the modular discriminant.

One can extend the above discussion to bounded symmetric domains with compact quotients which are algebraic varieties defined over number fields. These form one possible generalization of curves of higher genus, and one may even wonder if these varieties do not satisfy the Mordell property (of having only a finite number of rational points, or points rational over a given number field). Whenever one wants to conjecture such a statement about a variety, one must be sure that there is no obvious geometric reason why the variety could have many rational points. I know of three such reasons: An infinite group of automorphisms, the function field contained in a purely transcendental extension of the constants, and the variety containing "blown up" points, or straight lines. I am told by experts that the first two conditions are known not to prevail, and that the third is probably not known in general, but probably does not prevail either, for the varieties mentioned above.

Although I have omitted a discussion of values of zeta functions (principally for lack of competence) I would like to refer the reader to at least one paper which makes a connection between such values and values of functions of exponential and automorphic type, namely that of Damerell [30], who studies values of Hecke L-series at, say, $1/2$, in connection with the Birch-Swinnerton Dyer conjecture concerning the rank of the Mordell-Weil group of an elliptic curve.

In looking at 1-parameter subgroups of abelian varieties, and to prove certain estimates, especially from below for theta functions, I was led to conjecture [49]:

(4.8) *Let B be a 1-parameter subgroup of an abelian variety, embedded in projective space, and let H be a hyperplane section. Assume that B is Zariski dense in A. Then the intersection of B with H is infinite, unless B is algebraic (in which case Bezout's theorem applies).*

This was proved by Ax [10], who also discovered a general phenomenon concerning the intersection of analytic subgroups and algebraic subsets of a group variety, as follows:

(4.9) *Let A be an algebraic group and B an analytic subgroup. Let V be an algebraic subvariety of A. Assume that V, B pass through the origin. If locally at the origin the intersection of V and B has an analytic component W of excessive dimension, and if V is the Zariski closure of W in A, then there exists an analytic subgroup A' at the origin containing both V and B such that the intersection of V and B with respect to A' is not excessive.*

Of course, by excessive, we mean that if W is an analytic component of the intersection of V and B on A, then

$$\dim W > \dim V + \dim B - \dim A.$$

Ax relates this with the function theoretic formulation of Schanuel's conjecture. Thus we see throughout this section that the theory of transcendental numbers intermingles very closely with the theory of functions of several complex variables.

5. **E-functions.** Siegel [86], [87] defined an E-function to be a function which admits a power series expansion

$$f(z) = \sum \alpha_n z^n/n!$$

with complex coefficients α_n belonging to a number field K, satisfying the following conditions:

E1. There is some constant c such that all conjugates of α_n are bounded in absolute value by c^n.

E2. There exists a sequence of integers $d_n > 0$ such that d_n is a denominator for α_k ($k = 0, \cdots, n$) and $d_n \leq c^n$.

The ordinary exponential function e^z is an E-function, and so is the Bessel function

$$J_0(z) = \sum z^{2n}/(n!)^2.$$

Siegel [87] gives other examples similar to these. The product and sum of E-functions are E-functions.

(5.1) Let $Q=(Q_{ij})$ $(i, j=1, \cdots, s)$ be a matrix of rational functions in $K(z)$ over a number field K. Let F be the column vector formed by E-functions $f_1, \cdots, f_s$, and assume that it satisfies the linear differential equation $F'=QF$. Assume that the functions $f_1, \cdots, f_s$ are algebraically independent over $K(z)$. Let $\alpha \in K$ be distinct from 0 and from the poles of the rational functions Q_{ij}. Then the values $f_1(\alpha), \cdots, f_s(\alpha)$ are algebraically independent.

Siegel [86] originally proved this theorem for the Bessel function and its derivative, and extended it to the more general case under an extra condition. Shidlovsky showed how to eliminate this condition by formulating and proving the appropriate lemma [85]. An exposition is also given in [49].

Extensions have been given by Sprindzuk [89]. It is a problem both to weaken the conditions on the coefficients of the power series, and to prove that certain functions are E-functions (e.g. Bessel functions J_λ with algebraic λ).

In the applications, it is also necessary to prove that certain functions are algebraically independent. We refer to Siegel for this [87].

The main part of Siegel's arguments is linear. It remains an open problem to see to what extent one can replace "algebraic independence" by "linear independence" throughout the statement of the Siegel-Shidlovsky theorem.

Also, the p-adic analogue of the transcendence theorem for E-functions is not known. Siegel's arguments depend on an essential way on the factorials in the denominators, and no substitutes are known at present.

DIOPHANTINE APPROXIMATIONS

6. Metrical results. Let α be a real number, and assume that α is not rational. We denote by $\|\alpha\|$ the distance of α to the closest integer. If this distance is less than $1/2$, then there is a unique integer p such that $\|\alpha\| = |\alpha-p|$. Thus the norm which we have defined is essentially the distance on the circle. More generally, one can consider vectors X in R^n, and define their norm $\|X\|$ on the torus R^n/L, where L is a lattice. We concentrate mostly on single numbers.

We are interested in the distribution of the numbers $q\alpha$ on the circle, i.e. on R/Z, when q ranges over the positive integers, and we want to give quantitative results concerning this distribution. We look mainly at the homogeneous case, i.e. how close $\|q\alpha\|$ can come to the origin. The first observation is due to Dirichlet:

(6.1) *Let N be a positive integer. There exists an integer q, $0 < q \leq N$, such that $\|q\alpha\| < 1/N$.*

The proof is easy, so we give it. Cut up the interval $[0, 1]$ into N equal parts of length $1/N$, and consider the $N+1$ numbers 0α, 1α, 2α, $\cdots$, $N\alpha$ modulo $\mathbf{Z}$. Two of them must lie in the same segment (mod $\mathbf{Z}$), say $r\alpha$ and $s\alpha$ with $r < s$. We let $q = s - r$, and obtain

$$\|q\alpha\| < 1/N \leq 1/q,$$

as desired.

Note that $\|q\alpha\| = |q\alpha - p|$ for $\|q\alpha\|$ sufficiently small. The inequality $\|q\alpha\| < 1/q$ can also be written in the form

$$|q\alpha - p| < 1/q \quad \text{or} \quad |\alpha - p/q| < 1/q^2.$$

Thus depending on how we write these inequalities, we get an exponent of 1 or 2 on the q of the right-hand side. The last inequality shows that we are dealing with the approximation of α by rational numbers.

In higher dimensional space, we take vectors $A_1, \cdots, A_r$ in n-space. Then the same type of argument shows that there exist integers q_i such that

$$\|q_1 A_1 + \cdots + q_r A_r\| < 1/N \quad \text{and} \quad |q_i| \leq N^{n/r}.$$

Davenport and Schmidt [32] prove that such inequalities cannot be improved for almost all numbers. More generally, one considers linear forms $L_1, \cdots, L_m$ in n variables. We let

$$\delta(m, n) = (n - m)/m$$

be the "Dirichlet exponent," and we consider the simultaneous inequalities

$$|L_i(Q)| \ll 1/q^{\delta(m,n)}$$

where $Q = (q_1, \cdots, q_n)$ is a vector of integers, and $q = \max |q_i|$.

We recall that a set of numbers is said to have measure 0 if given $\epsilon > 0$ the set can be covered by a countable number of intervals such that the sum of the lengths of these intervals is $< \epsilon$. Khintchine proved:

(6.2) *Let ψ be a positive function such that $\sum_{q=1}^{\infty} \psi(q)$ converges.*

Then for almost all numbers α (i.e. outside a set of measure 0), there is only a finite number of solutions to the inequality

$$\|q\alpha\| < \psi(q).$$

If ψ is decreasing, and the above sum diverges, then for almost all numbers α, there exist infinitely many solutions to the inequality $\|q\alpha\| < \psi(q)$.

The proof of the second assertion is harder and we refer to Khintchine [46] for it. The proof of the first assertion is easy and we give it. We may restrict our attention to those numbers α lying in the interval [0, 1]. Consider those for which the inequality has infinitely many solutions. Given ϵ select q_0 such that

$$\sum_{q \geq q_0}^{\infty} \psi(q) < \epsilon.$$

For each $q \geq q_0$ consider the intervals of radius $\psi(q)/q$ surrounding the rational numbers

$$0/q, 1/q, \cdots, (q-1)/q.$$

Every one of our α will lie in one of these intervals because for such α we have

$$|\alpha - p/q| < \psi(q)/q.$$

The measure of the union of these intervals is bounded by the sum

$$\sum_{q \geq q_0} q \frac{2\psi(q)}{q} < 2\epsilon,$$

as was to be shown.

For example we can take $\psi(q) = 1/(\log q)^{1+\epsilon}$ for any $\epsilon > 0$.

Let ψ be a decreasing function with divergent sum. For each (real) number α, let $\lambda(N)$ (λ depends on α and ψ) be the number of solutions in integers p, q of the inequalities

$$0 < q\alpha - p < \psi(q) \quad \text{and} \quad 1 \leq q < N.$$

(6.3) *For almost all numbers α, we have the asymptotic relation*

$$\lambda(N) \sim \int_1^N \psi(x) \, dx.$$

A special case of (6.3) was first stated by LeVeque [55]. The general theorem was proved by Erdös [33] and Schmidt [73]. We note especially that Schmidt obtains important generalizations to the higher dimensional case, with certain error terms.

7. The type of a number. Results as in §6 which hold almost everywhere are said to be metrical results. They suggest a first order of magnitude for the typical behavior of numbers given as values of classical functions, suitably normalized. For instance, as expressed in [48], I would expect all such numbers to satisfy the property that

$$\|q\alpha\| > 1/q^{1+\epsilon}$$

for all but a finite number of q. However, let us call a number α of bounded type if there exists a constant $c>0$ such that

$$\|q\alpha\| > c/q$$

for all integers $q>0$. It can be shown that a number is of bounded type if and only if its continued fraction has bounded entries, and that the set of numbers of bounded type has measure 0. More importantly from our point of view, we have:

(7.1) *A number α is of bounded type if and only if for any positive function ψ with convergent sum $\sum \psi(q)$, the inequality*

$$\|q\alpha\| < \psi(q)$$

has only a finite number of solutions.

(For the proof, see [50]. Schanuel showed me how to prove one of the implications, namely that the convergent sum condition implies bounded type.) Consequently the metrical theorems cannot be held as models beyond this first sort of approximation, and one must look for a more subtle invariant, which will be associated with any particular number, or class of numbers, under considerations.

With this point of view, I defined the type of a number [50]. There are alternative definitions. Let f be a positive increasing function (not necessarily strictly). We say that α has type $\leq f$ if

$$\|q\alpha\| \geq 1/qf(q)$$

for all sufficiently large q. This condition implies that for all sufficiently large integers N, there exists a solution in relatively prime integers $p, q>0$ of the inequalities

$$|q\alpha - p| < 1/q \quad \text{and} \quad N/f(N) \leq q < N,$$

and the converse is almost true, when f does not grow too fast. See [50] for details. (The proofs use continued fractions.) One can then formulate a basic problem in diophantine approximations: Determine a type for the classical numbers.

To say that for all sufficiently large q we have

$$\|q\alpha\| \geq 1/q^{1+\epsilon}$$

amounts to saying that α has type $\leq q^\epsilon$. (Here, as always, we suppose that this holds for each $\epsilon > 0$.) However, it is more fruitful to work with the function f so that $f(q)$ appears as a factor of q, rather than with the ϵ in the exponent, because as shown in [50], the function f appears in an essential way in other estimates associated with the number. We recall two of these here.

If x is a real number, let $R(x)$ be the remainder of x modulo Z, i.e. the unique number such that $0 \leq x < 1$ and such that $x - R(x)$ is an integer. Let us form the sum

$$S_N(\alpha) = \sum_{n=1}^{N} R(n\alpha).$$

One expects the values of $R(n\alpha)$ to be somewhat evenly distributed around $1/2$. Using the type, one then finds the following estimate:

(7.2) *Let α be of type $\leq f$ and assume that the function $f(t)/t$ is decreasing. Then*

$$S_N(\alpha) = \tfrac{1}{2} N + O\left(\int_1^N \frac{f(t)}{t} dt \right).$$

Relations between rational approximations to α and the sum S_N had been noted by Behnke [19] and Ostrowski [65], but the above statement (which has a very simple proof) was first given in [50]. The essential thing here is the appearance of a canonical error term as an integral involving the type. A similar integral error term appears for the function λ discussed in (6.3). If α has type $\leq f$, let us write $\psi(t) = \omega(t)/t$. Under simple conditions on the type in relation to ω, but especially that ω tends to infinity faster than the type, one finds that

(7.3) $$\lambda(N) = \int_1^N \psi(t) \, dt + O\left(\int_1^N \frac{\omega(t)^{1/2} f(t)^{1/2}}{t} dt \right).$$

Thus if ω grows much faster than f, then the first integral dominates the second integral. In particular, if $\omega(t) = at$ with $0 < a \leq 1$, then this amounts to the usual "equidistribution" function, and we have

$$\lambda(N) = aN + O\left(\int_1^N \frac{f(t)^{1/2}}{t^{1/2}} dt \right).$$

If $f(t)/t$ tends to 0 as t becomes large, then the error term is $o(N)$. This gives a quantitative description of the equidistribution of the numbers $q\alpha$ on the circle, in terms of the type for α. For the proofs, see [50].

Because of Khintchine's convergence theorem (6.2) one sees that almost all numbers have type $(\log t)^{1+\epsilon}$ (for instance) so that the above asymptotic results apply for almost all numbers. But their formulation in terms of the type makes them applicable to specific numbers, and thus reduces the study of such asymptotic estimates to a determination of the type. However it is clear from the form of the error terms above that they are significant only for "good" types (e.g. satisfying the condition $f(t)/t \to 0$). Only for algebraic numbers or a few isolated transcendental ones is such a type known at present.

Adams [5] extended these theorems to the higher dimensional case, when the type is defined by an inequality

$$\|q_1\alpha_1 + \cdots + q_n\alpha_n\| > 1/q^n f(q)$$

and $q = \max |q_i|$. According to Schmidt [74], the measure theoretic expectancy is that $\lambda(N)$ is asymptotic to

$$\int_1^N \psi(t)^n \, dt$$

and Adams proves this with an integral error term generalizing (7.3). Schmidt [75] did it for the basis of a real algebraic number field.

When the function ω does not grow faster than the type f, then the above error terms break down and each number will be expected to exhibit its own peculiarities. We discuss some special types in the next two sections. Here we still mention the connection with the asymptotic function λ in the few known cases.

The Liouville inequality (see below) shows that quadratic numbers are of bounded type. In that case, I proved [50]:

(7.4) *Let α be a real quadratic number. Let c be a number such that the inequalities*

$$0 < q\alpha - p < 1/q \quad \text{and} \quad 1 \leq q \leq N$$

have infinitely many solutions. Let $\lambda(N)$ be the number of solutions. Then $\lambda(N) \sim c_1 \log N$ for some constant c_1.

This was extended to a basis of a real algebraic number field by Adams [6].

Adams [2] also gives the value of λ for e and numbers having the same kind of continued fraction, namely Hurwitz numbers (cf. Perron [66]). His result for e is typical:

(7.5) Let $g(x)$ be the inverse function of the function $4^x\Gamma(x+3/2)$, so that $g(x)$ is asymptotic to $\log x/\log \log x$. Let $\lambda(N)$ be the number of solutions of the inequalities

$$0 < qe - p < 1/q \quad \text{and} \quad 1 \leq q \leq N.$$

Then

$$\lambda(N) = \tfrac{1}{6}(2g(N))^{3/2} + O(g(N)).$$

Using the same technique as Adams, looking at the continued fraction, I determined the type of e as being $\leq 2g+O(1)$. This is best possible in the sense that subtracting a sufficiently large constant is not a type. Cf. [50]. In this case, the type is such that the function

$$\psi(q) = 1/qf(q)$$

has a divergent sum. The number e is behaving better than almost all numbers from the point of view of being badly approximable by rationals.

No other sharp statements like these are known at present. For instance, a type for e^3 similar to the above is not known. Computations confirm that a power of the log should be an expected upper bound for the types of classical numbers [8]. The computations actually suggest that the types differ from the log by a factor having lower order of magnitude.

I believe that the continued fraction for e, which has a "formula"

$$[2, 1, 2, 1, 1, 4, 1, 1, 6, 1, \cdots],$$

is accidental in the sense that, say for algebraic numbers of degree >2, the continued fraction should be essentially random. It would lead us too far to discuss continued fractions in detail here. We just mention a couple of their properties. If α is real, irrational, we let $a_0 = [\alpha]$ be the largest integer $\leq \alpha$. Then $\alpha - a_0 = 1/\alpha_1$ and $\alpha_1 > 1$. Write $\alpha_1 = a_1 + 1/\alpha_2$ with a positive integer a_1, and continue like this. Then $[a_0, a_1, a_2, \cdots]$ is called the continued fraction for α. What matters for us is that the partial fractions

$$[a_0, a_1, \cdots, a_n] = p_n/q_n$$

provide solutions of the fundamental inequality

$$|q_n\alpha - p_n| < 1/q_n,$$

and that $q_{n+1} \leq a_{n+1}q_n$. Thus the integers a_n ($n \geq 1$) in some way measure how far apart solutions of the fundamental inequality may be. The nth such solution, given by the continued fraction, has an order of magnitude equal to the product $a_1 \cdots a_n$. In the case of e

and numbers like it, this is something like $n!$ (the precise function is given by (7.5)). For proofs, see [2], or [50].

8. Algebraic numbers. If α is algebraic of degree $n>1$ (over $\mathcal{Q}$), then Liouville remarked that one has the trivial inequality

$$|\alpha - p/q| \geq c(\alpha)/q^n,$$

for some number $c(\alpha)$ which is easily computable in terms of the degree and discriminant of α. Indeed, let

$$f(X) = c \prod_{i=1}^{n} (X - \alpha_i)$$

be the irreducible polynomial of $\alpha = \alpha_1$ over $\mathcal{Q}$, taken with relatively prime integer coefficients, so that c is an integer ≥ 1. If p/q is close to α, then p/q is at a distance from any conjugate α_i of α approximately equal to $|\alpha_i - \alpha|$. Since $f(p/q) \neq 0$, we get the inequality

$$1/q^n \leq |f(p/q)| = |c| \prod_{i=1}^{n} |p/q - \alpha_i|.$$

The Liouville inequality follows at once from this factorization. (For generalizations of the Liouville inequality to polynomials in several variables, cf. Feldman [35], [36], [38], and [50] where some of Feldman's results are reproduced.)

In particular, if α is quadratic, then α is of bounded type. I would conjecture that no other "natural" number has bounded type, although random continued fractions with bounded entries provide random examples of such numbers. By "natural," I mean algebraic of degree >2, or transcendental in the field mentioned previously, generated by values of classical functions suitably normalized. It is however unknown except in a few cases like e whether any such numbers are of bounded type. In particular, it is unknown for any algebraic number of degree >2, say $2^{1/3}$, and for e^3.

Improvements on the exponent n in Liouville's theorem have proved to be very difficult to obtain. Thue and Siegel reduced this exponent significantly, in a manner depending on the degree, but Roth [72] gave the best exponent (conjectured by Siegel), namely:

(8.1) *If α is algebraic irrational, then there are only a finite number of solutions to the inequality $\|q\alpha\| < 1/q^{1+\epsilon}$.*

The first time that the correct exponent $1+\epsilon$ appeared in the literature was in Schneider's paper [81]. However, in parts of his proof, the arguments were still too weak to give the full result, and

Schneider could only prove that solutions q of the above inequality have to tend rapidly to infinity.

Following an idea of Siegel, Schneider considered a sequence of solutions p_i/q_i for the approximation

$$|\alpha - p_i/q_i| < 1/q_i^{2+\epsilon}$$

and constructed a polynomial in several variables $F(X_1, \cdots, X_m)$ such that the Taylor expansion

$$F(p/q) = F\left(\frac{p_1}{q_1}, \cdots, \frac{p_m}{q_m}\right)$$

$$= \sum F^{(j)}(\alpha, \cdots, \alpha)\left(\frac{p_1}{q_1} - \alpha\right)^{j_1} \cdots \left(\frac{p_m}{q_m} - \alpha\right)^{j_m}$$

starts only with a zero of high order, the order being measured by the special weights

$$\frac{j_1}{r_1} + \cdots + \frac{j_m}{r_m}$$

where $r_1, \cdots, r_m$ are the degrees of F in $X_1, \cdots, X_m$ respectively. By taking m large, solving appropriate linear equations, and taking sufficiently big gaps between the fractions p_i/q_i, one sees that the value $F(p/q)$ is small from the right-hand side if the p_i/q_i approximate α very closely. On the other hand, one can show that the linear equations achieving this can be solved with integer coefficients, and a denominator for $F(p/q)$ is at most

$$q_1^{r_1} \cdots q_m^{r_m}.$$

If $F(p/q) \neq 0$, then multiplying by such a denominator one gets an integer of absolute value ≥ 1, in other words, one gets an inequality on the left which contradicts the inequality on the right. Schneider's argument to find the polynomial F such that $F(p/q) \neq 0$ required large gaps between the fractions p_i/q_i. Roth improved this part of the proof, and showed that even if one gets a polynomial F such that $F(p/q) = 0$, then some suitable derivative of F will not vanish at p/q, thus getting his theorem. About 20 years had elapsed since Schneider's proof in 1936.

Even though the method of proof yields a bound on m (for complements on this, see Davenport and Roth [31]), it does not yield a bound on the size of the approximating fractions. Thus Roth's theorem is called "noneffective."

Contrary to statements sometimes made that Roth's theorem gives a "best possible approximation inequality," one can ask for a better one, namely one can ask for the determination of a type for algebraic numbers having a lower order of growth than epsilon in the exponent. As far as I know, this was first mentioned in [48], with the possibility of a type $(\log t)^{1+\epsilon}$, in line with the Khintchine convergence principle.

It is difficult to make precisely correct guesses, because within the range of the logarithm as a type, each number will exhibit its own peculiar behavior. It is conceivable that even a type $\log t$ occurs (even though the sum $\sum 1/q \log q$ diverges!). A few computations for some transcendental numbers definitely suggest such a low order of growth, fairly close to the log [8]. The Adams result for e, and further experience confirm that it is not likely that one gets precisely the logarithm as a type, but rather various small perturbations of it, depending on each particular number.

For algebraic numbers of degree >2, I expect basically a random behavior for the continued fraction, or at most a small departure from the random behavior. Some tables for continued fractions of a few algebraic numbers confirm this [96]. There occur some exceptionally large values among a generally uniform random behavior. This suggests the problem of determining whether such relatively large values continue to occur throughout the continued fraction, or whether they stop. If they do not stop, then the problem is to determine how they affect an otherwise rather smooth type. The existence of some exceptionally large integers in the continued fractions of numbers related to values of the modular function had already been observed by Brillhart (cf. Churchhouse and Muir [95], and also a forthcoming paper by Stark [97]).

Another regularity lies in the frequency count of the numbers in the continued fraction. According to a theorem of Kuzmin (see [46]) for almost all numbers α, the probability that the nth number a_n in the continued fraction for α is equal to a positive integer k is given by

$$\log_2 \frac{(k+1)^2}{k(k+2)}.$$

For $k=1$ and almost all numbers, this means that the probability for $a_n = 1$ is approximately .41. Among the first thousand a_n for $2^{1/3}$, $5^{1/3}$, $7^{1/3}$ (say) we find that 1 occurs respectively 422, 433, 409 times, which is rather close to the Kuzmin number.

The structure of Schneider's argument exhibiting the exponent $2+\epsilon$ in the inequality $|\alpha - p/q| < 1/q^{2+\epsilon}$ is purely combinatorial, and as such is applicable to more general contexts (e.g. function fields with algebraically closed constant fields). It is when dealing with algebraic numbers (or function fields over finite fields) that one can expect q^ϵ to be replaced by a function of q growing more like the log.

Recently Schmidt [76] gave a far reaching generalization of Roth's theorem, by extending it to vectors of algebraic numbers.

(8.2) Let $1, \alpha_1, \cdots, \alpha_n$ be algebraic, linearly independent over the rationals. Then the inequality

$$\|q_1\alpha_1 + \cdots + q_n\alpha_n\| < 1/q^{n+\epsilon}, \quad q = \max|q_i|,$$

has only a finite number of solutions.

A standard transference principle of Khintchine [24], [77], shows that (8.2) is equivalent with

(8.3) If $1, \alpha_1, \cdots, \alpha_n$ are linearly independent over the rationals, then the simultaneous inequalities

$$\|q_i\alpha\| < \frac{1}{q_n^{1+\epsilon}}$$

have only a finite number of solutions.

Schmidt's proof uses the same techniques as Schneider and Roth, but also relies heavily on Mahler's theory of compound convex bodies [62]. For more details, I refer the reader to Schmidt [76], [77]. He also proves theorems concerning the approximation of algebraic numbers by other algebraic numbers. If P is a polynomial with integer coefficients, we let $H = H(P)$ be the height of P, namely the maximum of the absolute values of its coefficients. If β is algebraic and P is the irreducible polynomial for β with relatively prime integer coefficients, we let $H(P)$ be the height of β. Schmidt proves:

(8.4) Let α be algebraic and let n be a positive integer. There are only finitely many algebraic numbers β of degree $\leq n$ such that

$$|\alpha - \beta| < \frac{1}{H(\beta)^{n+1+\epsilon}}.$$

This generalizes results of Wirsing [93], who had the weaker exponent $2n+\epsilon$. Schmidt's exponent is best possible. It is easily seen that (8.4) follows from (8.2). Schmidt's results, coming 15 years after Roth's theorem, again constitute an impressive advance in the subject.

Schmidt's theorem (8.4) has substance only if n is smaller than the degree d of α. Indeed, one has superficially at the level of a Liouville estimate the better result:

(8.5) *Let α be algebraic of degree d. Then there are only finitely many algebraic numbers β such that*

$$|\alpha - \beta| < \frac{1}{H(\beta)^{d+\epsilon}}.$$

A result of Wirsing gives the possibility of using the inequality (8.5), or even weaker ones, as a criterion for algebraicity or transcendence. Indeed, Wirsing proves [94]:

(8.6) *If w is a real transcendental number, then the inequality*

$$|w - \beta| < \frac{1}{H(\beta)^{(n+1)/2+1-\epsilon}}$$

has infinitely many solutions in algebraic numbers β of degree $\leq n$.

The essential thing here is that the exponent depending on n may be arbitrarily large, whereas for algebraic numbers, it is bounded as in (8.5). This result of Wirsing relates to the classification of numbers by Mahler and Koksma [47] (see also Schneider [80]), considering the lower bound for a polynomial $|P(w)|$.

Roth's theorem has p-adic analogues [71]. It can also be axiomatized to cover cases in algebraic geometry, and a finite number of absolute values, satisfying the product formula [52], e.g. in the function field case. However, when the constant field is, say, algebraically closed, one does not expect the type q^ϵ to be improvable. Only when the constant field is finite would I expect again an improvement in the type, to a power of the log (assuming that the irrationality α has only tame ramification).

The first improvement on the Liouville inequality which was effective is due to Baker [14], [15]. By using the same method as that for linear combinations of logarithms (2.3) he proves:

(8.7) *Let α be algebraic of degree $n \geq 3$ and let $k > n+1$. Then there is a computable constant $c = c(\alpha, k)$ such that*

$$\|q\alpha\| > c \frac{e^{(\log q)^{1/k}}}{q^n}$$

(This is improved to $k > 1$ in the joint paper with Stark [18].) Thus the effective result is still quite far from having the best exponent on q. In fact, it does not yet yield the possibility of lowering the exponent n

in the denominator by ϵ. However, it is significant because it allows us to give effective bounds for solutions of a wide class of diophantine equations.

Note that Coates gives p-adic extensions to Baker's results [27], also announced by Feldman [40].

9. **Some transcendence measures.** Given classical numbers $x_1, \cdots, x_n$ one is interested in a lower bound for linear combinations

$$|q_0 + q_1 x_1 + \cdots + q_n x_n| > F(q), \quad q = \max|q_i|,$$

with sufficiently large q. Such a function F is called a measure of linear independence for the numbers $x_1, \cdots, x_n$. Historically, there have been three levels of difficulty in determining the function F, namely:

Stage 1. When $F(q) = 1/q^{\phi(q)}$, and ϕ is increasing to infinity.

Stage 2. When $F(q) = 1/q^c$ for some fixed number c.

Stage 3. When $F(q) = 1/q^{n+\epsilon}$, the best possible expected exponent of q, in view of Dirichlet's theorem.

Ultimately, one wants the even better results involving the type, namely $1/q^n f(q)$, where f grows like the log, or a power of the log. No such results are known at present except those mentioned in §7. We could call this *Stage* 4. This is the point of view taken in [50].

In dealing with one number x which is transcendental, one considers the numbers $x_i = x^i$, so that the linear combination above can be written as a polynomial in x of degree n, namely

$$|q_0 + q_1 x + \cdots + q_n x^n| > F(q).$$

The function F is then called a measure of transcendence for x. We shall summarize some of the known results, which should be regarded as very tentative.

Even though the sharpest type for e became known only in 1966, it had been known long before that e is of type $\leq q^\epsilon$, in other words $\|qe\| < 1/q^{1+\epsilon}$ has only a finite number of solutions. In fact, Popken had proved [67]:

(9.1) *Let $P(X)$ be a polynomial of degree n with integer coefficients, and let $H = H(P)$ be the maximum of the absolute value of the coefficients. Then there is a number c depending only on n such that*

$$|P(e)| > \frac{1}{H^{n+c/\log \log H}}.$$

In particular,

$$\|qe\| > \frac{1}{q^{1+c/\log\log q}}.$$

For the proof and explicit dependence of c on n, due to Mahler, cf. Schneider's book [80]. In [48], I had already verified that the same argument also works for e^a, where a is rational. In this connection, Mahler [63] uses some of the original formulas by Hermite to get similar approximation estimates to exponential and logarithms of rational numbers. Even though these formulas have disappeared from the picture for transcendence proofs, it seems that they still contain germs of methods which would be useful for measures of approximation, more closely related to continued fractions.

When α is irrational algebraic, and one wants to study e^α from the present point of view, one meets difficulties analogous to those of Roth's theorem.

Siegel had a Roth type theorem for values of the Bessel function J_0, namely the inequality

$$\|q_1 J_0(\alpha) + q_2 J_0'(\alpha)\| < 1/q^{2+\epsilon}$$

has only a finite number of solutions whenever α is rational $\neq 0$, and similarly for a polynomial in $J_0(\alpha)$, $J_0'(\alpha)$ [86]. Here again, the problem is open for algebraic α, or when one deals with J_λ and λ is algebraic irrational. (Actually in this case it is unknown if J_λ is an E-function.) Once Shidlovsky had proved his transcendence result, it was easy to extend Siegel's estimates to arbitrary E-functions (see e.g. [49]).

It is unknown if a single value of the Bessel function $J_0(a)$, with a rational, is of type $\leq q^\epsilon$.

Popken also had the statement analogous to (9.1) for approximations with algebraic numbers, namely:

(9.2) *For all algebraic α of degree n and height $H \geq 3$, one has*

$$|e - \alpha| > 1/H^{n+1+c/\log\log H},$$

where c depends only on n.

Lower bounds which are somewhat worse for numbers α^β with α algebraic, β algebraic have been obtained by Gelfond and Feldman. Cf. Gelfond [44], Schneider [80], also for further reference to the literature. The Gelfond result does not show that α^β has type $\leq q^\epsilon$. It is weaker, roughly like

$$q^{(\log \log q)^\tau},$$

where τ is a low number like 5. The situation has been going through various improvements, so one must check the latest bulletins to know how much the 5 is reduced. From our point of view here, what matters is that the exponent of q is not even a fixed number, independent of q. The situation here is still in Stage 1.

For logarithms of algebraic numbers, the situation is slightly better. Mahler [58] had proved that for α algebraic,

$$\|q \log \alpha\| > 1/q^c$$

with some fixed number c, depending only on α. Feldman, following results of Gelfond, improved these results, and also obtained similar results for approximation by algebraic numbers [59], [60], [80]. For π, Mahler [60] had shown that for all positive q we have

$$\|q\pi\| > 1/q^{42}.$$

(The exponent 42 can be reduced if one allows a constant factor on the right.) None of these results has any semblance of finality.

The result of Baker, whose proof we sketched in §2, represented the first effective lower bound for linear forms in logarithms of several algebraic numbers. Current research is somewhere between Stage one and two. For instance, Feldman has reached Stage two for linear combinations of n logarithms of algebraic numbers, with $F(q) = q^{-c}$, where c depends on n and the logarithms [40].

Baker has also proved analogous results in Stage one for elliptic functions, and periods of elliptic functions [12], [13]. Through Baker's ideas, the whole theory is now in a considerable state of flux, with continuous extensions of Baker's method by Baker and Feldman [18], [41]. Here again one can formulate general conjectures for generalized logarithms on algebraic groups, especially toruses [51]. We shall mention these in connection with the applications to diophantine analysis.

Finally we mention that Baker type results can be used to give an upper bound for the discriminant of an imaginary quadratic field of class number 1 and 2 (again, consult the latest bulletins for improvements on this). It would take us here too far afield to give more details, and we refer the reader to the latest papers of Baker and Stark [18], [16], [91].

The Riemann Hypothesis would give a very good and very effective lower bound for the class number in terms of the discriminant via the Brauer-Siegel theorem. It is not clear (to me) from the present proofs of Baker and Stark if very good results in estimates for linear forms of logarithms of algebraic numbers would give as good a bound as the Riemann Hypothesis. It is also not clear to me if such results might in fact prove the absence of zeros of the zeta function of (quadratic) fields near 1.

10. A criterion of Gelfond. Gelfond gave a useful criterion to prove that two numbers are algebraically independent, namely:

(10.1) *Let x be a complex number. Let σ be a strictly monotone increasing real function tending to infinity, and assume that there is a number $a_0 > 1$ such that $\sigma(N+1) < a_0 \sigma(N)$ for all integers $N > N_0$. Assume that for each integer $N > N_0$ there exists a nonzero polynomial F_N with integer coefficients, such that*

$$|F_N(x)| < e^{-C\sigma(N)},$$

where $C = 50 a_0^2$, and

$$\max(\deg F_N, \log|F_N|) \leq \sigma(N).$$

Then x is algebraic.

(*Notation.* $|F_N|$ is the height of F_N.) In his book [44], he proved only a weaker version. I gave a proof in [48]. Waldschmidt has extended this theorem and proved interesting new applications by separating σ into two functions, describing independently how the degree and the height of the polynomials F_N grow [92].

It should be noted that the conjecture expressed at the end of [48], attempting to give a generalization to a criterion of algebraic independence for several numbers, is not valid. The expected exponent for polynomials F_N in several variables would be σ^{n+1}. However, Bombieri pointed out to me that Cassels [24] proves the existence of numbers whose linear combinations tend to 0 very rapidly. Indeed, by Theorem XIV of Chapter V, loc. cit, for every function $\psi(t)$, decreasing to 0, there exist numbers α_1, α_2 having the following property. For large positive integers q, there exist q_1, q_2 with $|q_i| < q$ such that

$$0 < \|q_1 \alpha_1 + q_2 \alpha_2\| < \psi(q).$$

Thus the polynomials F_q have degree 1. This is a phenomenon which appears only when $n \geq 2$. It is a problem to formulate those supplementary hypotheses which must be added to generalize the Gelfond criterion to several variables.

11. Applications to diophantine analysis. The Baker theorem (8.4) still corresponds to Stage 1 in the approximation theory. However, its effective improvement on the Liouville estimate allows affective bounds for the solutions in integers of a wide class of diophantine equations. For instance, let

$$F(X, Y) = a_n X^n + a_{n-1} X^{n-1} Y + \cdots + a_0 Y^n$$

be a binary form of degree ≥ 3 with integer coefficients. Let m be a positive integer. Any improvement of the Liouville inequality immediately shows that the equation

$$F(X, Y) = m$$

has only a finite number of solutions. Namely, we factor F into factors of degree 1, which are of the form $(X - \alpha^{(i)} Y)$ where the $\alpha^{(i)}$ are conjugates of an algebraic number α (if F is irreducible), otherwise split into families of conjugates. If p, q are integers such that $F(p, q) = m$, then it follows immediately from any improvement of Liouville's inequality that the number of such p, q is finite. With his effective improvement, Baker is led to an estimate [14]:

$$\log \max(|p|, |q|) < (nH)^{(10n)^5} + (\log m)^{2n+2},$$

where H is the height of F.

Using his theorem on diophantine approximations, Siegel [86] has shown:

(11.1) *If a curve of genus ≥ 1 is defined by an equation $f(X, Y) = 0$ with an irreducible polynomial f having algebraic coefficients, then this curve has only a finite number of points (x, y) with x, y lying in the ring of algebraic integers of a number field.*

Mahler extended this by using p-adic analogues of the Thue-Siegel theorem, and in fact the result holds for arbitrary finitely generated rings (without divisors of 0) over the integers $\mathbf{Z}$. Cf. [52]. Siegel's argument is based on a geometric version of the diophantine approximation inequality. If $P = (\alpha_0, \cdots, \alpha_n)$ is a point in projective space, represented with projective coordinates, one can define its height. For instance, if the α_i are relatively prime integers, then the height $H(P)$ is the maximum of the absolute values $|\alpha_i|$. The definition in number fields is similar. The inequality of Roth's theorem

$$|\alpha - p/q| < 1/q^{2+\epsilon}$$

then has a geometric analogue as follows.

(11.2) *Let V be a curve defined over a number field K. Let ϕ be a nonconstant rational function in $K(V)$. Let r be the maximum of the orders of zeros of ϕ. Then there are only finitely many rational points P of V in K satisfying the inequality*

$$|\phi(P)| < 1/H(P)^{2r+\epsilon}.$$

This is easily reduced to Roth's theorem. For instance, if V is a rational curve, with function field $K(x)$, then a rational function $R(x)$ has a factorization

$$R(x) = \prod (x - \alpha_i)^{r_i}$$

where the r_i are the multiplicities of the zeros and poles, and the α_i are algebraic numbers. Say $K = \mathbf{Q}$. A rational point is a fraction p/q, and $R(p/q)$ is small if and only if p/q approximates one of the roots of $R(x)$. In that case, such a fraction stays away from the other roots, and the equivalence between (11.2) and Roth's theorem is obvious. When V is not a rational curve, one reduces the proof of (11.2) to the case of a rational curve by projecting. Cf. [52].

Siegel [86] had an analogous theorem to (11.2), but corresponding to the weaker approximation result which he had available at that time.

In dealing with an integral point P, or say a point P such that $\phi(P)$ is an integer, one gets trivially an inequality like (11.2), except that the exponent of $H(P)$ on the right is much larger than the desired $2r+\epsilon$. For curves of genus 1, using the group law arising from the theory of elliptic functions, one can then reduce the exponent to an arbitrarily small one, whence the result. For curves of higher genus, Siegel [86] used abelian functions, and the analytic representation of the Jacobian variety of the curve. The arguments can be algebraicized, and in fact one can express the idea geometrically by saying that to reduce the exponent so as to apply (11.2) one must go to a covering of the given curve, by restricting to the curve the unramified covering of the Jacobian, obtained by division of the periods, i.e. the mapping

$$a \mapsto na$$

on the Jacobian, for a large integer n. Cf. [52].

These arguments are not effective in two ways: first in the non-effectiveness of the Roth theorem, and second in the use of the Mordell-Weil theorem when jacking up the diophantine approximation to unramified coverings of the curve.

For curves of genus 1, Baker and Coates [17] gave a different argument. Using the Riemann-Roch theorem, they reduce the study of integral points on a curve of genus 1 to that of a curve in standard form, $Y^2 = f(X)$, which can then be handled by a reduction to Baker's theorem (2.3), whence they get an effective upper bound for its integral points. So far, this method has no analogue to curves of higher genus, or abelian varieties, but it is worth emphasizing that the Coates-Baker argument represents a new approach to the finiteness problems we have been discussing.

By considering generalized logarithms on an elliptic curve, and using the idea that they should behave from the point of view of diophantine approximations so as to be of type $\leq q^\epsilon$, I conjectured that (11.2) should generalize as follows [51]:

(11.3) *Let A be an elliptic curve defined over a number field K. Let ϕ be a nonconstant function in $K(A)$. Let r be the maximum multiplicity of its zeros, and let m be the rank of the group of rational points A_K of A in K. Then there should be only a finite number of points P in A_K satisfying the inequality*

$$|\phi(P)| \leq 1/h(P)^{r(m+1)/2+\epsilon}$$

where $h(P) = \log H(P)$ is the logarithmic height.

Observe here that we are dealing with a lower order of magnitude on the right-hand side than in (11.2). Of course, a similar conjecture can be made in terms of a type for the logarithms of algebraic points, replace the $h(P)^\epsilon$ which corresponds to q^ϵ. The inequality of (11.3) is equivalent to an inequality involving logs of algebraic points. Indeed, let $P_1, \cdots, P_m$ be free generators of A_K modulo the torsion group. Let $u_j = \log P_j$. Let ω_1, ω_2 be fundamental periods. Let

$$P = q_1 P_1 + \cdots + q_m P_m + Q$$

where Q is in the torsion group. Since this torsion group is finite, we may assume that when we consider infinitely many P, the same Q appears. Such points P have a point of accumulation P_0. We let $u_0 = \log(P_0 - Q)$. Then the inequality of (11.3) amounts to

$$|-u_0 + q_1 u_1 + \cdots + q_m u_m + q_{m+1} \omega_1 + q_{m+2} \omega_2| \leq 1/q^{m+1+\epsilon},$$

which has the standard recognizable form. The same remark applies of course to ordinary logarithms.

Bibliography

1. W. W. ADAMS, *Asymptotic Diophantine approximations to e*, Proc. Nat. Acad. Sci. U.S.A. **55** (1966), 28–31. MR **32** #4085.
2. ———, *Asymptotic Diophantine approximations and Hurwitz numbers*, Amer. J. Math. **89** (1967), 1083–1108. MR **36** #5082.

3. ———, *Simultaneous asymptotic Diophantine approximations to a basis of a real cubic number field*, J. Number Theory **1** (1969), 179–194. MR **39** #1409.
4. ———, *A lower bound in asymptotic Diophantine approximations*, Duke Math. J. **35** (1968), 21–35. MR **36** #5083.
5. ———, *Simultaneous asymptotic Diophantine approximations*, Mathematika **14** (1967), 173–180. MR **36** #3730.
6. ———, *Simultaneous asymptotic Diophantine approximation to a basis of a real number field*, Nagoya Math. J. (to appear).
7. ———, *Transcendental numbers in the P-adic domain*, Amer. J. Math. **88** (1966), 279–308. MR **33** #5564.
8. W. Adams and S. Lang, *Some computations in Diophantine approximations*, J. Reine Angew. Math. **220** (1965), 163–173. MR **32** #91.
9. L. Alaoglu and P. Erdös, *On highly composite and similar numbers*, Trans. Amer. Math. Soc. **56** (1944), 448–469. MR **6**, 117.
10. J. Ax, *On Schannuel's conjecture*, Ann. of Math. (2) **93** (1971), 252–268 (and another paper to appear).
11. A. Baker, *Linear forms in the logarithms of algebraic numbers*. I, II, III, Mathematika **13** (1966), 204–216; ibid. **14** (1967), 102–107, 220–228. MR **36** #3732.
12. ———, *An estimate for the $\wp$-function at an algebraic point*, Amer. J. Math. (to appear).
13. ———, *On the quasi-periods of the Weierstrass zeta function*, Nachr. Akad. Wiss. Göttingen Math.-Phys. Kl. II **1969**, 145–157.
14. ———, *Contributions to the theory of Diophantine equations*. I. *On the representation of integers by binary forms*, Philos. Trans. Roy. Soc. London Ser. A **263** (1967/68), 173–191; II. *The Diophantine equation* $Y^2 = X^3 + k$, ibid., 193–208. MR **37** #4005; #4006.
15. ———, *The Diophantine equation* $Y^2 = ax^3 + bx^2 + cx + d$, J. London Math. Soc. **43** (1968), 1–9. MR **38** #111.
16. ———, *Imaginary quadratic fields with class number 2*, Ann. of Math. (to appear).
17. A. Baker and J. Coates, *Integer points on curves of genus* 1, Proc. Cambridge Philos. Soc. **67** (1970), 595–602. MR **41** #1638.
18. A. Baker and H. Stark, *On a fundamental inequality in number theory*, Ann. of Math. (to appear).
19. H. Behnke, *Über die Verteilung von Irrationalitaten mod 1*, Abh. Math. Sem. Univ. Hamburg **1** (1922), 252–267.
20. ———, *Zur Theorie der diophantischen Approximationen*, Abh. Math. Sem. Univ. Hamburg **3** (1924), 261–318.
21. E. Bombieri, *Algebraic values of meromorphic maps*, Invent. Math. **10** (1970), 267–287.
22. E. Bombieri and S. Lang, *Analytic subgroups of group varieties*, Invent. Math. **11** (1970), 1–14.
23. A. Brumer, *On the units of algebraic number fields*, Mathematika **14** (1967), 121–124. MR **36** #3746.
24. J. W. S. Cassels, *An introduction to Diophantine approximation*, Cambridge Tracts in Math. and Math. Phys., no. 45, Cambridge Univ. Press, New York, 1957. MR **19**, 396.
25. J. Coates, *An effective p-adic analogue of a theorem of Thue*, Acta Arith. **15** (1968/69), 279–305. MR **39** #4095.
26. ———, *Construction of rational functions on a curve*, Proc. Cambridge Philos. Soc. **68** (1970), 105–123.
27. ———, *An effective p-adic analogue of a theorem of Thue*. II. *The greatest prime factor*

of a binary form, Acta Arith. **17** (1970), 399–412; *The Diophantine equation* $Y^2 = X^3 + k$, ibid., 425–435.

28. ———, *The transcendence of linear forms in* $\omega_1, \omega_2, \eta_1, \eta_2, 2\pi i$ (to appear).
29. A. BAKER AND J. COATES, *Integer points on curves of genus* 1, Proc. Cambridge Philos. Soc. **67** (1970), 595–602. MR **41** #1638.
30. R. M. DAMERELL, *L-functions of elliptic curves with multiplication*. I, Acta Arith. **17** (1970), 287–301.
31. H. DAVENPORT AND K. ROTH, *Rational approximations to algebraic numbers*, Mathematika **2** (1955), 160–167. MR **17**, 1060.
32. H. DAVENPORT AND W. SCHMIDT, *Dirichlet's theorem on Diophantine approximations*. II, Acta Arith. **17** (1970), 413–424.
33. P. ERDÖS, *Some results on Diophantine approximation*, Acta Arith. **5** (1959), 359–369. MR **22** #12091.
34. N. I. FELDMAN, *Approximation of certain transcendental numbers*. I: *The approximation of logarithms of algebraic numbers*; II: *The approximation of certain numbers associated with the Weierstrass function*, Izv. Akad. Nauk SSSR Ser. Mat. **15** (1951), 53–74, 153–176; English transl., Amer. Math. Soc. Transl. (2) **59** (1966), 224–270. MR **12**, 595; MR **13**, 117.
35. ———, *Simultaneous approximation of the periods of an elliptic function by algebraic numbers*, Izv. Akad. Nauk SSSR Ser. Mat. **22** (1958), 563–576; English transl., Amer. Math. Soc. Transl. (2) **59** (1966), 271–284. MR **20** #5895.
36. ———, *Approximation of the logarithms of algebraic numbers by algebraic numbers*, Izv. Akad. Nauk SSSR Ser. Mat. **24** (1960), 475–492; English transl., Amer. Math. Soc. Transl. (2) **58** (1966), 125–142. MR **22** #5623b.
37. ———, *On the measure of transcendence of* π, Izv. Akad. Nauk SSSR Ser. Mat. **24** (1960), 357–368; English transl., Amer. Math. Soc. Transl. (2) **58** (1966), 110–124. MR **22** #5632a.
38. ———, *Arithmetic properties of the solutions of a transcendental equation*, Vestnik Moskov. Univ. Ser. I Mat. Meh. **1964**, no. 1, 13–20; English transl., Amer. Math. Soc. Transl. (2) **66** (1968), 145–153. MR **28** #2091.
39. ———, *Estimate for a linear form of logarithms of algebraic numbers*, Mat. Sb. **76** (118) (1968), 304–319 = Math. USSR Sb. **5** (1968), 291–307. MR **37** #4025.
40. ———, *Improved estimate for a linear form of the logarithms of algebraic numbers*, Mat. Sb. **77** (119) (1968), 423–436 = Math. USSR Sb. **6** (1968), 398–406. MR **38** #1059.
41. ———, *A certain inequality for a linear form in the logarithms of algebraic numbers*, Mat. Zametki **5** (1969), 681–689. (Russian) MR **40** #2610.
42. A. O. GELFOND AND N. I. FELDMAN, *On the measure of relative transcendentality of certain numbers*, Izv. Akad. Nauk SSSR Ser. Mat. **14** (1950), 493–500. (Russian) MR **12**, 679.
43. A. O. GELFOND, *Sur les propriétés arithmétiques des fonctions entières*, Tôhoku Math. J. **30** (1929), 280–285.
44. ———, *Transcendental and algebraic numbers*, GITTL, Moscow, 1952; English transl., Dover, New York, 1960. MR **15**, 292; MR **22** #2598.
45. C. HERMITE, "Sur la fonction exponentielle," in *Oeuvres*, Vol. III, pp. 150–181.
46. A. YA. KHINCHIN (A. JA. HINČIN), *Continued fractions*, Fizmatgiz, Moscow, 1961; English transl., Univ. of Chicago Press, Chicago, Ill., 1964. MR **28** #5037.
47. J. F. KOKSMA, *Über die Mahlersche Klasseneinteilung der transzendenten Zahlen und die Approximation komplexer Zahlen durch algebraische Zahlen*, Monatsh. Math. Phys. **48** (1939), 176–189. MR **1**, 137.

48. S. LANG, *Report on Diophantine approximations*, Bull. Soc. Math. France **93** (1965), 177–192. MR **33** #1286.
49. ———, *Introduction to transcendental numbers*, Addison-Wesley, Reading, Mass., 1966. MR **35** #5397.
50. ———, *Introduction to Diophantine approximations*, Addison-Wesley, Reading, Mass., 1966. MR **35** #129.
51. ———, *Diophantine approximations on toruses*, Amer. J. Math. **86** (1964), 521–533. MR **29** #2220.
52. ———, *Diophantine geometry*, Interscience Tracts in Pure and Appl. Math., no. 11, Interscience, New York, 1962. MR **26** #119.
53. ———, *Algebraic number theory*, Addison-Wesley, Reading, Mass., 1970.
54. H. W. LEOPOLDT, *Zur Arithmetik in abelschen Zahlkörpern*, J. Reine Angew. Math. **209** (1962), 54–71. MR **25** #3034.
55. W. J. LEVEQUE, *On the frequency of small fractional parts in certain real sequences.* II, Trans. Amer. Math. Soc. **94** (1960), 130–149. MR **22** #12089.
56. F. LINDEMANN, *Über die Zahl π*, Math. Ann. **20** (1882), 213–225.
57. K. MAHLER, *Über transzendente p-adische Zahlen*, Compositio Math. **2** (1935), 259–275.
58. ———, *Zur Approximation der Exponentialfunktion und des Logarithmus*, J. Reine Angew. Math. **66** (1932), 118–150.
59. ———, *On the approximation of logarithms of algebraic numbers*, Philos. Trans. Roy. Soc. London Ser. A **245** (1953), 371–398. MR **14**, 624.
60. ———, *On the approximation of π*, Nederl. Akad. Wetensch. Proc. Ser. A. **56** = Indag. Math. **15** (1953), 30–42. MR **14**, 957.
61. ———, *Ein Übertragungsprinzip für konvexe Körper*, Casopis Pěst. Mat. Fys. **68** (1939), 93–102. MR **1**, 202.
62. ———, *On compound convex bodies.* I, Proc. London Math. Soc. (3) **5** (1955), 358–379. MR **17**, 589.
63. ———, *Applications of some formulae by Hermite to the approximation of exponentials and logarithms*, Math. Ann. **168** (1967), 200–227. MR **34** #5754.
64. A. NÉRON, *Quasi-fonctions et hauteurs sur les variétiés abéliennes*, Ann. of Math. (2) **82** (1965), 249–331. MR **31** #3424.
65. A. OSTROWSKI, *Bemerkungen zur Theorie der Diophantischen Approximationen*, Abh. Math. Sem. Univ. Hamburg **1** (1921), 77–98.
66. O. PERRON, *Die Lehre von den Kettenbrüchen*, 2nd ed., Chelsea, New York, 1950. MR **12**, 254.
67. J. POPKEN, *Sur la nature arithmétique du nombre e*, C. R. Acad. Sci. Paris **186** (1928), 1505–1507.
68. ———, *Zur Transzendenz von e*, Math. Z, **29** (1929), 525–541.
69. ———, *Zur Transzendenz von π*, Math. Z. **29** (1929), 542–448.
70. K. RAMACHANDRA, *Some applications of Kronecker's limit formulas*, Ann. of Math. (2) **80** (1964), 104–148. MR **29** #2241.
71. D. RIDOUT, *Rational approximations to algebraic numbers*, Mathematika **4** (1957), 125–131. MR **20** #32.
72. K. ROTH, *Rational approximations to algebraic numbers*, Mathematika **2** (1955), 1–20; corrigendum, 168. MR **17**, 242.
73. W. SCHMIDT, *A metrical theorem in Diophantine approximation*, Canad. J. Math. **12** (1960), 619–631. MR **22** #9482.
74. ———, *Metrical theorems on fractional parts of sequences*, Trans. Amer. Math. Soc. **110** (1964), 493–518. MR **28** #3018.

75. ———, *Simultaneous approximation to a basis of a real numberfield*, Amer. J. Math. 88 (1966), 517–527. MR 34 #2529.
76. ———, *Simultaneous approximation to algebraic numbers by rationals*, Acta Math. 21 (1970), 189–201.
77. ———, *Lectures on Diophantine approximation*, University of Colorado, Boulder, Colo., 1970.
78. T. SCHNEIDER, *Zur Theorie der Abelschen Funktionen und Integrale*, J. Reine Angew. Math. 183 (1941), 110–128. MR 3, 266.
79. ———, *Ein Satz über ganzwertige Funktionen als Prinzip für Transzendenzbeweise*, Math. Ann. 121 (1949), 131–140. MR 11, 160.
80. ———, *Einführung in die transzendenten Zahlen*, Springer-Verlag, Berlin, 1957. MR 19, 252.
81. ———, *Über die Approximation algebraischer Zahlen*, J. Reine Angew. Math 175 (1936), 182–192.
82. J.-P. SERRE, *Abelian l-adic representations*, Benjamin, New York, 1968.
83. ———, "Dependence d'exponentielle p-adiques," in *Séminaire Delange-Pisot-Poitou*, 1965/66, Exposé 15, Secrétariat mathématique, Paris, 1967. MR 35 #6507.
84. A. BOREL, ET AL., *Seminar on complex multiplication*, Lecture Notes in Math., no. 21, Springer-Verlag, Berlin and New York, 1966. MR 34 #1278.
85. A. B. ŠIDLOVSKIĬ, *On criteria for algebraic independence of values of a class of integral functions*, Izv. Akad. Nauk SSSR Ser. Mat. 23 (1959), 35–66; English transl., Amer. Math. Soc. Transl. (2) 22 (1962), 339–370. MR 21 #1295.
86. C. L. SIEGEL, *Über einige Anwendungen diophantischer Approximationen*, Abh. Preuss. Akad. Wiss. 1929, 1–41.
87. ———, *Transcendental numbers*, Ann. of Math. Studies, no. 16, Princeton Univ. Press, Princeton, N. J., 1949. MR 11, 330.
88. ———, *Bestimmung der elliptischen Modulfunktion durch eine Transformationsgleichung*, Abh. Math. Sem. Univ. Hamburg 27 (1964/65), 32–38. MR 29 #2391.
89. V. G. SPRINDŽUK, *The irrationality of the values of certain transcendental functions*, Izv. Akad. SSSR Ser. Mat. 32 (1968), 93–107 = Math. USSR Izv. 2 (1968), 89–104. MR 36 #5087.
90. H. STARK, *A historical note on complex quadratic fields with class-number one*, Proc. Amer. Math. Soc. 21 (1969), 254–255. MR 38 #5743.
91. ———, *A transcendence theorem for class number problems*, Ann. of Math. (to appear).
92. M. WALDSCHMIDT, *Independence algebrique des valeurs de la fonction exponentielle*, Bull. Soc. Math. France (to appear).
93. E. WIRSING, *On approximations of algebraic numbers by algebraic numbers of bounded degree*, Proc. Sympos. Pure Math., vol. 20, Amer. Math. Soc., Providence, R. I., 1971, pp. 213–247.
94. ———, *Approximation mit algebraischen Zahlen beschränkten Grades*, J. Reine Angew. Math. 206 (1960), 67–77, MR 26 #79.
95. CHURCHOUSE AND MUIR, *Continued fractions, algebraic numbers and modular invariants*, J. Inst. Math. Appl. (1969), 318–328.
96. S. LANG AND H. TROTTER, *Continued fractions of some algebraic numbers*, J. Reine Angew. Math. (to appear).
97. H. STARK, *An explanation of some exotic continued fractions found by Brillhart* (to appear).

On the Zeta Function of Number Fields

SERGE LANG (Princeton)

The standard and systematic development of basic analytic number theory is given in Titchmarsh for the rational zeta function [10]. A number of papers (listed in the bibliography, but I don't claim completeness) have extended some of these results to the zeta function, or L-functions in number fields. There is a simple principle, however, which helps in understanding how the Titchmarsh proofs can be extended to the general case of a number field k, namely: Whenever you see a T in an estimate, with a logarithm, then replace it by $d_k T^{N_k}$, where d_k is the absolute value of the discriminant, and N_k is the absolute degree, $[k:\mathbf{Q}]$. The proofs will then go through essentially automatically. If you deal with L-series, say with characters of finite period and conductor $\mathfrak{f}$, then use $d_\chi = d_k |\mathfrak{f}|$ instead of d_k. Here, $|\mathfrak{f}|$ is the absolute norm of $\mathfrak{f}$. Thus it is structurally rather simple to keep track of the constants depending on k, even though the literature is usually loose about this. The constants are essentially not kept track of in Landau [6], the dependence on N_k is not given explicitly in Fogels [2]. Papers like those of Hooley [4] and Goldstein [3] give some explicit dependence on N_k and d_k, but the $d_k T^{N_k}$ principle does not come out clearly in these papers, where the matter is considered as subsidiary to other considerations. In dealing with L-series, Davenport [1] is careful to keep track of the conductor. In his case, $N=1$, so that T is replaced by qT (q = conductor). The noteworthy aspect of the dT^N principle is that the discriminant occurs only to the power 1, not N.

I shall give a systematic summary of the various basic estimates of analytic number theory to show how the principle applies. As an application, one sees on the Riemann Hypothesis that an asymptotic formula is valid for the number of zeros of ζ_k in a box of height T fixed, but with variable k. Namely, if $N_k/\log d_k \to 0$, then

$$\mathbf{N}_k(T) \sim \frac{1}{\pi} T \log d_k.$$

Note that the usual terms like $T \log T$ occurring when k is fixed and $T \to \infty$ are now absorbed by the error terms. The situation here is similar to that of the Brauer-Siegel theorem, cf. [7]. In [9] Siegel deals with a similar problem for L-series, letting the characters vary, and the conductor also. By averaging over the characters, he gets an asymptotic estimate without using the Riemann Hypothesis, essentially for the zeta function of cyclotomic fields. The general case has to remain a conjecture for the moment.

We now describe things in more detail, putting the emphasis on how the standard treatments are to be adjusted in the case of a number field k. The O notation, also expressed by $\ll$, refers to absolute constants. Although these are not given explicitly, they could all be easily computed, and the proofs show that they are not especially large. Numbers depending on k, only through d_k and N_k, are given explicitly. Since k will not change, we write $d = d_k$ and $N = N_k$. Also, we shall want to deal with small values of T, and so we write

$$T_2 = \max(2, T), \qquad t_2 = \max(2, t).$$

We let $\zeta = \zeta_k$, and write $\zeta_\mathbf{Q}$ for the Riemann zeta function. We let

$$A = A_k = 2^{-r_2} \pi^{-N/2} d^{\frac{1}{2}}.$$

The standard proof for the *Riemann-Von Mangoldt formula* yields:

$$\frac{\pi}{2} \mathbf{N}(T) = T \log A + r_2 (T \log T - T) + r_1 \left(\frac{T}{2} \log \frac{T}{2} - \frac{T}{2} \right) \qquad (1)$$

$$+ \pi - r_1 \frac{\pi}{8} + \Delta_V \zeta + \Delta_H \zeta + O\left(\frac{N}{T+1} \right)$$

where $\mathbf{N}(T)$ is the number of zeros of ζ in the box with vertical sides T between 0 and 1; and Δ_V, Δ_H are the variations of argument of the zeta function on the vertical segment from 2 to $2 + iT$, and on the horizontal segment from $2 + iT$ to $\frac{1}{2} + iT$ respectively.

By using the logarithm of the zeta function on the vertical line $\sigma = 2$, and comparing with the log of $\zeta_\mathbf{Q}$, one sees at once that

$$\Delta_V \zeta \ll N. \qquad (2)$$

To get a first estimate for the horizontal variation, one uses the function

$$\varphi(s) = (s-1) \zeta(s),$$

and the function

$$g(s) = \frac{1}{2} \left[\frac{\varphi(s)}{\varphi(s_0)} + \frac{\varphi(s - 2Ti)}{\varphi(s_0 - 2Ti)} \right]$$

which is the real part of $\varphi(s)/\varphi(s_0)$ on the horizontal segment from $s_0 = 2 + iT$ to $\frac{1}{2} + iT$. The variation of argument is estimated by the number of changes of signs of this real part, and this is estimated by

Jensen's formula, as usual, to give:

$$\Delta_H \zeta \ll \log(d\, T_2^N). \tag{3}$$

Note here the canonical error term which now appears, in lieu of the standard log T. The proofs as in Ingham [5] or Titchmarsh [10] work, and one needs no minimum for the real part of the function on the line 2. Note that the proof in Landau [6] does not work, because Landau goes too far out to the right to select the center of his circle for Jensen's formula, and hence the dependence of the constants under these circumstances would not be the desired one. Normalizing the zeta function with s_0 on the line $\sigma = 2$ avoids this. Cf. Siegel [9], dealing with ordinary L-series.

From this last estimate, we deduce at once the estimate

$$N(T+1) - N(T) \ll \log(d\, T_2^N). \tag{4}$$

If, as usual, we let ρ denote the zeros of the zeta function in the strip between 0 and 1, then from (4) we conclude at once absolute estimates for the sums over reciprocals of zeros, namely for $T \geq \frac{3}{2}$:

$$\sum_{1 \leq |\rho| \leq T} \frac{1}{|\rho|} \ll (\log T)(\log d\, T^N) \tag{5}$$

and

$$\sum_{|\rho| \geq T} \frac{1}{|\rho|^2} \ll \frac{\log d\, T^N}{T}. \tag{6}$$

Observe that in (5), the canonical replacement of T by $d\, T^N$ occurs only in one of the factors.

The explicit formula is given in Landau [6]. Let

$$\psi(x) = \psi_k(x) = \sum_{|\mathfrak{n}| \leq x} \Lambda(\mathfrak{n}),$$

where $\mathfrak{n}$ ranges over the ideals $\neq 0$, $|\mathfrak{n}|$ is the absolute norm, and

$$\Lambda(\mathfrak{n}) = \log |\mathfrak{p}| \quad \text{if } \mathfrak{n} = \mathfrak{p}^\nu$$
$$= 0 \quad \text{otherwise}.$$

Then for $x \geq \frac{3}{2}$ we find:

$$\psi(x) = x - (r \log x + b) - \frac{r_1}{2} \log\left(1 - \frac{1}{x^2}\right) - r_2 \log\left(1 - \frac{1}{x}\right) - \sum_\rho \frac{x^\rho}{\rho}. \tag{7}$$

Here, $b = b_k$ is the constant term in the expansion of ζ'/ζ at 0, and $r = r_1 + r_2 - 1$ as always. We let $\rho = \beta + i\gamma$.

With the Riemann Hypothesis, one sees that the estimate for b comes out as it should. Namely, we select x to be exactly between two integers, and use a standard trick, cf. for instance Hooley [4]. We take $x \leq y \leq x + \frac{1}{4}$. On RH, we get using (5):

$$\sum_{|\gamma| \leq x} \frac{y^\rho}{\rho} \ll y^{\frac{1}{2}} \sum_{|\gamma| \leq x} \frac{1}{|\rho|} \ll x^{\frac{1}{2}} \log x \log dx^N.$$

We use RH to signify that the Riemann Hypothesis is being assumed. Let

$$E(x) = O(x^{\frac{1}{2}} \log x \log dx^N).$$

Then

$$\psi(x) = \psi(y) = y - b - \sum_{|\rho| > x} \frac{y^\rho}{\rho} + E(x). \qquad (8.\text{RH})$$

Integrating from x to $x + \frac{1}{4}$ and using the uniform convergence of the series (Landau [6], §23, Satz 199), as well as (6), we find

$$\psi(x) = x - b + E(x). \qquad (9.\text{RH})$$

For $x = 2$, say, this yields

$$b \ll \log d\, 2^N, \qquad (10.\text{RH})$$

and therefore we can finally write an estimate for ψ, namely for $x \geq \frac{3}{2}$,

$$\psi(x) = x + O(x^{\frac{1}{2}} \log x \log dx^N). \qquad (11.\text{RH})$$

Estimates for ψ give rise to estimates for other sums involving Λ. We don't go into this here, but just mention a trivial upper bound for such a sum. Let $\sigma > 0$. Then

$$\sum_{|\mathfrak{n}| \leq x} \frac{\Lambda(\mathfrak{n})}{|\mathfrak{n}|^\sigma \log |\mathfrak{n}|} \leq N \sum_{n \leq x} \frac{\Lambda_\mathbf{Q}(n)}{n^\sigma \log n}. \qquad (12)$$

This is easily seen as follows. The sum on the left hand side is equal to

$$\sum_{p^m \leq x} \sum_{|\mathfrak{p}|^\nu = p^m} \frac{f(\mathfrak{p})}{f(\mathfrak{p}) \nu\, p^{f(\mathfrak{p})\nu\sigma}}$$

where the inner sum above is taken over those pairs $(\mathfrak{p}, \nu)$ with $\mathfrak{p} | p$ and $|\mathfrak{p}|^\nu = p^m$. For each $\mathfrak{p} | p$ there is at most one value of ν such that $|\mathfrak{p}|^\nu = p^m$, and we have $\sum f(\mathfrak{p}) \leq N$. Hence our estimate (12) follows.

We then estimate the right hand side of (12), getting

$$\sum_{n \leq x} \frac{\Lambda_\mathbf{Q}(n)}{n^{\frac{1}{2}} \log n} \ll \frac{x^{\frac{1}{2}}}{\log x}. \qquad (13)$$

To see this, we sum the left hand side of (13) by parts, and use the trivial estimate $\psi_\mathbf{Q}(x) \ll x$. The inequality falls out. The estimate (12) with $\sigma = \frac{1}{2}$ will be used below. Summing by parts, one also obtains an estimate for $\sigma = 1$, with a main term of $\log \log x$, plus a bounded term easily estimated using RH. More complicated arguments (known for the rational zeta function) also allow a determination of the bounded term, as being a constant, plus an error term tending to 0 with x. The constant involves the Euler constant, and the residue of the zeta function at 1.

We also note that the usual formula for the logarithmic derivative of the zeta function yields for $t > 0$, using (10.RH):

$$\zeta'/\zeta(s) = \sum_\rho \left[\frac{1}{s-\rho} + \frac{1}{\rho} \right] + O(\log dt_2^N). \tag{14.RH}$$

From (4) and (6) we get

$$\sum_\rho \frac{1}{|\rho|^2} \ll \log d\, 2^N, \tag{15.RH}$$

and combining this with (14.RH) yields

$$\operatorname{Re} \zeta'/\zeta(s) = \sum_\gamma \frac{\sigma - \frac{1}{2}}{(\sigma - \frac{1}{2})^2 + (t - \gamma)^2} + O(\log dt_2^N). \tag{16.RH}$$

In order to estimate the variation of argument of ζ on the horizontal segment we use a formula of Selberg [8] designed for that purpose. By means of this formula, Selberg was able to prove elegantly the estimate for the rational zeta function due to Littlewood from $\log T$ to

$$\frac{\log T}{\log \log T}.$$

As above, the Selberg argument extends to number fields with a similar result using our general principle. I am much indebted to Selberg for pointing out his formula to me. Selberg's arguments are also reproduced in Titchmarsh [10], 14.20 and 14.21, which can be followed essentially without change. We recall the formula and mention only those points where the number fields affect the arguments.

Define

$$\Lambda_x(\mathfrak{n}) = \Lambda(\mathfrak{n}) \qquad \text{if } |\mathfrak{n}| \leq x$$
$$= \frac{\Lambda(\mathfrak{n}) \log(x^2/|\mathfrak{n}|)}{\log x} \qquad \text{if } x \leq |\mathfrak{n}| \leq x^2.$$

Then Selberg's formula states:

$$\zeta'/\zeta(s) = -\sum_{|n|\leq x^2} \frac{\Lambda_x(n)}{|n|^s} + \sum_\rho \frac{x^{\rho-s} - x^{2(\rho-s)}}{(\rho-s)^2 \log x}$$
$$+ r \frac{x^{-s} - x^{-2s}}{s^2 \log x} + \frac{x^{1-s} - x^{2(1-s)}}{(1-s)^2 \log x}$$
$$+ (r_1 + r_2) \sum_{q=1}^{\infty} \frac{x^{-2q} - x^{-2(2q+s)}}{(2q+s)^2 \log x} + r_1 \sum_{q=1}^{\infty} \frac{x^{-(2q-1+s)} - x^{-2(2q-1+s)}}{(2q-1+s)^2 \log x}.$$

The essential thing about this formula is the appearance of log x in the denominators of all terms except the first. For simplicity we denote the first term by

$$F(s, x) = \sum_{|n|\leq x^2} \frac{\Lambda_x(n)}{|n|^s}.$$

We let $\sigma_1 = \frac{1}{2} + \frac{1}{\log x}$. Trivial estimates show that for $\sigma \geq \sigma_1$ we have

$$\zeta'/\zeta(s) = F(s, x) + O\left(\frac{x^{2(1-\sigma)}}{|1-s|^2 \log x}\right) + 2\alpha x^{\frac{1}{2}-\sigma} \sum_\gamma \frac{\sigma_1 - \frac{1}{2}}{(\sigma_1 - \frac{1}{2})^2 + (t-\gamma)^2},$$

for some complex number α with $|\alpha|\leq 1$. Note that

$$x^{\frac{1}{2}-\sigma_1} = \frac{1}{e}.$$

Arguing as in Titchmarsh, substituting $s=\sigma_1$, and using (16.RH), one finds

$$\sum_\gamma \frac{\sigma_1 - \frac{1}{2}}{(\sigma_1-\frac{1}{2})^2 + (t-\gamma)^2} \ll F(\sigma_1, x) + \log dt_2^N + \frac{x}{\log x}. \quad (17.\text{RH})$$

Let $g(x)$ be the right hand side of (17.RH). Then for $\sigma \geq \sigma_1$ we find that there is some number α' with $|\alpha'|<1$ such that

$$\zeta'/\zeta(s) = F(s, x) + O\left(\frac{x^{2(1-\sigma)}}{|1-s|^2 \log x}\right) + 2\alpha' x^{\frac{1}{2}-\sigma} g(x). \quad (18.\text{RH})$$

We find the variation of argument between $\sigma_1 + it$ and $2+it$ by integrating (18.RH). The integral of $F(s, x)$ can be estimated by (12) and (13), where a log x appears in the denominator, so that

$$\sum_{|n|\leq x^2} \frac{\Lambda(n)}{|n|^\sigma \log |n|} \ll \frac{Nx}{\log x}.$$

The integrals of the other terms are easily estimated. We just note that we can take the integral over the segment except for a small semicircle of fixed radius surrounding 1 from above. We take such a semicircle in order to avoid the $|1-s|^2$ in the denominator from giving a large contribution near 1 (after integrating the contribution would be of the order of $1/t$). We find

$$\Delta_{\sigma_1,2} \ll \frac{Nx}{\log x} + \frac{\log dt_2^N}{\log x}. \tag{19.RH}$$

The variation of argument between $\frac{1}{2}$ and σ_1 is done again as in Titchmarsh, and similar estimates arise, with the same type of terms involved in (18.RH). Thus finally we find that (19.RH) also applies to the variation of argument from $\frac{1}{2}$ to 2. We now select

$$x = \frac{1}{N} \log dt_2^N,$$

and find our final estimate for the horizontal variation of argument for the zeta function, between $\frac{1}{2}+it$ and $2+it$, namely:

$$\Delta_H \zeta \ll \frac{\log dt_2^N}{\log x}. \tag{20.RH}$$

Observe that if $N_k/\log d_k$ tends to 0 then $\log x$ tends to infinity for our value of x, and hence this variation is of a smaller order of magnitude in the Riemann-Von Mangoldt formula than the main term $t \log d_k$, provided that t stays away from 0. Of course, when $t=0$, the problem becomes much deeper, namely determine some sort of criterion when ζ_k actually has a zero at $\frac{1}{2}$. This was in fact my original motivation for studying this problem. It arose in connection with the Birch-Swinnerton Dyer conjecture on the rank of the Mordell-Weil group, which should be the order of the zero at $\frac{1}{2}$ for a Hecke L-function in the case of complex multiplication. It is already unknown if the zeta functions of imaginary quadratic fields have such a zero.

One possible approach to the problem of a zero at $\frac{1}{2}$, is to consider the Hecke formula for the function

$$F_k(s) = A^s \Gamma\left(\frac{s}{2}\right)^{r_1} \Gamma(s)^{r_2} \zeta_k(s),$$

whose residue at 1 is given by

$$\lambda = \frac{2^{r_1} h R}{w}.$$

Using this formula (reproduced in [7], Ch. XIII, §3, Th. 3), say for quadratic imaginary fields, with $s = \frac{1}{2}$, and assuming that the only roots of unity in k are ± 1, we see that $F_k(\frac{1}{2}) = 0$ if and only if

$$h = \sum_{\mathfrak{a}} \int_1^\infty \exp(-2\pi d_k^{-\frac{1}{2}} \mathbf{N}\mathfrak{a} \cdot y) y^{\frac{1}{2}} \frac{dy}{y},$$

where the sum is taken over all integral ideals $\neq 0$. From the functional equation, it is clear that any zero at $\frac{1}{2}$ is of even order. From the Riemann Hypothesis on the segment $[\frac{1}{2}, 1]$, one sees that the right hand side in the above equation is always $\leq h$. From the behavior of the zeta function at 1, as in the Brauer-Siegel theorem, one also sees that the right hand side cannot be very much smaller than $d_k^{\frac{1}{2}}$, so that the behavior at 1 is affecting the behavior at $\frac{1}{2}$, and there is some obstruction to the sum on the right being too small compared with h. The problem is to determine whether it can ever reach the integer h.

Estimates for the zeta function on the segment between $\frac{1}{2}$ and 1 should turn out better than estimates like (20.RH). One sees this by going back over the original proof of Siegel for the Brauer-Siegel theorem, assuming RH. Siegel's arguments are reproduced in [7], Chapter XVI, § 2. Taking $s_0 = 1 - \frac{1}{\log d}$ one finds

$$\frac{1}{2^{r_1}(2\pi)^{r_2} 2^N e^{4\pi N} 2e} \frac{1}{\log d} \leq \frac{hR}{w\sqrt{d}} \leq (c \log d)^N$$

for some absolute constant c. This suggests that similar low estimates for the zeta function hold on the segment mentioned above.

Finally, in a more appropriate place, one should develop the whole theory axiomatically for functions of type

$$A^s \prod \Gamma(\alpha_i s + \beta_i) \varphi(s)$$

where $\varphi(s)$ is a suitable Dirichlet series, including L-series, and those series coming from Modular forms with Euler products. A forthcoming paper of Goldstein carries out the analogue of the Riemann-Von Mangoldt formula in this context, but again does not keep track of the constants. Rather than extend the language and length of this paper, I preferred to keep it brief and simply state the main point concerning the constants.

References

1. Davenport, H.: Multiplicative number theory. Chicago: Markham Publishing Co. 1967.
2. Fogels, E.: On the zeros of Hecke's L-functions, I, II, III. Acta Arithmetica **7**, 87–106 (1962); **7**, 131–148 (1962); **8**, 307–309 (1963).
3. Goldstein, L.: A generalization of the Siegel-Walfisz theorem. Trans. AMS (to appear).

4. Hooley, C.: On Artin's conjecture. J. Reine Angew. Math. **225**, 209–222 (1967).
5. Ingham, A.: The distribution of prime numbers. Cambridge Tract No. 30. New York: Stechert-Hafner 1964.
6. Landau, E.: Einführung in die elementare und analytische Theorie der Algebraischen Zahlen und der Ideale. New York: Chelsea 1949.
7. Lang, S.: Algebraic number theory. Reading: Addison Wesley 1970.
8. Selberg, A.: Contributions to the theory of the Riemann zeta function. Arch. f. Math. og Naturv. **48**, Nr. 5 (1945).
9. Siegel, C. L.: On the zeros of Dirichlet L-functions. Ann. of Math. **46**, 409–422 (1945).
0. Titchmarsh, E. C.: The theory of the Riemann zeta function. Oxford 1951.

> Serge Lang
> Department of Mathematics
> Fine Hall
> Princeton, N. J. 08540
> USA

(Received November 15, 1970)

The Group of Automorphisms of the Modular Function Field

SERGE LANG (Princeton)

Let F be the field of modular functions of all levels on the upper half plane, with, say, Fourier coefficients in the field of all roots of unity. One can define in an obvious way two types of automorphisms of F. First we note that F is a Galois extension of $\mathbf{Q}(j)$, where j is the modular function. We have therefore the Galois group U of automorphisms of F over $\mathbf{Q}(j)$. Second, if α is an element of $G_\mathbf{Q}^+$ (rational 2×2 matrices with positive determinant), then the map $f \mapsto f \circ \alpha$ gives an automorphism of F. Shimura [4] has determined the structure of the group of automorphisms of F, and in particular has shown that it is equal to $U G_\mathbf{Q}^+$. We shall give another proof of this fact, based on an entirely different principle than Shimura's arguments.

Let σ be an automorphism of F. If $\sigma j = j$, then $\sigma \in U$ and we are done. The whole point is to prove that if σ moves j, then we can compose σ with some α so as to fix j. Let A be an abelian curve having invariant j, defined over $\mathbf{Q}(j)$, say by a Weierstrass equation $y^2 = 4x^3 - g_2 x - g_3$. As is standard (cf. [4]), we can identify F with the field obtained by adjoining to $\mathbf{Q}(j)$ all the x-coordinates of the points of finite order on A. For any positive integer N let A_N be the group of points of period N on A. Let p be a prime number, and let

$$F_A^{(p)} = \bigcup_{\nu=1}^{\infty} F_1(x(A_{p^\nu}))$$

be the subfield of F obtained by adjoining to $F_1 = \mathbf{Q}(j)$ the x-coordinates of all points of period a power of p. Then $F_A^{(p)}$ is a p-extension of $F_1(x(A_p))$, and $\sigma F_A^{(p)}$ is the corresponding tower over $\mathbf{Q}(j^\sigma)$ for the curve A^σ.

We now make the constant field extension to the complex numbers $\mathbf{C}$, and let $E = \mathbf{C}(j, j^\sigma, A_p, A_p^\sigma)$, so that E is a finite extension of $\mathbf{C}(j)$. If we translate F over $\mathbf{C}$, then the Galois group of $F\mathbf{C}$ over $\mathbf{C}(j)$ is simply the product

$$G = \prod_l SL_2(\mathbf{Z}_l)/\pm 1.$$

The field of points of period a power of p on A has a Galois group V_p over E, which is open of finite index in $SL_2(\mathbf{Z}_p)$, and is a p-group. There exists a subgroup W of G, which contains almost all factors $SL_2(\mathbf{Z}_l)$, and maps surjectively on V_p. For any l, the group $SL_2(\mathbf{Z}_l)$ is an extension of $SL_2(\mathbf{Z}/l\mathbf{Z})$ by an l-group, and $SL_2(\mathbf{Z}/l\mathbf{Z})/\pm 1$ is simple. It follows that an open subgroup V'_p of finite index in V_p must be the homomorphic image of the p-factor of W. Expressed in terms of fields, this means that up to a finite extension, the fields of p-primary points on A and A^σ over E are the same. It follows from a theorem of Deligne [1], §4, (the generic analogue of results of Serre [2], [3]) that if for two generic elliptic curves A_{j_1}, A_{j_2} the fields of p-primary points are the same over a finite extension of $\mathbf{C}(j_1, j_2)$, then they are isogenous. Consequently there exists an integral matrix α such that $j^\sigma = j \circ \alpha$. Thus finally the map $\alpha^{-1} \circ \sigma$ is an automorphism of F leaving j fixed, as was to be shown.

Added in Proof. The generic analogue of Serre's result mentioned above can be proved like the "elementary case" of the elliptic curves over number fields having a j-invariant which is not integral. With only minor modifications, the arguments given at the end of Serre's book [2] apply, if one works over the ring $\mathbf{Z}[1/j]$, with maximal ideal generated by $(p, 1/j)$. Details will appear in a forthcoming paper.

Appendix
by P. DELIGNE

Soient S une variété algébrique complexe connexe, munie d'un point base s_0, et E, F deux courbes elliptiques sur S, avec j_E non constant. Dans [1], on prouve que si les représentations de $\pi_1(S, s_0)$ sur $H_1(E_{s_0}, \mathbf{Q})$ et $H_1(F_{s_0}, \mathbf{Q})$ sont isomorphes, alors E et F sont isogènes. Pour que ces représentations soient isomorphes, il suffit que les représentations du groupe de Galois du corps des fonctions de S sur $T_p(E) \otimes \mathbf{Q}$ et $T_p(F) \otimes \mathbf{Q}$ soient isomorphes. Pour verifier que tel est le cas ici, on utilise le lemme suivant.

Lemme. *Soient G un sous-groupe fermé d'indice fini de $SL_2(\mathbf{Z}_p)$, et ρ_1, ρ_2 deux représentations fidèles continues $\rho_i: G \to SL_2(\mathbf{Q}_p)$. Alors il existe un sous-groupe d'indice fini G' de G et $g \in GL_2(\mathbf{Q}_p)$ tels que $\rho_1 = g \rho_2 g^{-1}$ sur G'.*

References
1. Deligne, P.: Hodge structures II, to appear, Publications IHES.
2. Serre, J. P.: Abelian l-adic representations and elliptic curves. New York: W. A. Benjamin 1968.
3. — Cours au College de France, to appear, 1971.
4. Shimura, G.: Introduction to the arithmetic theory of automorphic functions. Princeton and Tokyo University Presses 1971.

Serge Lang
Princeton University
Department of Mathematics
Fine Hall
Princeton, New Jersey 00540
USA

(Received June 15, 1971)

ISOGENOUS GENERIC ELLIPTIC CURVES.

By Serge Lang.

Serre has shown that two elliptic curves over a number field, whose j-invariants are not integral, and whose p-adic representations are isomorphic (the terminology will be recalled below), mus be isogenous [7]. The result was shown to be also valid for elliptic curves with transcendental j-invariants over the complex numbers by Deligne [1], who uses Hodge structures. The present paper shows that the "elementary" technique of the Tate parametrization used by Serre can also be applied to this case, by considering the ring $Z[1/j]$, and elliptic curves defined over this ring (or its integral closure in a finite extension of $Q(j)$). We shall emphasize those points where Serre's proof must be adjusted to deal with this case, but we reproduce many arguments for the convenience of the reader.

1. The Tate parametrization. Let K be complete under a discrete valuation. Let q be an element of K such that $0 < |q| < 1$. For any $w \in K^*$ let

$$X(w) = \frac{w}{(1-w)^2} + \sum_{n=1}^{\infty} \left[\frac{q^n w}{(1-q^n w)^2} + \frac{q^n w^{-1}}{(1-q^n w^{-1})^2} - 2 \frac{q^n}{(1-q^n)^2} \right]$$

and

$$Y(w) = \frac{w}{(1-w)^3} + \sum_{n=1}^{\infty} \left[\frac{q^{2n} w^2}{(1-q^n w)^3} - \frac{q^n w^{-1}}{(1-q^n w^{-1})^3} + \frac{q^n}{(1-q^n)^2} \right].$$

The map $w \mapsto \phi(w) = (X(w), Y(w))$ is a homomorphism of K^* into the K-rational points of an elliptic curve A defined over K, by the equation

$$Y^2 - XY = X^3 - h_2 X - h_3,$$

where h_2, h_3 are given by correspondingly classical power series in q with integer coefficients, and with invariant $j(q)$ given by the usual power series in q of classical analysis,

$$j(q) = \frac{1}{q} + 744 + \cdots$$

having integral coefficients, and starting with a polar term of order 1. The

---

Received December 17, 1971.

kernel of the above homomorphism is q^Z, the infinite cyclic group generated by q in K^*. For proofs, cf. [5].

The same parametrization is then also valid over any Galois extension E of K, and maps E^* into A_E. The nature of the power series (with integer coefficients) shows that the homomorphism commutes with the operations of the Galois group $G(E/K)$.

The elliptic curve parametrized as above may be denoted by $A(q)$. The general theory shows that for any positive integer m, the curves $A(q)$ and $A(q^m)$ are isogenous. This is analogous to the fact that over complex numbers, two elliptic curves having commensurable lattices are isogenous. In particular, for any rational number r, the curves $A(q)$ and $A(q^r)$ are isogenous.

2. Complete local rings. Let R be a complete local ring, Notherian, without divisors of zero, integrally closed in its quotient field K. Let $\mathfrak{m}$ be the maximal ideal and let $j \in K$ be such that $j^{-1} \in \mathfrak{m}$. Then we can get an element $q \in \mathfrak{m}$ such that

$$q = \frac{1}{j} + f\left(\frac{1}{j}\right)$$

where f is the power series which inverts the classical series $j(q)$, and which will converge in R. Conversely, given $q \in \mathfrak{m}$, the series $j(q)$ converges in R.

We can aways find a discrete valuation on K which induces the topology on R such that the powers of $\mathfrak{m}$ tend to 0. For instance, if R is regular, for any element $a \in R$ we define $\mathrm{ord}(a)$ to be the largests exponent e such that $a \in \mathfrak{m}^e$, and extend this order function to the quotient field so as to make it a homomorphism. In general, we know from the Cohen structure theorems that R is always a finite module over a complete local ring R_0 which is regular. We can then apply the above construction to R_0, and extend the valuation to the quotient field of R to obtain what we want. We can also visualize the above procedure as blowing up the closed point of $\mathrm{spec}(R)$ to a prime divisor. It is known quite generally that the discrete valuation thus obtained induces the desired topology on R (cf. Zariski [10]). Such a discrete valuation will be said to be admissible.

Let v be admissible. Let R_N be the integral closure of R in $K_N = K(\zeta_N, q^{1/N})$ for any positive integer N, where ζ_N is a primitive N-th root of unity. Then v extends to K_N and is admissible on R_N. If $w = \zeta q^{s/N}$ where ζ is an N-th root of unity, and s is an integer, then all but a finite number of terms of the series $X(w)$ and $Y(w)$ lie in R_N, and these series converge in R_N.

Each ring R_N is closed in the completion $(K_N)_v$. Let $D_q{}^{1/N}$ be the multiplicative subgroup of $K_N{}^*$ consisting of all elements whose N-th power lies in $D_q = q^z$. This subgroup is generated, modulo D_q, by the N-th roots of unity, and $q^{1/N}$. The factor group is isomorphic to a direct product of cyclic groups of order N, provided N is not divisible by the characteristic of K. In this case, the image of $D_q{}^{1/N}$ on the elliptic curve $A = A(q)$ is precisely A_N because we know that A_N has order N^2. Thus the Tate parametrization induces a $\mathrm{Gal}(K_N/K)$-isomorphism

$$D_q{}^{1/N}/D_q \approx A_N.$$

3. The p-adic spaces. Let A be an elliptic curve defined over a field K. Let p be a prime number, and let A_n denote the subgroup of A (in some fixed algebraic closure) consisting of all points of order p^n. Let $A^{(p)}$ be the group of points of p-power order, i.e. the union of A_n for all n. Let $T_p(A)$ be the group of vectors

$$(a_1, a_2, \cdots)$$

such that $a_n \in A_n$ and $pa_{n+1} = a_n$ for $n \geq 1$. The p-adic integers $\mathbf{Z}_p$ operate in the obvious way on $T_p(A)$. Cf. [4], Chapter VIII. One can also form the vector space $V_p(A)$ over $\mathbf{Q}_p$ consisting of all vectors

$$(a_0, a_1, a_2, \cdots),$$

where $a_0 \in A^{(p)}$ is an arbitrary point of p-power order on A, and $pa_{n+1} = a_n$. Let G_K be the Galois group $\mathrm{Gal}(K_a/K)$, where K_a is the algebraic closure of K. Then G_K operates continuously both on $T_p(A)$ and $V_p(A)$. We call this operation the p-adic representation of G_K associated with A.

The Galois representation of G_K on $V_p(A)$ factors through the Galois group leaving $K(A^{(p)})$ fixed. Hence in studying such representations, we are really concerned with the representation of the Galois group of $K(A^{(p)})$ over K. In particular, if A, A' are two elliptic curves defined over K, and $V_p(A)$, $V_p(A')$ are G_K-isomorphic, then $K(A^{(p)}) = K(A'^{(p)})$.

We observe that the T_p and V_p construction applies to any p-divisible group, e.g. the p-primary part of the roots of unity, which we denote by $\mu^{(p)}$.

If p is prime to the characteristic of K, then $T_p(A)$ is a free module of dimension 2 over $\mathbf{Z}_p$ and $V_p(A)$ is a vector space of dimension 2 over $\mathbf{Q}_p$ (cf. [4], Chapter VIII). If p is equal to the characteristic of K, then the dimension is 1 or 0 according as the elliptic curve is singular or supersingular.

4. Results of Kummer theory. Let us now assume that K has characteristic 0. Let μ_n be the group of p-th roots of unity in the algebraic closure K_a of K. Suppose that K is the quotient field of a ring R as above, complete, local, Noetherian, integrally closed, and assume that p lies in the maximal ideal $\mathfrak{m}$.

Let q, q' be elements of $\mathfrak{m}$, and let $A = A(q)$, $A' = A(q')$ be the elliptic curves as in the Tate parametrization, defined over K. Let $G = \operatorname{Gal}(K_a/K)$ throughout this section. We use the p-logarithmic notation, and so we let $A_n = A_{p^n}$ be the group of points of period p^n on A. We have the G-isomorphism

$$D_q^{1/p^n}/D_q \approx A_n.$$

We shall identify these two groups from now on.

As in Kummer theory, if $z \in D_q^{1/p^n}$ then z^{p^n} lies in D_q and there is an integer c such that $z^{p^n} = q^c$. The association

$$z \mapsto \text{class of } c \bmod p^n \mathbf{Z}\ .$$

defines a homomorphism of A_n onto $\mathbf{Z}/p^n\mathbf{Z}$, and hence gives rise to the exact sequence

(1) $$0 \to \mu_n \to A_n \to \mathbf{Z}/p^n\mathbf{Z} \to 0$$

of G-modules, with trivial action on $\mathbf{Z}/p^n\mathbf{Z}$. Taking the limit, we obtain an exact sequence

(2) $$0 \to T_p(\mu) \to T_p(A) \to \mathbf{Z}_p \to 0.$$

Tensoring with Q_p yields an exact sequence of G-modules,

(3) $$0 \to V_p(\mu) \to V_p(A) \to \mathbf{Q}_p \to 0.$$

LEMMA 1. *The above sequence does not split.*

To prove Lemma 1, we introduce an invariant x which belongs to the group

$$\varprojlim H^1(G, \mu_n).$$

Let d be the coboundary homomorphism

$$d \colon H^0(G, \mathbf{Z}/p^n\mathbf{Z}) \to H^1(G, \mu_n)$$

with respect to the exact sequence (1), and let $x_n = d(1)$. We define x to be the element of $\varprojlim H^1(G, \mu_n)$ defined by the family $\{x_n\}$, $n \geq 1$.

LEMMA 2. (i) *The isomorphism*

$$\delta \colon K^*/K^{*p^n} \to H^1(G, \mu_n)$$

of Kummer theory transforms the class of $q \mod K^{*p^n}$ into x_n.

(ii) *The element x is of infinite order.*

Proof. Recall that δ is induced by the coboundary map relative to the exact sequence
$$1 \to \mu_n \to K^*_a \xrightarrow{p^n} K^*_a \to 1.$$

The first assertion of Lemma 2 is immediate from the definitions, because the isomorphism of Kummer theory transforms an element $\alpha \in K^*$ into the class of the cocycle α^σ/α, $\sigma \in G$.

To prove the second assertion, let v be an admissible discrete valuation on K. Then the valuation defines a homomorphism
$$f_n \colon K^*/K^{*p^n} \to \mathbf{Z}/p^n\mathbf{Z},$$
and hence a homomorphism
$$f \colon \lim K^*/K^{*p^n} \to \mathbf{Z}_p.$$
If wqe identify x with the corresponding element of $\lim K^*/K^{*p^n}$, as in (i), then we have $f(x) = v(q)$, and hence x is of infinite order, proving Lemma 2.

To get Lemma 1, suppose the sequence (3) splits. There is a G-subspace W of $V_p(A)$ which is mapped isomorphically onto $\mathbf{Q}_p$. Let $W_T = W \cap T_p(A)$. The image of W_T in $\mathbf{Z}_p$ is $p^m \mathbf{Z}_p$ for some $m \geq 0$. But then it follows immediately that $p^m x = 0$, contradicting the fact that x has infinite order.

LEMMA 3. *Let R_∞ be the integral closure of R in*
$$K_\infty = K(\mu^{(p)}, q^{1/p^\infty}) = K(A^{(p)}).$$
Let $\mathfrak{M}$ be the maximal ideal of R_∞ lying above $\mathfrak{m}$. Let I be the inertia group of $\mathfrak{M}$ in $\mathrm{Gal}(K_\infty/K)$. Then I is of finite index in $\mathrm{Gal}(K_\infty/K)$.

Proof. Let v be an admissible discrete valuation on K. We denote an extension of this valuation to K_∞ by the same letter. Let I_v be the inertia group for this extended valuation. It will suffice to prove that I_v is of finite index in $\mathrm{Gal}(K_\infty/K)$, because $I \supset I_v$. Without loss of generality, we may therefore assume that R is a discrete valuation ring. Let K_v be the completion of K at v, and let L be the completion of the maximal unramified extension of K_v. Then L again has a discrete valuation v. It will suffice to prove that $\mathrm{Gal}(L(A^{(p)})/L)$, identified in the usual manner with a subgroup of $\mathrm{Gal}(K(A^{(p)})/K)$, is of finite index. The picture of Galois theory is as follows.

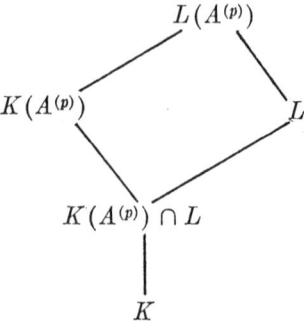

It is known from elementary algebraic number theory that if ζ is a primitive p^n-th root of unity, then $1-\zeta$ has order $1/\phi(p^n) = 1/(p-1)p^{n-1}$ at the p-adic valuation giving p order 1. Since v is a discrete valuation, it follows that there is a constant c such that for all n,

$$[L(\mu_n) : L] \geqq cp^n,$$

and in fact the ramification index of $L(\mu_n)$ over L satisfies a similar inequality. The operation of the Galois group on $T_p(A)$ is represented, relative to a basis, by matrices

$$\begin{pmatrix} a & b \\ 0 & d \end{pmatrix}$$

with components in $\mathbf{Z}_p$. There exists some positive integer r such that the equation

$$x^{p^r} - q = 0$$

has no root in $L(\mu^{(p)})$. For otherwise, we obtain

$$L(\mu^{(p)}) = L(\mu^{(p)}, q^{1/p^\infty}) = L(A^{(p)}),$$

and the Galois group of $L(A^{(p)})$ over L is abelian, which means that the above matrices must be diagonal, whence the representation is reducible, contradicting Lemma 1, applied to the field L, with its discrete valuation ring. By an elementary irreducibility criterion [3], this implies that the degree of

$$L(\mu^{(p)}, q^{1/p^n})$$

over $L(\mu^{(p)})$ satisfies an inequality of the same kind as above, i. e. it is at least equal to cp^n for some constant c. Hence there is a constant c such that for all n,

$$[L(A_n) : L] \geqq cp^{2n}.$$

Since the Galois groups of $K(A_n)$ over K and $L(A_n)$ over L have an order of magnitude at most equal to $c'p^{2n}$ for some constant c', it follows that $\mathrm{Gal}(L(A^{(p)})/L)$ is of finite index in $\mathrm{Gal}(K(A^{(p)})/K)$. Since L is maximal unramified, it follows that $L(A^{(p)})$ is totally ramified over L, thereby proving our lemma.

5. The local isogeny theorems.

THEOREM 1. *Let R be a Noetherian complete local ring, without divisors of zero, integrally closed, with quotient field K of characteristic 0. Assume that the maximal ideal $\mathfrak{m}$ of R contains the prime number p, and that $R/\mathfrak{m}$ is finite. Let A, A' be elliptic curves defined over K, with invariants j, j' such that $1/j$ and $1/j'$ are contained in $\mathfrak{m}$. Suppose that $V_p(A)$ and $V_p(A')$ are G_K-isomorphic. Then A and A' are isogenous.*

Proof. It will suffice to prove that there exist integers i, i' such that $q^i = q'^{i'}$. Let

$$\phi: V_p(A) \to V_p(A')$$

be an isomorphism. By Lemma 1, we know that $V_p(\mu)$ is the only 1-dimensional subspace of $V_p(A)$ (resp. $V_p(A')$) which is stable by G_K. Hence ϕ maps $V_p(\mu)$ into itself. Moreover, after multiplying ϕ be some p-adic integer, we may suppose that ϕ maps $T_p(A)$ into $T_p(A')$. We then have a commutative diagram

(4)
$$\begin{array}{ccccccccc} 0 & \to & T_p(\mu) & \to & T_p(A) & \to & \mathbf{Z}_p & \to & 0 \\ & & r\downarrow & & \downarrow & & s\downarrow & & \\ 0 & \to & T_p(\mu) & \to & T_p(A') & \to & \mathbf{Z}_p & \to & 0 \end{array}$$

where the vertical arrows on the ends are multiplication by p-adic integers r, s respectively. Let x, x' be the elements of $\lim H^1(G_K, \mu_n)$ associated to A and A' as in the preceding section. Then the commutativity of (4) shows that

$$rx = sx'.$$

Using again our discrete admissible valuation v, we get a homomorphism

$$\lim H^1(G_K, \mu_n) = \lim K^*/K^{*p^n} \to \mathbf{Z}_p,$$

and we have seen that the image of x is $v(q)$, while the image of x' is $v(q')$. Hence

$$rv(q) = sv(q').$$

It will now suffice to prove that

$$\alpha = q^{v(q')}/q'^{v(q)}$$

is a root of unity.

We look at the image of α in $\lim K^*/K^{*p^n}$. This image is

$$v(q')x - v(q)x',$$

and multiplying by s, we find 0 by using the above relations. Hence the image of α in $\lim K^*/K^{*p^n}$ is 0.

We are thus reduced to proving that the kernel of the canonical map

$$K^* \to \lim K^*/K^{*p^n}$$

is finite. If an element α lies in the kernel, then α must be a p^n-th power in K for all n. If α does not lie in R, then $1/\alpha$ does not generate the unit ideal in $R[1/\alpha]$, for otherwise α would be integral over R, whence in R, a contradiction. A minimal prime over the ideal $(1/\alpha)$ in the integral closure of $R[1/\alpha]$ would give rise to a discrete valuation where α has a pole, and hence could not be a p^n-th power for large n. So α lies in R. Similarly α cannot lie in $\mathfrak{m}$, otherwise $1/\alpha$ does not lie in R. Hence α is a unit in R. Since the residue class field is finite, and R is complete, there is a finite subgroup k^* in R representing the non-zero elements of $R/\mathfrak{m}$, and the group of units U of R is isomorphic to a product

$$k^* \times U_1,$$

where U_1 consists of the units congruent to 1 mod $\mathfrak{m}$. If $w \in \mathfrak{m}$, then $(1+w)^{p^n}$ lies in $1 + \mathfrak{m}^n$. From this it is clear that α must lie in k^*. This concludes the proof of Theorem 1.

THEOREM 2. *Let R be a complete Noetherian local ring, without divisors of zero, integrally closed, with maximal ideal $\mathfrak{m}$, and quotient field K of characteristic 0. Assume that $R/\mathfrak{m}$ is finite. Let A be an elliptic curve defined over K, with invariant $j \in R$, and let A' be defined over K, with invariant j' such that $1/j' \in \mathfrak{m}$. Then the representations of G_K on $V_p(A)$ and $V_p(A')$ for any prime p are not isomorphic.*

Proof. Passing to a finite extension of K and the integral closure of R in this extension if necessary, we may assume that A has non-degenerate reduction mod $\mathfrak{m}$. Furthermore, A' becomes isomorphic over a finite extension of K to the curve having the Tate parametrization in terms of q', and hence again without loss of generality, we may assume that A' is the Tate curve. We now distinguish two cases.

The reduction $\bar{A}$ of A mod $\mathfrak{m}$ is not supersingular, i.e. has a point of order p in the algebraic closure of the residue class field $\bar{R} = R/\mathfrak{m}$. Then $K(A^{(p)})$ contains an infinite unramified part, corresponding to the infinite residue class field extension

$$\bar{R}(\bar{A}^{(p)}).$$

On the other hand, by Lemma 3, we know that $K(A'^{(p)})$ is almost totally ramified, in the sense of that lemma. Hence A and A' cannot be isogenous. (If $p \neq$ characteristic of $R/\mathfrak{m}$, then all of $K(A^{(p)})$ is unramified, and the argument works even more strongly.)

The reduction $\bar{A}$ of A mod $\mathfrak{m}$ is supersingular, i.e. has no point of order p, so that $\bar{A}^{(p)} = 0$. In that case, we use an admissible discrete valuation v. The repesentation of G_K on $V_p(A')$ is triangular, and has in particular an invariant subspace of dimension 1, corresponding to $V_p(\mu)$. On the other hand, Serre has proved that the representation of G_K on $V_p(A)$ is irreducible, [7], Appendix A.2.2, Theorem, first line of the proof, referring to [8], p. 128, Prop. 8. Hence the representations cannot be isomorphic, and the curves are not isogenous, as was to be proved.

Remark. The assumption that the residue class field is finite can be weakened to finitely generated over the prime field, since it is known that for such field k, the extension $k(A^{(p)})$ of k has an infinite separable part. However, we shall not use this in the sequel.

6. The global isogeny theorems.

THEOREM 3. *Let A, A' be elliptic curves over a field K, finitely generated over the rationals. Assume that they have transcendental j-invariants. Let p be a prime number, and assume that $V_p(A)$ and $V_p(A')$ are G_K-isomorphic. Then the curves are isogenous.*

Proof. It is trivial that j, j' must be algebraically dependent. Hence K can be selected to be a finite extension of $\mathbf{Q}(j, j')$, of transcendence degree 1 over $\mathbf{Q}$.

Next we prove that j' is integral over $\mathbf{Z}[j]$ and vice versa. Suppose this is not the case. There exists a homomorphism of $\mathbf{Z}[j]$ which extends to $\mathbf{Z}[j, 1/j']$ sending $1/j'$ to 0. Let R be the integral closure of $\mathbf{Z}[j, 1/j']$ in K. Extend the homomorphism to R. By composing our homomorphism with another one if necessary, we may assume that our homomorphism takes on its values in a finite field. Let $\mathfrak{m}$ be the kernel in R. The completion $\hat{R}_\mathfrak{m}$ has no divisors of 0 by EGA, Chapter IV, 7.8.3. and 7.8.6. The Galois representations being isomorphic on G_K, they are isomorphic with respect to any closed subgroup, in particular the subgroup arising from the extension $K_\mathfrak{m}(A^{(p)}) = K_\mathfrak{m}(A'^{(p)})$, where $K_\mathfrak{m}$ is the quotient field of $\hat{R}_\mathfrak{m}$. This is a contradiction in view of Theorem 2. Hence j' is integral over $\mathbf{Z}[j]$.

Consider the ring $\mathbf{Z}[1/j, 1/j']$. We contend that the ideal generated by p, $1/j$, $1/j'$ is not the unit ideal. Let $\mathfrak{o}$ be the local ring in $\mathbf{Q}(j)$ of the homomorphism of $\mathbf{Z}[1/j]$ which sends p and $1/j$ to 0. Any place of $\mathbf{Q}(j)$ over this homomorphism must send $1/j'$ to 0. Otherwise, suppose $1/j'$ goes to a finite element $c \neq 0$. Then j' goes to $1/c$, and j goes to infinity, which we have already seen is impossible. Similarly, $1/j'$ cannot go to infinity. This proves our contention.

Let R be the integral closure of $\mathbf{Z}[1/j, 1/j']$ in K and let $\mathfrak{m}$ be a maximal ideal of R containing p, $1/j$, $1/j'$. We now argue as in the first part of the proof, with $\hat{R}_\mathfrak{m}$, reducing our problem to the local case, and cite Theorem 1 to conclude the proof.

THEOREM 4. *Let K be a finitely generated field over an algebraically closed field k of characteristic 0. Let A, A' be elliptic curves defined over K, with invariants j, j' which are transcendental over k. Assume that $V_p(A)$ and $V_p(A')$ are G_K-isomorphic. Then the elliptic curves are isogenous.*

Proof. As in Theorem 3, the invariants j, j' must be algebraically dependent over k, and we can assume K finite over $k(j, j')$. Without loss of generality, we can assume that A is defined by a Weierstrass equation

$$y^2 = 4x^3 - gx - g,$$

and that A' is defined by

$$y^2 = 4x^3 - g'x - g',$$

after replacing K with a finite extension if necessary. Then $k(g) = k(j)$ and $k(j') = k(g')$. There exists a function field K_0 with constant field k_0, such that k_0 is contained in k, is finitely generated over $\mathbf{Q}$, K_0 is a finite extension of $k_0(j, j')$, and K is obtained from K_0 by extending the constants from k_0 to k. The picture is as follows.

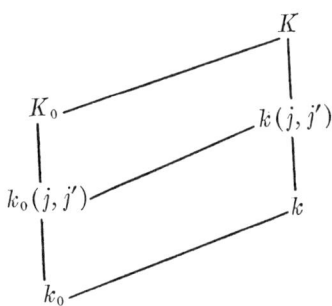

The constant field (i. e. algebraic closure of $\mathbf{Q}$) in $\mathbf{Q}(j, A^{(p)})$ is the field of p-power roots of unity (this was already known to Weber and Fricke, for a recent references, cf. Shimura [9]). Let k_1 be the constant field of $K_0(A^{(p)})$, i. e. the algebraic closure of k_0 in $K_0(A^{(p)})$. Let $K_1 = k_1 K_0$ be the corresponding constant field extension. We must then have

$$K_1(A^{(p)}) = K_1(A'^{(p)}).$$

Indeed, if we make the constant field extension to k, the two fields

$$K_1(A^{(p)}) \quad \text{and} \quad K_1(A'^{(p)})$$

become equal. Let

$$E = K_1(A^{(p)}) \cap K(A'^{(p)}).$$

If E is a proper subfield of $K_1(A^{(p)})$, then there is an element $\neq 1$ of the Galois group of $K_1(A^{(p)})$ over E which extends to an element of the Galois group of $K_1(A^{(p)}, A'^{(p)})$ over $K_1(A'^{(p)})$, thus acting trivially on $A'^{(p)}$, contradicting the hypothesis that the Galois representations over K on $V_p(A)$ and $V_p(A')$ are isomorphic.

We now conclude that

$$K_0(A^{(p)}) = K_0(A'^{(p)}).$$

Let G be the Galois group of the extension $K_0(A^{(p)})$ over K_0. We have two representations

$$\rho: G \to \operatorname{Aut} T_p(A) \quad \text{and} \quad \rho': G \to \operatorname{Aut} T_p(A').$$

onto open subgroups of these automorphism groups, each one of which is isomorphic to $GL_2(\mathbf{Z}_p)$ after a choice of basis over $\mathbf{Z}_p$. Let S be the Galois group of $K_1(A^{(p)})$ over K_1. The the image of S under both ρ and ρ' is an open subgroup of the special linear subgroup of $\operatorname{Aut} T_p(A)$ and $\operatorname{Aut} T_p(A')$ respectively, that is this image is open in $SL_2(\mathbf{Z}_p)$ under both representations. The center of S maps onto open subgroups of the diagonal groups, formed with units in $\mathbf{Z}_p$. An open subgroup W of the center, not containing -1, is then such that $W \cap S = 1$, whence $S \times W$ is open in G.

The representations of W on $V_p(A)$ and $V_p(A')$ give rise to two characters ψ, ψ' of W into the group of p-adic units, such that if $\sigma \in W$, then the matrix representation of σ on $T_p(A)$ is a diagonal matrix

$$\begin{pmatrix} \psi(\sigma) & 0 \\ 0 & \psi(\sigma) \end{pmatrix},$$

and similarly for σ'. The effect of σ on a p-power root of unity ζ is classically known to be

$$\sigma(\zeta) = \zeta^{\det \rho(\sigma)}.$$

This implies that $\psi(\sigma^2) = \psi'(\sigma^2)$ for all $\sigma \in W$. Since $(W:W^2)$ is finite, passing to an open subgroup of W if necessary, we may assume without loss of generality that $\psi = \psi'$ on W.

By hypothesis, we know that there is a $\mathbf{Q}_p$-isomorphism

$$h: V_p(A) \to V_p(A')$$

which is also an S-isomorphism. Since W acts on $V_p(A)$ and $V_p(A')$ as the same group of p-adic multiplications, it follows that h is also a W-isomorphism, in other words, h is a G-isomorphism for the group $G = S \times W$. The fixed field of G is a finite extension of K_0, finitely generated over the rationals, and we are therefore reduced to the situation of Theorem 4, thus concluding the proof of Theorem 5.

The argument given at the end also shows:

THEOREM 5. *Let A, A' be elliptic curves defined over a field K. Assume that the representations*

$$\rho: G_K \to \operatorname{Aut} V_p(A) \quad \text{and} \quad \rho': G_K \to \operatorname{Aut} V_p(A')$$

map G_K onto open subgroups of $\operatorname{Aut} V_p(A)$ and $\operatorname{Aut} V_p(A')$ respectively. Let $L = K(\mu^{(p)})$ be the field obtained by adjoining all p-power roots of unity to K. If the restrictions of ρ and ρ' to G_L are isomorphic, then ρ and ρ' are isomorphic.

7. Applications. The need for a generic version of the isogeny theorem arose for a new proof that the group of automorphisms of the modular function field is no bigger than expected. I include here the auxiliary results which allow us to apply the isogeny theorem in the form needed for the applications.

THEOREM 6. *Let A, A' be elliptic curves defined over a field K, and assume that $K(A^{(p)}) = K(A'^{(p)})$. Let G be the Galois group of $K(A^{(p)})$ over K and assume that the representations of G on $T_p(A)$ and $T_p(A')$ map G onto open subgroups of $SL_2(\mathbf{Z}_p)$. Then the representations on $V_p(A)$ and $V_p(A')$ are isomorphic on an open subgroup of G.*

The theorem follows from the next lemma, which just drops the hypotheses where the p-adic spaces came from, but preserves the other hypotheses of Theorem 6.

LEMMA 4. *Let G be an open subgroup of $SL_2(\mathbf{Z}_p)$ and let*

$$\rho_1: G \to SL_2(\mathbf{Z}_p) \quad \text{and} \quad \rho_2: G \to SL_2(\mathbf{Z}_p)$$

be continuous injective representations. Then there exists $g \in GL_2(\boldsymbol{Q}_p)$ *such that* $g^{-1} \rho_2 g = \rho_1$ *on an open subgroup of* G.

Proof. Without loss of generality we may assume that ρ_1 is the identity and $\rho_2 = \rho$. Thus ρ induces a local isomorphism of $SL_2(\boldsymbol{Z}_p)$ into itself, because its differential at the origin must be injective. Since the Lie algebra of SL_2 is simple, one knows that the induced map on the Lie algebra comes from an inner derivation, and it then follows that ρ is induced by a conjugation, in some neighborhood of the identity. (For the foundations of p-adic Lie groups necessary for the above, cf. Serre's *Lie Algebras and Lie groups* [6].)

The statement of Theorem 6 can even be slightly strengthened.

COROLLARY. *Let A, A' be elliptic curves over K, and assume that $K(A^{(p)})$ and $K(A'^{(p)})$ have an intersection which is of infinite degree over K. Assume that the representations of G_K on $T_p(A)$ and $T_p(A')$ map G_K onto open subgroups of $SL_2(\boldsymbol{Z}_p)$. Then there is a finite extension E of K such that*

$$E(A^{(p)}) = E(A'^{(p)}),$$

and the theorem applies.

Proof. Since the Lie algebra of $SL_2(\boldsymbol{Z}_p)$ is simple, there exists an open subgroup W of $\mathrm{Gal}(K(A^{(p)})/K)$ having the following properties:

(i) W has no finite subgroup other than 1.
(ii) Any closed normal, non-trivial subgroup of W is also open, and hence of finite index.

Let K_1 be the fixed field of W. We consider the inclusion of fields

$$K_1 \subset K_1(A^{(p)}) \cap K_1(A'^{(p)}) \subset K_1(A^{(p)}).$$

The intermediate field is of infinite degree over K_1, and is the fixed field of a closed normal subgroup of W. By the above two properties, it must be equal to $K_1(A^{(p)})$. Arguing in the same way with respect to A', i.e. selecting an open subgroup W' in a similar way, we can find a finite extension K_2 of K such that

$$K_2(A^{(p)}) = K_2(A'^{(p)}).$$

This proves our corollary, with $E = K_2$.

The assumptions of the theorem and corollary are easily satisfied in the "generic" case, and so these results apply. That they are also satisfied in the case of number fields is a much deeper result of Serre [7].

REFERENCES.

[1] P. Deligne, *Hodge Structures*, to appear, Publications IHES.
[2] A. Grothendieck, *Elements de Geometrie Algebrique* (EGA), Chapter IV.
[3] S. Lang, *Algebra*, Addison Wesley, Reading, 1966, Chapter VIII, § 9.
[4] ———, *Abelian varieties*, Interscience, 1959.
[5] P. Roquette, "Analytic theory of elliptic functions over local fields," *Hamb. Math. Einzelschriften*, Neue Folgen, Heft 1, 1970.
[6] J. P. Serre, *Lie Algebras and Lie Groups*, W. A. Benjamin, 1965.
[7] ———, *Abelian l-adic representations and elliptic curves*, W. A. Benjamin, 1968.
[8] ———, " Sur les groupes de Galois attachés aux groupes p-divisibles," *Proceed. Conf. on Local Fields*, Springer-Verlag, 1967, pp. 113-131.
[9] G. Shimura, *Introduction to the arithmetic theory of automorphic functions*, Iwanami Shoten and Princeton University Press, 1971.
[10] O. Zariski, "A simple analytical proof of a fundamental property of birational transformations," *Proc. Nat. Acad. Sci. USA* (1949), pp. 62-66.

Journal für die reine und angewandte Mathematik

Herausgegeben von **Helmut Hasse** und **Hans Rohrbach**

Sonderdruck aus Band 255, Seite 112 bis 134

Verlag Walter de Gruyter · Berlin · New York 1972

Continued fractions for some algebraic numbers

By Serge Lang and Hale Trotter at Princeton

Continued fractions for some algebraic numbers

By *Serge Lang* and *Hale Trotter* at Princeton

The following tables at the end of the paper contain the continued fractions and some related information for a few algebraic numbers, viz:

$$\sqrt[3]{2},\ \sqrt[3]{3},\ \sqrt[3]{4},\ \sqrt[3]{5},\ \sqrt[3]{7},\ \alpha_1, \alpha_2, \alpha_3,$$

where

$\alpha_1 = 2\cos\dfrac{2\pi}{7}$ is a root of $x^3 + x^2 - 2x - 1$,

α_2 is a root of $x^5 - x - 1$,

$\alpha_3 = \sqrt[3]{2} + \sqrt{3}$ is a root of $x^6 - 9x^4 - 4x^3 + 27x^2 - 36x - 23$.

The first numbers were chosen as cubic irrationalities, α_1 as being totally real, and we picked α_2, α_3 at random to be non cubic. The original motivation was to see if computations for algebraic numbers would be in line with some conjectures made in [3] and [4], and to observe whatever else might come up.

1. Table I

In each case, this table is to be read horizontally, and gives the first thousand terms in the continued fraction. For instance, for $\sqrt[3]{2}$, the continued fraction begins

$$[1, 3, 1, 5, 1, 1, 4, 1, 1, 8, \ldots].$$

The last five terms among these first thousand are $[\ldots, 2, 1, 1, 1, 1]$. Thus the continued fraction $[a_1, a_2, \ldots]$ is laid out in rows of twenty integers, and there are fifty such rows. The position of any a_n is therefore easily determined.

For clarity, and reasons of space distribution, the table is filled only with the 2-digit integers which arose. Whenever a larger integer occurred, it is indicated by the symbols *a, *b, *c, ..., and these are then listed separately at the bottom. For instance, for $\sqrt[3]{2}$, we have

$$*a = 534, \quad *b = 121, \quad \ldots, \quad *k = 4941, \quad \ldots$$

2. Table II

This table gives the frequency count for the digits appearing in the continued fraction. According to a theorem of Kuzmin [2], for almost all numbers α, the probability that the n-th integer a_n in the continued fraction for α is equal to a positive integer k is given by

$$\log_2 \frac{(k+1)^2}{k(k+2)}.$$

For $k = 1$, and almost all numbers, this means that the probability for $a_n = 1$ is approximately .41. In each case of the computations, we find that the numbers behave very closely to this generic expectation, thus confirming that in most respects, the continued fractions for algebraic numbers of degree > 2 should be essentially like those of almost all numbers. For instance, among the first thousand a_n for $\sqrt[3]{2}$, $\sqrt[3]{4}$, and α_3 we find that 1 occurs respectively 422, 412, and 418 times. The biggest divergence from the Kuzmin number is for $\sqrt[3]{5}$, and even then 433 is still in line with the expected asymptotic estimate. The Kuzmin probability that 2 occurs is approximately .17, and again in this case we find that 2 occurs 165 times, resp. 159 times, resp. 168 times for $\sqrt[3]{2}$, $\sqrt[3]{3}$, α_1 respectively. One should also note that this type of regularity is exhibited throughout, among the first thousand terms.

3. The occurrence of large integers

Aside from what appear to be routine low numbers, there occur larger numbers which seem to be of two kinds: Some which are only somewhat larger than the ordinary ones, and some which appear to be exceptionally large. For instance, for $\sqrt[3]{2}$, we meet

$$a_{572} = 7\,451 \text{ and } a_{620} = 4\,941.$$

For $\sqrt[3]{5}$, we meet

$$a_{19} = 3\,052, \quad a_{691} = 13\,977, \quad a_{813} = 49\,968.$$

The occurrence of large numbers in the continued fractions of certain cubic irrationalities has already been observed by Brillhart, see Churchhouse and Muir [1]. A theoretical explanation for some of them has been proposed by Stark [6], but the following problems remain: Determine whether there is a basic theoretical distinction between what seem to be only medium large numbers, and very large ones. More importantly, determine whether exceptionally large integers will continue to occur throughout the continued fraction, or whether they will stop from occurring. The explanation given by Stark depends on some class numbers being equal to 1, and thus would account for only a finite number of them. In general, the appearance of such large integers may depend on the arithmetic properties of the field obtained from the square root of the discriminant, e. g. its class number. The tables seem to indicate that they stop.

To discuss the statistical significance of exceptionally large values of a_n occurring near the beginning of the sequence of partial quotients, we need an estimate of the probability $q_{N,K}$ that the first N partial quotients of a "random" number are all less than a given integer K. It is perhaps most natural to consider a random number as distributed uniformly on $(0, 1)$, but in this context the distribution given by

$$Pr\{X \leq c\} = \log_2(1 + c), \quad \text{for } c \in (0, 1),$$

is more appropriate, because if X has this distribution, then the distribution of the partial quotient $a_n(X)$ is independent of n. In fact,

$$Pr\{a_n(X) < K\} = \gamma_K,$$

where
$$\gamma_K = Pr\{X^{-1} < K\} = 1 - \log_2\left(1 + \frac{1}{K}\right),$$
which is the Kuzmin theorem already alluded to.

To see that this is so, observe that $a_n(X) = [X_n^{-1}]$, where $X_1 = X$, and for $n > 0$,
$$X_{n+1} = X_n^{-1} - [X_n^{-1}].$$
As usual, $[x]$ is the largest integer $\leq x$. It is then an exercise in calculus to show that if f_n is a density function for X_n, so that
$$Pr\{X_n \leq c\} = \int_0^c f_n(x)\,dx,$$
then $f_{n+1} = T f_n$ is a density function for X_{n+1}, where T is the linear operator on $L^1(0, 1)$ defined by
$$(Tf)(x) = \sum_{k=1}^\infty (x+k)^{-2} f((x+k)^{-1}).$$
It follows that $f_{n+1} = T^n f_1$. It is easy to verify that the function
$$\frac{1}{(\log 2)(1+x)}$$
is a density function and is invariant under T, so that if X has this density, in which case, by integration,
$$Pr\{X \leq c\} = \log_2(1+c),$$
all the X_n have the same distribution. In fact Kuzmin's theorem states that if f is any smooth probability density, then
$$\lim_{n \to \infty} T^n f(x) = \frac{1}{(\log 2)(1+x)}.$$
It follows that as $n \to \infty$ the distribution of $a_n(X)$ tends to the one given above if X has any smooth distribution. For a discussion of all these ideas and a proof of Kuzmin's theorem, see [2].

If the random variables $a_n(X)$ were independent, then we would have
$$q_{N,K} = \gamma_K^N.$$
This is not strictly correct, but can be expected to give a good approximation for large values of N and K. A combination of theoretical and numerical analysis indicated strongly that the relative error is bounded by $\lambda N K^{-2}$, with $\lambda < 1$, and we are confident that the approximation is entirely adequate for our purposes.

The short Table A at the end of this section shows, for each of the numbers investigated, the maximum value A of the first 1000 partial quotients, and the value M for which $a_M = A$. The third column gives
$$p_{1000,A} = 1 - q_{1000,A},$$
the probability that a "random" number would have a value as large as A among its first 1000 partial quotients. The smaller the value of p, the stronger the evidence that

the number is unusual. The fourth column gives

$$p_{M,A} = 1 - q_{M,A},$$

the probability of getting a value as large as A among the first M quotients. The value is of course smaller than that of p, and its statistical meaning less clear since M is taken a posteriori to make the probability small.

The table shows that if one goes by the maximum quotient found, only $\sqrt[3]{5}$ appears highly unusual, although one might question $\sqrt[3]{4}$ and α_2. If one also takes into account the second largest quotient, then $\sqrt[3]{5}$ with $a_{691} = 13977$ appears even more unusual, and $\sqrt[3]{2}$ with $a_{620} = 4941$ perhaps comes to be of interest also.

Table A

Number	A	M	$p = 1 - q_{1000,A}$	$p_M = 1 - q_{M,A}$
$\sqrt[3]{2}$	7451	572	.18	.10
$\sqrt[3]{3}$	3502	916	.34	.31
$\sqrt[3]{4}$	14902	579	.09	.05
$\sqrt[3]{5}$	49968	813	.03	.02
$\sqrt[3]{7}$	689	611	.88	.72
α_1	904	830	.80	.73
α_2	11644	588	.12	.07
α_3	1446	54	.63	.05

4. Table III

In each case, this table begins with the columns labelled n, a_n, and q_n. The n indicates n-th position in the continued fraction. The a_n means the n-th partial quotient. The q_n means the denominator in the approximating fraction p_n/q_n (classical notation). For instance, in the case of $\sqrt[3]{2}$, we have

$$a_{36} = 534, \quad q_{36} = 3.06 \times 10^{19}$$
$$a_{42} = 121, \quad q_{42} = 8.95 \times 10^{22},$$

and so forth. In machine language, $E\,19$ means multiplication by 10^{19}, and $E\,486$ means multiplication by 10^{486} (the last line in Table III).

Table III includes these data for all n among the first thousand such that $a_n \geq 50$. We picked 50 as a cutting point after looking at preliminary computations, because it included all the numbers a_n which could be labelled as somewhat large, and at the same time provided only a rather small table.

The last column r_n in Table III gives (up to three decimals) the quotient

$$\frac{q_n}{q_{n-1} \log q_{n-1}}$$

for those values of n when $a_n \geq 50$. The reason for this quotient to be interesting are as follows. According to a theorem of Roth, if α is algebraic, there is only a finite number of integers $q > 0$, p such that

$$|q\alpha - p| < \frac{1}{q^{1+\varepsilon}}.$$

It was suggested in [3] and [4] that this theorem should be improvable by an inequality

$$|q\alpha - p| < \frac{1}{qf(q)},$$

where f is a function close to the logarithm, for instance $(\log q)^{1+\varepsilon}$, or perhaps even $\log q$ itself, up to a constant factor of course. Such a function is called a type in [4]. If some a_n is small, then q_n/q_{n-1} being approximately equal to a_n shows that the quotient

$$\frac{q_n}{q_{n-1} \log q_{n-1}}$$

is approximately like $1/\log q_{n-1}$. Thus to investigate the possibility of a type f, we look at those n for which a_n is comparatively large. Again n such that $a_n \geq 50$ seemed to give the most information for the least amount of space used. It is even unsolved for any algebraic number of degree > 2 whether it is of bounded type, but the tables seem to fall fairly well in line with expectations, e. g. differing from the log by a function with a lower order of magnitude (above or below).

We have also programmed the same data for the first 3000 terms of the continued fractions of the cubic numbers listed. In every case, exceptionally large integers did not seem to recur, and generally speaking, the ratio r_n seems to decrease. We thought it pointless to reproduce these more extensive tables in full, but we give in Table B the portion of Table III for $n > 1000$ when $r_n > 1$, rounding off r_n to one decimal.

The tables therefore suggest that the type may in fact not be bigger than a constant times the logarithm, and may even be of an order of magnitude smaller than the logarithm. Following certain asymptotic estimates of Adams, who looked at the continued fraction of e, it was shown (cf. [4]) that the type of e is asymptotic to $\log q/\log\log q$. Thus one is beginning to be accustomed to such small types. Note that for a function essentially not bigger than the log, the series

$$\Sigma \frac{1}{qf(q)}$$

diverges, so that these cases go very slightly against the Khintchine convergence principle: If ψ is such that $\Sigma \psi(q)$ diverges (resp. converges), then for almost all numbers, the inequality

$$|q\alpha - p| < \psi(q)$$

has infinitely many solutions (resp. only a finite number). However, this statistical result is delicate to use for specific numbers with a type in the range of the log, because one also

Table B

	n	a_n	q_n	r_n
$\sqrt[3]{2}$	1191	12737	7.74 E 1010	5.5
	2248	2897	2.97 E 1136	1.1
$\sqrt[3]{3}$	1988	2967	3.47 E 1024	1.3
	2407	9559	1.25 E 1242	3.3
$\sqrt[3]{4}$	1974	6368	4.88 E 1010	2.7
	2248	4157	6.92 E 1146	1.6
$\sqrt[3]{5}$	1196	18905	1.47 E 600	13.8
$\sqrt[3]{7}$	None			
$2 \cos \dfrac{2\pi}{7}$	1102	1374	6.84 E 576	1.0

knows that if α is a number such that for every function ψ (decreasing) having convergent sum, the above inequality has only a finite number of solutions, then α must be of bounded type. For all this, cf. [4].

5. Computational method

The computations were done by the following algorithm, which uses integer arithmetic only, and thus involves no rounding error.

Given a polynomial $p_n(x)$, of degree d, with positive leading coefficient and a unique positive root y_n which is simple, irrational, and greater than 1, we construct a polynomial $P_{n+1}(x)$ as follows. Let $a_n = [y_n]$ be the greatest integer such that $P_n(a_n) < 0$. Define

$$Q_n(x) = P_n(x + a_n) \text{ and } P_{n+1}(x) = -x^d Q_n(x^{-1}).$$

Then $Q_n(x)$ has exactly one root between 0 and 1, and since the roots of $P_{n+1}(x)$ are the reciprocals of the roots of $Q_n(x)$, we see that $P_{n+1}(x)$ has a unique positive root y_{n+1}. Obviously y_{n+1} is also a simple root, irrational, and greater than 1. Note that the constant term of $Q_n(x)$ is $P_n(a_n) < 0$, so that $P_{n+1}(x)$ also has a positive leading coefficient. Thus $P_{n+1}(x)$ has the properties assumed for $P_n(x)$, and starting from any $P_1(x)$ with these properties, we can define an infinite sequence $P_1(x), P_2(x), \ldots$ with associated positive roots $y_1, y_2, \ldots$

We have

$$a_n = [y_n] \text{ and } y_{n+1} = (y_n - a_n)^{-1}.$$

This is precisely equivalent to saying that $a_1, a_2, \ldots$ is the sequence of partial quotients in the continued fraction expansion of y_1. If $P_1(x)$ has integer coefficients, then so has every $P_n(x)$, and the calculation involves only addition and multiplication of integers.

Table I

$$\sqrt[3]{2}$$

	1	2	3	4	5	6	7	8	9	10	1	2	3	4	5	6	7	8	9	10
0	1	3	1	5	1	1	4	1	1	8	1	14	1	10	2	1	4	12	2	3
20	2	1	3	4	1	1	2	14	3	12	1	15	3	1	4	*a	1	1	5	1
40	1	*b	1	2	2	4	10	3	2	2	41	1	1	1	3	7	2	2	9	4
60	1	3	7	6	1	1	2	2	9	3	1	1	69	4	4	5	12	1	1	5
80	15	1	4	1	1	1	1	1	89	1	22	*c	6	2	3	1	3	2	1	1
100	5	1	3	1	8	9	1	26	1	7	1	18	6	1	*d	3	13	1	1	14
120	2	2	2	1	1	4	3	2	2	1	1	9	1	6	1	38	1	2	25	1
140	4	2	44	1	22	2	12	11	1	1	49	2	6	8	2	3	2	1	3	5
160	1	1	1	3	1	2	1	2	4	1	1	3	2	1	9	4	1	4	1	2
180	1	27	1	1	5	5	1	3	2	1	2	2	3	1	4	2	2	8	4	1
200	6	1	1	1	36	9	13	9	3	6	2	5	1	1	1	2	10	21	1	1
220	1	2	1	2	6	2	1	6	19	1	1	18	1	2	1	1	1	27	1	1
240	10	3	11	38	7	1	1	1	3	1	8	1	5	1	5	4	4	4	7	2
260	1	21	1	1	5	10	3	1	72	6	9	1	3	3	2	1	4	2	1	1
280	1	1	2	1	7	8	1	2	1	8	1	8	3	1	1	3	2	1	8	1
300	1	1	1	1	6	1	4	3	4	1	1	1	4	30	39	2	1	3	8	1
320	1	2	1	3	1	9	1	4	1	2	2	1	6	2	1	1	3	1	4	1
340	2	1	1	5	1	2	10	1	5	4	1	1	4	1	2	1	1	2	12	2
360	1	8	3	2	6	1	3	10	1	2	20	1	6	1	2	*e	2	2	1	2
380	47	1	19	2	2	1	1	1	2	1	1	3	2	8	1	18	3	5	39	1
400	2	1	1	1	1	4	1	5	2	6	3	1	1	1	4	2	1	6	1	1
420	*f	1	3	1	3	1	4	5	1	2	1	13	2	2	2	1	1	1	1	7
440	2	1	7	1	3	1	1	11	1	2	2	4	2	33	3	1	1	2	6	3
460	1	1	3	6	8	3	4	84	1	1	2	1	10	2	2	20	1	3	1	7
480	13	14	1	29	1	1	5	1	7	1	1	2	1	56	1	3	2	1	13	2
500	1	2	2	2	1	1	1	1	1	1	*g	2	4	5	1	1	1	3	1	3
520	3	1	6	1	1	6	1	71	1	9	1	2	1	11	5	1	25	1	6	67
540	2	9	6	1	5	2	15	1	2	48	2	7	1	3	1	4	21	1	1	2
560	1	27	3	26	2	1	1	2	5	7	3	*h	2	29	4	3	8	17	3	8
580	2	3	1	1	1	5	*i	1	3	4	1	4	1	1	13	1	34	1	2	7
600	1	3	3	7	1	3	1	1	4	2	69	1	3	12	34	1	2	*j	1	*k
620	4	1	1	12	3	4	2	3	1	1	1	1	1	2	1	1	6	16	1	2
640	27	2	13	4	1	1	1	3	11	1	1	3	1	53	2	15	1	1	1	1
660	1	1	2	1	1	1	3	3	1	9	1	1	10	3	1	1	2	1	2	2
680	1	10	9	1	2	5	1	2	2	1	1	2	4	7	1	5	1	1	1	1
700	4	2	25	16	5	4	1	3	2	3	13	1	49	6	2	5	1	1	2	7
720	3	2	1	1	1	4	1	1	1	5	1	2	1	2	1	1	1	1	2	2
740	4	1	2	1	10	5	4	8	10	2	4	1	1	1	4	1	41	1	3	1
760	56	3	1	1	3	1	3	1	5	6	6	3	1	2	1	1	1	12	1	10
780	2	1	1	1	1	50	5	1	2	6	5	1	2	5	6	5	2	77	1	4
800	2	1	1	1	1	1	4	2	1	2	1	1	1	1	1	6	2	1	1	7
820	1	5	1	1	1	1	2	2	1	1	5	2	1	5	1	1	1	4	1	2
840	17	1	20	7	4	2	1	1	1	2	1	4	7	3	4	3	3	5	31	1
860	1	2	2	6	1	1	1	1	1	1	1	1	6	1	6	1	1	23	20	1
880	22	16	4	2	1	3	2	1	1	2	5	5	1	1	15	3	1	1	2	1
900	1	1	4	2	1	2	23	6	10	3	2	3	6	2	1	1	1	1	1	1
920	4	3	2	1	2	1	4	10	7	1	1	1	1	3	3	2	*m	1	1	11
940	2	6	1	4	1	2	2	9	3	1	1	3	22	4	1	93	1	3	1	4
960	2	1	2	3	2	1	2	11	1	1	3	1	2	1	28	23	4	11	1	9
980	1	4	3	1	6	1	2	1	12	2	6	19	1	4	4	2	1	1	1	1

a = 534   b = 121   c = 186   d = 372   e = 186   f = 220
g = 255   h = 7451 i = 113   j = 151   k = 4941 m = 108

Table II

$$\sqrt[3]{2}$$

FREQUENCY COUNTS

1	422	22	4	56	2
2	165	23	3	67	1
3	91	25	3	69	2
4	66	26	2	71	1
5	40	27	4	72	1
6	37	28	1	77	1
7	20	29	2	84	1
8	16	30	1	89	1
9	15	31	1	93	1
10	15	33	1	108	1
11	8	34	2	113	1
12	9	36	1	121	1
13	8	38	2	151	1
14	4	39	2	186	2
15	5	41	2	220	1
16	3	44	1	255	1
17	2	47	1	372	1
18	3	48	1	534	1
19	3	49	2	4941	1
20	4	50	1	7451	1
21	3	53	1		

Table III

$$\sqrt[3]{2}$$

n	a_n	q_n	r_n
36	534	3.06 E 19	13.844
42	121	8.95 E 22	2.530
73	69	4.48 E 39	0.799
89	89	3.61 E 48	0.835
92	186	1.56 E 52	1.618
115	372	6.69 E 65	2.560
269	72	1.90 E 146	0.219
376	186	4.78 E 194	0.421
421	220	4.51 E 216	0.447
468	84	1.19 E 239	0.154
494	56	1.01 E 253	0.098
511	255	3.59 E 260	0.430
528	71	3.38 E 268	0.117
540	57	1.32 E 276	0.106
572	7451	8.64 E 297	11.005
587	113	4.07 E 308	0.160
611	69	7.64 E 320	0.095
618	151	5.97 E 326	0.202
620	4941	2.97 E 330	6.568
654	53	1.50 E 347	0.068
761	56	5.13 E 395	0.063
786	50	7.54 E 406	0.054
798	77	4.14 E 414	0.082
937	108	3.81 E 475	0.099
956	93	1.37 E 486	0.084

Table I

$$\sqrt[3]{3}$$

	1	2	3	4	5	6	7	8	9	10	1	2	3	4	5	6	7	8	9	10
0	1	2	3	1	4	1	5	1	1	6	2	5	8	3	3	4	2	6	4	4
20	1	3	2	3	4	1	4	9	1	8	4	3	1	3	2	6	1	6	1	3
40	1	1	1	1	12	3	1	3	1	1	4	1	6	1	5	1	2	1	3	3
60	11	8	1	*a	8	2	8	5	1	2	2	2	2	3	1	1	2	1	1	1
80	52	2	46	2	2	3	3	1	5	1	6	1	6	1	1	1	6	1	20	10
100	4	8	1	1	1	2	1	2	*b	1	2	5	6	1	4	3	1	1	*c	2
120	3	3	1	6	1	3	1	2	1	24	2	1	5	4	1	1	2	1	1	1
140	3	1	1	20	5	60	1	1	37	6	3	28	1	1	2	1	31	1	3	3
160	1	2	58	1	11	1	31	1	8	1	11	2	2	3	1	4	1	4	37	1
180	2	1	2	82	1	2	6	11	5	1	1	1	7	54	2	27	1	2	1	24
200	3	3	1	2	1	1	*d	1	14	2	89	1	2	4	1	5	2	1	2	2
220	3	2	1	5	1	4	2	1	15	2	2	3	10	2	1	1	1	1	1	1
240	9	2	67	1	1	1	9	5	2	1	3	1	60	1	3	1	1	3	1	9
260	2	18	1	3	1	1	3	3	2	1	3	1	5	4	2	1	1	3	1	6
280	2	39	1	3	4	1	2	1	1	2	1	1	4	3	1	1	4	2	3	1
300	4	1	1	1	1	1	5	*e	1	84	3	1	2	1	3	7	3	1	1	1
320	8	1	7	1	1	1	11	1	1	2	1	1	5	1	1	1	3	1	1	2
340	2	1	7	24	5	4	1	1	1	2	2	1	1	95	1	3	1	2	1	6
360	2	1	1	6	1	1	1	3	1	16	*f	6	4	1	9	4	3	*g	1	2
380	3	4	2	1	1	*h	2	6	4	2	1	1	5	2	4	1	1	2	1	7
400	6	1	2	10	*i	3	1	2	1	1	34	1	1	2	2	1	1	10	4	15
420	1	2	1	1	2	4	1	3	1	7	1	42	1	3	1	2	4	6	2	1
440	2	1	28	3	1	5	3	1	1	1	3	2	4	4	11	1	1	3	2	10
460	6	2	1	1	26	3	2	1	1	1	1	2	26	2	3	1	66	6	1	8
480	2	1	4	1	10	3	1	1	2	1	1	1	24	4	1	2	23	4	8	1
500	41	1	1	1	1	25	1	4	1	4	6	23	1	5	2	23	1	4	3	1
520	1	5	16	1	8	2	1	11	2	2	1	1	10	3	58	4	1	34	1	1
540	1	19	1	1	1	1	1	1	1	1	1	1	7	3	3	1	11	1	16	1
560	6	5	19	7	2	4	2	2	7	4	1	3	1	1	1	3	1	1	3	*j
580	8	1	1	10	6	2	8	23	5	2	1	17	2	2	15	1	1	1	20	1
600	3	6	1	1	3	1	1	1	4	2	26	1	2	12	2	8	1	15	2	3
620	54	3	2	22	1	1	3	1	2	1	92	1	1	1	4	1	3	4	4	1
640	1	1	1	1	12	2	1	18	1	5	9	1	1	5	3	1	1	1	9	2
660	1	4	1	2	1	12	1	1	15	1	1	1	3	1	6	2	2	2	12	21
680	1	3	1	15	3	4	4	6	1	10	1	1	1	1	1	5	2	4	25	2
700	3	1	2	2	2	3	9	71	14	2	1	1	1	2	4	1	1	2	*k	1
720	1	*m	1	1	1	1	1	6	3	1	*n	5	12	2	2	*p	4	9	1	1
740	1	3	1	1	3	7	1	1	5	1	*q	1	34	1	12	2	1	1	2	1
760	69	1	2	3	2	*r	1	1	4	1	2	1	1	10	1	5	3	1	1	1
780	1	4	2	1	2	3	1	3	10	21	1	1	1	1	2	1	13	1	10	2
800	1	4	21	39	1	19	1	1	5	1	38	1	2	1	28	56	5	3	1	1
820	2	11	2	2	1	1	1	1	5	2	11	5	3	1	2	1	6	4	1	4
840	1	3	18	1	1	1	1	6	3	1	5	1	2	1	1	1	54	1	16	2
860	1	1	20	1	1	1	9	2	2	21	1	12	2	2	1	1	1	1	4	1
880	1	5	2	1	1	1	2	3	4	1	13	2	1	8	1	*s	2	1	60	1
900	1	6	10	8	4	1	3	1	2	5	1	3	16	20	5	*t	1	1	7	8
920	1	4	1	1	2	7	4	1	33	1	1	1	12	1	11	3	2	1	11	2
940	41	5	1	1	2	17	1	1	1	2	7	2	2	2	2	1	1	1	1	1
960	15	2	28	2	3	1	3	13	4	1	1	1	3	72	1	13	8	1	2	1
980	2	2	3	1	6	1	1	3	1	*u	1	2	4	4	3	15	7	2	39	2

a = 139   b = 249   c = 612   d = 220   e = 123   f = 131
g = 196   h = 729   i = 164   j = 396   k = 343   m = 137
n = 139   p = 268   q = 247   r = 1232   s = 116   t = 3502
u = 164

Table II

$$\sqrt[3]{3}$$

FREQUENCY COUNTS

1	425	13	4	25	2	46	1	84	1	247	1
2	159	14	2	26	3	52	1	89	1	249	1
3	97	15	8	27	1	54	3	92	1	268	1
4	64	16	5	28	4	56	1	95	1	343	1
5	37	17	2	31	2	58	2	116	1	396	1
6	32	18	3	33	1	60	3	123	1	612	1
7	14	19	3	34	3	66	1	131	1	729	1
8	18	20	5	37	2	67	1	137	1	1232	1
9	10	21	4	38	1	69	1	139	2	3502	1
10	13	22	1	39	3	71	1	164	2		
11	12	23	4	41	2	72	1	196	1		
12	9	24	4	42	1	82	1	220	1		

Table III

$$\sqrt[3]{3}$$

n	a_n	q_n	r_n
64	139	6.85 E 30	2.118
81	52	9.42 E 38	0.614
109	249	3.42 E 54	2.077
119	612	8.70 E 60	4.575
146	60	7.46 E 73	0.363
163	58	1.94 E 84	0.307
184	82	2.18 E 96	0.379
194	54	3.33 E 102	0.233
207	220	3.57 E 110	0.885
211	89	9.94 E 113	0.347
243	67	3.00 E 128	0.231
253	60	4.08 E 133	0.200
308	123	1.53 E 157	0.345
310	84	1.31 E 159	0.235
354	95	5.96 E 177	0.236
371	131	9.09 E 185	0.310
378	196	7.22 E 191	0.450
386	729	1.24 E 197	1.631
405	164	4.22 E 207	0.347
477	66	2.34 E 242	0.121
535	58	2.72 E 274	0.093
580	396	9.29 E 296	0.585
621	54	5.56 E 320	0.074
631	92	2.84 E 326	0.124
708	71	2.10 E 363	0.085
719	343	5.43 E 369	0.406
722	137	1.49 E 372	0.161
731	139	4.61 E 376	0.162
736	268	3.91 E 381	0.307
751	247	3.47 E 389	0.278
761	69	2.10 E 395	0.077
766	1232	6.02 E 399	1.349
816	56	2.01 E 425	0.057
857	54	6.13 E 443	0.054
896	116	1.77 E 462	0.110
899	60	3.23 E 464	0.057
916	3502	3.38 E 477	3.209
974	72	1.22 E 507	0.062
990	164	1.91 E 515	0.139

Table I

$$\sqrt[3]{4}$$

	1	2	3	4	5	6	7	8	9	10	1	2	3	4	5	6	7	8	9	10
0	1	1	1	2	2	1	3	2	3	1	3	1	30	1	4	1	2	9	6	4
20	1	1	2	7	2	3	2	1	6	1	1	1	25	1	7	7	1	1	1	1
40	*a	1	3	2	1	3	60	1	5	1	8	5	6	1	4	20	1	4	1	1
60	14	1	4	4	1	1	1	1	7	3	1	1	2	1	3	1	4	4	1	1
80	1	3	1	34	8	2	10	6	3	1	2	31	1	1	1	4	3	44	1	45
100	93	12	1	7	1	1	5	12	1	1	2	4	19	1	12	1	16	1	8	1
120	1	2	1	*b	1	1	1	6	3	1	6	1	2	2	2	3	2	6	1	5
140	20	1	2	1	78	2	1	12	2	2	4	22	2	11	4	6	23	99	1	12
160	4	4	1	1	2	7	2	1	4	1	1	2	1	2	1	9	7	1	2	4
180	1	1	1	1	10	2	56	11	2	1	7	1	2	1	4	1	1	9	1	4
200	4	9	1	2	1	4	1	17	1	1	4	26	4	1	1	1	12	1	11	3
220	1	20	10	1	4	2	5	3	5	1	2	1	1	9	3	1	8	1	6	3
240	13	1	3	5	6	5	1	1	18	1	1	3	3	8	1	3	1	12	1	2
260	8	2	8	3	1	2	44	11	5	7	1	35	1	1	2	1	1	4	2	1
280	1	1	1	5	1	1	1	2	4	6	1	3	17	2	18	1	3	1	1	1
300	3	1	1	5	1	3	1	4	3	3	2	2	6	2	3	9	15	78	1	2
320	1	1	1	3	1	3	1	2	1	1	20	1	1	1	5	1	2	3	5	8
340	1	1	1	6	12	2	1	4	1	11	2	3	1	1	1	6	5	6	5	1
360	3	1	1	1	4	3	2	1	1	1	4	1	5	10	2	3	2	1	*c	1
380	5	2	1	23	2	9	1	2	2	4	2	3	1	1	2	1	3	1	37	1
400	1	1	2	79	2	4	10	1	2	4	3	7	3	2	5	1	2	1	3	*d
420	2	1	1	8	1	1	1	1	2	2	1	2	6	1	2	2	2	4	15	1
440	2	3	1	8	24	2	2	1	1	1	2	1	16	7	5	3	7	7	3	16
460	1	1	1	1	1	1	41	1	3	1	2	5	4	1	41	1	1	2	3	1
480	1	6	29	1	14	3	1	2	2	3	1	3	1	2	28	2	1	1	2	27
500	1	2	1	4	1	3	4	*e	1	8	2	1	4	1	1	2	1	1	1	1
520	2	3	3	1	2	1	*f	1	4	2	1	2	5	1	1	2	2	12	1	13
540	33	1	2	1	4	13	1	2	4	7	1	5	24	4	3	1	8	1	1	1
560	1	10	3	2	55	1	1	1	12	1	2	3	1	10	3	1	1	1	*g	1
580	58	2	6	4	34	1	1	1	3	1	2	1	1	3	11	56	1	7	2	2
600	2	3	1	6	2	17	2	1	15	1	1	6	3	1	8	9	1	*h	1	1
620	24	17	2	1	*i	1	*j	9	25	1	1	1	1	1	2	1	1	3	4	2
640	3	3	33	2	1	13	4	6	1	1	1	1	4	1	1	23	8	1	26	4
660	7	1	4	4	2	2	3	1	1	1	1	2	4	1	3	5	7	6	2	2
680	21	4	1	5	2	1	5	1	3	2	1	1	1	1	3	2	2	1	4	9
700	1	50	8	10	2	2	1	1	2	1	1	27	1	24	12	1	11	5	3	1
720	1	1	5	3	2	3	12	2	6	4	2	2	1	1	1	6	1	4	1	1
740	2	8	4	20	1	9	3	2	2	20	1	8	1	27	1	1	1	3	1	1
760	2	1	1	11	3	12	1	1	6	3	6	2	5	5	4	1	24	1	1	2
780	2	1	12	2	1	5	2	1	1	2	1	1	2	4	38	1	9	1	3	4
800	1	1	1	2	6	4	13	1	3	1	3	2	2	1	4	5	1	3	1	2
820	5	1	2	3	10	2	1	8	2	10	14	2	5	3	1	2	2	14	1	1
840	1	1	1	1	1	6	2	1	1	15	3	2	2	1	2	1	4	4	3	3
860	2	3	3	1	11	41	1	10	1	1	7	1	1	1	1	2	7	1	3	2
880	1	2	11	31	1	1	3	1	3	9	1	2	1	46	3	20	1	1	2	1
900	1	12	1	3	4	9	1	1	2	6	1	1	1	1	4	1	1	3	3	3
920	1	1	1	1	4	54	3	1	5	4	3	2	2	2	1	4	4	1	1	1
940	3	1	1	44	2	2	46	1	8	1	1	1	2	5	1	1	2	5	5	1
960	3	1	1	6	1	13	1	1	11	8	5	1	20	1	1	1	1	1	2	3
980	2	1	2	6	4	3	39	1	1	1	1	1	1	2	4	3	5	6	2	10

a = 266   b = 745   c = 372   d = 110   e = 511   f = 144
g = 14902   h = 139   i = 303   j = 2470

Table II

$$\sqrt[3]{4}$$

FREQUENCY COUNTS

1	412	22	1	54	1
2	164	23	3	55	1
3	100	24	5	56	2
4	69	25	2	58	1
5	39	26	2	60	1
6	32	27	3	78	2
7	19	28	1	79	1
8	20	29	1	93	1
9	14	30	1	99	1
10	12	31	2	110	1
11	11	33	2	139	1
12	15	34	2	144	1
13	6	35	1	266	1
14	4	37	1	303	1
15	4	38	1	372	1
16	3	39	1	511	1
17	4	41	3	745	1
18	2	44	3	2470	1
19	1	45	1	14902	1
20	8	46	2		
21	1	50	1		

Table III

$$\sqrt[3]{4}$$

n	a_n	q_n	r_n
41	266	1.93 E 19	6.868
47	60	5.59 E 22	1.248
101	93	9.85 E 51	0.808
124	745	8.43 E 65	5.136
145	78	2.00 E 76	0.460
158	99	5.29 E 86	0.508
187	56	1.53 E 100	0.249
318	78	2.14 E 167	0.205
379	372	6.02 E 194	0.842
404	79	1.26 E 207	0.168
420	110	2.84 E 216	0.223
508	511	4.53 E 260	0.861
527	144	4.29 E 268	0.236
565	55	1.14 E 289	0.084
579	14902	1.09 E 298	22.025
581	58	6.43 E 299	0.086
596	56	2.54 E 308	0.079
618	139	9.69 E 320	0.191
625	303	7.55 E 326	0.407
627	2470	1.87 E 330	3.283
702	50	2.38 E 367	0.060
926	54	2.40 E 475	0.050

Table I

$$\sqrt[3]{5}$$

	1	2	3	4	5	6	7	8	9	10	1	2	3	4	5	6	7	8	9	10
0	1	1	2	2	4	3	3	1	5	1	1	4	10	17	1	14	1	1	*a	1
20	1	1	1	1	1	2	2	1	3	2	1	13	5	1	1	1	13	2	41	1
40	4	12	1	5	2	7	1	1	3	33	2	1	1	1	1	1	1	3	2	2
60	1	15	12	8	10	48	1	2	1	1	3	4	1	*b	1	13	2	4	1	1
80	49	3	10	1	8	1	1	1	1	4	1	60	7	2	2	2	3	3	2	1
100	3	2	1	61	1	10	2	8	3	4	4	2	33	2	1	1	1	7	3	4
120	2	1	1	1	42	1	6	1	3	1	2	1	3	2	1	1	1	1	1	1
140	2	1	2	3	2	2	1	3	1	1	3	12	2	45	3	7	31	1	6	5
160	2	1	1	3	1	1	5	2	3	1	1	4	1	1	1	4	1	6	1	1
180	2	3	2	1	8	1	7	1	35	2	8	6	1	4	2	1	3	1	2	1
200	12	1	7	20	1	2	42	1	1	2	5	1	3	1	81	2	21	1	5	1
220	1	4	8	9	1	52	1	2	1	2	2	5	1	13	2	3	1	1	1	9
240	2	1	3	13	2	17	2	2	10	1	11	1	2	1	2	9	2	1	1	1
260	2	11	1	5	2	1	9	1	3	1	17	1	6	1	1	1	51	2	*c	2
280	8	1	1	1	1	1	25	1	5	6	3	1	2	2	3	4	1	4	9	4
300	1	1	2	1	1	1	1	1	1	1	2	4	1	2	5	1	2	11	1	1
320	1	1	1	2	1	3	3	2	1	21	1	2	1	2	1	1	38	1	4	2
340	1	5	19	4	1	1	1	2	1	2	1	1	2	4	1	2	1	3	1	1
360	2	1	1	1	4	48	5	1	6	1	1	*d	1	3	*e	23	1	51	12	2
380	35	1	1	2	3	2	1	59	1	1	2	2	3	1	1	2	3	5	2	1
400	2	5	2	1	3	*f	1	5	2	1	4	3	2	5	61	2	1	5	1	2
420	1	5	2	1	2	1	3	6	3	2	9	8	1	1	8	13	1	1	2	1
440	5	1	1	21	9	1	2	1	1	3	1	1	2	2	4	1	*g	1	1	2
460	1	6	1	50	3	2	1	4	1	1	20	5	4	2	1	1	1	1	1	1
480	7	9	10	2	9	1	*h	1	12	2	5	2	2	1	*i	1	6	1	5	1
500	9	1	4	1	23	1	3	15	3	1	2	2	2	10	2	1	25	3	2	55
520	1	1	2	1	3	2	1	7	2	2	5	1	18	3	1	2	8	3	1	3
540	1	1	5	5	1	1	1	1	4	1	1	21	1	8	1	3	2	1	5	1
560	4	3	1	2	1	13	1	1	1	2	2	1	13	2	1	5	72	1	1	4
580	1	2	1	1	1	10	2	1	3	1	3	3	1	4	2	2	1	6	6	1
600	1	1	1	1	1	29	2	2	3	1	1	3	2	1	1	1	1	2	6	1
620	1	3	1	28	20	3	1	6	2	2	1	6	1	29	1	2	15	5	3	3
640	1	5	1	1	2	1	2	2	5	3	1	1	3	2	6	7	2	1	1	4
660	4	3	1	1	3	1	3	1	3	1	37	1	8	15	1	2	4	1	10	9
680	1	2	3	1	4	1	3	7	1	1	*j	1	1	1	3	1	4	1	1	2
700	5	9	3	8	2	2	5	2	6	4	1	3	7	2	2	1	4	1	8	1
720	2	1	5	1	1	1	2	1	1	8	4	4	2	13	3	1	7	1	45	2
740	2	1	5	3	3	2	1	2	5	13	1	2	1	7	10	1	3	1	1	1
760	1	75	7	1	63	1	2	1	3	5	1	1	5	1	1	1	3	1	1	3
780	1	1	5	2	2	1	1	1	1	16	2	1	3	1	1	3	1	1	1	9
800	1	37	6	1	1	1	40	2	*k	1	5	1	*m	7	4	1	1	1	7	2
820	1	2	2	2	2	4	1	1	2	2	4	1	5	2	8	2	3	5	1	5
840	3	7	5	1	1	1	53	4	12	1	3	2	7	8	2	9	2	47	1	1
860	2	6	3	1	37	2	1	1	1	1	*n	13	1	6	2	1	2	3	2	1
880	1	3	4	1	1	4	1	2	1	1	3	1	1	2	1	2	2	3	1	1
900	3	1	6	1	7	1	1	1	4	2	1	4	1	1	2	23	1	1	1	1
920	5	1	3	1	10	16	*p	1	6	1	9	1	1	2	1	13	1	8	2	1
940	3	8	23	4	2	1	2	9	1	11	1	1	3	1	2	4	1	6	2	1
960	1	13	14	1	12	2	1	6	1	53	1	3	3	5	1	1	1	1	2	1
980	1	6	1	1	25	2	1	1	1	34	4	1	10	1	40	1	3	1	3	1

a = 3052   b = 474   c = 854   d = 131   e = 170   f = 1051
g = 182   h = 326   i = 135   j = 13977   k = 451   m = 49968
n = 739   p = 121

Table II

$\sqrt[3]{5}$

FREQUENCY COUNTS

1	433	17	3	41	1	75	1
2	180	18	1	42	2	81	1
3	95	19	1	45	2	121	1
4	50	20	3	47	1	131	1
5	46	21	4	48	2	135	1
6	25	23	4	49	1	170	1
7	19	25	3	50	1	182	1
8	19	28	1	51	2	326	1
9	16	29	2	52	1	451	1
10	12	31	1	53	2	474	1
11	4	33	2	55	1	739	1
12	8	34	1	59	1	854	1
13	13	35	2	60	1	1051	1
14	2	37	3	61	2	3052	1
15	4	38	1	63	1	13977	1
16	2	40	2	72	1	49968	1

Table III

$\sqrt[3]{5}$

n	a_n	q_n	r_n
19	3052	4.28 E 11	162.730
74	474	1.23 E 40	5.511
92	60	9.82 E 49	0.548
104	61	2.43 E 56	0.491
215	81	1.88 E 109	0.331
226	52	2.20 E 116	0.200
277	51	3.15 E 140	0.162
279	854	5.44 E 143	2.636
372	131	1.57 E 183	0.315
375	170	1.07 E 186	0.402
378	51	1.34 E 189	0.120
388	59	4.04 E 195	0.134
406	1051	1.00 E 205	2.260
415	61	2.21 E 210	0.127
457	182	2.35 E 230	0.348
464	50	6.48 E 233	0.095
487	326	5.54 E 246	0.581
495	135	8.42 E 251	0.236
520	55	9.33 E 265	0.091
577	72	2.06 E 291	0.108
691	13977	2.20 E 345	17.792
762	75	1.46 E 379	0.087
765	63	7.47 E 381	0.073
809	451	4.93 E 401	0.491
813	49968	1.73 E 407	53.913
847	53	3.15 E 423	0.055
871	739	8.92 E 438	0.737
927	121	7.42 E 462	0.114
970	53	9.37 E 484	0.048

Table I

$$\sqrt[3]{7}$$

	1	2	3	4	5	6	7	8	9	10	1	2	3	4	5	6	7	8	9	10
0	1	1	10	2	16	2	1	4	2	1	21	1	3	5	1	2	1	1	2	11
20	5	1	3	1	2	27	4	1	*a	8	1	2	1	1	3	1	3	2	6	4
40	1	2	1	5	1	1	2	1	1	1	3	2	8	1	2	2	4	5	1	1
60	36	1	1	1	1	2	1	2	31	2	1	1	7	1	1	1	1	6	7	6
80	5	7	1	6	1	6	9	6	*b	5	33	8	2	1	1	1	6	2	1	1
100	1	1	2	13	7	1	1	1	17	3	2	1	1	1	23	*c	1	*d	1	72
120	1	4	6	6	1	11	6	12	7	3	5	1	4	6	2	2	1	2	4	4
140	1	1	9	11	6	2	1	15	6	2	1	1	2	1	3	5	1	2	1	1
160	1	1	1	15	1	35	1	2	1	2	4	2	1	11	2	1	1	1	1	15
180	4	1	1	2	22	2	1	31	2	2	1	1	3	1	10	2	1	2	1	1
200	2	1	1	8	1	5	1	1	6	2	1	14	1	4	7	5	2	6	6	1
220	4	1	1	3	2	10	1	3	2	16	2	34	1	1	1	18	3	1	7	1
240	3	12	1	1	15	3	2	1	2	2	1	2	2	17	3	2	2	3	5	8
260	1	56	1	2	1	25	1	3	1	3	3	8	20	8	1	3	1	1	1	8
280	1	4	12	1	3	6	3	1	3	7	19	3	1	1	13	1	1	1	1	10
300	2	2	1	1	3	15	1	4	1	*e	1	7	1	8	2	1	1	2	13	3
320	7	56	3	2	4	2	4	2	5	1	1	9	3	1	1	7	4	1	6	1
340	47	1	2	31	6	2	4	1	4	7	9	1	1	3	9	2	1	14	1	3
360	1	5	5	1	6	15	1	5	1	*f	6	1	1	11	4	1	6	8	19	1
380	1	1	8	2	6	4	19	3	14	14	3	1	11	1	6	1	1	2	2	1
400	1	1	1	5	2	1	1	1	1	4	2	3	2	1	1	1	1	1	2	3
420	1	2	1	4	1	2	12	1	1	13	3	1	1	2	*g	4	1	1	1	14
440	1	10	1	*h	7	60	9	32	1	6	2	13	1	1	1	2	1	98	1	1
460	1	2	1	2	1	*i	1	4	3	2	2	4	4	1	2	35	2	25	8	3
480	3	1	1	6	1	1	1	10	1	2	3	2	4	2	2	5	2	6	2	2
500	1	1	1	10	1	1	1	6	1	35	30	1	2	1	1	1	3	2	1	3
520	4	1	9	1	1	9	6	1	1	15	2	1	5	1	3	3	1	1	2	5
540	2	60	3	23	1	2	1	1	1	4	11	9	13	1	4	1	*j	2	1	11
560	3	2	1	1	5	2	52	1	4	1	8	2	3	2	3	25	1	14	1	1
580	1	7	8	1	2	11	15	1	2	14	2	1	6	1	2	7	1	1	2	2
600	1	1	1	2	1	1	1	1	1	9	*k	3	3	2	5	1	1	1	1	3
620	1	8	1	1	1	11	2	3	2	1	1	3	4	1	3	1	4	1	1	6
640	19	1	1	2	1	1	1	1	1	3	8	7	99	1	41	1	11	51	1	1
660	7	17	2	1	7	1	5	1	4	15	1	1	1	3	1	6	2	1	69	1
680	3	1	18	2	2	2	4	2	13	11	1	6	1	2	1	14	3	3	1	3
700	7	1	12	2	78	2	1	1	1	1	*m	5	1	10	5	2	1	2	5	1
720	6	3	2	2	10	13	4	27	1	2	1	1	44	19	4	3	1	3	1	1
740	2	2	5	3	1	1	1	1	1	1	1	2	1	1	1	11	1	8	6	1
760	7	1	1	2	19	1	5	3	1	1	3	1	3	1	1	44	1	34	1	5
780	3	8	1	2	3	*n	11	1	4	10	3	1	1	2	2	2	18	16	6	3
800	63	1	1	5	1	19	24	1	1	7	53	1	1	4	2	1	1	*p	2	1
820	5	2	7	5	4	1	1	3	1	30	1	1	1	2	1	7	1	7	1	1
840	24	2	1	4	1	1	3	2	1	3	21	1	1	1	73	1	4	2	1	1
860	4	9	1	2	4	2	3	1	1	3	7	2	11	2	1	4	8	8	15	11
880	18	17	1	1	2	7	43	2	3	1	1	6	2	1	10	1	5	3	3	3
900	8	22	4	3	1	4	4	4	6	1	7	1	1	9	1	18	3	4	1	1
920	2	1	2	7	1	2	2	1	15	1	1	2	5	1	2	52	3	1	87	1
940	3	1	1	1	6	10	1	1	1	3	2	4	1	6	1	2	1	3	2	3
960	1	1	1	5	2	1	1	8	1	1	8	1	2	4	1	*q	2	8	7	2
980	84	4	1	11	2	2	12	3	1	1	1	3	4	12	2	1	9	1	72	2

$a = 282$ $b = 104$ $c = 277$ $d = 429$ $e = 303$ $f = 341$
$g = 110$ $h = 197$ $i = 118$ $j = 133$ $k = 689$ $m = 115$
$n = 111$ $p = 202$ $q = 628$

Table II

$$\sqrt[3]{7}$$

FREQUENCY COUNTS

1	409	18	5	43	1	104	1
2	161	19	7	44	2	110	1
3	88	20	1	47	1	111	1
4	55	21	2	51	1	115	1
5	34	22	2	52	2	118	1
6	40	23	2	53	1	133	1
7	29	24	2	56	2	197	1
8	24	25	3	60	2	202	1
9	13	27	2	63	1	277	1
10	12	30	2	69	1	282	1
11	17	31	3	72	2	303	1
12	7	32	1	73	1	341	1
13	8	33	1	78	1	429	1
14	8	34	2	84	1	628	1
15	11	35	3	87	1	689	1
16	3	36	1	98	1		
17	4	41	1	99	1		

Table III

$$\sqrt[3]{7}$$

n	a_n	q_n	r_n
29	282	6.14 E 15	9.209
89	104	2.74 E 44	1.066
116	277	7.19 E 59	2.096
118	429	3.10 E 62	3.120
120	72	2.27 E 64	0.507
262	56	1.28 E 135	0.185
310	303	4.11 E 160	0.834
322	56	7.57 E 167	0.147
370	341	2.07 E 195	0.770
435	110	2.82 E 226	0.214
444	197	1.46 E 232	0.374
446	60	6.16 E 234	0.112
458	98	4.21 E 242	0.178
466	118	2.06 E 246	0.211
542	60	2.86 E 283	0.093
557	133	1.21 E 293	0.200
567	52	4.71 E 298	0.077
611	689	1.20 E 321	0.940
653	99	6.74 E 339	0.127
658	51	1.79 E 344	0.065
679	69	1.36 E 355	0.086
705	78	9.95 E 369	0.093
711	115	1.50 E 373	0.135
786	111	4.29 E 411	0.118
801	63	9.92 E 421	0.065
811	53	5.06 E 428	0.054
818	202	5.07 E 432	0.204
855	73	2.51 E 450	0.071
936	52	3.44 E 494	0.046
939	87	1.21 E 497	0.077
976	628	2.69 E 513	0.535
981	84	5.88 E 517	0.071
999	72	1.14 E 528	0.060

Table I

$$\alpha_1 = 2\cos\frac{2\pi}{7}, \text{ Root of } x^3 + x^2 - 2x - 1$$

	1	2	3	4	5	6	7	8	9	10	1	2	3	4	5	6	7	8	9	10
0	1	4	20	2	3	1	6	10	5	2	2	1	2	2	1	18	1	1	3	2
20	1	2	1	2	1	39	2	1	1	1	13	1	2	1	30	1	1	1	3	2
40	5	4	1	5	1	5	1	2	1	1	94	6	2	19	11	1	60	1	1	50
60	2	1	1	8	53	1	3	1	6	3	2	1	5	1	1	3	4	*a	1	2
80	1	3	3	7	9	1	2	10	3	1	22	1	*b	3	32	1	2	1	2	4
100	2	1	2	2	62	2	1	1	8	1	14	5	6	5	1	1	8	1	7	*c
120	1	1	2	2	1	2	2	2	2	30	3	1	13	1	19	3	1	4	1	1
140	2	1	33	1	10	1	13	2	26	1	1	1	9	1	9	1	6	13	1	5
160	1	1	1	6	1	9	1	7	1	3	1	1	1	12	1	1	4	3	1	8
180	1	2	2	2	1	2	4	1	1	1	1	3	1	19	1	3	3	2	2	1
200	13	3	4	1	1	1	3	1	1	1	2	1	3	19	1	3	9	3	2	4
220	1	3	1	6	1	25	20	1	1	2	1	2	5	1	1	1	14	1	13	1
240	1	1	55	8	1	1	24	17	1	11	17	1	4	1	1	9	1	1	1	3
260	1	1	1	1	20	4	1	45	1	2	4	1	1	2	1	1	2	11	*d	1
280	4	1	2	1	2	*e	1	3	5	3	7	3	25	2	3	1	1	2	1	1
300	1	1	5	1	1	5	2	1	1	1	1	1	1	3	2	1	4	2	1	5
320	1	1	6	1	30	1	62	1	36	11	7	1	21	1	19	1	15	1	2	12
340	5	5	9	2	1	5	1	1	1	2	6	1	3	1	1	3	*f	8	30	2
360	1	1	1	3	1	3	1	1	44	2	19	1	3	1	1	17	20	1	4	3
380	97	2	1	10	4	2	1	5	1	1	4	1	3	18	1	1	1	4	14	2
400	4	13	3	2	9	2	1	2	1	2	1	4	1	22	1	1	3	1	16	61
420	1	2	3	2	5	1	1	2	1	1	7	3	11	2	1	4	2	3	1	2
440	2	1	27	4	1	3	5	17	2	2	10	2	1	2	15	1	37	5	7	1
460	24	4	56	2	2	4	1	1	1	4	3	1	2	8	3	3	2	4	1	1
480	2	14	*g	1	16	1	5	1	1	1	1	*h	1	3	1	*i	1	1	13	2
500	2	3	2	1	1	1	2	1	11	1	8	4	2	72	3	1	5	7	1	3
520	7	2	1	4	4	2	9	2	1	2	3	10	1	1	2	1	13	1	5	1
540	4	4	2	1	1	6	3	1	3	1	1	5	2	1	49	1	10	2	1	11
560	1	3	4	1	6	1	2	1	2	1	2	2	1	13	1	3	2	2	3	2
580	1	2	*j	1	3	2	2	1	3	3	9	1	3	*k	3	2	3	12	6	11
600	3	2	1	1	1	2	2	2	1	4	1	3	4	7	1	1	3	1	1	1
620	7	1	16	19	9	1	6	8	1	75	1	2	3	7	3	1	1	2	4	2
640	2	4	2	1	1	27	1	1	1	9	15	1	9	3	12	1	13	3	2	16
660	1	1	18	2	3	1	6	1	7	4	1	1	1	1	2	2	4	4	1	3
680	8	1	19	3	21	1	3	1	4	3	3	1	2	2	1	*m	4	7	1	*n
700	1	1	2	1	6	1	2	1	1	3	4	1	2	2	2	22	3	1	9	4
720	3	1	5	1	1	3	5	20	1	12	2	1	1	1	1	2	87	2	2	2
740	59	1	1	2	1	4	17	3	1	1	1	1	2	3	2	1	1	2	3	3
760	1	*p	4	11	19	2	2	1	5	2	3	1	4	1	1	1	1	4	3	1
780	6	1	12	2	7	1	5	4	9	3	3	2	2	1	1	2	4	2	4	3
800	5	3	1	1	5	1	1	4	1	1	20	1	1	1	6	1	6	2	3	1
820	3	1	2	9	1	1	1	10	10	*q	1	1	10	1	3	1	14	1	2	2
840	5	1	1	2	3	2	6	1	3	1	12	1	25	9	*r	1	1	3	1	26
860	34	2	2	5	3	1	2	5	2	3	2	1	1	2	1	8	2	1	1	5
880	3	1	2	9	32	1	1	3	4	1	3	1	1	3	1	1	20	2	2	1
900	3	10	57	1	*s	20	1	1	1	66	1	26	1	4	4	6	5	50	1	1
920	5	3	1	1	6	21	4	4	1	1	1	1	2	7	5	3	9	5	3	1
940	4	2	1	2	1	17	1	59	3	1	8	1	1	10	3	4	2	5	1	1
960	1	2	1	14	5	7	1	1	6	46	1	2	4	6	3	1	3	8	24	7
980	1	1	2	1	3	11	4	1	14	1	13	2	1	2	2	1	7	2	2	1

a = 636	b = 119	c = 425
g = 424	h = 165	i = 114
n = 108	p = 704	q = 904

d = 202	e = 136	f = 699
j = 283	k = 267	m = 716
r = 124	s = 152	

Table II

$$\alpha_1 = 2\cos\frac{2\pi}{7}, \text{ Root of } x^3 + x^2 - 2x - 1$$

FREQUENCY COUNTS

1	401	15	3	33	1	59	2	136	1
2	168	16	4	34	1	60	1	152	1
3	109	17	6	36	1	61	1	165	1
4	60	18	3	37	1	62	2	202	1
5	40	19	9	39	1	66	1	267	1
6	23	20	8	44	1	72	1	283	1
7	19	21	3	45	1	75	1	424	1
8	13	22	3	46	1	87	1	425	1
9	19	24	3	49	1	94	1	636	1
10	12	25	3	50	2	97	1	699	1
11	10	26	3	53	1	108	1	704	1
12	7	27	2	55	1	114	1	716	1
13	12	30	4	56	1	119	1	904	1
14	7	32	2	57	1	124	1		

Table III

$$\alpha_1 = 2\cos\frac{2\pi}{7}, \text{ Root of } x^3 + x^2 - 2x - 1$$

n	a_n	q_n	r_n
51	94	1.34 E 24	1.854
57	60	2.49 E 29	0.958
60	50	2.53 E 31	0.739
65	53	5.83 E 34	0.698
78	636	2.51 E 42	6.978
93	119	3.20 E 51	1.054
105	62	1.96 E 58	0.480
120	425	1.61 E 68	2.815
243	55	7.83 E 124	0.196
279	202	4.61 E 144	0.617
286	136	3.97 E 148	0.404
327	62	3.30 E 166	0.166
357	699	1.22 E 185	1.665
381	97	1.44 E 200	0.213
420	61	2.64 E 220	0.121
463	56	4.85 E 244	0.101
483	424	6.15 E 255	0.728
492	165	6.02 E 260	0.278
496	114	3.47 E 263	0.191
514	72	4.03 E 272	0.116
583	283	3.59 E 305	0.406
594	267	1.45 E 312	0.375
630	75	4.20 E 331	0.100
696	716	2.26 E 366	0.856
700	108	8.11 E 369	0.129
737	87	2.72 E 387	0.098
741	59	1.95 E 390	0.066
762	704	2.68 E 400	0.770
830	904	5.77 E 434	0.909
855	124	2.41 E 448	0.121
903	57	3.80 E 472	0.053
905	152	5.92 E 474	0.141
910	66	2.45 E 478	0.061
918	50	2.34 E 484	0.045
948	59	1.04 E 500	0.052

Table I

α_2, Root of $x^5 - x - 1$

	1	2	3	4	5	6	7	8	9	10	1	2	3	4	5	6	7	8	9	10
0	1	5	1	42	1	3	24	2	2	1	16	1	11	1	1	2	31	1	12	5
20	1	7	11	1	4	1	4	2	2	3	4	2	1	1	11	1	41	12	1	8
40	1	1	1	1	1	9	2	1	5	4	1	25	4	6	11	1	4	1	6	1
60	1	1	2	2	2	4	11	1	4	1	3	2	8	1	3	3	6	21	11	2
80	1	1	10	2	1	3	2	8	1	10	4	3	1	1	1	1	2	1	1	18
100	7	4	1	2	2	6	1	1	2	2	6	20	1	43	3	2	4	2	1	1
120	2	3	1	4	4	1	3	1	1	12	1	2	2	3	2	14	1	5	1	7
140	1	2	2	10	2	2	2	1	38	1	59	2	1	4	1	4	2	4	23	13
160	1	1	1	3	1	32	2	1	3	6	4	1	4	1	1	1	1	91	1	7
180	2	8	1	18	2	2	1	1	28	2	1	12	1	3	4	55	1	1	2	2
200	6	1	3	1	16	1	2	1	1	*a	1	1	4	3	2	1	2	1	4	1
220	1	2	43	3	3	1	55	1	1	2	10	1	2	1	12	12	36	1	8	1
240	18	1	3	1	1	1	6	1	24	2	1	1	1	6	1	3	1	1	11	1
260	4	2	1	1	1	2	2	1	73	1	1	4	*b	54	1	8	4	2	3	4
280	1	1	1	1	99	4	2	2	1	4	1	1	1	10	1	1	1	1	1	1
300	1	2	1	5	3	21	5	1	6	2	1	3	2	1	1	4	1	1	1	85
320	8	1	1	30	1	3	2	3	1	1	3	1	1	1	85	6	1	1	3	2
340	1	10	1	1	*c	3	1	1	17	1	12	1	1	3	5	2	1	3	1	3
360	1	5	1	1	1	*d	1	2	1	2	1	9	3	1	1	3	1	4	3	7
380	1	69	5	1	2	2	3	1	1	8	17	13	1	2	3	2	11	1	8	2
400	4	5	1	2	1	5	1	1	1	5	2	1	10	1	1	1	1	1	6	3
420	3	1	5	2	2	6	4	1	10	1	2	1	1	1	4	2	1	1	2	1
440	4	1	2	4	5	1	1	3	1	5	6	1	4	1	*e	3	18	2	11	9
460	9	2	20	1	10	2	4	1	1	1	5	3	2	2	2	4	3	1	1	8
480	1	7	4	1	3	12	16	1	1	2	2	2	2	3	5	2	1	3	1	16
500	2	1	1	2	4	1	3	5	2	12	1	1	1	12	1	2	26	21	7	2
520	1	2	8	2	2	1	1	1	2	2	1	1	1	3	1	1	1	1	39	4
540	1	29	18	1	8	13	3	1	1	1	1	1	8	1	4	1	3	2	2	2
560	1	5	2	5	1	5	2	8	8	2	8	5	1	4	3	2	2	2	3	3
580	7	2	4	4	2	18	6	*f	6	32	5	13	2	3	6	1	5	2	1	1
600	1	1	1	3	12	2	1	1	2	1	1	48	1	1	1	13	1	5	1	4
620	1	1	5	1	1	3	1	2	1	21	2	2	3	12	1	3	1	1	3	2
640	3	1	*g	11	5	1	1	12	2	2	2	2	3	3	14	1	42	17	1	1
660	1	2	1	2	1	1	5	2	8	1	2	18	2	27	1	14	1	1	3	1
680	1	4	2	3	3	3	1	2	1	9	1	1	1	1	4	1	17	4	3	12
700	1	25	15	5	1	2	2	6	1	7	7	5	1	5	1	7	1	2	1	1
720	1	*h	1	1	1	2	2	1	1	34	4	5	4	16	3	4	1	1	1	10
740	46	2	1	1	1	5	1	1	2	1	7	10	1	3	2	1	1	2	1	7
760	4	*i	1	4	6	1	4	1	1	8	1	1	15	2	3	16	7	1	6	1
780	3	1	1	1	7	1	1	1	1	3	4	1	1	10	1	1	4	2	1	47
800	1	3	3	6	1	*j	74	14	2	24	22	*k	3	9	2	5	3	2	4	2
820	1	19	2	1	2	8	4	2	5	7	4	1	1	4	3	25	1	1	*m	*n
840	2	44	3	1	50	1	1	2	1	10	1	15	1	1	*p	5	2	1	3	37
860	5	7	1	4	4	1	26	2	2	*q	1	7	2	1	11	2	1	2	1	1
880	1	5	1	6	1	1	4	14	3	1	2	1	1	12	1	9	52	1	9	6
900	2	2	4	1	33	3	3	*r	1	1	23	7	1	2	9	1	7	1	2	1
920	1	1	7	4	1	1	1	17	9	2	3	1	14	35	1	1	1	6	9	12
940	1	4	2	*s	1	5	3	1	1	1	2	5	3	7	1	32	8	1	6	1
960	1	3	*t	1	25	1	1	1	26	3	1	3	1	1	7	2	17	6	1	1
980	6	4	1	6	2	23	1	3	2	5	8	1	3	10	1	30	1	13	1	2

a = 761   b = 195   c = 166   d = 264   e = 701   f = 11644
g = 169   h = 673   i = 457   j = 409   k = 274   m = 174
n = 124   p = 172   q = 1033   r = 110   s = 684   t = 1292

Table II

α_2, Root of $x^5 - x - 1$

FREQUENCY COUNTS

1	406	16	6	31	1	48	1	169	1
2	162	17	6	32	3	50	1	172	1
3	89	18	7	33	1	52	1	174	1
4	67	19	1	34	1	54	1	195	1
5	41	20	2	35	1	55	2	264	1
6	28	21	4	36	1	59	1	274	1
7	23	22	1	37	1	69	1	409	1
8	21	23	3	38	1	73	1	457	1
9	11	24	3	39	1	74	1	673	1
10	14	25	4	41	1	85	2	684	1
11	11	26	3	42	2	91	1	701	1
12	16	27	1	43	2	99	1	761	1
13	6	28	1	44	1	110	1	1033	1
14	6	29	1	46	1	124	1	1292	1
15	3	30	2	47	1	166	1	11644	1

Table III

α_2, Root of $x^5 - x - 1$

n	a_n	q_n	r_n
151	59	2.97 E 77	0.344
178	91	3.13 E 91	0.444
196	55	1.29 E 102	0.239
210	761	5.27 E 109	3.096
227	55	1.58 E 118	0.208
269	73	2.00 E 138	0.235
273	195	3.54 E 141	0.609
274	54	1.91 E 143	0.166
285	99	1.35 E 149	0.294
320	85	8.07 E 163	0.230
335	85	3.32 E 171	0.219
345	166	1.87 E 177	0.413
366	264	4.47 E 187	0.621
382	69	1.92 E 195	0.157
455	701	1.44 E 229	1.347
588	11644	6.58 E 300	17.042
643	169	7.49 E 327	0.226
722	673	8.41 E 369	0.797
762	457	1.47 E 391	0.511
806	409	1.39 E 413	0.434
807	74	1.03 E 415	0.078
812	274	4.43 E 421	0.284
839	174	8.98 E 437	0.174
840	124	1.11 E 440	0.123
845	50	2.03 E 444	0.050
855	172	9.54 E 449	0.167
870	1033	3.09 E 461	0.979
897	52	3.27 E 474	0.048
908	110	2.20 E 482	0.100
944	684	3.68 E 503	0.594
963	1292	1.91 E 515	1.096

Table I

$$\alpha_3 = \sqrt[3]{2} + \sqrt{3}, \quad x^6 - 9x^4 - 4x^3 + 27x^2 - 36x - 23$$

	1	2	3	4	5	6	7	8	9	10	1	2	3	4	5	6	7	8	9	10
0	2	1	*a	1	1	3	1	1	5	7	2	3	2	4	1	18	5	1	13	3
20	3	3	4	1	69	2	1	1	7	1	1	3	1	1	13	2	5	2	1	3
40	1	2	38	3	1	2	1	1	2	1	5	1	1	*b	1	1	6	1	2	5
60	1	1	9	4	1	5	2	1	4	5	1	1	18	3	3	2	24	3	1	1
80	1	2	74	3	2	4	3	1	1	10	1	1	1	1	4	1	1	1	3	7
100	8	*c	4	1	4	1	1	2	1	5	1	2	3	1	23	18	4	1	2	1
120	85	1	2	1	2	1	8	1	1	1	22	3	1	3	1	1	8	3	15	30
140	1	7	1	1	1	11	4	1	19	1	1	1	3	6	1	44	3	8	3	1
160	1	1	10	4	1	8	3	5	16	6	3	1	2	12	1	2	3	2	1	3
180	9	1	5	2	4	1	3	2	26	1	2	1	1	2	2	4	2	1	3	2
200	4	5	1	4	1	2	1	4	*d	1	1	8	4	1	1	9	1	1	2	5
220	1	4	2	2	1	29	4	1	*e	3	61	1	4	15	1	3	23	1	5	1
240	1	1	2	3	2	3	6	1	8	1	2	2	1	1	10	1	1	3	3	*f
260	1	1	1	1	1	3	7	8	1	42	1	3	1	1	2	1	2	6	2	1
280	1	1	1	7	16	1	1	1	1	3	37	1	7	38	63	3	1	14	6	1
300	1	4	1	1	2	4	1	6	1	1	1	1	3	1	30	4	1	4	1	8
320	10	4	3	*g	25	2	1	2	1	1	*h	1	1	1	2	2	6	2	9	1
340	13	1	2	4	4	1	1	19	1	1	3	2	3	1	2	1	1	1	4	1
360	7	2	1	*i	8	4	20	1	2	1	3	1	1	94	2	1	3	4	1	3
380	1	7	1	3	1	9	9	1	4	2	4	2	35	1	2	2	1	2	1	1
400	3	12	4	1	1	2	6	1	1	1	1	2	3	13	1	1	5	1	3	7
420	7	2	4	4	3	1	1	1	1	3	12	10	2	2	1	1	1	1	1	3
440	6	1	2	6	28	2	1	1	1	2	1	1	1	2	1	1	10	1	5	3
460	2	3	1	1	3	2	2	8	1	13	4	1	1	7	1	1	2	2	10	4
480	1	1	5	1	8	2	1	4	4	1	1	1	8	12	2	5	3	18	4	27
500	2	3	1	1	1	1	4	3	2	1	7	1	1	1	8	5	1	5	1	5
520	2	1	4	1	14	14	1	1	1	1	2	1	1	22	7	1	1	5	17	1
540	2	4	1	15	1	3	1	1	1	11	1	2	16	1	1	3	1	1	1	9
560	1	3	30	4	3	4	36	1	6	2	2	1	36	2	2	1	5	7	1	1
580	99	2	16	1	37	1	2	1	1	3	1	4	1	1	5	2	2	1	3	3
600	1	3	2	4	1	2	2	31	1	11	1	1	24	1	2	8	3	1	*j	1
620	*k	3	19	2	2	8	1	1	11	2	1	7	1	1	1	7	1	1	5	3
640	3	1	25	1	8	2	2	1	6	1	1	2	1	2	1	1	4	1	1	10
660	22	1	37	1	19	2	1	17	1	38	2	3	8	1	8	2	30	1	2	2
680	5	2	2	3	1	2	1	7	1	3	1	1	12	1	11	3	3	*m	6	1
700	30	1	1	2	1	1	7	1	5	7	1	75	1	12	1	2	1	1	7	1
720		1	2	2	1	1	7	1	3	1	1	27	10	4	1	6	2	1	*n	1
740	2	2	1	1	4	5	7	3	17	21	1	1	58	13	33	2	4	1	5	3
760	12	1	16	3	3	7	*p	1	1	13	2	*q	1	7	2	1	3	1	1	1
780	1	1	2	1	2	1	1	2	7	3	1	3	34	13	10	1	1	1	3	1
800		1	32	1	*r	3	55	3	2	1	6	1	3	1	2	2	1	2	5	1	7
820		1	34	1	5	1	13	1	2	8	1	9	5	1	21	3	2	4	1	1	1
840		1	1	1	5	7	1	1	2	1	2	3	5	28	1	1	11	1	4	1	3
860		2	47	2	3	14	1	1	2	29	1	1	1	7	1	3	1	3	1	2	1
880		8	1	1	1	2	1	4	2	2	2	1	1	2	1	2	13	2	1	50	13
900		23	1	2	5	6	1	2	1	2	1	53	1	6	3	3	3	23	1	1	1
920		5	1	11	1	4	5	*s	2	9	1	27	1	15	2	1	29	1	3	2	2
940		2	3	6	2	80	3	1	9	1	3	9	1	1	2	1	3	11	8	17	1
960		3	1	1	4	8	2	3	1	*t	6	1	1	2	1	*u	3	38	2	1	2
980		3	1	2	2	3	1	9	5	1	8	2	7	2	1	1	1	5	8	1	10

a = 123   b = 1446   c = 126   d = 121   e = 154   f = 452
g = 315   h = 135   i = 103   j = 120   k = 331   m = 184
n = 133   p = 430   q = 298   r = 150   s = 208   t = 186
u = 138

Table II

$$\alpha_3 = \sqrt[3]{2} + \sqrt{3}, \quad x^6 - 9x^4 - 4x^3 + 27x^2 - 36x - 23$$

FREQUENCY COUNTS

1	418	20	1	42	1	123	1
2	156	21	2	44	1	126	1
3	105	22	3	47	1	133	1
4	56	23	4	50	1	135	1
5	38	24	2	53	1	138	1
6	20	25	2	55	1	150	1
7	30	26	1	58	1	154	1
8	25	27	3	61	1	184	1
9	12	28	2	63	1	186	1
10	11	29	3	69	1	208	1
11	8	30	5	74	1	298	1
12	7	31	1	75	1	315	1
13	11	32	1	80	1	331	1
14	4	33	1	85	1	430	1
15	4	34	2	94	1	452	1
16	5	35	1	99	1	1446	1
17	4	36	2	103	1		
18	4	37	3	120	1		
19	4	38	4	121	1		

Table III

$$\alpha_3 = \sqrt[3]{2} + \sqrt{3}, \quad x^6 - 9x^4 - 4x^3 + 27x^2 - 36x - 23$$

n	a_n	q_n	r_n
3	123	1.24 E 2	************
25	69	1.45 E 14	2.462
54	1446	3.93 E 28	24.700
83	74	2.19 E 43	0.779
102	126	1.22 E 53	1.074
121	85	1.36 E 63	0.608
209	121	8.04 E 107	0.497
229	154	3.52 E 118	0.578
231	61	6.49 E 120	0.224
260	452	9.82 E 135	1.473
295	63	1.06 E 154	0.180
324	315	6.59 E 169	0.818
331	135	4.31 E 174	0.341
364	103	4.51 E 189	0.240
374	94	1.00 E 196	0.212
581	99	2.74 E 294	0.148
619	120	6.52 E 313	0.168
621	331	2.18 E 316	0.459
698	184	9.56 E 356	0.226
712	75	7.91 E 364	0.091
739	133	9.57 E 377	0.154
753	58	3.86 E 386	0.066
767	430	2.47 E 398	0.472
772	298	4.13 E 402	0.324
804	150	1.49 E 418	0.158
806	55	2.48 E 420	0.057
899	50	2.45 E 463	0.048
911	53	6.77 E 470	0.050
927	208	5.44 E 480	0.189
945	80	9.75 E 491	0.071
969	186	4.09 E 505	0.161
975	138	2.61 E 509	0.119

The program to do the calculation was written in Fortran, using machine-language subroutines for multiple-precision integer arithmetic to handle the coefficients of the polynomials. The calculation of a_n was done in floating-point arithmetic (approximately 14 significant digits), using a floating-point approximation to $P_n(x)$ (suitably scaled). This procedure avoids the use of multiple presicion arithmetic in any trial-and-error steps, and so makes for greater efficiency. One could be even more efficient, using an idea suggested by Lehmer [5], and compute several successive partial quotients from an approximation to $P_n(x)$. It is possible to find $P_{n+m}(x)$ from $P_n(x)$ and $a_n, a_{n+1}, \ldots, a_{n+m-1}$ with less multiple-precision calculation than is needed to find all the intervening polynomials explicitly. The additional complication in the program, however, did not seem worth while, since the results given here were obtained by the simpler method in about 6 minutes on an IBM 360/91. A listing of the actual program may be obtained on request from Trotter.

References

[1] *Churchhouse* and *Muir*, Continued fractions, algebraic numbers, and modular invariants, J. Inst. Math. App. **5** (1969), 318—328.

[2] *A. Y. Khinchin*, Continued fractions, Chicago 1964.

[3] *S. Lang*, Report on diophantine approximations, Bull. Math. Soc. France **93** (1965), 177—192.

[4] *S. Lang*, Introduction to diophantine approximations, Reading 1967.

[5] *D. Lehmer*, Euclid's algorithm for large numbers, Am. Math. Monthly **45** (1938), 227—233.

[6] *H. Stark*, An explanation of some exotic continued fractions found by Brillhart, to appear.

Fine Hall, Princeton University, Princeton, N. J. 08540 USA

Eingegangen 3. Juni 1971

FROBENIUS AUTOMORPHISMS OF MODULAR FUNCTION FIELDS.

By Serge Lang.

Ihara [2] has determined the Frobenius automorphism of primes in modular function fields of characteristic p. We shall carry out this determination in the generic case of characteristic 0, over $\mathbf{Z}[j]$. More detailed comments on the relation to Ihara's work and Shimura's work [6] will be made later. One reason why the generic case is important is that it allows us to get the analogous result for extensions of a number field by points of finite order on an elliptic curve, giving rise to a correspondence from certain non-abelian extensions to abelian extensions arising from complex multiplication.

1. Generalities on Frobenius automorphisms and points of finite period. Let R_1 be a local ring, without divisors of 0, and integrally closed in its quotient field F_1. Let F be a Galois extension of F_1 (possibly infinite). Let $\mathfrak{m}$ be the maximal ideal of R_1. Let R be the integral closure of R_1 in F and G the Galois group of F over F_1. Then G operates transitively on the set of maximal ideals of R lying above $\mathfrak{m}$. If $\mathfrak{M}$ is one such ideal, then it has a decomposition group (isotropy group) $G_\mathfrak{M}$ in G, which maps surjectively on the Galois group of the residue class field extension. For finite extensions this is elementary algebra ([4], Chapter 1, §5), and the infinite case follows by a trivial application of Zorn's lemma. In our applications, the residue class field $R_1/\mathfrak{m}$ will be finite of characteristic p, and hence modulo the inertia group, there is a Frobenius automorphism σ in the decomposition group of $\mathfrak{M}$, characterized modulo the inertia group by the condition

$$\sigma x \equiv x^q \pmod{\mathfrak{M}},$$

if q is the cardinality of $R_1/\mathfrak{m}$. Frobeniuses associated with different $\mathfrak{M}$ over $\mathfrak{m}$ differ by a conjugation.

An abelian curve A will be said to be defined over R_1 if it has non degenerate reduction modulo the maximal ideal. If the curve is in Weierstrass form

---

Received August 24, 1971.

$$y^2 = 4x^3 - g_2 x - g_3,$$

this means that g_2, g_3 lie in R_1, and that its discriminant is a unit in R_1, provided the characteristic is not 2 or 3.

Suppose that A is defined over R_1. Let $K_N = F_1(A_N)$ be the extension generated by the coordinates of the division points A_N of period N and let K be the union of all K_N. Then each K_N and K are Galois over F_1. The first thing to note is that these extensions are unramified over $\mathfrak{m}$ if $p \nmid N$. Indeed, denote reduction mod $\mathfrak{M}$ by a bar. Let $\sigma \in G_{\mathfrak{M}}$. We know that reduction mod $\mathfrak{M}$ induces an injection on the points of finite period prime to p, the characteristic of $R_1/\mathfrak{m}$. Suppose that $\bar{\sigma}$ is the identity on the residue class field. Since the action of σ is determined by its effect of the points of A_N, it follows that σ is the identity. This also shows that the residue class field is generated by the coordinates of the points of $\bar{A}_N$. The same argument applies to the subfield generated say by the x-coordinate if the curve is in non-degenerate Weierstrass form as above.

On the other hand, suppose that $R_1/\mathfrak{m}$ has q elements, and that τ is an element of $\mathrm{Gal}(F/F_1)$ such that

$$\tau w \equiv w^q \pmod{\mathfrak{M}}$$

for every coordinate function w of all points of period prime to p, but we do not know if τ lies in the decomposition group of $\mathfrak{M}$. Let

$$\sigma = (\mathfrak{M}, R/R_1)$$

be the Frobenius automorphism of $\mathfrak{M}$ in $G_{\mathfrak{M}}$. Then

$$\sigma^{-1} \tau w \equiv w \pmod{\mathfrak{M}}.$$

Hence from the injectivity of the reduction of points of finite period prime to p, we conclude that $\sigma^{-1} \tau = \mathrm{id}$, so that $\sigma = \tau$.

The above is in a sense standard, but we shall apply it to rings which are not discrete valuation rings.

2. Coordinate representations. We recall briefly what they are. For a good recent exposition, cf. Shimura [6]. Let $z \in \mathfrak{H}$ (the upper half plane). We let $L_z = [z, 1]$ be the lattice generated by z and 1, and we denote by A^z an abelian curve whose j-invariant is $j(z)$. If $a = (a_1, a_2)$ is in $\mathbf{Q}^2$, then the dot product

$$a \begin{pmatrix} z \\ 1 \end{pmatrix} = a_1 z + a_2$$

lies in QL_z, and an analytic representation of A^z as a quotient C/L_z induces a homomorphism

$$\phi_z: Q^2 \to QL_z/L_z \to A^z.$$

If

$$\alpha = \begin{pmatrix} a & b \\ c & d \end{pmatrix}$$

is an integral matrix with positive determinant, and $\mu = cz + d$, then there is an isogeny $\lambda: A^{\alpha(z)} \to A^z$ such that the following diagram is commutative.

$$\begin{array}{ccccc}
Q^2 & \longrightarrow & QL_{\alpha(z)} & \longrightarrow & A^{\alpha(z)} \\
\alpha \downarrow & & \mu \downarrow & & \downarrow \lambda \\
Q^2 & \longrightarrow & QL_z & \longrightarrow & A^z.
\end{array}$$

In other words,

$$\lambda \circ \phi_{\alpha(z)}(a) = \phi_z(a\alpha).$$

(we view a as a row veceor.)

3. Deuring representatives. If K is a field, we denote its algebraic closure by ^aK. Consider a fixed place $\mathfrak{P}$ of ^aQ into ^aF_p, denoted by a bar,

$$x \mapsto \bar{x}.$$

As in Ihara, we start with Deuring's canonical lifting [1]. Let J_p be the set of singular invariants $j(z)$ in characteristic 0, such that the order $\mathfrak{o}$ of $[z,1]$ has conductor not divisible by p, and such that p splits completely in the imaginary quadratic field $k = Q(z)$, these two conditions being denoted by the symbol $(\mathfrak{o}/p) = 1$. Then the association

$$j(z) \mapsto \overline{j(z)}$$

gives a bijection between J_p and the set of non-supersingular invariants in ^aF_p. A point $z \in \mathfrak{H}$ such that $\overline{j(z)} = \bar{j}$ will be called *Deuring representative* of $\bar{j}$ in $\mathfrak{H}$.

Consider such a point. Let $\mathfrak{p} = \mathfrak{P} \cap \mathfrak{o}_k$, where $\mathfrak{o}_k$ is the ring of algebraic integers of k. Let $M^p = M_2^p(Z)$ be the set (monoid) of 2×2 rational integral matrices whose determinant is equal to a power of p. Then M^p is a subset of G^+_Q (rational matrices with positive determinant), and acts in the usual way on $\mathfrak{H}$. We let M_z^p denote the subset of M^p which leaves z fixed. We shall prove that there exist two elements α, α' of M_z^p such that M_z^p is a disjoint union of two direct products

$$M_z^p = \{\alpha\} \times p^N \times T \cup \{\alpha'\} \times p^N \times T,$$

where $\{\alpha\}$, $\{\alpha'\}$ are positive powers of α, α' respectively, p^N consists of all power of p with natural numbers, and T is isomorphic to the group of units in the order $\mathfrak{o}$ of $[z, 1]$.

Remark. Ihara factors out by powers of p, and is thus led to a group, consisting of all elements in $GL_2(\mathbf{Z}[\frac{1}{p}])$ whose determinants are powers of p, modulo $\pm p^{\mathbf{Z}}$, this losing something of the full structure in the present situation. However, his arguments work just as well in general, and even with some simplifications, as we now see.

Let $\mathfrak{p}_0 = \mathfrak{p} \cap \mathfrak{o}$. Then $\mathfrak{p}_0$ has finite order D in the group of $\mathfrak{o}$-ideal classes. Let $\mathfrak{p}_0^D = (\mu)$. Then there is a unique matrix α in M^p such that

$$\alpha \begin{pmatrix} z \\ 1 \end{pmatrix} = \mu \begin{pmatrix} z \\ 1 \end{pmatrix},$$

and $\alpha(z) = z$, i.e. z is a fixed point of α. By the Kronecker congruence relation, we have

$$\overline{j(\mathfrak{p}_0^{-1}\mathfrak{a})} = \overline{j(\mathfrak{a})}^p,$$

where $\mathfrak{a} = [z, 1]$. Hence the invariants

$$\overline{j(\mathfrak{a})}, \overline{j(\mathfrak{p}_0 \mathfrak{a})}, \cdots, \overline{j(\mathfrak{p}_0^{D-1}\mathfrak{a})}$$

form a complete set of conjugates of $\bar{j}$ over $\mathbf{F}_p$ (no repetitions because of the injectivity in Deuring's reduction mapping on J_p). Therefore D is also the degree of $\bar{j}$ over $\mathbf{F}_p$. In the present situation, the commutative diagram of §2 yields

$$\bar{\lambda} \, \overline{\phi_z(\mathfrak{a})} = \overline{\phi_z(\mathfrak{a}\alpha)}.$$

Furthermore, $\bar{\lambda}$ is purely inseparable (the standard situation of complex multiplication as originated by Deuring). Hence $\bar{\lambda}$ differs from the Frobenius map $\pi_p{}^D$ by an automorphism ϵ of $\bar{A}$, and consequently we get the relation

$$\overline{\phi_z(\mathfrak{a}\alpha)} = \epsilon \overline{\phi_z(\mathfrak{a})}^{p^D}.$$

The matrix α has infinite period modulo $p^N \times T$ ($T = $ torsion) because μ does not lie in $\mathfrak{p}\mathfrak{o}$ (not divisible by the conjugate of $\mathfrak{p}_0$). Let $\beta \in M_z^p$. Dividing out a positive power of p, we may assume that β is primitive. Then

$$\beta \begin{pmatrix} z \\ 1 \end{pmatrix} = \mu_1 \begin{pmatrix} z \\ 1 \end{pmatrix}$$

with some $\mu_1 \in \mathfrak{o}$, because z is a fixed point of β. But $\mu_1 \notin \mathfrak{po}$. Hence if $\mathfrak{p}'$ is the conjugate of $\mathfrak{p}$, then

$$\mu_1 \mathfrak{o} = \mathfrak{p}_0^m \quad \text{or} \quad \mu_1 \mathfrak{o} = \mathfrak{p}_0'^m$$

for some positive integer m. Since D is the period of $\mathfrak{p}_0$ in the ideal class group of $\mathfrak{o}$, we must have $D \mid m$. Hence

$$\mu_1 = \mu^{m/D}\zeta, \quad \text{or} \quad \mu_1 = \mu'^{m/D}\zeta,$$

where ζ is a unit of $\mathfrak{o}$. Hence $\beta = \alpha^{m/D}\gamma$, where γ has finite period, and corresponds to a unit of $\mathfrak{o}$, or $\beta = \alpha'^{m/D}\gamma$, where α' relates to μ' as α relates to μ.

We have now proved what we wanted about the structure of M_z^p. Observe that the distinction between α and α' can be made by the choice of $\mathfrak{p}$ in $\mathfrak{o}_k$. We then call α a *positive generator* of M_z^p with respect to $\mathfrak{p}$ (or the generator of M_z^p relative to $\mathfrak{p}$). It is then well defined modulo elements of T.

Remarks. By factorizing out $p^{\mathbf{Z}}$, Ihara is led to a coarser group, and those points in $\mathfrak{H}$ which have infinite isotropy group form a larger orbit than the orbit of a Deuring representative for $SL_2(\mathbf{Z})$. Ihara's arguments had to go through the above steps, and although by factoring out $p^{\mathbf{Z}}$ he gets a function field over a finite field in characteristic p, one should not give that part up in the generic case, or over number fields. Ihara is exclusively interested in the applications to characteristic p, and in his conjective, now proved, concerning the unramified extensions of the modular function fields into which supersingular primes split completely, thus eliminating extensions by roots of unity.

Also the property for the eigenvalues of α to be in $\mathbf{Q}_p$, and to be distinct, arises here naturally from the complex multiplication situation. In [2], Ihara gets such a property in a more general set-up, depending on the discreteness of a subgroup of $GL_2(\mathbf{R}) \times GL_2(\mathbf{Q}_p)$. As we see, this can be avoided.

4. The generic situation. Let j be the modular function, normalized to take the values $j=0$ and $j=1728$ when the carresponding abelian curve has extra automorphisms. Let $F_1 = \mathbf{Q}(j)$, and let F_N be the field of modular functions of level N, identified with the field of "x-coordinates" of the points of period N. Let F be the union of all F_N, with N prime to p. Let $\bar{j} \in {}^a\mathbf{F}_p$ be a non-supersingular element of the algebraic closure of $\mathbf{F}_p$, and let z be a Deuring representative. Let $k = \mathbf{Q}(z)$. Then k is contained in F (because p splits completely in k, whose discriminant is therefore prime to p, and hence k is contained in the field of N-th roots of unity, which is contained in F_N).

We have a homomorphism called the bar,

$$Z[j] \to {}^aF_p$$

reducing Z mod p, and sending j on $\bar{j}$. We let R_1 be the local ring of this homomorphism in $Q(j)$, and we let R be the integral closure of R_1 in F.

Let k_{ab} be the maximal abelian extension of k, and let $\mathfrak{P}$ be a place of k_{ab} in ${}^aF_p$. Let $\mathfrak{p} = \mathfrak{P} \cap \mathfrak{o}_k$, so that $p\mathfrak{o}_k = \mathfrak{p}\mathfrak{p}'$. We denote $\mathfrak{P}$ by a bar. Observe that for $f \in R_1$, we can obtain $\bar{f}$ either by the direct homomorphism $Z[j] \to {}^aF_p$, or by the succession

$$f \mapsto f(z) \mapsto \overline{f(z)}.$$

Thus the bar homomorphism on R_1 extends to a homomorphism on R given by the same successsion of mappings. We let $\mathfrak{M}$ be the kernel of the map $R \to \bar{R}$, so that $\mathfrak{M}$ is a maximal ideal in R.

Let $\phi: Q^2 \to A$ be a coordinatization of the points of finite order on a generic curve A with invariant j, defined over $Q(j)$.

THEOREM. *The Frobenius automorphism* $(\mathfrak{M}, F/F_1)$ *is that automorphism which is induced by the automorphism*

$$\phi(a) \mapsto \phi(a\alpha),$$

on the field of all points of finite order, where α is the element of $M_z^\mathfrak{p}$ described in § 3, i.e. the positive generator of $M_z^\mathfrak{p}$ with respect to $\mathfrak{p}$.

Proof. If z is not equivalent to i or ρ under the modular group, then the assertion of the theorem is now essentially immediate from the considerations of § 1, and the relation

$$\overline{\phi_z(a\alpha)} = \pm \overline{\phi_z(a)}^{p^D}$$

of § 3. Indeed, a coordinate representation of A specializes to a coordinate representation of A^z, because we can select A defined over R_1 to specialize to A^z, even by an equation of the form

$$y^2 = 4x^3 - cx - c,$$

with $c/(c-27) = j/1728$. The exceptional cases have always proved to be an additional pain in the literature, whether in Ihara [2], or through the additional arguments of Shimura's [6], Theorem 6.31. However, for simplicity now we quote this theorem of Shimura, i.e. his "reciprocity law" at fixed points, which says that if s is an idele of k and $\sigma = (s, k)$ is the associated element of the Galois group of the maximal abelian extension of k, then for any function $f \in F$ we have

$$f(z)^\sigma = f^{u^{-1}}(z)$$

where u is a certain element of the group of automorphisms of F, corresponding to the embedding of the idele s in the adelized matrix group of 2×2 matrices. In our case, σ is simply the Frobenius automorphism of k_{ab} over $k(j(z))$, associated with a prime divisor of $\mathfrak{p}$ in $k(j(z))$. By the standard formalism of Frobenius automorphisms, this is the same as the Frobenius automorphism

$$(\mathfrak{p}^D, k_{ab}/k),$$

because $\mathfrak{p}^D$ is the norm of such a prime divisor. Thus for our idele we can take the idele having $\mathfrak{p}$-component μ, and 1 everyhere else, or equivalently, $s = (\cdots, \mu^{-1}, \mu^{-1}, 1, \mu^{-1}, \mu^{-1})$ having 1 at $\mathfrak{p}$ and μ^{-1} everywhere else, where μ is such that $\mathfrak{p}_o{}^D = (\mu)$, as described in §3. It is then clear that the embedding of this idele in the adelized matrix group yields the matrix α of §3. This proves the theorem.

Of course, the Frobenius automorphism is only determined up to the inertia group. On these fields F_N with N prime to p, it is easy to see that the inertia group is precisely T (in the product decomposition of §3). For extenensions with points of period a power of p, we get an unramified part corresponding to the points which don't vanish under reduction, and a ramified part corresponding to the others. Ihara's description of what happens when $\bar{j}$ is supersingular also works in the present context, and is in fact much easier. For instance when $\bar{j} \in \mathbf{F}_p$ and when there are no automorphisms other than ± 1 on the corresponding abelian curve, then $\pi_p{}^2 = \pm p\delta$, represented by the diagonal matrix pI_2. We note that $\mathfrak{m}$ splits completely in a large extension (up to an extension of degree 2, it is the field of all functions $j \circ \psi$, with $\psi \in G_{\mathbf{Q}^+}$).

Having factored out p^z Ihara had to consider a smaller field than F, but it is no harder to consider the general case.

Having the Frobenius automorphism over $\mathbf{Z}[j]$ now allows us to get it in the field abtain by reduction modulo a prime ideal. Alegbraically, the situation amounts to selecting a prime q of dimension 1 in the maximal ideal $\mathfrak{m}$. We take a prime $\mathfrak{Q}$ in $\mathfrak{M}$ lying above q. Those elements of $G_{\mathfrak{M}}$ which leave $\mathfrak{Q}$ invariant then induce a Frobenius automorphism of $R/\mathfrak{Q}$ over R_1/q. To get Ihara's theorem in characteristic p, we select q to be the ideal generated by p. We can, however, take a prime q which yields extensions of a number field, proceeding as follows.

Take a value $z \in \mathfrak{H}$ such that $j(z)$ is algebraic, and such that, for simplicity, there is no degeneracy in the extension generated by the points of finite order on an abelian curve defined over $\boldsymbol{Q}(j(z))$, having invariant $j(z)$. Examples of many such cases (they happen most of the time) are due to the recent work of Serre [5]. Take a prime p for which $j(z)$ is p-integral. Then the theorem can be used to describe how a divisor $\mathfrak{p}_z$ of p in $\boldsymbol{Q}(j(z))$ decomposes in the field $F(z)$, i.e. the residue class field of the place

$$f \mapsto f(z)$$

for $f \in F$. Reducing $j(z)$ mod $\mathfrak{p}_z$ yields a value $\bar{j}$. Suppose that $\bar{j}$ is not supersingular. The Galois group of $F(z)$ over $\boldsymbol{Q}(j(z))$ can be identified with the Galois group of F over $\boldsymbol{Q}(j)$, and thus the theorem gives us a description of the Frobenius automorphism of $\mathfrak{p}_z$ in $F(z)$. The process consists in finding a Deuring lifting z' of $\bar{j}$ in $\mathfrak{H}$, and then applying the theorem to z'. Thus we obtain a *correspondence* from certain *non-abelian* extensions of $\boldsymbol{Q}(j(z))$ to *abelian extensions* $\boldsymbol{Q}(z', j(z'))$ over $\boldsymbol{Q}(z')$, and the non-abelian Frobenius automorphism can be reduced to the study of an abelian one. For example, suppose that we want to know which primes split completely in some $F_N(z)$. We just compute the matrix $\alpha(z')$ and see whether $\alpha(z')$ is the identity (mod ± 1) on the group

$$\frac{1}{N} \boldsymbol{Z}^2 / \boldsymbol{Z}^2 \subset \boldsymbol{Q}^2 / \boldsymbol{Z}^2.$$

i.e. whether a prime divisor of p in $k(j(z'))$ splits completely in $F_N(z')$. Note that the computation takes place "in the ground field." It involves $\bar{j} = j(z) \bmod p$, and then only finding the μ in $\boldsymbol{Q}(z')$ such that $\mathfrak{p}_0{}^D = (\mu)$.

Finally I would like to recall the problem which I had already encountered many years ago [3], for abelian class field theory over finitely generated rings over $\boldsymbol{Z}$, namely give an appropriate equivalence among the maximal ideals to determine which ones have the same Artin symbol in an abelian extension. It turns out here that we have some determination of the Frobenius automorphism in a non-abelian situation of Kronecker dimension 2, i.e. a situation where both p and j vary (not only fixed j, variable p as in ordinary complex multiplication, or fixed p, variable j, as in Ihara).

PRINCETON UNIVERSITY.

REFERENCES.

[1] M. Deuring, " Die Typen der Multiplikatorenringe elliptischer Funktionenkörper," *Abh. Math. Sem. Hamburg* (1941), pp. 197-272.
[2] Y. Ihara, *On congruence monodromy problems*, University of Tokyo, 1968, escpecially Chapter 5.
[3] S. Lang, " Unramified class field theory over function fields," *Annals of Mathematics*, vol. 64, No. 2 (1956), pp. 285-325.
[4] ———, *Algebraic Number Theory*, Addison Wesley (1970).
[5] J. P. Serre, *Abelian l-adic representations and elliptic curves*, Benjamin,
[6] G. Shimura, *Introduction to the arithmetic theory of automorphic functions*, Tokyo University Press and Princeton University Press, 1971.

HIGHER DIMENSIONAL DIOPHANTINE PROBLEMS

BY SERGE LANG

As on previous occasions [L 5], [L 6], I shall discuss some general conjectures concerning diophantine analysis on varieties. These involve rational points, integral points, and the possibility of treating by diophantine methods questions which in the past have been handled by congruence methods.

1. **Rational points.** A classical conjecture of Mordell states that a curve of genus ≥ 2 over the rational numbers has only a finite number of rational points. Let K be a finitely generated field over the rational numbers. Then the same statement should hold for a curve defined over K, and a specialization argument due to Néron shows in fact that this latter statement is implied by the corresponding statement over number fields (cf. [L 1, Chapter VII, §6]).

Let V be a variety in projective space, defined over the complex numbers, and therefore over some finitely generated field over the rationals. We shall say that V has the **Mordell property** if it has only a finite number of rational points in any finitely generated field over Q. One possibility to extend Mordell's conjecture to higher dimensional varieties is as follows.

(1.1) *Let D be a bounded domain in C^n (it should be irrelevant whether D is symmetric or not). Let Γ be a discrete group of automorphisms, acting freely on D, and assume that the quotient $\Gamma \backslash D$ is compact, and embedded as a variety V in projective space. Then V has the Mordell property.*

One must assume that Γ operates freely (the isotropy group at each point is the identity), otherwise the quotient may have singularities, whose effect is analogous to decreasing the genus in the case of curves. Similarly, one must assume that the quotient is compact, otherwise one is faced with a situation which may be like that of modular curves which may have a low genus 0 or 1. The Mordell conjecture is a special case of the above, because a curve of genus ≥ 2 is a quotient of the disc. On the other hand, it has always been useful to regard a curve of genus ≥ 2 as a

---

An address delivered before the New York meeting of the Society on April 13, 1974 by invitation of the Committee to Select Hour Speakers for Eastern Sectional Meetings; received by the editors March 12, 1973.

AMS (MOS) subject classifications (1970). Primary 32H20, 14G99, 53C99, 14K05.

Copyright © American Mathematical Society 1974

subvariety of its Jacobian, and one had the conjecture:

(1.2) *Let V be a subvariety of an abelian variety which does not contain the translation of an abelian subvariety of dimension ≥ 1. Then V has the Mordell property.*

I therefore looked for a way to unify this with the previous one, and it seems that the most natural hypothesis on V is hyperbolicity. There are alternative conditions defining this property, not known at present to be equivalent. The following two conditions are the most relevant for us.

H 1. *The Kobayashi metric is strictly positive and V is complete.*

(For the Kobayashi metric, see [**K 1**].)

H 2. *The sectional holomorphic curvature is bounded above by a negative constant, and V is complete.*

(Cf. Kobayashi [**K 1**, Chapter III, Theorem 4.11].) The second condition could be weakened slightly to require the property only of an unramified covering. The first condition is usually taken as the definition. Then we conjecture that:

(1.3) *If V is a projective hyperbolic variety, then V has the Mordell property.*

A variety satisfying the hypotheses of (1.1) is always hyperbolic. It does not seem to be known if the universal covering space of a hyperbolic variety is a bounded domain. Note that (1.2) is related to (1.3). I owe to Griffiths the remark that if a subvariety of an abelian variety does not contain the translate of an abelian subvariety of dimension ≥ 1, then its curvature is strictly negative. Indeed, the holomorphic curvature is decreasing on submanifolds [**K 1**, Chapter III, Theorem 1.1], and if the curvature were to be 0 at a point, then this would imply that there is a flat torus passing through that point, i.e. an abelian subvariety. Thus (1.3) implies (1.1) and (1.2), assuming all varieties nonsingular. The question also arises whether the universal covering space of a subvariety of an abelian variety as in (1.2) is a bounded domain.

There are "obvious" geometric ways of generating rational points on some varieties, for instance, if they contain a straight line or an elliptic curve. Neither is possible for hyperbolic varieties, cf. [**K 1**, Chapter V, Theorem 1.1]. If V admits a bounded domain as universal covering space, then we can see this from the Liouville theorem that a bounded holomorphic function is constant. Another way of generating rational points is by means of an infinite group of automorphisms, also impossible for hyperbolic varieties [**K 1**, Chapter V, Corollary 2.3].

As usual, the absolute Mordell property has a relative formulation for algebraic families of hyperbolic varieties: If there is an infinity of sections, then the family contains split subfamilies, and almost all sections are due to constant sections. In the split case, i.e. when one deals with the product of a hyperbolic variety and a fixed variety, we are thus led to consider the following conjecture, whose proof would give a generalization of a theorem of De Franchis for curves of genus ≥ 2 (cf. [L 1, Chapter VII, Historical Note]):

(1.4) *Let V be a projective hyperbolic variety and W any algebraic variety. Then there is only a finite number of surjective rational maps of W onto V.*

2. Integral points. Let V be an affine variety. We shall say that V has the **Siegel property** if V has only a finite number of points in any finitely generated ring (without divisors of zero) over the integers Z. Siegel's theorem is that a curve of genus ≥ 1 has this property. Some time ago, I conjectured that affine subsets of abelian varieties have this property [L 3], [L 1]. One possible approach is through the methods which have been developed in connection with the theory of transcendental numbers, as suggested in [L 3]. Let A be an abelian variety defined over a number field K. We suppose that A is embedded in projective space. Let A_K be the group of points on A rational over K. The Mordell-Weil theorem states that A_K is finitely generated. The notion of height of a point can be defined in general. For simplicity let us assume further that $K = Q$. If $(x_0, \cdots, x_{d'})$ are projective coordinates for a point P in A_K, with $x_i \in Z$ $(i = 0, \cdots, m)$ relatively prime to each other, then the height is defined by

$$H(P) = \max_i |x_i(P)|.$$

Let $P^1, \cdots, P^n$ be a basis for the Mordell-Weil group A_K, modulo torsion. Given $P \in A_K$ there exists a torsion point Q and integers q_j such that

$$P = q_1 P^1 + \cdots + q_n P^n + Q.$$

By the quadraticity of the Néron-Tate height [N], [L 2], there exist constants C_1, C_2 such that $C_1^{q^2} \leq H(P) \leq C_2^{q^2}$ for all $P \in A_K$. More precisely, $\log H(P)$ is equal to a quadratic function of P, plus a linear function, plus a bounded function. If φ, ψ are two positive functions, let us use the Vinogradov notation, and write $\varphi \ll \psi$ if there exists a constant C such that $\varphi \leq C\psi$. We write $\varphi \gg\ll \psi$ if $\varphi \ll \psi$ and $\psi \ll \varphi$. Then we have $\log H(P) \gg\ll q^2$.

We may view the complex points A_C as parametrized by abelian functions on C^d ($\dim A = d$), relative to a suitably normalized exponential map, represented by theta functions, $\exp: \mathbf{C}^d \to A_\mathbf{C}$, normalized so that the differential at the origin is algebraic. For $i = 1, \cdots, d'$ let

$$f_i(z) = x_i(\exp(z)), \quad z \in \mathbf{C}^d.$$

If $\exp u$ is an algebraic point on A, we call u an algebraic point of the exponential map, or also an abelian logarithm of an algebraic point on A. Let

$$\exp w = Q, \quad \exp u^j = P^j, \quad \text{with } w, u^j \in \mathbf{C}^d.$$

Then w is a division point of a period. We have

$$\exp(q_1 u^1 + \cdots + q_n u^n + w) = P.$$

Suppose that A_K contains infinitely many integral points with respect to the affine coordinates $y_i = x_i/x_1$, so that $y_1 = 1$. If P is such an integral point, then for some coordinate, say y_0, we have

$$|y_0(P)| = H(P) \geq C^{q^2}.$$

Let $x_i = y_i/y_0$, so that $x_1 = 1/y_0$. Then $|x_1(P)| \leq C^{-q^2}$.

Suppose that A_K contains infinitely many integral points. Selecting a subsequence of these if necessary, we may assume without loss of generality that the following condition holds. In their expression as a linear combination of a basis of the Mordell-Weil group, the same torsion point Q occurs. For all such points, we have $|y_0(P)| = H(P)$. These integral points converge to a point on the divisor of zeros of f_1, say to a point $P^0 = \exp u^0$. In the case of dimension 1, as already pointed out in [L 4], Siegel's theorem that there is only a finite number of integral points on an elliptic curve follows from an approximation statement of type

$$|q_1 u^1 + \cdots + q_n u^n - r\omega - u^0| > e^{-\tau(q)},$$

where τ is a function of q which is $o(q^2)$, r is a fixed rational number, and ω is a period. Indeed, after a projective linear transformation over $\mathbf{Z}$, we may assume that $f = (f_1, \cdots, f_d)$ gives an analytic isomorphism in a neighborhood of u^0. Then $|f(z) - f(u^0)|$ and $|z - u^0|$ have the same order of magnitude, so that the inequality in terms of the algebraic function on A can be transferred to an inequality on the universal covering space, in terms of the abelian logarithms.

Recently, Masser was able to prove the desired diophantine inequality, with a function $\tau(q) = q^\varepsilon$, for elliptic curves with complex multiplication [M 2].

In the higher dimensional case one does not have much information about the point u^0. First, it may be that $\exp u^0$ is not algebraic. Second, in order to prove the finiteness of integral points, one needs a conjecture of the following type.

(2.1) *Let x be a nonconstant abelian function. If $\tau(q) = o(q^2)$, then we have*

$$|x(P)| \geq e^{-\tau(q)}$$

for all $P \in A_K$ not lying in the divisor of zeros of x, and $q \gg\!\!\ll \log H(P)$ sufficiently large.

Of course, one conjectures the much stronger inequality with a function $\tau(q) = C \log q$ with a sufficiently large constant C, or even a constant C only epsilon larger than the "Dirichlet exponent" which guarantees that points can always be found to satisfy the inequality. Cf. [L 4].

Thus the conjecture applies to a single function. It is then a problem to prove diophantine inequalities first simultaneously for all the coordinates $(f_1, \cdots, f_d)$, and then to eliminate one after the other to get similar inequalities with one function. I was able to generalize part of Masser's results for *all* the coordinates in [L 9], although the measure function $\tau(q)$ which I obtain is poor. It is enough to prove the transcendence of $\exp(\alpha^1 u^1 + \cdots + \alpha^n u^n)$ when the α^j are algebraic, and when, for each i, some component α_i^j does not lie in the field of complex multiplication. This is a small beginning in the desired direction.

From the point of view of integral points, we are also led to relationships between the values of functions and their heights. For simplicity, let A be a simple abelian variety defined over K, and let φ be a nonconstant abelian function. I expect that the height of $\varphi(P)$ tends to infinity, for P ranging over any infinite subset of A_K. This is implied by (1.2). Indeed, if the height of $\varphi(P)$ is bounded, then φ takes on only a finite number of values, and the points P lie in the divisors of such values. Another problem here is whether one can extend to one coordinate the quadraticity of the height.

However, even as the height goes to infinity rapidly, from the point of view of Mordell-type conjectures, I also would expect a rather strong limitation on the speed with which the absolute value of the coordinate goes to infinity, in line with (2.1). For instance one expects an inequality $|\varphi(P)| \geq q^C$, for $P \in A_K$ such that $\varphi(P)$ is defined, and $q^2 \gg\!\!\ll \log H(P)$ sufficiently large, as mentioned above.

The absolute value $|\varphi(P)|$ can also be interpreted geometrically as being of the order of magnitude of a power of the distance of P to the divisor of zeros of φ, when $|\varphi(P)|$ tends to 0. Thus the above inequality can be interpreted as giving a limitation to the closeness between a point in A_K and the divisor of zeros of (φ). Considering φ^{-1} instead of φ gives an interpretation in terms of poles.

Even on an elliptic curve without complex multiplication, if ω_1, ω_2 are fundamental periods and u^0 is an algebraic point for the Weierstrass $\wp$

function, it is still a problem to prove an inequality of type

(2.2) $\quad |q_0 u^0 + q_1 \omega_1 + q_2 \omega_2| > \exp(-q^\varepsilon), \quad q = \max |q_j|.$

A similar inequality can be asked p-adically. It is known and easy to prove that if f is an elliptic function and $f(u)$ is algebraic, then the denominators of $f(u/p^n)$ are bounded when p is a prime number and $n \to \infty$. It is a problem to generalize this to abelian varieties, and to give inequalities (lower bounds) for the closeness of u/p^n to the divisor of zeros or poles of f if u/p^n does not lie in such a divisor.

Thus one is led to consider single functions rather than a set of local uniformizing parameters in proving desired diophantine inequalities. In this line, when an abelian variety does not have complex multiplication, or when it is not strongly normalized (for the definition, cf. [L 4]), it is still unknown that

(2.3) *If u is an algebraic point for the exponential map, $u \neq 0$, then every coordinate of u is transcendental.*

In the special case when A has complex multiplication and the exponential map is strongly normalized, I proved it as a corollary of a theorem of Bombieri [B 1], [L 9].

For inequalities giving measures of transcendence in cases simpler than (2.2) and (2.3), cf. Baker [Ba 1]] [Ba 2], Feldman [F], Coates [Co 1], [Co 2], Masser [M 1].

3. **Algebraic points.** There is a certain class of results which in the past has been obtained essentially by congruence methods. For instance, the Mordell-Weil theorem. On the other hand, one knows that two algebraically independent functions cannot take algebraic values at any point when suitably normalized and when they satisfy a differential equation [Sch], [L 8]. This suggests that the Mordell-Weil theorem, which concerns one function, the Weierstrass $\wp$-function, and its derivative, should be provable by methods related to those used in the theory of transcendental numbers and diophantine approximations.

Furthermore, it is also reasonable to expect that one can attack the isogeny theorem (still a conjecture due to Serre) by these methods. It states:

(3.1) *Let A, B be elliptic curves defined over a number field, K, and without complex multiplication. If their Galois representations $V_p(A)$ and $V_p(B)$ are isomorphic, then the curves are isogenous.*

Let us recall briefly the definitions involved in this statement. For more details, see Serre [Se] and [L 7, Chapter XVI]. Let $A^{(p)}$ denote the group of points on A whose order is a power of a prime p. Let A_{p^n} denote the group of points on A of order dividing p^n. Let $V_p(A)$ be the set of all infinite vectors $(a_0, a_1, a_2, \cdots)$ where $a_0 \in A^{(p)}$ and $pa_{i+1} = a_i$ for all i. Then $V_p(A)$ is a vector space over $\mathbf{Q}_p$, and the Galois group $G = \text{Gal}(\bar{K}/K)$ operates on $V_p(A)$ in a natural way. If $\sigma \in G$ then $\sigma(a_0, a_1, \cdots) = (\sigma a_0, \sigma a_1, \cdots)$. Serre proved the theorem when one of the two curves has a j-invariant which is not integral for some prime number.

One can approach the problem by disregarding the representation aspects and concentrating on the degrees of the fields of division points $K(A_{p^n})$. (One reason for this lies in [L 7, Theorem 1, Chapter 16, §1]. If $K(A^{(p)}) = K(B^{(p)})$, then the p-adic Galois representations on $V_p(A)$ and $V_p(B)$ become isomorphic over a finite extension of K. Thus enough knowledge about the degree of the fields of division points implies automatically the isomorphism of the representations.) If A does not have complex multiplication, a theorem of Serre asserts that the degree of p^n-division points satisfies the inequality

$$[K(A_{p^n}):K] \gg p^{4n}.$$

The Galois group $\text{Gal}(K(A^{(p)})/K)$ is closed in $\text{GL}_2(\mathbf{Z}_p)$, and is a Lie subgroup. Therefore it is a priori clear that the degrees above have order of increase p^n, or p^{2n}, or p^{3n}, or p^{4n}. The fields $K(A^{(p)})$ and $K(B^{(p)})$ have the field of all p^nth roots of unity in common.

A very simple argument based on the fact that $\text{SL}_2(\mathbf{Z}_p)$ does not contain a closed, normal, nontrivial subgroup of infinite index shows that either

$$[K(A_{p^n}, B_{p^n}):K] \ll p^{4n},$$

or the fields $K(A^{(p)})$ and $K(B^{(p)})$ are linearly disjoint over a finite extension of the field of all p^nth roots of unity (see [L 7, Corollary of Theorem 1, Chapter XVI, §1]). We must then have

$$[K(A_{p^n}, B_{p^n}):K] \gg p^{7n}.$$

Thus to prove the isogeny theorem, it suffices to prove:

(3.2) *Let A, B be elliptic curves without complex multiplication, defined over a number field K. Let $\varepsilon > 0$. Then*

$$[K(A_{p^n}, B_{p^n}):K] \gg p^{n(4+\varepsilon)}.$$

The number 4 in the exponent reflects "dimension 4".

The proofs of transcendence for the classical numbers, values of the exponential or Weierstrass function at algebraic numbers, actually have nothing to do with transcendental numbers. Assuming that these values are algebraic, one derives a contradiction by juggling with arithmetic and analytic inequalities. In fact, a basic theorem ([Sch] and [L 8]) gives an upper bound for the number of points where certain functions satisfying a differential equation can take values in a fixed number field K. The degree $[K:Q]$ is bounded from below by this number of points and other factors. (Cf. [L 8, Theorem 1, Chapter III, §1, and Theorem 2, Chapter II, §2].)

The situation here is completely analogous. Suppose that the elliptic curves A, B are not isogenous. We consider the *five* algebraically independent functions

$$e^{2\pi i z}, \quad \wp_A(\omega_1 z), \quad \wp_A(\omega_2 z), \quad \wp_B(\omega_3 z), \quad \wp_B(\omega_4 z),$$

where $\wp_A$, $\wp_B$ are the Weierstrass functions associated with A and B, and where $[\omega_1, \omega_2]$, $[\omega_3, \omega_4]$ are fundamental periods of $\wp_A$ and $\wp_B$, respectively. One can form the usual approximating function with coefficients in the field of division points,

$$\sum \alpha_{(\lambda)} \wp_A(\omega_1 z)^{\lambda_1} \wp_A(\omega_2 z)^{\lambda_2} \wp_B(\omega_3 z)^{\lambda_3} \wp_B(\omega_4 z)^{\lambda_4} e^{2\pi i z \lambda_0},$$

and activate the standard arguments which have previously been used in proving transcendence results (possibly with variations, using several variables). Although I obtain not completely trivial estimates (lower bounds) for the degrees of the fields of p^nth division points, I fall short of the desired $4+\varepsilon$. The difficulty here is exactly the same as that which one meets when trying to prove a statement like (2.2). Even for one elliptic curve the difficulty arises in trying to prove by these methods the known Serre theorem giving the lower bound $[K(A_{p^n}):K] \gg p^{4n}$. Actually, in this case, a simple Lie theoretic argument (bottom of p. IV-11 in Serre's book [Se]) shows that it suffices to prove

$$[K(A_{p^n}):K] \gg p^{n(2+\varepsilon)}$$

showing that the Galois group of $K(A^{(p)})$ has dimension >2 in order to jump immediately to 4. Indeed, if the dimension were 3, one would be led to a contradiction of the Šafarevič theorem, through the irreducibility of the Galois representation on $V_p(A)$. Cf. [L 7, Chapter XVII, §§1, 2].

Thus instead of *using* Serre type theorems as in Coates [Co 1] to prove transcendence and approximation results, one would *prove* lower bounds on degrees of fields of division points by the methods of diophantine approximations-transcendental numbers. It may also be that these two types of results generate feedback on each other, and by a suitable recursive procedure, one can use one after the other to strengthen results in both directions.

Bibliography

[Ba 1] A. Baker, *On the periods of the Weierstrass $\wp$-function*, Symposia Mathematica, vol. IV (INDAM, Rome, 1968/69), Academic Press, New York, 1970, pp. 155–174.

[Ba 2] ———, *An estimate for the $\wp$-function at an algebraic point*, Amer. J. Math. 92 (1970), 619–622. MR **43** #7409.

[Bo] E. Bombieri, *Algebraic values of meromorphic maps*, Invent. Math. **10** (1970), 267–287. MR **46** #5328.

[Co 1] J. Coates, *An application of the division theory of elliptic functions to diophantine approximation*, Invent. Math. **11** (1970), 167–182. MR **44** #3963.

[Co 2] ———, *Linear forms in the periods of the exponential and elliptic functions*, Invent. Math. **12** (1971), 290–299. MR **45** #3330.

[F] N. Feldman, *An elliptic analogue of an inequality of Gelfond*, Trudy Moskov. Mat. Obšč. **18** (1968), 65–76 = Trans. Moscow Math. Soc. **18** (1968), 71–84. MR **40** #1345.

[K 1] S. Kobayashi, *Hyperbolic manifolds and holomorphic mappings*, Pure and Appl. Math., 2, Dekker, New York, 1970. MR **43** #3503.

[K 2] ———, *Some problems on intrinsic distances and measures*, Carathéodory Centennial (to appear).

[L 1] S. Lang, *Diophantine geometry*, Interscience Tracts in Pure and Appl. Math., no. 11, Interscience, New York, 1962. MR **26** #119.

[L 2] ———, *Les formes bilinéaires de Néron et Tate*, Séminaire Bourbaki: 1963/64, Exposé 274, fasc. 3, Secrétariat mathématique, Paris, 1964. MR **31** #1252.

[L 3] ———, *Integral points on curves*, Inst. Hautes Études Sci. Publ. Math. No. 6 (1960), 27–43. MR **24** #A86.

[L 4] ———, *Diophantine approximation on toruses*, Amer. J. Math. **86** (1964), 521–533. MR **29** #2220.

[L 5] ———, *Some theorems and conjectures in diophantine equations*, Bull. Amer. Math. Soc. **66** (1960), 240–249. MR **22** #9469.

[L 6] ———, *Transcendental numbers and diophantine approximations*, Bull. Amer. Math. Soc. **77** (1971), 635–677. MR **44** #6615.

[L 7] ———, *Elliptic functions*, Addison-Wesley, Reading, Mass., 1974.

[L 8] ———, *Introduction to transcendental numbers*, Addison-Wesley, Reading, Mass., 1966. MR **35** #5397.

[L 9] ———, *Diophantine approximations on abelian varieties with complex multiplication* (to appear).

[M 1] D. Masser, *Transcendence properties of elliptic functions* (to appear).

[M 2] ———, *Algebraic points of an elliptic function* (to appear).

[N] A. Néron, *Quasi-fonctions et hauteurs sur les variétés abéliennes*, Ann. of Math. (2) **82** (1965), 249–331. MR **31** #3424.

[Sch] T. Schneider, *Einführung in die transzendenten Zahlen*, Springer-Verlag, Berlin, 1957. MR **19**, 252.

[Se] J.-P. Serre, *Abelian l-adic representations and elliptic curves*, Benjamin, New York, 1968. MR **41** #8422.

Journal für die reine und angewandte Mathematik

Herausgegeben von **Helmut Hasse** und **Hans Rohrbach**

Sonderdruck aus Band 267, Seite 219 bis 220

Verlag Walter de Gruyter · Berlin · New York 1974

Addendum to
"Continued fractions of some algebraic numbers"
(This Journal, vol. 255, 1972)

By *S. Lang* at New Haven and *H. Trotter* at Princeton

References [2], [3], and [4] came to our attention after the proof-sheets of [1] had been corrected. It is clear that the computational method we used is essentially the same as that used in [2] and [4], and is presumably the same as that used in [3] (which does not give details of the computation).

Reference [4] gives the results of a χ^2-test comparing the observed frequencies of partial quotients of certain algebraic numbers with the theoretical frequencies for a "random" number. Results are reported for nine algebraic numbers, for each of which between 700 and 800 partial quotients were calculated. Nothing was found to suggest non-randomness except for a very low value of χ^2 (indicating unusually *good* agreement between expected and observed frequencies) for the expansion of the cube root of 2. After some discussion the authors remark ". . . the impression persists that the expansion of $2^{\frac{1}{3}}$ is peculiar. Probably the expansion will have to be carried to many more terms to verify or contradict this impression."

It therefore occurred to us that it might be worth while to exhibit the results of applying a similar χ^2-test to the expansions that we had calculated. Following [4], we divided the partial quotients into ten groups consisting of 1, 2, 3, 4, 5 and 6, 7 and 8, 9 through 12, 13 through 19, 20 through 40, and over 40. For each of the eight numbers for which we obtained expansions, we give the χ^2 value obtained from the distribution of the first 1000 partial quotients, and in the column headed P, the approximate probability that the χ^2 for a random sample would be no larger. (The probabilities are computed for the ordinary χ^2-distribution on nine degrees of freedom. This is not strictly correct because the partial quotients of a "random" number are not independent. The error involved is assessed in [4], and we agree with the authors that it is negligible for present purposes.) For the first six numbers (the numbers of degree 3) we give the same information for the distribution of the first 3000 partial quotients.

The rows of the table correspond to the numbers reported on in [1]. Thus the first five are the cube roots of 2, 3, 4, 5, and 7, and the last three are the positive roots of

$$x^3 + x^2 - 2x - 1, \quad x^5 - x - 1,$$

and

$$x^6 - 9x^4 - 4x^3 + 27x^2 - 36x - 23.$$

28*

The results do not suggest any significant departure from random behaviour. In particular the anomaly observed in [4] for the cube root of 2 appears not to persist when the expansion is carried further.

χ^2	P	χ^2	P
\multicolumn{2}{c}{$N = 1000$}	\multicolumn{2}{c}{$N = 3000$}
4.61	.13	5.59	.22
8.41	.51	10.33	.68
8.47	.51	7.71	.44
8.07	.47	9.48	.61
10.22	.67	13.32	.85
8.08	.48	7.72	.44
4.08	.09	—	—
12.73	.83	—	—

References

[1] S. *Lang* and H. *Trotter*, Continued fractions of some algebraic numbers, J. reine u. angew. Math. **255** (1972), 112—134.

[2] A. D. *Bryuno*, Continued Fraction Expansion of Algebraic Numbers, Zh. Vychisl. Mat. i Mat. Fiz. **4** nr. 2 (1964), 211—221. English translation, U. S. S. R. Comput. Math. and Math. Phys. **4** (1964), 1—15.

[3] J. *von Neumann* and B. *Tuckerman*, Continued Fraction Expansion of $2^{\frac{1}{3}}$, Math. Tables Aids Comput. **9** (1955), 23—24.

[4] R. D. *Richtmyer*, M. *Devaney* and N. *Metropolis*, Continued fraction expansions of algebraic numbers, Numer. Math. **4** (1962), 68—84.

Yale University, Mathematics Department, Box 2155, Yale Station, New Haven, Connecticut 06520, USA

Fine Hall, Princeton University, Princeton, New Jersey, 08540, USA

Eingegangen 17. September 1972

Diophantine Approximation on Abelian Varieties with Complex Multiplication

SERGE LANG*

Received January 1974

INTRODUCTION

Let A be an abelian variety defined over a number field K. We suppose that A is embedded in projective space. Let A_K be the group of points on A rational over K. In section 1 the definition of the height of a point is $P \in A_K$. For simplicity here, let $K = \mathbf{Q}$. If $(x_0, x_1, ..., x_{d'})$ are projective coordinates for P, with $x_i \in \mathbf{Z}$ ($i = 0,..., d'$) relatively prime to each other, then

$$H(P) = \max |x_i(P)|.$$

Let $P^1,..., P^n$ be a basis for the Mordell–Weil group A_K, modulo torsion. Given $P \in A_K$, there exists a torsion point Q and integers q_j such that

$$P = q_1 P^1 + \cdots + q_n P^n + Q.$$

By the quadraticity of the Néron–Tate height [17, 25], there exist constants C_1, C_2 such that

$$C_1^{q^2} \leqslant H(P) \leqslant C_2^{q^2}; \qquad q = \max |q_j|$$

for all $P \in A_K$. More precisely, $\log H(P)$ is equal to a quadratic function of P, plus a linear function, plus a bounded function.

We may view the complex points $A_\mathbf{C}$ as parametrized by abelian functions on $\mathbf{C}^d$ (dim $A = d$), relative to a suitably normalized exponential map (recalled in Section 2, cf. [16]) represented by theta functions,

$$\exp: \mathbf{C}^d \to A_\mathbf{C}.$$

For $i = 1,..., d'$ let

$$f_i(z) = x_i(\exp(z)); \qquad z \in \mathbf{C}^d.$$

* Supported by NSF grant.

If exp u is an algebraic point on A, we call u an *algebraic point of the exponential map*, or also an *abelian logarithm of an algebraic point on A*. Let

$$\exp w = Q; \quad \exp \exp u^j = P^j; \quad \text{with} \quad w, u^j \in \mathbf{C}^d.$$

Then w is a division point of a period. We have

$$\exp(q_1 u^1 + \cdots + q_n u^n + w) = P.$$

We can select the projective coordinates sufficiently generally that, say, $f_1, \ldots, f_d$ give a local analytic isomorphism in a neighborhood of the origin. Let $f = (f_1, \ldots, f_d)$. Then near the origin, $|f(z)|$ and $|z|$ have the same order of magnitude. Therefore, if $H = \max |q_i|$ we have an inequality of type

$$\frac{1}{C^{H^2}} \leqslant |q_1 u^1 + \cdots + q_n u^n + \omega|,$$

for any period ω. The first basic problem of diophantine approximations on abelian varieties is to improve this inequality. For reasonably strong conjectures concerning such lower bounds for linear combinations of abelian logarithms of algebraic points on A, cf. [15].

A first possibility is to replace the function C^{H^2} by $C^{o(H^2)}$, for instance e^{H^ϵ}. In increasing order of improvement one can then try for $e^{(\log H)^\kappa}$ for some positive number κ, then H^c for some positive number c (large), and then for the best possible expected value of c in line with the Roth theorem. Finally, one can ask for a type along the lines discussed in my book [18].

Just obtaining $C^{o(H^2)}$ has significance for applications to diophantine problems, in the following manner. Let $y_i = x_i/x_1$ so that $y_1 = 1$. If P is an integral point with respect to the affine coordinates y_i, then for some coordinate, say y_0, we have

$$|y_0(P)| = H(P) \geqslant C^{q^2}.$$

Let $x_i = y_i/y_0$, so that $x_1 = 1/y_0$. Then

$$|x_1(P)| \leqslant C^{-q^2}.$$

Suppose that A_K contains infinitely many integral points. Selecting a subsequence of these if necessary, we may assume without loss of generality that the following conditions hold. In their expression as a linear combination of a basis of the Mordell–Weil group, the same torsion point Q occurs. For all such points, we have

$$|y_0(P)| = H(P).$$

These integral points converge to a point on the divisor of zeros of f_1, say to a point

$$P^0 = \exp u^0.$$

In the case of dimension 1, as already pointed out in [15], Siegel's theorem that there is only a finite number of integral points on an elliptic curve is equivalent to an approximation statement of type

$$|q_1 u^1 + \cdots + q_n u^n - \alpha \omega - u^0| > e^{-\epsilon q^2},$$

where α is rational, ω is a period, and the inequality holds for all ϵ and q sufficiently large (depending on ϵ). In the case of abelian varieties, an inequality like this one is equivalent to a similarly inequality for

$$|f(P) - f(P^0)|,$$

where $f = (f_1, \ldots, f_d)$ give an analytic isomorphism in a neighborhood of, say, the origin and u^0. The inequality does not say anything about one fixed abelian function, which is what is needed to yield the finiteness of integral points on affine open subsets of abelian varieties, which I conjectured some time ago [11, 21]. A difficulty appears when $d > 1$, and one is led to other conjectures as follows.

For simplicity, let A be a simple abelian variety defined over a number field K, and let φ be a nonconstant abelian function. I expect that the height of $\varphi(P)$ tends to infinity, for P ranging over any infinite subset of A_K. This is implied by the conjecture that there is only a finite number of points of A_K on any proper subvariety of A (which does not contain an abelian subvariety, since A is simple). Indeed, if the height of $\varphi(P)$ is bounded, then φ takes on only a finite number of values, and the points P lie in the divisors of such values.

Another problem here is to extend to one coordinate the quadraticity of the height. It is conceivable that the height of every coordinate (nonconstant) will go to infinity more or less as rapidly as any other.

However, even as the height goes to infinity rapidly, from the point of view of Mordell-type conjectures, I also would expect a rather strong limitation on the speed with which the absolute value of the coordinate goes to infinity, say that the inequality

$$|\varphi(P)| \leq e^{-q^\epsilon}$$

has only a finite number of solutions, or various improvements along the lines discussed in [16] and [18], e.g., replacing the right-hand side by q^{-c} with an appropriate constant c. This would mean that the numerator and denominator of the coordinate go to infinity much more rapidly than the absolute value of the coordinate, and at about the same speed, up to a much lower order of magnitude.

The absolute value of $|\varphi(P)|$ can also be interpreted geometrically as being of the order of magnitude of a power of the distance of P to the divisor of zeros of φ, when $|\varphi(P)|$ tends to 0. Thus, the above inequality can be interpreted as giving a limitation to the closeness between a point in A_K and the divisor of zeros of (φ). Considering φ^{-1} instead of φ gives an interpretation in terms of poles.

The above considerations concerning algebraic points provide much of the motivation for considering linear combinations of algebraic points of the exponential map with algebraic coefficients. The same methods apply to such "algebraic" questions as they do to prove transcendence results, and we shall obtain in this paper a generalization of a theorem of Masser from elliptic curves to abelian varieties with complex multiplication, namely:

Let A be an abelian variety with complex multiplication, and an exponential map normalized as in Section 2. Let $u^1,..., u^n$ be algebraic points of the exponential map, linearly independent over the field k of complex multiplication. Let $\alpha^1,..., \alpha^n$ be algebraic elements of $\mathbf{C}^d$ such that for each $i = 1,..., d$, some coordinate α_i^j does not lie in k. Then the point

$$\exp(\alpha^1 u^1 + \cdots + \alpha^n u^n)$$

is transcendental on A.

We shall also prove:

Under the same normalization of the exponential map, if u is an algebraic point of this map, $u \neq 0$, then each coordinate u_i of u is transcendental.

I view these as a small step in the higher dimensional program just outlined.

We turn to the method of proof. Aside from the original Hermite procedure dealing with e, and extended by Lindemann to deal with e^α (α algebraic), there is essentially one known method of proof dealing with transcendence questions about exponential mappings. Beyond the original work of Siegel, Gelfond, and Schneider on ordinary logarithms and elliptic curves, some very substantial variations and deepenings of this method have been given recently.

Baker [2-4] discovered how to expand the extrapolation by diminishing slightly the derivatives and expanding the set of points, using several variables. Feldman [10] gave a measure of transcendence for quotients of elliptic logarithms using a clever set of points for his interpolation. Coates, in an unpublished set of comments, suggested that when there is complex multiplication, the extrapolation method becomes more efficient, and may succeed even though failing (as far as we know today) when there is not. Whereas Baker with his method could handle algebraic linear combinations of ordinary logarithms, he was able to deal only with combinations of ordinary logarithms, he was able to deal with only combinations of two periods of elliptic functions without complex multiplication.

Masser [23, 24] recently made very important contributions that allowed him to deal with the general case on elliptic curves having complex multiplication. Before him, fairly explicit arguments about determinants of various functions involved were used to derive the final contradiction after the extrapolation. Masser gives a much more general and much more powerful way of making the elimination, essentially based on the Lagrange interpolation method, bounding the coefficients of a polynomial by its values at suitable points. He was thus able to prove an inequality of the form

$$|\alpha^1 u^1 + \cdots + \alpha^n u^n - u^0| > e^{-H^\epsilon}$$

on elliptic curves with complex multiplication. The main object of the present paper is to generalize Masser's theorem to abelian varieties, still with complex multiplication. However, I obtain a weaker inequality then he does (cf. Theorem 2), not in the desired range $e^{-o(H^2)}$. On the other hand, I do not enter into Masser's elimination and induction procedure, which allowed him to get the better measure with H^ϵ. The principal difficulty to carry out this induction is to prove the case $n = 1$! The other difficulties seem to me to be manageable, and I may deal separately with them elsewhere.

In [6], Bombieri and I discuss the role of quantitative results concerning the equidistribution of abelian logarithms in certain questions of transcendence that had arisen in [14] (see also [16]). A negative approximation condition at the origin implies a positive approximation condition everywhere. Masser uses such equidistribution and transference in a new manner to prove his theorem. The method used here follows Masser to a large extent.

For the convenience of the reader, I reproduce the proofs of general transference theorems in a self-contained appendix.

As for the extension of transcendence results from elliptic curves to abelian varieties, the first paper is due to Schneider, who dealt with the periods of a curve [27]. I gave further results and techniques [12–14, 16] along the lines both of the Hermite–Lindemann and Gelfond–Schneider theorems on group varieties.

Other questions suggest themselves, and will be discussed in Section 2, after we have set up the terminology more formally.

For the basic reference to the algebraic theory of abelian varieties, we refer to [19]. For the connection with theta functions, see [20].

I am much indebted to Masser for letting me have his manuscript before publication.[1]

The main contents of the paper are:

Contents

Part I. The Theorems
1. Heights
2. Abelian varieties
3. The reduction theorem

Part II. The Approximation Function and the Main Proof
4. Special points
5. Determination of the basic parameters
6. Estimate of differences
7. Estimates on heights and sizes
8. The Baker–Coates Lemma
9. Extrapolating on integral multiples
10. Extrapolation on division points

[1] *Added in proof:* Masser's papers [22], [23], [24] have appeared in his *Elliptic Functions and Transcendence*, Springer Lecture Notes 437. The present paper was delayed at Masser's request, pending the appearance of the above.

Appendix 1. Geometry of Numbers and Transference
 1. Successive minima
 2. General transference theorem
 3. Applications to abelian varieties

Appendix 2. Interpolation
 1. Polynomials with many zeros
 2. Estimates for a holomorphic function

Appendix 3. Good Estimates for meromorphic maps

PART 1. THE THEOREMS

1. *Heights*

Let $\alpha = (\alpha_1, ..., \alpha_d)$ be a vector of algebraic numbers in an algebraic number field K. Let $\{v\}$ be the set of absolute values on K that induce either the p-adic absolute value $|\ |_p$ or the ordinary absolute valute $|\ |_\infty$ on the rational numbers $\mathbf{Q}$. The *height* of α, *relative* to K, is defined to be

$$H_K(\alpha) = \prod_v \max(1, \|\alpha_1\|_v, ..., \|\alpha_d\|_v),$$

where

$$\|x\|_v = |x|_v^{N_v}$$

is normalized by using the local degree $N_v = [K_v : \mathbf{Q}_v]$ in such a way as to satisfy the product formula. Note that for two vectors α, β we have trivially

$$H(\alpha\beta) \leq H(\alpha) H(\beta).$$

The product $\alpha\beta = (\alpha_1\beta_1, ..., \alpha_d\beta_d)$ is taken componentwise.

The height above is defined relative to K, but letting

$$h(\alpha) = H_K(\alpha)^{1/[K:\mathbf{Q}]}$$

defines an *absolute height*, independent of the field in which the components of α are embedded. For properties of heights, cf. [11].

Suppose that in K, the element α_i has an ideal factorization

$$(\alpha_i) = \frac{\mathfrak{a}_i}{\mathfrak{b}_i}$$

with ideals $\mathfrak{a}_i$, $\mathfrak{b}_i$ relatively prime. Let $\mathfrak{b} = $ l.c.m. $\mathfrak{b}_i$. Then it is immediate from the definition of the height that

$$H_K(\alpha) = \prod_{v \text{ arch}} \max_i(1, \|\alpha_i\|_v) \cdot \mathbf{N}\mathfrak{b},$$

where $\mathbf{N}$ is the absolute norm. Observe that $\mathbf{N}\mathfrak{b}$ is a denominator for each α_i, i.e. a positive integer b such that $b\alpha_i$ is an algebraic integer for all i. Thus, the height is decomposed into an estimate for the conjugates of α_i and a denominator for these numbers.

We use the notation $\|\alpha\|$ to denote the maximum of the absolute values of the conjugates of an algebraic number α, or a family of algebraic numbers $\alpha = (\alpha_1, \ldots, \alpha_d)$. Since $\| \ \|$ occurs without subscript, no confusion can arise with the normalized absolute value $\| \ \|_v$ mentioned above. If $P(T_1, \ldots, T_d) = P(T)$ is a polynomial,

$$P(T) = \sum \alpha_{(\lambda)} T_1^{\lambda_1} \cdots T_d^{\lambda_d},$$

with coefficients in a number field, we let $\|P\| = \max \|\alpha_{(\lambda)}\|$.

By the *size* of an algebraic number, or of a family of algebraic numbers, we mean the maximum of a smallest denominator for it, and the absolute values of the conjugates. It is clear how the size gives a bound for the height and conversely.

2. Abelian varieties

Let A be an abelian variety defined over a number field K. Then the complex torus $A_\mathbf{C}$ (set of complex points of A) can be represented as a quotient

$$\exp: \mathbf{C}^d \to A_\mathbf{C},$$

by means of theta functions, which even given a projective embedding of $A_\mathbf{C}$. Say $(\theta_0, \ldots, \theta_{d'})$ are such functions. Then each function

$$f_i = \frac{\theta_i}{\theta_0},$$

is an abelian function, periodic with respect to a period lattice Λ. The exponential map above can be changed by a linear automorphism of $\mathbf{C}^d$. If we assume that the differential at the origin is algebraic, then this linear automorphism is determined up to linear algebraic automorphisms and the exponential is then called *weakly normalized*. This is the normalization used in [12]. If this differential is defined over K, if $z_1, \ldots, z_d$ are the d complex variables of $\mathbf{C}^d$, and if f is an abelian function defined over K, then $\partial f/\partial z_i$ is also defined over K. We assume from now on that the exponential map is weakly normalized. We shall soon impose a stronger normalization property for the applications we have in mind.

The affine ring
$$K[f_1, \ldots, f_{a'}],$$
is mapped into itself by the derivations $\partial/\partial z_i$ because the partial derivative of a function is holomorphic at every point where the function is holomorphic, and the affine ring is integrally closed, because A is nonsingular.

If $u \in \mathbf{C}^d$ and $\exp(u)$ is an algebraic point of A, then we shall call u an *algebraic point of the exponential map*.

Let $u^1, \ldots, u^n$ be algebraic points for this exponential map.

Extending K by a finite extension, we may assume that the algebraic points $\exp u^j$ are all rational over K. We let
$$(\theta_0, \ldots, \theta_{d'})$$
be the projective coordinates of our exponential map. Changing the projective embedding by a projective linear transformation over $\mathbf{Z}$, we may assume without loss of generality that θ_0 does not vanish at u^j ($j = 0, \ldots, n$), and also does not vanish at the origin. We let
$$f_i = \frac{\theta_i}{\theta_0}.$$

Without loss of generality, we may also assume that the following conditions are satisfied.

The functions $f_1, \ldots, f_d$ are algebraically independent, and $f_{d+1}, \ldots, f_{a'}$ are integral over $K[f_1, \ldots, f_d]$.

The functions $f_1, \ldots, f_d$ give an analytic isomorphism of the ball $\mathbf{D}_{3\rho}(u^j)$ of radius 3ρ centered at each of the points u^j and the origin. We may also assume that $f(0) = 0$.

Finally, for technical convenience, the projective embedding can be selected in such a way that it corresponds to a linear system in which one hyperplane is invariant under the mapping $P \mapsto -P$ on A. This is useful to get a purely quadratic function from the height. See Section 7.

The above conditions set up the exponential map in a convenient form. More seriously, *we shall assume from now on that A has complex multiplication*. By this we mean that
$$\text{End}(A)_{\mathbf{Q}} = \text{End}(A) \otimes \mathbf{Q} = k,$$
is a totally imaginary number field k, which is a quadratic extension of a totally real field, and
$$[k : \mathbf{Q}] = 2d.$$

This is what Shimura calls the *primitive case of sufficiently many complex multiplications* [28]. We assume that the reader is acquainted with the basic facts concerning this situation, e.g. [28] Chapter II, Section 5 and 6. We recall some of these facts and their proofs briefly for convenience.

To begin with, we note that A is simple. Suppose that A is defined over a number K containing k and all its conjugates. The complex representation of $\mathrm{End}(A)_Q$ is equivalent to the representation on differential forms of first kind. Since k is semisimple (being a commutative field!) we can find a basis for the differentials of first kind on A over K, say

$$\Omega^1, \ldots, \Omega^d,$$

such that if $\gamma \in k$, then

$$\Omega^i \circ \gamma = \gamma_i \Omega^i,$$

where γ_i is a conjugate of γ in $\mathbf{C}$. Since the rational representation is equivalent to the direct sum of the complex representation and its conjugate (see for instance [20], Chapter V, Section 3, Theorem 4), it follows that $\gamma_1, \ldots, \gamma_d$ represent distinct pairs of complex conjugates. These differentials of first kind then determine a normalization of the exponential map, such that their pull back to $\mathbf{C}^d$ is $dz_1, \ldots, dz_d$. We call this a *strong normalization*.

If f is an abelian function defined over K, then $\partial f/\partial z_i$ is also defined over K. In particular, a strongly normalized exponential map is also weakly normalized.

If $\alpha, u \in \mathbf{C}^d$ then we write the product

$$\alpha u = (\alpha_1 u_1, \ldots, \alpha_d u_d).$$

If $\gamma \in k$, then we also view γ as an element of $\mathbf{C}^d$,

$$\gamma \mapsto (\gamma_1, \ldots, \gamma_d),$$

where γ_i is the conjugate mentioned above. We say that an element $\alpha \in \mathbf{C}^d$ is *proper* if no coordinate α_i is 0. Observe that the image of an element of k in $\mathbf{C}^d$ is proper.

We denote by $\| \ \|$ or $| \ |$ the sup norm in $\mathbf{C}^d$, so that if $z = (z_1, \ldots, z_d)$ then $|z| = \max |z_i|$.

From now on, unless otherwise specified, we assume that A is an abelian variety with complex multiplication, defined over a number field K containing k, and all the conjugates of k. We assume that the exponential map is strongly normalized.

We begin by giving a transcendence result on each coordinate of an algebraic point for the exponential map. Given a lattice in $\mathbf{C}^d$ (of maximal rank $2d$), we first note that it cannot be contained in a hypersurface. In fact, given an integer b, there exists a ball of radius R such that some element of the lattice in that ball is not contained in a hypersurface of degree $\leqslant b$. Otherwise, for each integer R say that the lattice elements in the ball of radius R are contained in the hypersurface H_R of degree $\leqslant b$. Without loss of generality, we may assume that all the coefficients of the equation defining the hypersurface are of absolute value $\leqslant 1$, and some coefficient has absolute value $=1$. The set of such equations is compact. Hence the hypersurfaces H_R have a limit, which is also a hypersurface, whence a contradiction.

THEOREM 1. *Let $u \neq 0$ be an algebraic point of the exponential map, with coordinates*

$$u = (u_1, ..., u_d).$$

Then each coordinate u_i is transcendental.

Proof. We shall first prove that no coordinate can be equal to 0. After reordering the coordinates, suppose that

$$u_1 = \cdots = u_r = 0 \quad \text{but} \quad u_{r+1}, ..., u_d \neq 0.$$

The set of all points

$$(\gamma_1, ..., \gamma_r, \gamma_{r+1} u_{r+1}, ..., \gamma_d u_d)$$

with $\gamma = (\gamma_1, ..., \gamma_d)$ in $\mathfrak{o}$ is a lattice in $\mathbf{C}^d$. Let b be a large constant. Let S be a finite subset of the points of this lattice which is not contained in a hypersurface of degree $\leqslant b$. Without loss of generality, after a projective change of coordinates on $(\theta_0, ..., \theta_{d'})$, we may assume that θ_0 does not vanish at any point

$$(0, ..., 0, \gamma_{r+1} u_{r+1}, ..., \gamma_d u_d),$$

with $(\gamma_1, ..., \gamma_r, \gamma_{r+1} u_{r+1}, ..., \gamma_d u_d)$ in S. We let as before $f_i = \theta_i/\theta_0$, and we let

$$g_i(z) = f_i(0, ..., 0, z_{r+1}, ..., z_d), \quad i = 1, ..., d'.$$

Let $\mathbf{C}^{d-r}$ be the space of the last $d-r$ coordinates. Then $\exp(\mathbf{C}^{d-r})$ is an analytic subgroup of $A_{\mathbf{C}}$. Since A is simple, it follows that the transcendence degree of the functions g_i with $i = 1, ..., d'$ is also d, because this analytic subgroup must be Zariski dense in $A_{\mathbf{C}}$, and the Zariski closure of a subgroup is a subgroup. Hence the ring

$$K[g_1,...,g_{d'},z_1],$$

has transcendence degree at least $d+1$. Furthermore, the functions $g_i(z)$, z_1 take on values in K at all points of the set S. The ring $K[g_1,...,g_{d'},z_1]$ is mapped into itself by the partial derivatives. A theorem of Bombieri [5] now yields a contradiction if we chose the constant b to sufficiently large with respect to d. This proves that no coordinate can be 0.

Suppose that one coordinate u_1 is algebraic. We repeat essentially the same argument, using the ring of functions

$$K[f_1,...,f_{d'},z_1],$$

taking values in K at the lattice points γu, for $\gamma \in \mathfrak{o}$. This proves our theorem.

To prove that the coordinate of an algebraic point is transcendental when the exponential map is only weakly normalized, one is led to consider linear combinations

$$\alpha_1 u_1 + \cdots + \alpha_d u_d$$

with algebraic coefficients, and to estimate their difference with algebraic numbers (an inhomogeneous problem, cf. [4, 7, 8].

Similarly one also wants to determine the transcendence of linear forms of coordinates, in particular linear combinations

$$\alpha^1 u^1 + \cdots + \alpha^n u^n$$

of algebraic points. From this point of view, as well as the point of view of integral points on the abelian variety, it is important to have results concerning one coordinate, whether taken in $\mathbf{C}^d$ or on the abelian variety. In this paper, we only obtain a result for all coordinates simultaneously.

THEOREM 2. *Let $u^0, u^1,..., u^n$ be algebraic points for the strongly normalized exponential map, linearly independent over the field of complex multiplication k. Given a positive integer d_0, there exists a number H' having the following property. Let $\alpha^1,..., \alpha^n$ be algebraic elements of $\mathbf{C}^d$, with components of degree $\leq d_0$. Let H be a number $\geq H'$ such that $H \geq H(\alpha^j)$ for all j. Then*

$$|\alpha^1 u^1 + \cdots + \alpha^n u^n - u^0| > e^{-\tau^*(H)},$$

where

$$\log \tau^*(H) = (\log H)^{2(2/\kappa)^n} \quad \text{and} \quad \kappa = \frac{1 - \left(\frac{2+1}{d}\right)}{6(n+1)(2d+1)}.$$

We observe that κ is a number <1, and that our measure of inequality given by $\tau^*(H)$ is lousy.

The proof of Theorem 2 will be given in the next section, using a third result stated as Theorem 4. The remaining sections will be devoted to the proof of Theorem 4.

Theorem 2 implies a transcendence result.

THEOREM 3. *Assume that $u^1,..., u^n$ are algebraic points for the strongly normalized exponential map, linearly independent over the ring of endomorphisms. Let $\alpha^1,..., \alpha^n$ be algebraic elements of $\mathbf{C}^d$, and assume that for each $i = 1,..., d$ there exists one of these vectors α^j such that the ith components α_i^j does not lie in any conjugate of the field of complex multiplication k. Then*

$$\exp(\alpha^1 u^1 + \cdots + \alpha^n u^n)$$

is a transcendental point on A.

Proof. Assume otherwise, so that

$$\alpha^1 u^1 + \cdots + \alpha^n u^n = u^0$$

where u^0 is an algebraic point for the exponential map. If $u^0,..., u^n$ are linearly independent over k, then the theorem applies and we are done. Otherwise, there exists a linear combination

$$\gamma^1 u^1 + \cdots + \gamma^n u^n = \gamma^0 u^0$$

with $\gamma^j \in \mathfrak{o}$, and by hypothesis, $\gamma^0 \neq 0$. Dividing by γ^0 yields

$$\beta^1 u^1 + \cdots + \beta^n u^n = u^0; \qquad \beta^j \in k.$$

We then obtain

$$(\alpha^1 - \beta^1) u^1 + \cdots + (\alpha^n - \beta^n) u^n = 0.$$

Suppose that no component of α^1 lies in a conjugate of k. Then we can divide by $\alpha^1 - \beta^1$ and get a contradiction from the theorem. The hypothesis made on $\alpha^1,..., \alpha^n$ allows us to assume this extra condition without loss of generality. Indeed, suppose that α_1^1 lies in a conjugate of k but say α_1^2 does not. Then for a suitably large integer q we consider the linear combination

$$(\alpha^1 + q\alpha^2) u^1 + \alpha^2(u^2 - qu^1) + \cdots + \alpha^n u^n = u^0,$$

i.e., we replace u^2 by $u^2 - qu^1$ and change α^1, α^2 accordingly. Then the first coordinate of $\alpha^1 + q\alpha^2$ does not lie in a conjugate of k. We proceed similarly for the other coordinates, observing that if some coordinate α_i^1 does not lie in k, then for q sufficiently large, the ith coordinate of $\alpha^1 + q\alpha^2$ also does not lie in a conjugate of k. In other words, adding such a multiple of α^2 preserves the desired property. This concludes the proof of Theorem 3.

Naturally, it is a problem to get rid of the condition on the coordinates. It should suffice that some α^j should not lie in a conjugate of k. The problem is analogous to those mentioned above.

3. The Reduction Theorem

Let $\mathfrak{o} = \text{End}(A)$. Let B be a positive number. Let $\mathfrak{o}(B)$ denote the set of elements $\gamma \in \mathfrak{o}$ that $\|\gamma\| \leq B$. If φ, ψ are two positive functions on a set, we write

$$\varphi \ll \psi$$

to mean that there is a constant C such that $\varphi \leq C\psi$. We write

$$\varphi \mathrel{>\!\!\!<} \psi$$

to mean $\varphi \ll \psi$ and $\psi \ll \varphi$.

We let

$$\delta = \frac{1 - \left(\dfrac{2+1}{d}\right)^{\!\!\!\cdot\, 4}}{2(n+1)(2d+1)} \quad \text{and} \quad \kappa = \frac{\delta}{3}.$$

We see that δ is a number <1, depending only on n and d. The reason for its particular shape is given in the constructions of Section 5, see (7) below.

We let

$$M = M(H) = (\log H)^4.$$

We let

$$\tau(H) = e^{(\log H)^{1/4}}.$$

We shall devote the entire second part of this paper to the proof of the following result.

MAIN LEMMA. *Given the algebraic points $u^0, u^1,..., u^n$ linearly independent over k, and positive numbers d_0, b, there exists a number H' having the following property. Let $\alpha^1,..., \alpha^n$ be algebraic elements of $\mathbf{C}^d$, with components of degrees $\leqslant d_0$. Let H be a number $\geqslant H'$ such that $H \geqslant H(\alpha^j)$ for all j, and such that*

$$| \alpha^1 u^1 + \cdots + \alpha^n u^n - u^0 | < e^{-\tau(H)}.$$

Then there exists a nonzero polynomial P in $(n+1)d$ variables, of degree $\leqslant L$ in each variable, such that

$$P\left[f\left(\frac{\gamma}{q} u^1\right),...,f\left(\frac{\gamma}{q} u^n\right), f\left(\frac{\gamma}{q} u^0\right)\right] = 0,$$

for all quotients γ/q, where $f = (f_1,..., f_d)$ and:

  q *is a positive integer $\leqslant M^b$;*

  $\gamma \in \mathfrak{o}$ *and* $\|\gamma\| \leqslant e^{M^\kappa}$;

  $(\gamma/q)u^j \in \mathbf{D}_{2\rho}(0)$ *modulo the period lattice;*

  L *is equal to a lower power of $\log H$ than M.*

We shall now assume the Main Lemma, and derive some consequences. We need the lemma only with $b = 1/2$, to be able to apply Masser's criterion, Appendix 2, Section 1.

THEOREM 4. *Given the points $u^0, u^1,..., u^n$ and a positive integer d_0, there exists a number H' having the following property. Let $\alpha^1,..., \alpha^n$ be algebraic elements of $\mathbf{C}^d$ with components of degrees $\leqslant d_0$. Let H be a number $\geqslant H'$ such that $H \geqslant H(\alpha^j)$ for all j, and such that*

$$| \alpha^1 u^1 + \cdots + \alpha^n u^n - u^0 | < e^{-\tau(H)}.$$

Then there exist elements $\gamma^j \in \mathfrak{o}(B)$, with $B = (\log H)^2$, such that

$$| \gamma^1 u^1 + \cdots + \gamma^n u^n + \gamma^0 u^0 | < e^{-B^\kappa}.$$

Proof. Suppose otherwise. By the transference theorem, Theorem 6 of Appendix 1, given elements

$$\xi^j = f(\zeta^j), \quad \text{with} \quad \zeta^j \in \mathbf{D}_\rho(0)$$

there exists $z^0 \in \mathbf{C}^d$ such that

$$|f(z^0 u^j) - \xi^j| \ll 1/B \quad \text{for} \quad j = 0,\ldots, n$$

and

$$|z^0| \leqslant e^{B^\kappa}; \quad \|z^0 u^j - \zeta^j\|_\Lambda \ll 1/B,$$

We take $q = [B]$. We let γ/q be the element of the divided lattice $(1/q)\mathfrak{o}$ closest to z^0. Then

$$\left| f\left(\frac{\gamma}{q} u^j\right) - f(z^0 u^j) \right| \ll \frac{1}{B}$$

because f satisfies a Lipschitz condition in a neighborhood of the origin, and both $(\gamma/q) u^j$, $z^0 u^j$ are close to 0 modulo the period lattice. It follows that

$$\left| f\left(\frac{\gamma}{q} u^j\right) - \xi^j \right| \ll \frac{1}{B}.$$

The main lemma shows that the polynomial P has a zero sufficiently close to $(\xi^1,\ldots, \xi^n, \xi^0)$ to apply Masser's criterion, whose statement is reproduced in Appendix 2, Section 1. Therefore P is identically zero, a contradiction which proves Theorem 4.

We now make repeated use of Theorem 4. Suppose that we start with an inequality

$$|\alpha^1 u^1 + \cdots + \alpha^n u^n - u^0| < e^{-\tau(H_0)},$$

with H_0 sufficiently large, and with algebraic vectors $\alpha^\nu = \alpha^{0\nu}$, of heights $\leqslant H_0$. For any positive number B, abbreviate

$$e^{B^\kappa} = \psi(B).$$

For each $j = 0,\ldots, n-1$ let B_{j+1} satisfy

Condition 1. $\quad B_{j+1} = (C^* B_j^2)^{1/\kappa}$

where C^* is a sufficiently large constant, depending only on $u^0,\ldots, u^n, A$. Actually, we could take C^* much larger, going to infinity slowly with H. Let H_j be determined by the condition

$$(\log H_j)^2 = B_j, \quad \text{so that} \quad H_j = e^{B_j^{1/2}}.$$

Suppose that for each $j = 1,\ldots, n$ we can find a solution of the inequality

$$\left| \sum_{\nu=1}^n \alpha^{j\nu} u^\nu - u^0 \right| < e^{-\tau(H_j)}$$

with algebraic vectors $\alpha^{j\nu}$ of degrees $\leq d_0$ and height $\leq H_j$. We apply Theorem 4 to find elements $\gamma^j \in \mathfrak{o}(B_j)$ such that

$$\left|\sum_{\nu=0}^{n} \gamma^{j\nu} u^\nu\right| < e^{-B_j^\kappa}; \quad j = 0, \ldots, n.$$

We shall derive a contradiction. Let $\Delta = \det(\gamma^{j\nu})$. Then Δ is an algebraic integer of degree $\leq 2d$, and

$$\|\Delta\| \ll B_n^{n+1}.$$

If $\Delta \neq 0$, then

$$\frac{1}{B_n^{(n+1)2d}} \ll |\Delta|.$$

By Cramer's rule,

$$|\Delta u^1| \ll \sum_j \frac{B_n^{n-1}}{\psi(B_j)} \ll \frac{B_n^{n-1}}{\psi(B_0)}.$$

Therefore

$$\|\Delta\| \ll \frac{B_n^{n-1}}{\psi(B_0)}.$$

However, we have

Condition 2. $\psi(B_0)$ is much larger than $B_n^{(n+1)2d+n-1}$.

Therefore $\Delta = 0$. Consider the matrix

$$\Gamma = (\gamma^{j\nu}).$$

Let $\Gamma_0, \ldots, \Gamma_n$ be its rows. Say $\Gamma_{r+1}, \ldots, \Gamma_n$ are linearly independent, and Γ_r depends linearly on $\Gamma_{r+1}, \ldots, \Gamma_n$. Then

$$\Gamma_r = \beta^{r+1}\Gamma_{r+1} + \cdots + \beta^n \Gamma_n,$$

with some elements $\beta^{r+1}, \ldots, \beta^n$ in k of heights $\leq B_n^{2n}$. This yields

$$\left|\sum \gamma^{r\nu} u^\nu\right| \ll \sum_{j=r+1}^{n} \frac{B_n^{2n}}{\psi(B_j)} \ll \frac{B_n^{2n}}{\psi(B_{r+1})}. \qquad (*)$$

On the other hand, since $\gamma^{r\nu}$ are in $\mathfrak{o}(B_r)$, we get a left inequality

$$\frac{1}{C^{B_r^2}} \leq \left|\sum \gamma^{r\nu} u^\nu\right|, \qquad (**)$$

with some constant C. Thus we have a contradiction between (*) and (**), because from Condition 1 we have

Condition 3. $\psi(B_{j+1})$ is much larger than $C^{B_j^2} B_n^{2n}$.

This final contradiction shows that for some $j \geq 1$ we cannot apply Theorem 4. It follows that for any vectors of algebraic numbers

$$\alpha^{j0}, \ldots, \alpha^{jn}$$

of degrees $\leq d_0$ and heights $\leq H_j$ we have an inequality

$$e^{-\tau(H_n)} \leq e^{-\tau(H_j)} < \left| \sum_{\nu=1}^{n} \alpha^{j\nu} u^\nu - u^0 \right|.$$

This applies in particular to the original vectors $\alpha^1, \ldots, \alpha^n$. Therefore, we get the inequality

$$e^{-\tau(H_n)} < |\alpha^1 u^1 + \cdots + \alpha^n u^n - u^0|.$$

It is trivial to express H_n as a power of H_0, and we see that Theorem 2 is proved, with $\tau^*(H_0) = \tau(H_n)$.

Observe the logic: Given one very close approximation with numbers of height $\leq H$, *either* we can find other close approximations with numbers of height H^* fairly close to H, in which case an elimination yields a contradiction; *or* we cannot find other close approximations with numbers of height H^* fairly close to H, in which case we get a lower bound for the linear combination under consideration, with numbers of height $\leq H$. A method of proof is all the better as it allows H^* to be closer to H. This particular type of argument is reminiscent of the formalism of type functions and their relations with continued fractions and best possible approximations (cf. [18]). I believe that it should be applicable to the study of algebraic numbers to improve the Thue–Siegel–Roth theorem, both to make it effective and also to reach the improvements suggested in [18]. The poor function $\tau^*(H)$ is due only to the difficulty in reconciling Condition 3 above with (7) below, used in Lemma 9.1.

Remark 1. We have been careful to formulate the results and their proofs in such a way that the necessary and sufficient conditions on the functions involved are readily apparent. For instance, this is one reason why we extract Conditions 1, 2, 3 explicitly. If one can improve the main lemma with a better function $\tau(H)$, equal to a power of $\log H$, then Theorem 4 remains valid as stated. If Theorem 4 is valid for some function $\tau(H)$, then the preceding proof shows that Theorem 2 is valid with the function

$$\tau^*(H) = \tau(H_n),$$

where

$$\log H_n = (\log H)^c, \quad \text{and} \quad c = 2\left(\frac{2}{\kappa}\right)^n.$$

Thus again, $\tau^*(H)$ is a power of $\log H$.

Remark 2. Theorem 4 reduces the proof of the diophantine approximation inequality from arbitrary coefficients to coefficients in k, and even with a better approximation. In other words, B^κ is a larger function of B then $\tau(H)$ is a function of H.

Remark 3. I feel that not all the juice has been squeezed out of Theorem 4. For instance, instead of making a descent by one step, one might descend several steps, i.e., go to $(\log \log B)^2$, and so on, taking further logs. Although this improves somewhat the final estimate $\tau(H_n)$, it is still not sufficient to yield a measure with $o(H^2)$.

PART 2. THE APPROXIMATING FUNCTION AND THE MAIN PROOF

4. *Special Points*

For any set of elements $\{\ \}$, we let $\#\{\ \}$ denote the cardinality of this set.

Let $\mathfrak{o} = \text{End}(A)$. For any positive integer S, we denote by $\mathfrak{o}(S)$ the subset of elements $\gamma \in \mathfrak{o}$ such that $\|\gamma\| \leq S$. We note that $\mathfrak{o}$ is a lattice in $\mathbf{C}^d$, and has a basis over $\mathbf{Z}$ consisting of $2d$ elements. It follows at once that

$$\#\{\mathfrak{o}(S)\} \gtrless S^{2d}.$$

(The constants implicit in the symbol $\gtrless$ depend on $\mathfrak{o}$ but not on S.) We let Λ be the lattice of periods.

> Let $u^1, \ldots, u^n \in \mathbf{C}^d$ and let $\rho > 0$. Let $\mathbf{D}_\rho(0)$ be the ball of radius ρ centered at the origin in $\mathbf{C}^d$. Then
>
> $$\#\{\gamma \in \mathfrak{o}(S), \gamma u^j \in \mathbf{D}_\rho(0) \bmod \Lambda \text{ for all } j\} \gg S^{2d}.$$

Proof. Decompose $\mathbf{C}^d/\Lambda$ into parallelotopes of width approximately equal to ρ. The number of such parallelotopes is $\gg\!\!\ll 1/\rho^{2d}$. There exists a subset Γ_1 of $\mathfrak{o}(S)$ such that

$$\#(\Gamma_1) \gg \frac{S^{2d}}{\rho^{2d}}$$

and such that all elements γu^1 with $\gamma \in \Gamma_1$ lie in the same small parallelotope. There exists a subset $\Gamma_2 \subset \Gamma_1$ such that

$$\#(\Gamma_2) \gg \frac{S^{2d}}{(\rho^{2d})^{2d}}$$

and such that all elements γu^2 with $\gamma \in \Gamma_2$ lie in the same small parallelotope. Inductively, we obtain a subset Γ_n of $\mathfrak{o}(S)$ such that γu^j lie in the same small parallelotope for all $\gamma \in \Gamma_n$ and all j. Pick $\gamma^0 \in \Gamma_n$. Then $(\gamma - \gamma^0) u^j$ lie in a small parallelotope with a corner at 0, thus proving the lemma.

We also want to deal with conjugates. Let $P^j = \exp u^j$. For $\gamma \in \mathfrak{o}$, let γ_A be the corresponding algebraic endomorphism of A (we have viewed γ as operating on $\mathbf{C}^d$, as distinguished from $A_\mathbf{C}$). Then for any embedding σ of the algebraic numbers in $\mathbf{C}$,

$$(\gamma_A P^j)^\sigma = \gamma_A^\sigma P^{j\sigma}.$$

We may now repeat the procedure to find a subset of $\mathfrak{o}(S)$ having $\gg S^{2d}$ elements γ, such that $\gamma_A^\sigma P^{j\sigma}$ lies in a disk of radius centered at the origin in $(A^\sigma)_\mathbf{C}$.

Let $u^0, \ldots, u^n$ be the algebraic points as in the main theorem to be proved, and $P^j = \exp u^j$, $j = 0, \ldots, n$. We shall use the following notation. We let $\mathfrak{o}_\rho(u^0, \ldots, u^n; S) = \mathfrak{o}_\rho(u; S)$ consist of those elements $\gamma \in \mathfrak{o}(S)$ such that $(\gamma_A P^j)^\sigma$ lies in a disk of radius ρ centered at the origin in $(A^\sigma)_\mathbf{C}$ for all σ, all j. We have proved:

LEMMA 4.1. $\mathfrak{o}_\rho(u, S) \gg S^{2d}$, *where the constant implicit in $\gg$ depends on ρ, u, A but not on S.*

In the applications, we shall deal with a finite set of abelian functions $f_1, \ldots, f_{d'}$ whose polar divisors do not intersect the origin. If f is one such function, let f_A be the corresponding function on A, so that $f(z) = f_A(\exp z)$. Then $f_A(0)$ is defined, whence for all σ, $f_{A^\sigma}(O^\sigma)$ is also defined, and O^σ does not lie in a pole of f_{A^σ}. Thus we obtain:

LEMMA 4.2. *Let f_i ($i = 1,..., d'$) be abelian functions defined over K. Assume that none of the points u^j lie in the poles of the functions f_i for all i, j, and also that these functions are defined at the origin. Let ρ be a positive number such that the ball of radius 3ρ in each abelian variety A_C^σ does not intersect the polar divisor of any function f_{iA}^σ for all i, σ. Then the values*

$$(f_i(\gamma u^j))^\sigma$$

are bounded for all i, j, σ, and $\gamma \in \mathfrak{o}_\rho(u, S)$, independently of S.

Remark. With the arrangement of the proof of Theorem 4, only points in a neighborhood of 0 where f is holomorphic will be considered, as in [14]. In other contexts, points approaching the divisor of poles of f also will have to be considered.

In the sequel, we let ρ have the meaning assigned in Lemma 2.

5. Determination of the Approximating Function and Basic Parameters

It seems to me clearer not to plug in the specific values of the parameters used in the proof too early, thus bringing out somewhat more sharply the interrelationships between these parameters. Thus we let τ be a positive function strictly increasing to infinity, and we suppose that we have an inequality

$$| \alpha^1 u^1 + \cdots + \alpha^n u^n - u^0 | < e^{-\tau(H)}; \quad H(\alpha^j) \leqslant H$$

with sufficiently large H (depending on $u^0 u^1,..., u^n$, A and τ). We shall then derive consequences of such an inequality. In various estimates, we get upper bounds for certain expressions that must be compared with $e^{-\tau(H)}$, and must go to infinity slower than $e^{\tau(H)}$. Thus, we let $\tau'(H)$ be a function that is also increasing to infinity and is $o(\tau(H))$, i.e., such that the limit of $\tau'(H)/\tau(H)$ is 0 as H goes to infinity.

We assume that the reader is acquainted with the estimates of [16], Chapter III, Section 2, Lemma 1. It is helpful to have read the proof of the main theorem of that chapter, as giving in a much simpler situation the basic techniques of the beginnings of the proof carried out here. We recall explicitly the basic lemma used for estimating.

LEMMA 5.1. *Let K be a number field. Let $f_1,..., f_{d'}$ be functions on $\mathbf{C}^d$ such that the ring $K[f_1,..., f_{d'}]$ is mapped into itself by the partial derivatives $D_1,..., D_d$. There exists a number C_1 having the following property. If*

$Q(T_1,...,T_{d'})$ is a polynomial in d' variables with coefficients in $\mathfrak{o}_K$, of total degree $\leq L$, and if

$$D = D_1^{m_1} \cdots D_d^{m_d}$$

is a differential operator of order $M = m_1 + \cdots + m_d$, then

$$D(Q(f_1,...,f_{d'})) = Q_D(f_1,...,f_{d'})$$

where $Q_D \in K[T_1,...,T_{d'}]$ is a polynomial satisfying:

(i) $\deg Q_D \ll M + L$
(ii) $\|Q_D\| \leq \|Q\| M! C_1^{M+L}$
(iii) There exists a denominator for Q_D bounded by $\text{den}(Q) C_1^{M+L}$.

The formulation is slightly different, but the proof is the same as that for Lemma 1 of the references cited above.

Evaluating $D(Q(f_1,...,f_{d'}))$ at an algebraic point u of the functions $f_1,...,f_{d'}$ then yields an algebraic number whose absolute values and denominator satisfy a corresponding estimates, just by plugging in the values $f_i(u)$ in Q_D. The constants appearing in the estimate then depend on the point u.

We let $z^1,...,z^n$ be independent vectors of d complex variables, so that

$$z^j = (z_1^j,...,z_d^j).$$

Thus $\{z_i^j\}$ is a family of nd variables. We then form the function

$$F(z^1,...,z^n) = \sum_{(\lambda)}^{L-1} a_{(\lambda)} \prod_{i=1}^{d} \prod_{j=1}^{n} f_i(z^j)^{\lambda_{ij}} \prod_{i=1}^{d} f_i(\alpha^1 z^1 + \cdots + \alpha^n z^n)^{\lambda_{i0}},$$

where the sum is taken over the indices λ_{ij} ($i = 1,...,d$; $j = 0,...,n$) satisfying

$$0 \leq \lambda_{ij} \leq L - 1,$$

and the coefficients $a_{(\lambda)}$ are integers to be determined in such a way that certain conditions to be stated are satisfied.

For each $(m) = (m_{ij})$ ($i = 1,...,d$; $j = 1,...,n$) we have a differential operator

$$D^{(m)} = \prod_{i,j} \left(\frac{\partial}{\partial z_i^j}\right)^{m_{ij}}.$$

The *order of the operator* is denoted by
$$|m| = \sum m_{ij}.$$
By Lemma 5.1, for any differential operator $D = D^{(m)}$,
$$DF(z^1,\ldots,z^n)$$
is a polynomial
$$P_D[f(z^1),\ldots,f(z^n), f(\alpha^1 z^1 + \cdots + \alpha^n z^n)],$$
where $f = (f_1,\ldots,f_{d'})$. Certain power products of the components α_i^j occur in the coefficients of this polynomial. Let z^0 be a new family of variables. We denote by
$$D^{(m)} F(z^1,\ldots, z^n; z^0) = P_D[f(z^1),\ldots, f(z^n), f(z^0)]$$
the expression obtained by substituting z^0 for $\alpha^1 z^1 + \cdots + \alpha^n z^n$. In particular, if $\eta \in k$ and ηu^j does not lie in any pole of any function f_i then
$$D^{(m)} F(\eta u^1,\ldots, \eta u^n; \eta u^0)$$
is an algebraic number. If in addition $\eta \in \mathfrak{o}$ is an endomorphisms of A, then this algebraic number lies in K.

We require that
$$D^{(m)} F(\gamma u^1,\ldots, \gamma u^n; \gamma u^0) = 0, \tag{1}$$
for all $\gamma \in \mathfrak{o}_\rho(u, N)$ and all (m) satisfying $|m| \leq M$, where M, N are parameters still to be selected. The number ρ is selected as in Lemma 4.2. The vanishing of these derivatives with the substitution of u^0 amounts to a system of linear equations for the $a_{(\lambda)}$, in the field $K(\alpha^1,\ldots,\alpha^n)$ If we wanted uniformity only with respect to $\alpha^1,\ldots, \alpha^n$ lying in a fixed number field rather than over a variable field of bounded degree, we could then pick a fixed basis of the field, and reduce our system of linear equations to another system over the integers, the number of equations being multiplied by $[K(\alpha^1,\ldots, \alpha^n) : \mathbf{Q}]$, as well as the number of variables. For simplicity, we shall carry out the proof in this context, which allows us to use the ordinary Siegel's lemma [16], Chapter I, Section 2. To treat the general case, one has to use the analog of Siegel's lemma for inequalities, cf. for instance [16], Chapter VI, Section 3. This has the effect of introducing only a minor perturbation in the choice of M, N below.

We have:

Number of variables $\gg L^{(n+1)d}[K(\alpha) : \mathbf{Q}] \gg L^{(n+1)d}$
Number of equations $\ll M^{nd} N^{2d}$
Size of coefficients $\leq M^M C_1^{M+L} C_1^{N^2 M} H^{cM}$

where C_1, c are constants depending on the degrees of α^j. We solve for $a_{(\lambda)}$ by Siegel's lemma. Ultimately we want

$$|a_{(\lambda)}| \leq C_2^{MN} \tag{2}$$

which will be somewhat smaller than the size of the coefficients of the linear equations. This smallness will be used in Section 8 below. Thus we want the exponent

$$\frac{\# \text{ equations}}{\# \text{ variables} - \# \text{ equations}}$$

in Siegel's lemma to be at most $1/N$. For this purpose we impose the sufficient condition that

$$L^{(n+1)d} = M^{nd} N^{2d} N \tag{3}$$

i.e.,

$$L = M^{1-1/(n+1)} N^{(2+1/d)/(n+1)} \tag{3'}$$

We let $N = \log H$, and we let

$$M = N^a \tag{4}$$

where a is a positive number that we select to satisfy two further conditions. For convenience of notation, suppose that we deal with quantities X, Y such that

$$X = N^b \quad \text{and} \quad Y = N^c$$

We write $X \prec Y$ to mean $b < c$. With this notation, the first condition is that, for some $\delta > 0$, we have

$$L \prec M^{1-(2d+1)\delta} \tag{5}$$

The second condition is that

$$MN \leq N^2 L; \quad \text{i.e.,} \quad M \leq NL. \tag{6}$$

The first of these will be used in the proof in §9. It amounts to

$$\delta(2d+1)(n+1) < 1 - \frac{2 + \frac{1}{d}}{a}; \quad \text{or equivalently} \quad a > 2 + \frac{1}{d}. \tag{5'}$$

The second of these conditions amounts to

$$a \leqslant n + 1 + 2 + \frac{1}{d}. \tag{6'}$$

As we have $n \geqslant 1$ we pick $a = 4$. We then select δ such that

$$\delta < \frac{1 - \frac{2+1}{a}}{(n+1)(2d+1)}, \quad \text{say} \quad \delta = \frac{1 - \frac{2+1}{a}}{2(n+1)(2d+1)}.$$

Then $a\delta < 1/2$. We note that δ is defined only in terms of d and n. In view of our choices, we have

$$M = N^4 = (\log H)^4.$$

It is clearer to carry out the proofs leaving M, N essentially as variables, so that the reader can see just what is needed to make the method work.

We let

$$\tau(H) = e^{(\log H)^{1/4}}.$$

Then

$$M^{\delta/2} < \log \tau(H). \tag{7}$$

This is roughly the lowest possible value of $\tau(H)$ needed to carry out the proof of Lemma 9.1.

We let

$$\tau'(H) = (\log H)^5.$$

Then

$$MN \leqslant \tau'(H). \tag{8}$$

This is of significance in Lemma 6.1. One figures out these functions precisely to make the methods of proof of these lemmas applicable. We recommend that the reader treat M, N, τ, τ' as variables subject to the necessary conditions arising from the lemmas.

6. *Estimate of Differences*

We shall deal with the two expressions

$$D^{(m)}F\left(\frac{\gamma}{q} u^1, \ldots, \frac{\gamma}{q} u^n\right) \quad \text{and} \quad D^{(m)}F\left(\frac{\gamma}{q} u^1, \ldots, \frac{\gamma}{q} u^n; \frac{\gamma}{q} u^0\right),$$

and we need an estimate for their difference. *For this paper*, only values of γ/q such that $(\gamma/q)\, u^j$ lie in a fixed neighborhood of 0 will occur, that is in a ball of radius ρ, such that the functions $f_1, ..., f_d$ give an analytic isomorphism on the ball of radius 3ρ centered at the origin. For other applications one needs to come closer to the divisor of poles of (f). This introduces a slight technical complication (which I can deal with), but I prefer to keep the present proof as easy as possible.

From now on, all assertions are meant to hold for H sufficiently large.

In the next lemma, what matters is that $\tau'(H)$ is much smaller than $\tau(H)$.

LEMMA 6.1. *Let S be a positive number, and q a positive integer. Let $\gamma \in \mathfrak{o}(S)$, and assume that $(\gamma/q)\, u^j$ lies in the ball of radius ρ, modulo the period lattice. Also assume that*

$$MN \leqslant \tau'(H), \qquad \log S \leqslant \tau'(H).$$

Then for $|m| \leqslant M$ we have

$$\left| D^{(m)} F\left(\frac{\gamma}{q} u^1, ..., \frac{\gamma}{q} u^n; \frac{\gamma}{q} u^0\right) - D^{(m)} F\left(\frac{\gamma}{q} u^1, ..., \frac{\gamma}{q} u^n\right) \right| < C_3^{-\tau(H)}.$$

Proof. Write

$$w^0 = \frac{\gamma}{q} u^0 \quad \text{and} \quad w^\alpha = \frac{\gamma}{q}(\alpha^1 u^1 + \cdots + \alpha^n u^n).$$

From the original approximation assumption, we have

$$|w^0 - w^\alpha| \leqslant S e^{-\tau(H)} \leqslant C_4^{-\tau(H)}.$$

Since f satisfies a Lipschitz condition on the ball of radius 3ρ, we obtain

$$|f_i(w^0) - f_i(w^\alpha)| \leqslant C_5^{-\tau(H)}. \tag{*}$$

The difference of two monomials involving $f_i(w^\alpha)$ and $f_i(w^0)$ is easily estimated. Indeed, for a positive integer $\mathfrak{o}$,

$$X^\lambda - Y^\lambda = (X - Y)(X^{\lambda-1} + \cdots + Y^{\lambda-1}).$$

We put $X = f_i(w^\alpha)$ and $Y = f_i(w^0)$. By assumption these values are bounded independently of γ, q. Using (*) we obtain from Lemma 5.1 and (2) the estimate

$$\left| P_D\left[f\left(\frac{\gamma}{q}u\right); f(w^0)\right] - P_D\left[f\left(\frac{\gamma}{q}u\right); f(w^\alpha)\right] \right|$$

$$\leqslant \max_i |f_i(w^0) - f_i(w^\alpha)| \, C_6^{MN}$$

$$\leqslant C_7^{-\tau(H)},$$

thus proving Lemma 6.1.

7. Estimates on Heights and Sizes

Let $f = (1, f_1, ..., f_{d'})$ denote the embedding of A in projective space. Let h denote the absolute height. It is known that the map

$$w \mapsto \log h(f(w))$$

defined on the group of algebraic points of the exponential map modulo the period lattice, is equal to a quadratic function, plus a bounded function. This is the Neron–Tate theorem; [11], [17], [25], Property 3 of Chapter IV, Section 2. Observe the uniformity with respect to algebraic extensions of the fixed field K, obtained from division points of algebraic points. Such uniformity had not been previously used in the extrapolation procedures of transcendence and approximation methods.

Let w be a point such that $f(qw)$ is rational over K. Then

$$[K(f(w)) : \mathbf{Q}] \leqslant q^{2d}[K : \mathbf{Q}].$$

In particular, let $\gamma \in \mathfrak{o}(S)$ be such that $(\gamma/q)u = w$ does not lie in the polar divisors $(f_i)_\infty$ for all $i = 1, ..., d'$. Let H denote the height with respect to the field $K(f(w))$. We obtain

$$H(f(w))^{1/q^{2d}} = \prod_v \max_i(1, |f_i(w)|_v)^{N_v/q^{2d}} \leqslant C_8^{S^2/q^2} C_8$$

for a suitable constant C_8. In particular, if u is an algebraic of the exponential map such that $f(u)$ is rational over K, then we get a bound

$$\text{size } f_i\left(\frac{\gamma}{q}u\right) \leqslant C_8^{S^2 q^{2d-2}} C_8^{q^{2d}} \tag{9}$$

The constant C_8 also depends on u, and the applications, u is equal to some u^j.

The derivative $D^{(m)}F$ is a polynomial in the functions

$$f_i(z^j); \quad f_i(\alpha^1 z^1 + \cdots + \alpha^n z^n),$$

of degree $\ll M + L \ll M$, and with coefficients bounded as in Lemma 5.1. From (9) we obtain an estimate for a derivative $D^{(m)}F$ with $|m| \leq M$.

LEMMA 7.1. *Let $\gamma \in \mathfrak{o}(S)$, and let q be a positive integer such that $(\gamma/q) u^j$ does not lie in $(f)_\infty$. Then the algebraic number*

$$D^{(m)}F\left(\frac{\gamma}{q} u^1, \ldots, \frac{\gamma}{q} u^n; \frac{\gamma}{q} u^0\right)$$

is either equal to 0, or its size is bounded by $C_9^{MS^2 q^{2d-2}} C_9^{M q^{2d}}$.

The next lemma will be used only in Section 8.

LEMMA 7.2. *Let $f = (f_1, \ldots, f_{d'})$. For each i there exist polynomials $\Phi_{i\gamma}$, $\Psi_{i\gamma}$ of degrees $\ll S^2$, with coefficients in K of heights $\ll C_{10}^{S^2}$ such that for $z \in \mathbf{C}^d$ we have*

$$f_i \circ \gamma = \frac{\Phi_{i\gamma}(f)}{\Psi_{i\gamma}(f)}.$$

These polynomials can be so chosen that $\Psi_{i\gamma}(f(u^j)) \neq 0$ for all j.

Proof. The existence of the polynomials $\Phi_{i\gamma}$ and $\Psi_{i\gamma}$ is a special case of a result of Altman [1], Theorem 3.5, applied to a generic point. For the convenience of the reader, we recall briefly the argument. We can assume that the projective coordinates

$$(\theta_0, \ldots, \theta_{d'})$$

from which we obtained affine coordinates $(f_1, \ldots, f_{d'})$ are those of a linear system $\mathscr{L}(X)$, where X is a positive divisor such that $X = X^-$, and X is sufficiently ample. The expression in $\gamma, \eta \in \mathfrak{o}$ given by

$$D_X(\gamma, \eta) = (\gamma + \eta)^{-1} X - \gamma^{-1}(X) - \eta^{-1}(X)$$

is bilinear modulo linear equivalence. This can be proved algebraically almost immediately from the theorem of the square, but for our purposes in characteristic zero over the complex numbers, it is obvious by using the functional equation of theta functions. One proves that the expression

$$D_\theta(\gamma, \eta) = \frac{\theta \circ (\gamma + \eta)}{(\theta \circ \gamma)(\theta \circ \eta)}$$

is bimultiplicative in the group of theta functions modulo trivial theta functions and abelian functions; cf. [20], Chapter IV.

By induction, it then follows at once that for any positive integer s, we have

$$(s\delta)^{-1} X \sim \left(\frac{s^2 + s}{2}\right) X + \left(\frac{s^2 - s}{2}\right) X^-,$$

which is just $s^2 X$ because of our assumption $X = X^-$. Having taken X sufficiently ample, the linear system $L(s^2 X)$ is linearly equivalent to $L(X)^{s^2}$, i.e., differs from it by the product with a fixed function, by elementary considerations. See for instance my *Introduction to Algebraic Geometry*, Chapter V, Theorem 6, Section 5.

Let $\eta^1, \ldots, \eta^{2d}$ be a basis of $\mathfrak{o}$ over $\mathbf{Z}$. Any element $\eta \in \mathfrak{o}(S)$ can be written in the form

$$\eta = s_1 \eta^1 + \cdots + s_{2d} \eta^{2d},$$

with integers s_i satisfying $|s_i| \ll S$. The addition of $2d$ terms and the multiplication by η^i ($i = 1, \ldots, 2d$) are polynomial functions in the homogeneous coordinates of fixed degree and fixed coefficients. From this the existence of the polynomials $\Phi_{i\gamma}$, $\Psi_{i\gamma}$ follows at once.

To see that $\Psi_{i\gamma}(f)$ can be so chosen as to be nonzero at the points u^j, we could have assumed from the beginning that the projective linear transformation over $\mathbf{Z}$ made on the coordinates was such that the function $g = f_{d+1}$ (a sufficiently general linear combination of $f_1, \ldots, f_{d'}$) is integral over $K[f_1, \ldots, f_d]$, and generates $K(f_1, \ldots, f_{d'})$. Furthermore, the dual basis to

$$1, g, g^2, \ldots, g^{r-1},$$

for the field extension $K(f_1, \ldots, f_{d'}) = K(A)$ of $K(f_1, \ldots, f_d)$, lies in the local ring of the points u^j. Let us write $f_i \circ \gamma$ as a polynomial

$$f_i \circ \gamma = \sum R_\nu(f_1, \ldots, f_d) g^\nu,$$

where R_ν are rational functions, in $K(f_1, \ldots, f_d)$. From the bounds on the degrees and coefficients of $\Phi_{i\gamma}$, $\Psi_{i\gamma}$ obtained above, and for instance

[16], Lemmas 1 and 2, Chapter V, Section 2, we conclude that these rational functions satisfy similar bounds. Since $R_\nu(f_1,...,f_d)$ is expressible as a trace of g^ν times an element of the dual basis, this means that the denominator of the rational function $R_\nu(f_1,...,f_d)$ is defined at the prescribed points u^j for all j. This proves the lemma.

Remark. Lemma 7.2 can be expressed in the obvious fashion to apply to abelian varieties without complex multiplication, and to functions f_i applied to points of A in some finitely generated field over the rationals. The proof is the same, but I preferred to stick to the notation used in the immediate applications rather than give a reformulation with new notation for the general case.

8. Baker–Coates Lemma

The next lemma is intended to get an estimate of type C^{S^2L} rather than the weaker C^{S^2M} obtained without further hypotheses in Lemma 7.1.

LEMMA 8.1 (Baker–Coates lemma). *Let (m) be such that $|m| \leqslant M$. Let $\gamma \in \mathfrak{o}_p(u, S)$. Assume that*

$$D^{(\mu)}F(\gamma u^1,...,\gamma u^n; \gamma u^0) = 0$$

for all (μ) with $|\mu| < |m|$. Then either

$$D^{(m)}F(\gamma u^1,...,\gamma u^n; \gamma u^0) = 0$$

or:

(i) *Its conjugates are bounded by $H^{cM}M^MC^{M+L} \leqslant C_{11}^{MN}$.*

(ii) *Its denominator is bounded by $C^{S^2L}H^{cM}C^MM^M \leqslant C_{11}^{S^2L}$ with some constants c, C, C_{11}. This is $\leq C_{ii}^{S^2L}$ if $S \geq N$.*

Proof. We know that $D^{(m)}F$ is a polynomial P_D as at the beginning of the proof, and we have an estimate for the degree and coefficients of this polynomial. Under our special assumption on γ, we know that the values $f_i(\gamma u^j)$ ($j = 0,..., n$) are bounded, so that when we substitute these values in the polynomial, we get a number for which estimate (i) holds. Note that we are using the estimate (2), that $|a_{(\lambda)}| \leqslant C_2^{MN}$ here.

To deal with the denominators, let

$$E(z^1,..., z^n) = F(\gamma z^1,..., \gamma z^n)$$

$$= \sum a_{(\lambda)} \prod_{i=1}^{d} \prod_{j=1}^{n} f_i(\gamma z^j)^{\lambda_{ij}} \prod_{i=1}^{d} f_i(\gamma \alpha^1 z^1 + \cdots + \gamma \alpha^n z^n)^{\lambda_{i0}}.$$

Lemma 7.2 gives us an expression for $f_i(\gamma z^j)$ as a rational function in $f(z)$. It allows us to construct a generic denominator by taking a product of the $\Psi_{i\gamma}$. Let

$$G(z^1,\ldots,z^n) = \prod_{i=1}^{d}\prod_{j=1}^{n} \Psi_{i\gamma}^L(f(z^j)) \prod_{i=1}^{d} \Psi_{i\gamma}^L(f(\alpha^1 z^1 + \cdots + \alpha^n z^n)).$$

Then GE is a polynomial in the functions

$$f_i(z^j); \quad f_i(\alpha^1 z^1 + \cdots + \alpha^n z^n),$$

$i = 1,\ldots, d'$. Its degree is $\ll S^2 L$, and its coefficients have size $\ll C^{S^2 L}$.

The derivative $D^{(m)} GE(z^1,\ldots,z^n)$ can be expressed as a sum of terms $D^{(\mu)} G D^{(m-\mu)} E$, with binomial coefficients. Furthermore,

$$D^{(\mu)} E(u^1,\ldots, u^n; u^0) = \gamma^{(\mu)} D^{(\mu)} F(\gamma u^1,\ldots, \gamma u^n; \gamma u^0),$$

where $\gamma^{(\mu)}$ is a monomial of order $|\mu|$ in the components γ_i ($i = 1,\ldots, d$). This latter expression vanishes by assumption if $|\mu| < |m|$. Consequently,

$$D^{(m)}(GE)(u^1,\ldots, u^{n-1}; u^0)$$
$$= G(u^1,\ldots, u^n; u^0)\, D^{(m)} E(u^1,\ldots, u^n; u^0)$$
$$= G(u^1,\ldots, u^n; u^0)\, \gamma^{(\mu)} D^{(m)} F(\gamma u^1,\ldots, \gamma u^n; \gamma u^0).$$

Therefore, the denominator that we want to estimate for (ii) is a product of:

(a) a denominator for $D^{(m)}(GE)(u^1,\ldots, u^n; u^0)$;
(b) $G(u^1,\ldots, u^n; u^0)$; and
(c) $\gamma^{(\mu)}$.

By Lemma 5.1, $D^{(m)}(GE)(u^1,\ldots, u^n; u^0)$ is a polynomial in the numbers

$$f_i(u^j); \quad i = 1,\ldots, d'; \; j = 0,\ldots, n$$

satisfying

$$\text{degree} \ll S^2 L + M$$

$$\text{size of coefficients} \ll C^{S^2 L} M^M C^M H^{cM}.$$

As the values $f_i(u^j)$ are fixed from the beginning, we see that term (a) introduces a denominator with the required bound. Using the fact that

$H(\xi) = H(\xi^{-1})$ for any algebraic number ξ, we see that a similar bound prevails for term (b). Finally, we note that

$$H(\gamma^{(\mu)}) \leqslant H(\gamma)^M,$$

and $H(\gamma) \ll S^{\deg \gamma}$. Thus term (c) contributes at most $S^{Md \deg \gamma} C^M$ to the denominator, which is within the desired bound. This proves Lemma 8.1.

Observe that (2) and (5) have been used in the above estimates.

9. Extrapolating on Integral Multiples

We shall make use of the interpolation lemma of Appendix 2. The set to which we apply this lemma will be a subset of $\lambda(S)$ of cardinality $\gg S^{2d}$, for various values of S.

LEMMA 9.1. *For each positive integer ν, let $S_\nu = NM^{\nu\delta}$. If*

$$\nu \leqslant M^{\delta/2} \ll \frac{\tau'(M)}{\log M}; \quad |m| \leqslant M - \nu M^{1-\delta}; \quad \gamma \in \mathfrak{o}_\rho(u, S_\nu)$$

then

$$D^{(m)} F(\gamma u^1, \ldots, \gamma u^n; \gamma u^0) = 0.$$

Proof. Recall that $\gamma \in \mathfrak{o}_\rho(u, S_\nu)$ implies that γu^j is uniformly away from the poles of the functions f_i for all i, j and that $\|\gamma\| \ll S_\nu$. By Lemma 4.2, the set of such γ has cardinality $\gg S_\nu^{2d}$. Since $\mathfrak{o}$ is a lattice in $\mathbf{C}^d$, the distance between any two of its points is bounded from below.

We prove the assertion by induction on ν. For $\nu = 0$ the assertion is merely a property of the constructed function F. Assume the assertion true up to some integer ν in the given range. We prove it for $\nu + 1$ in the given range. Suppose the assertion for $\nu + 1$ is false for some $\eta \in \lambda_\rho(u, S_{\nu+1})$. Let (m') be such that $|m'|$ is the smallest value

$$\leqslant M - (\nu + 1) M^{1-\delta}$$

such that

$$D^{(\mu)} F(\eta u^1, \ldots, \eta u^n; \eta u^0) = 0 \quad \text{if} \quad |\mu| < |m'|$$

but

$$D^{(m')} F(\eta u^1, \ldots, \eta u^n; \eta u^0) \neq 0.$$

Let

$$\Theta(z^1, \ldots, z^n) = [\theta_0(z^1) \cdots \theta_0(z^n) \theta_0(\alpha^1 z^1 + \cdots + \alpha^n z^n)]^L.$$

Then ΘF is an entire function. Let $z = (z_1, \ldots, z_d)$ and let

$$G(z) = D^{(m')}(\Theta F)(zu^1, \ldots, zu^n).$$

Thus G is a function of d complex variables. If $(r) = (r_1, \ldots, r_d)$ is a d-tuple of integers ≥ 0, then

$$D^{(r)}G(z) = \sum_{|\kappa| \leq |r|} b(\kappa, u) D^{(m')+(\kappa)}(\Theta F)(zu^1, \ldots, zu^n),$$

where $b(\kappa, u)$ is a polynomial in the components u_i^j and binomial coefficients, easily estimated. We take (r) such that $|r| \leq M^{1-\delta}$, so that the multiplicity to which we apply the interpolation estimate is $M^{1-\delta}$.

The derivative $D^{(m')+(\kappa)}(\Theta F)$ is also expressible as a sum,

$$\sum c(\mu, \mu') D^{(\mu)} \Theta D^{(\mu')} F.$$

By the difference Lemma 6.1 and a trivial estimate for $D^{(\mu)}\Theta$ obtained by iteration of Cauchy's formula in one variable, we obtain

$$|D^{(r)}G(\gamma)| \leq C^{-\tau(H)} \quad \text{all} \quad \gamma \in \mathfrak{o}_p(u, S_\nu). \tag{*}$$

The coefficients $a_{(\lambda)}$ satisfy (2), namely

$$|a_{(\lambda)}| \leq C_2^{MN}.$$

We take the radius R to be a large constant times $S_{\nu+1}$. Again using Cauchy's formula on derivatives of F, we obtain the estimate

$$|G|_R \leq C^{S_{\nu+1}^2 L} C^{MN} M^{cM} \leq C_{12}^{S_{\nu+1}^2 L}.$$

We apply Lemma 3 of Appendix 2, Section 2 to the function G. We get

$$|G(\eta)| \leq \frac{|G|_R}{C^{MS^{2d}/R^{2d-2}}} + (C'S_{\nu+1})^{5MS_\nu^{2d}} \max |D^{(r)}G(\gamma)|.$$

The second term on the right is small with an estimate essentially the same as that for the derivatives (*), because of the range for S_ν and (7). The denominator of the first term has an exponent equal to

$$\gg \frac{M^{1-\delta}S_\nu^{2d}}{R^{2d-2}} = M^{1-\delta}\left(\frac{S_\nu}{S_{\nu+1}}\right)^{2d} S_{\nu+1}^2$$

$$= M^{1-\delta}M^{-2d\delta}S_{\nu+1}^2.$$

The original determination of parameters (5) was designed so that the exponent of the denominator is substantially larger than the exponent of the numerator $S^2_{\nu+1}L$ at this stage of the proof. Consequently we obtain

$$\log |G(\eta)| \ll -M^{1-(2d+1)\delta}S^2_{\nu+1}. \qquad (**)$$

This is an upper bound showing that $|G(\eta)|$ is quite small. From it we obtain an upper bound for $D^{(m')}F$. We can write

$$G(\eta) = \Theta(\eta u^1,\ldots,\eta u^n)\, D^{(m')}F(\eta u^1,\ldots,\eta u^n)$$
$$+ \sum_{|\mu|<|m'|} c(\mu)\, D^{(m')-(\mu)}\Theta(\eta u^1,\ldots,\eta u^n)\, D^{(\mu)}F(\eta u^1,\ldots,\eta u^n);$$

By Lemma 6.1 again, since $D^{(\mu)}F(\eta u^1,\ldots,\eta u^n;\eta u^0) = 0$, we conclude that the second sum on the right is bounded by

$$\exp[-\tfrac{1}{2}\tau(H)].$$

On the other hand, since ηu^j is away from the zeros of θ_0, it follows from the functional equation of the theta function that

$$\log |\Theta(\eta u^1,\ldots,\eta u^n)| \gg -S^2_{\nu+1}L.$$

Hence we obtain the estimate

$$\log |D^{(m')}F(\eta u^1,\ldots,\eta u^n)| \ll -M^{1-(2d+1)\delta}S^2_{\nu+1}.$$

By Lemma 6.1 once more, the substituted derivative satisfies

$$\log |D^{(m')}F(\eta u^1,\ldots,\eta u^n;\eta u^0)| \ll -M^{1-(2d+1)\delta}S^2_{\nu+1}. \qquad (***)$$

By the Baker–Coates lemma we find the lower inequality

$$-S^2_{\nu+1}L \ll |D^{(m')}F(\eta u^1,\ldots,\eta u^n;\eta u^0)| \ll -M^{1-(2d+1)\delta}S^2_{\nu+1}. \qquad (****)$$

The extreme inequalities are contradictory to each other by (5), and Lemma 9.1 is proved.

10. Extrapolation on Division Points

LEMMA 10.1. *Let b be a fixed positive integer, given a priori, depending only on n and d. Let q be a positive integer, and $\gamma \in \mathfrak{o}$. Suppose that $(\gamma/q)\, u^j$ lies in $\mathbf{D}_{2\rho}(0)$ modulo the period lattice for all j. If*

$$q \leq M^b; \qquad |m| \leq \tfrac{1}{4}M; \qquad \|\gamma\| \leq e^{M\delta/3},$$

DIOPHANTINE APPROXIMATION

then

$$D^{(m)}F\left(\frac{\gamma}{q}u^1,\ldots,\frac{\gamma}{q}u^n;\frac{\gamma}{q}u^0\right) = 0.$$

Proof. Suppose that the lemma is false for some fraction η/q, satisfying the required conditions. Let (m') be such that m' is the smallest value in the given range such that

$$D^{(\mu)}F\left(\frac{\eta}{q}u^1,\ldots,\frac{\eta}{q}u^n;\frac{\eta}{q}u^0\right) = 0 \quad \text{if} \quad |\mu| < |m'|$$

but

$$D^{(m')}F\left(\frac{\eta}{q}u^1,\ldots,\frac{\eta}{q}u^n;\frac{\eta}{q}u^0\right) \neq 0.$$

Let Θ be as in the proof of Lemma 9.1, and also let again

$$G(z) = D^{(m')}(\Theta F)(zu^1,\ldots,zu^n).$$

We consider the derivatives $D^{(r)}G(s)$ as before, with $|r| \leq (1/4)M$. This will give the multiplicity in the application of Lemma 3, Appendix 2, to the set of points $\Gamma = o_p(u, S)$, where

$$S = e^{M\delta/2}.$$

This choice of S is essentially the maximal value obtainable from Lemma 9.1.

The difference Lemma 6.1 and a trivial estimate for $D^{(\mu)}\Theta$ obtained by iteration of Cauchy's formula in one variable show that

$$|D^{(r)}G(\gamma)| \leq C^{-\tau(H)}, \quad \text{all} \quad \gamma \in \Gamma. \qquad (*)$$

We take a circle of radius R equal to a large constant times S, and get the estimate

$$\left|G\left(\frac{\eta}{q}\right)\right| \leq \frac{|G|_R}{C^{MS^{2d}/(qR)^{2d-2}}} + (C'qS)^{5MS^{2d}} \max |D^{(r)}G(\gamma)|.$$

The second term on the right is small, with an estimate essentially the same as $(*)$. The first term is easier to handle than in the preceding section, and gives us the estimate

$$\log\left|G\left(\frac{\eta}{q}\right)\right| \ll -MS^2/q^{2d-2}. \qquad (**)$$

Actually, we could even drop the factor of M for our purposes.

Arguing as before to get rid of the theta function (this amounts to replacing η by η/q without any further change) we obtain the corresponding estimate for the algebraic number under consideration, i.e.

$$\log \left| D^{(m')} F \left(\frac{\eta}{q} u^1, \ldots, \frac{\eta}{q} u^n; \frac{\eta}{q} u^0 \right) \right| \ll -MS^2/q^{2d-2} \qquad (***)$$

This time we do not have to use Coates' lemma, but merely the universal estimate of Lemma 7.1. Let $T = e^{M^{\delta/3}}$. Since the algebraic number we deal with has degree $\ll q^{2d(n+1)}$, we obtain the lower estimate

$$-MT^2 q^{2d-2} q^{2d(n+1)} \ll \log \left| D^{(m')} F \left(\frac{\eta}{q} u^1, \ldots, \frac{\eta}{q} u^n; \frac{\eta}{q} u^0 \right) \right|. \qquad (****)$$

We picked T sufficiently smaller than S so that the two inequalities (***) and (****) would be contradictory, thereby proving our lemma.

We need the lemma only with $b = 1$, and The Main Lemma stated in Section 3 is the special case with $(m) = (0)$, with the polynomial

$$P_F(X) = \sum a_{(\lambda)} \prod_{i=1}^{d} X_{i1}^{\lambda_{i1}} \cdots X_{in}^{\lambda_{in}} X_{i0}^{\lambda_{i0}}.$$

Remark. For other applications, I can extend Lemma 10.1 to points coming close to the divisor of poles $(f)_\infty$. This merely requires a corresponding strengthening of Lemma 6.1, and some estimates on the functions f_i taken at points near the poles. See Appendix 3.

Appendix 1

Since I wish to make this paper as accessible as possible to those in several complex variables, I reproduce for their convenience the basic transference theorem of Khintchine, and the Minkowski–Mahler theorems concerning successive minima on which the proofs are based. Although Cassels in his *Introduction to Diophantine Approximations* obtains quite good constants in his theorems, he perversely (in my view) advises the reader to omit the technique of successive minima in first reading, whereas this seems the most natural way of doing things. Therefore, I follow Davenport [9] and Schmidt [26].

A corollary is stated for nonintegral lattices, which in the applications of this paper are lattices in fields of complex multiplication of abelian varieties. These are treated in Section 3, where we generalize a theorem of Masser to abelian varieties.

The notation and numbering of theorems is self-contained.

1. Successive Minima

Let S be a convex set in $\mathbf{R}^n$. We say that S is *symmetric* if S is symmetric with respect to the origin. In other words, $X \in S$ implies $-X \in S$.

A *lattice* Λ is a discrete subgroup of $\mathbf{R}^n$ that has a basis over $\mathbf{Z}$ consisting of n elements, which are also linearly independent over $\mathbf{R}$. If $X_1, ..., X_n$ is such a basis then the set of elements

$$t_1 X_1 + \cdots + t_n X_n\,; \quad 0 \leqslant t_i < 1$$

is a fundamental domain for the lattice. Its volume is equal to

$$|\det(X_1, ..., X_n)|,$$

This volume is independent of the choice of basis for Λ, and is called the *determinant* of the lattice, denoted also by $\det(\Lambda)$.

MINKOWSKI'S THEOREM. *Let S be convex symmetric in $\mathbf{R}^n$. Let Δ be a lattice in $\mathbf{R}^n$. If*

$$\operatorname{Vol}(S) > 2^n \det(\Lambda)$$

then S contains a nonzero element of Λ.

Proof. Suppose otherwise. Under the map $\mathbf{R}^n \to \mathbf{R}^n/\Lambda$, the set $\tfrac{1}{2}S$ cannot map injectively, otherwise $\mathbf{R}^n/\Lambda$ would have volume $> \det(\Lambda)$, which is against the definition of $\det(\Lambda)$. Hence there exist $X, Y \in S$ such that $X \neq Y$ and $\tfrac{1}{2}(X - Y) \in \Lambda$. By symmetry, $-Y \in S$ and $\tfrac{1}{2}(X - Y) \in S$ by convexity. This proves the theorem.

Let λ_1 be the radius of the smallest ball around the origin O such that this ball contains a nonzero element of Λ. In general, let λ_k $(1 \leqslant k \leqslant n)$ be the radius of the smallest ball around O containing k linearly independent of Λ. We call

$$\lambda_1 \leqslant \lambda_2 \leqslant \cdots \leqslant \lambda_n$$

the *successive minima* of Λ. We have just given an invariant definition of these. It is useful to have a noninvariant characterization. Let X^1 be and element of Λ such that

$$|X^1| = \lambda_1\,.$$

[We use | | for the Euclidean norm.] Let X^2 be an element of Λ linearly independent from X^1 and at smallest distance from 0. Then

$$|X^1| = \lambda_1 \leq |X^2|,$$

and the ball of radius $|X^2|$ has two linearly independent vectors of Λ in it, so that

$$\lambda_2 \leq |X^2|.$$

On the other hand, suppose the ball of radius λ_2 contains the linearly independent elements X^1, Y^2 of Λ, with $|Y^2| = \lambda_2$. Then by the minimality property of X^2 we get

$$|X^2| \leq |Y^2| = \lambda_2,$$

so $|X^2| = \lambda_2$. Inductively, one sees:

Let $X^1,..., X^k$ be elements of Λ such that $|X^i| = \lambda_i$ for $i = 1,..., k$ and $X^1,..., X^k$ are linearly independent. Let X^{k+1} be linearly independent from $X^1,..., X^k$ and at smallest distance from O. Then $\lambda_{k+1} = |X^{k+1}|$.

Indeed, the minimality property of λ_{k+1} shows that $\lambda_{k+1} \leq |X^{k+1}|$. On the other hand, the ball of radius λ_{k+1} has the linearly independent vectors $X^1,..., X^k$, Y^{k+1} with $|Y^{k+1}| = \lambda_{k+1}$. The minimality property of X^{k+1} yields

$$|X^{k+1}| \leq |Y^{k+1}| = \lambda_{k+1},$$

as desired.

Thus, the successive minimal are taken on by elements of Λ constructed by the above inductive procedure. We call such elements $X^1,..., X^n$ *minimal points for Λ*.

Remark. If $A_1,..., A_n$ are linearly independent vectors of Λ such that

$$|A_1| \leq |A_2| \leq \cdots \leq |A_n|,$$

then for $k = 1,..., n$ we have

$$\lambda_k \leq |A_k|.$$

This is immediate from the invariant definition of successive minima.

Let $\det(\Lambda) = r^n$ for some positive number r. Then

$$\det\left(\frac{1}{r}\Lambda\right) = 1.$$

If $X^1,..., X^n$ are minimum points for Λ, then $(1/r) X^1,..., (1/r) X^n$ are minimum points for $(1/r)\Lambda$. Thus

$$\lambda_k\left(\frac{1}{r}\Lambda\right) = \frac{1}{r}\lambda_k(\Lambda).$$

Because of this homogeneity property, in some results to follow, we deal with the case when $\det(\Lambda) = 1$, the general case resulting by homogeneity from this special case.

THEOREM 1. *Assume* $\det(\Lambda) = 1$. *Then*

$$1 \leqslant \lambda_1 \cdots \lambda_n \leqslant 2^n v_n,$$

where v_n *is the volume of the unit ball in* $\mathbf{R}^n$.

Proof. Let $X^1,..., X^n \in \Lambda$ be such that $|X^k| = \lambda_k$ for $k = 1,..., n$, and also linearly independent. We can find an orthogonal basis of $\mathbf{R}^n$ such that the coordinates of $X^1,..., X^n$ are

$$X^1 = (X_1^1, 0,..., 0)$$
$$X^2 = (X_1^2, X_2^2,..., 0)$$
$$\vdots$$
$$X^n = (X_1^n, X_2^n,..., X_n^n).$$

Then $\det(X^1,..., X^n) = X_1^1 \cdots X_n^n$ is an integral multiple of $\det(\Lambda)$, and therefore it is a nonzero integer, of absolute value $\geqslant 1$. Since $|X_k^k| \leqslant |X^k|$, we get

$$1 \leqslant \lambda_1 \cdots \lambda_n.$$

Conversely, consider the symmetric convex body defined by

$$\frac{x_1^2}{\lambda_1^2} + \cdots + \frac{x_n^2}{\lambda_n^2} < 1,$$

whose volume is $\lambda_1 \cdots \lambda_n v_n$. If $\lambda_1 \cdots \lambda_n v_n > 2^n$ then there exists a lattice point $X = (x_1,..., x_n) \neq 0$ in this body. Say X is linearly dependent on $X^1,..., X^k$ but not $X^1,..., X^{k-1}$. Then

$$X = (x_1,..., x_k, 0,..., 0); \qquad x_k \neq 0,$$

and $|X| \geq \lambda_k$ by the minimality property of X^k. Thus

$$1 > \frac{x_1^2}{\lambda_1^2} + \cdots + \frac{x_k^2}{\lambda_k^2} \geq \frac{x_1^2 + \cdots + x_k^2}{\lambda_k^2} = \frac{|X|^2}{\lambda_k^2} \geq 1,$$

a contradiction that proves the right-hand side inequality and also proves the theorem.

Let f, g be two functions on some set, with g positive. We write $f \ll g$ if there is a number $C > 0$ such that $|f(x)| \leq Cg(x)$ for all x in the set. We specify in each case on what extra parameters the constant C will depend. *In the present situation, with lattices, unless otherwise specified, the constant depends only on n, the dimension of the space in which we operate.*

With this notation, we can express Theorem 1 in the weaker form

$$1 \ll \lambda_1 \cdots \lambda_n \ll 1.$$

We write $f \gg\ll g$ to mean $f \ll g$ and $g \ll f$.

An orthonormal basis of $\mathbf{R}^n$ such that linearly independent vectors $X^1, \ldots, X^n$ have coordinates as in the proof of Theorem 1 will be said to be *associated* with this sequence of vectors.

THEOREM 2. *Assume* $\det(\Lambda) = 1$. *There exists a basis* $\{Y^1, \ldots, Y^n\}$ *of Λ such that in the orthonormal system associated with a sequence of minimum points we have*

$$|Y^k| \gg\ll \lambda_k \quad \text{and} \quad |Y_k^k| \gg\ll \lambda_k.$$

Proof. We let $Y^1 = X^1$. Let $(X^1, \ldots, X^k)_\mathbf{Q} = V_k$ be the vector space over $\mathbf{Q}$ generated by $X^1, \ldots, X^k$. Suppose inductively that $\{Y^1, \ldots, Y^\nu\}$ have been selected to be a basis of $\Lambda \cap V_\nu$ satisfying

$$|Y^\nu| \gg\ll \lambda_\nu \quad \text{and} \quad |Y_\nu^\nu| \gg\ll \lambda_\nu$$

for $1 \leq \nu \leq k$. Under the homomorphism

$$V_{k+1} \to V_{k+1}/V_R = \bar{V}_{k+1}$$

the lattice $\Lambda \cap V_{k+1}$ goes to a one-dimensional lattice generated by a vector $\bar{Y}^{k+1}$. We lift $\bar{Y}^{k+1}$ to a vector Y^{k+1} in V_{k+1} such that

$$Y^{k+1} = a_1 Y_1 + \cdots + a_k Y^k + a_{k+1} X^{k+1}$$

with rational $a_1, \ldots, a_{k+1}$. By induction hypothesis, after adding integral multiples of $Y^1, \ldots, Y^k$ we may assume that

$$|a_i| \leqslant 1 \quad \text{for} \quad i = 1, \ldots, k.$$

Write $a_{k+1} = b/d$ where b, d are relatively prime integers. Then

$$dY^{k+1} - bX^{k+1} = da_1 Y_1 + \cdots + da_k Y^k \in \Lambda \cap V_k.$$

Since X^{k+1} can be expressed as a linear combination of $Y^1, \ldots, Y^{k+1}$ it follows that $b = \pm 1$. Hence

$$|Y^{k+1}| \ll (k+1) |X^{k+1}| \ll \lambda_{k+1}.$$

The minimality property of X^{k+1} shows the opposite inequality,

$$\lambda_{k+1} \ll |Y^{k+1}|.$$

Finally,

$$1 = \det(Y^1, \ldots, Y^n) = |Y_1^1| \cdots |Y_n^n|$$
$$\ll |Y^1| \cdots |Y_k^k| \cdots |Y^n|$$
$$\ll \lambda_1 \cdots |Y_k^k| \cdots \lambda_n.$$

It follows that $|Y_k^k| \gg \lambda_k$, thus proving the theorem.

The *dual lattice* Λ^* of Λ is defined to be the lattice of points $X^* \in \mathbf{R}^n$ such that

$$X \cdot X^* \in \mathbf{Z} \quad \text{for all} \quad X \in \Lambda.$$

Let $\lambda_1^*, \ldots, \lambda_n^*$ be the successive minima of Λ^*. Note that

$$\det(\Lambda^*) = (\det \Lambda)^{-1}.$$

THEOREM 3 (Mahler). *We have* $\lambda_i \lambda_{n-i+1}^* \gtrless 1$ *for* $i = 1, \ldots, n$.

Proof. By homogeneity, we may assume without loss of generality that $\det(\Lambda) = 1$. Let $X_1, \ldots, X_n$ be minimal points for Λ, and let $X_1^*, \ldots, X_n^*$ be minimal points for Λ^*. One of $X_1^*, \ldots, X_{n-k+1}^*$ is not perpendicular to $X_1, \ldots, X_k$, say X_i^*. Then

$$1 \leqslant |X_i^* \cdot X_j| \leqslant |X_i^*| |X_j| \leqslant \lambda_i^* \lambda_j \leqslant \lambda_{n-k+1}^* \lambda_k.$$

By Theorem 1, we then get

$$1 \ll \lambda_1 \cdots \lambda_n \lambda_1^* \cdots \lambda_n^* \ll 1.$$

Theorem 3 follows at once.

2. General Transference Theorem

Let the *maximum* $\mu = \mu(\Lambda)$ of a lattice be defined as

$$\mu = \max_P \text{dist}(P, \Lambda),$$

the maximum distance between points $P \in \mathbf{R}^n$ and the lattice.

THEOREM 4. *Let $\lambda_1, \ldots, \lambda_n$ be the successive minima of Λ. Then*

$$\lambda_n \leqslant 2\mu \leqslant \lambda_1 + \cdots + \lambda_n.$$

Hence $\lambda_n \gg\ll \mu$.

Proof. Let $X^1, \ldots, X^n$ be linearly independent in Λ such that $|X^k| = \lambda_k$. The diameter of the parallelogram spanned by $X^1, \ldots, X^n$ is at most

$$\lambda_1 + \cdots + \lambda_n,$$

and therefore any point of $\mathbf{R}^n$ is at distance $\leqslant \frac{1}{2}(\lambda_1 + \cdots + \lambda_n)$ from the lattice, whence

$$2\mu \leqslant \lambda_1 + \cdots + \lambda_n.$$

Conversely, let $X^1, \ldots, X^n$ be a basis for Λ. There exists Y^k such that

$$|\tfrac{1}{2}X^k - Y^k| \leqslant \mu, \qquad k = 1, \ldots, n.$$

The point $X^k - 2Y^k$ lies in Λ. Since $\det(X^1, \ldots, X^n) = \det(\Lambda)$ we get

$$\det(X^1 - 2Y^1, \ldots, X^n - 2Y^n) \equiv \det(\Lambda) \bmod 2 \det(\Lambda).$$

Hence the points $X^k - 2Y^k$ are linearly independent, and also

$$|X^k - 2Y^k| \leqslant 2\mu; \qquad k = 1, \ldots, n.$$

Hence $\lambda_n \leqslant 2\mu$, as desired.

THEOREM 5. (Transference theorem.) *There is a constant $c_1 = c_1(m, n)$ having the following property. Let $B, \delta > 0$. Let*

$$L: \mathbf{R}^m \to \mathbf{R}^n$$

be a linear map. Assume that

$$\text{if} \quad Y \in \mathbf{Z}^n, \quad |Y| \leqslant c_1 B \quad \text{then} \quad \|{}^t L(Y)\| \geqslant c_1 \delta.$$

where $\| \ \|$ is the distance from a point to the integral lattice. Then for all $A \in \mathbf{R}^n$ there exists $X \in \mathbf{Z}^m$ such that

$$|X| \leqslant \frac{1}{\delta} \quad \text{and} \quad \|L(X) - A\| \leqslant \frac{1}{B}.$$

Proof. It suffices to prove the existence of $X \in \mathbf{Z}^m$ such that

$$|X| \ll \frac{1}{\delta} \quad \text{and} \quad \|L(X) - A\| \ll \frac{1}{B},$$

where $\ll$ now depends only on m, n. Let

$$U = \begin{pmatrix} u_{11} & \cdots & u_{1n} \\ \vdots & & \vdots \\ u_{m1} & \cdots & u_{mn} \end{pmatrix}$$

be the matrix such that

$${}^t L(Y) = y_1 U_1 + \cdots + y_n U_n$$

if $Y = (y_1, \ldots, y_n)$, and $U_1, \ldots, U_n$ are the column vectors of U. Define lattices Λ and Λ^* in $\mathbf{R}^{m+n}$ to have as bases the column vectors in the following $(m+n) \times (m+n)$ matrices.

$$\begin{array}{c} 1/\delta \\ \vdots \\ 1/\delta \\ 1/B \\ \vdots \\ 1/B \end{array} \begin{bmatrix} u_{11} & \cdots & u_{1n} & 1 & \cdots & 0 \\ \vdots & & \vdots & \vdots & & \vdots \\ u_{m1} & \cdots & u_{mn} & 0 & \cdots & 1 \\ 1 & \cdots & 0 & 0 & \cdots & 0 \\ \vdots & & \vdots & \vdots & & \vdots \\ 0 & \cdots & 1 & 0 & \cdots & 0 \end{bmatrix} \qquad \begin{array}{c} \delta \\ \vdots \\ \delta \\ B \\ \vdots \\ B \end{array} \begin{bmatrix} 0 & \cdots & 0 & 1 & \cdots & 0 \\ \vdots & & \vdots & \vdots & & \vdots \\ 0 & \cdots & 0 & 0 & \cdots & 1 \\ 1 & \cdots & 0 & -u_{11} & \cdots & -u_{m1} \\ \vdots & & \vdots & \vdots & & \vdots \\ 0 & \cdots & 1 & -u_{1n} & \cdots & -u_{mn} \end{bmatrix}$$

$$\Lambda_1^* \ \cdots \ \Lambda_n^* \ \Lambda_{n+1}^* \ \cdots \ \Lambda_{n+m}^* \qquad \Lambda_1 \ \cdots \ \Lambda_n \ \Lambda_{n+1} \ \cdots \ \Lambda_{n+m}$$

The indications $1/\delta$ on the left and δ on the right mean that the first m rows on the left should be multiplied by $1/\delta$, and the first m rows on the right should be multiplied by δ. Similarly, the last n rows on the left and right should be multiplied by $1/B$ and B, respectively. Then, trivially, we have

$$\Lambda_i^* \cdot \Lambda_j = \delta_{ij},$$

so that Λ and Λ^* are dual lattices.

We assert that $\lambda_1^* \gg 1$. Indeed, consider an element

$$q_1 \Lambda_1^* + \cdots + q_{m+n} \Lambda_{m+n}^* \in \Lambda^*, \quad \text{with} \quad q_i \in \mathbf{Z}.$$

If $|q_i/B| > c_1$ for some i, then this element has norm $> c_1$. On the other hand, if $|q_i/B| \leq c_1$ for all i then the assumption of the theorem implies that one of the first m rows has norm ≥ 1. Hence $\lambda_1^* \gg 1$.

By Theorem 3 it follows that $\lambda_n \ll 1$, and by Theorem 4, it follows that $\mu \ll 1$.

This means: Given

$$A = \begin{pmatrix} \alpha_1 \\ \vdots \\ \alpha_n \end{pmatrix} \quad \text{and} \quad A_0 = \begin{pmatrix} \alpha_1 \\ \vdots \\ \alpha_n \\ 0 \\ \vdots \\ 0 \end{pmatrix},$$

there exist integers $p_1, \ldots, p_{m+n}$ such that

$$|p_1 \Lambda_1 + \cdots + p_{m+n} \Lambda_{m+n} - BA_0| \ll 1.$$

Looking at the top m rows, we conclude that

$$|p_j| \ll \frac{1}{\delta}$$

Looking at the bottom n rows, we conclude that

$$\|L(X) - A\| \ll \frac{1}{B}.$$

This proves our theorem.

COROLLARY. *Let $V_1 = \mathbf{R}^m$ and $V_2 = \mathbf{R}^n$. Let J_i $(i = 1, 2)$ be the integral lattice in V_i. Let Λ_i be a lattice in V_i, and let T_i be a linear automorphism of V_i sending Λ_i onto J_i. Assume that there exist positive integers d_i such that ${}^tT_i$ maps J_i into $d_i^{-1} \Lambda_i$ for $i = 1, 2$. There exists a constant c depending only on Λ_1, Λ_2 such that, if*

$$L \colon V_1 \to V_2$$

is a linear map, and B, $\delta > 0$ satisfy

$$\|{}^tL(Y)\|_{\Lambda_1} \geq c\delta \quad \text{for all} \quad Y \in \Lambda_2, \quad |Y| \leq cB, \quad Y \neq O,$$

then given $A \in V_2$ there exists $X \in \Lambda_1$ such that

$$\|L(X) - A\|_{\Lambda_2} \leqslant \frac{1}{B} \quad \text{and} \quad |X| \leqslant \frac{1}{\delta}$$

Proof. By assumption, for some $\Omega_1 \in \Lambda_1$ we have

$$|{}^tL(Y) - \Omega_1| \gg \delta \quad \text{and} \quad |Y| \ll B$$

We can write $Y = T_2^{-1}Y'$ with $Y' \in J_2$, and also $\Omega_1 = T_1^{-1}\Omega_1'$ with $\Omega_1' \in J_1$. Then

$$|T_1^t L T_2^{-1}(Y') - \Omega_1'| \gg \delta \quad \text{and} \quad |Y'| \ll B$$

We apply the theorem in euclidean space to get some $X' \in J_1$ such that

$$|{}^tT_2^{-1}L\,{}^tT_1(X') - A - \Omega_2'| \ll \frac{1}{B} \quad \text{and} \quad X'| \ll \frac{1}{\delta}$$

We can rewrite this as

$$|L\,{}^tT_1(X') - A - {}^tT_2(\Omega_2')| \ll \frac{1}{B} \quad \text{and} \quad |{}^tT_1(X')| \ll \frac{1}{\delta}.$$

Our assumption that ${}^tT_i$ maps the integral lattice into an integral fraction of Λ_i, we can clear denominators by $d_1 d_2$, and our corollary follows at once, replacing L by $(d_1 d_2)^{-1} L$ and A by $(d_1 d_2)^{-1} A$.

3. Applications to Abelian Varieties

Let k be an imaginary quadratic extension of a totally real number field. Let $\mathfrak{o}$ be a sublattice of the ring of algebraic integers in k, closed under complex conjugation. Let $2d = [k : \mathbf{Q}]$. We have an embedding

$$k \to \mathbf{C}^d \quad \text{by} \quad \xi \mapsto (\xi_1, ..., \xi_d).$$

Viewing $\mathbf{C}^d$ as $\mathbf{R} \otimes k$, the Euclidean product on $\mathbf{C}^d = \mathbf{R}^{2d}$ can be written in the form

$$\langle z, w \rangle = \sum_{i=1}^{d} \operatorname{Re}(z_i \bar{w}_i) = \tfrac{1}{2} \operatorname{Tr}(z\bar{w}),$$

where the trace Tr is extended to $\mathbf{C}^d$ from the trace $\operatorname{Tr}_{k/\mathbf{Q}}$. We note that $\mathfrak{o}$ is a lattice in $\mathbf{C}^d$. Let $v = (v_1, ..., v_d)$. Then we have a linear map

$$L: \mathbf{C}^d \to \mathbf{C}^d \quad \text{given by} \quad L(z) = zv = (z_1 v_1, ..., z_d v_d).$$

Since
$$\langle L(z), w \rangle = \sum_{i=1}^{d} \operatorname{Re}(z_i v_i \bar{w}_i) = \langle z, \bar{v}w \rangle,$$
it follows that
$$^t L(w) = w\bar{v}.$$

Observe that the field k has the property that $\langle \xi, \eta \rangle \in \mathbf{Q}$ for all $\xi, \eta \in k$. Furthermore an element $z \in \mathbf{C}^d$ lies in k if and only if $\langle z, \eta \rangle \in \mathbf{Q}$ for all $\eta \in k$. From this we conclude that the property of Theorem 5 concerning the lattice $\mathfrak{o}$ holds, namely:

Let $T: \mathfrak{o} \to \mathbf{Z}^{2d}$ be a linear map of $\mathfrak{o}$ onto $\mathbf{Z}^{2d}$. Then its transpose $^t T$ maps $\mathbf{Z}^{2d}$ into $a^{-1}\mathfrak{o}$, for some positive integer a.

Proof. Let $m \in \mathbf{Z}^{2d}$. For any $\eta \in k$ we have
$$\langle {}^t T(m), \eta \rangle = \langle m, T(\eta) \rangle \in \mathbf{Q}.$$
This implies that $^t T$ maps m into k, and since $\mathbf{Z}^{2d}$ is finitely generated, it follows that $^t T$ maps $\mathbf{Z}^{2d}$ into a lattice in k. Any two lattices in k are commensurable, i.e., there exists a positive integer a such that $a\,{}^t T(\mathbf{Z}^{2d}) \subset \mathfrak{o}$, whence our assertion follows.

LEMMA 1. *There exists a number $c > 0$ having the following property. Let $v \in \mathbf{C}^d$. If B, δ are positive numbers such that*
$$\|\gamma v\|_\mathfrak{o} \geq c\delta \quad \text{for all} \quad \gamma \in \mathfrak{o}, \quad \|\gamma\| \leq cB,$$
Then, given $w \in \mathbf{C}^d$ there exists $\eta \in \mathfrak{o}$ such that
$$\|\eta v - w\|_\mathfrak{o} \leq \frac{1}{B} \quad \text{and} \quad \|\eta\| \leq \frac{1}{\delta}.$$

Proof. This is a special case of Theorem 5, taking into account the preceding remarks. Watch out for the notation: The sign $\|\ \|$ means maximum of the absolute values, whereas the sign $\|\ \|_\mathfrak{o}$ with the subscript $\mathfrak{o}$ means distance from the lattice $\mathfrak{o}$.

We can generalize this situation. Let $v^1, \ldots, v^n$ be elements of $\mathbf{C}^d$, and consider the linear map
$$L: \mathbf{C}^d \to \mathbf{C}^{dn} \quad \text{given by} \quad L(z) = (zv^1, \ldots, zv^d).$$

Then an argument similar to the preceding one shows that

$$
{}^t L(w^1, \ldots, w^n) = \sum_{j=1}^{n} w^j \bar{v}^j.
$$

Furthermore, let $T^n \colon \mathfrak{o}^n \to \mathbf{Z}^{2dn}$ be the linear map obtained by taking the nth product of T with itself. Then we see as before that $\mathfrak{o}^n$ in $\mathbf{C}^{dn}$ satisfies the hypothesis of the transference theorem with respect to the integral lattice. Thus we obtain a generalized version of Lemma 1.

LEMMA 2. *There exists a number $c > 0$ having the following property. Let $v^1, \ldots, v^n \in \mathbf{C}^d$. If B, δ are positive numbers such that*

$$\| \gamma^1 v^1 + \cdots + \gamma^n v^n \|_{\mathfrak{o}} \geq c\delta \quad \text{for all} \quad \gamma^j \in \mathfrak{o}, \quad \| \gamma^j \| \leq cB,$$

then given $w^1, \ldots, w^n \in \mathbf{C}^d$ there exists $\eta \in \mathfrak{o}$ such that

$$\| \eta v^j - w^j \|_{\mathfrak{o}} \leq \frac{1}{B} \quad \text{and} \quad \| \eta \| \leq \frac{1}{\delta}.$$

THEOREM 6. *Let A be a simple abelian variety admitting k as algebra of complex multiplications. Let $f = (f_1, \ldots, f_d)$ be abelian functions on $\mathbf{C}^d$ with respect to a strongly normalized exponential map. Assume that they give an analytic isomorphism on a ball of radius 2ρ centered at the origin. Let $u^1, \ldots, u^n$ be algebraic points of the exponential map. There exists a constant $C = C(u, f)$ having the following property. Let B, δ be positive numbers such that*

$$| \gamma^1 u^1 + \cdots + \gamma^n u^n | \geq C\delta \quad \text{for all} \quad \gamma^j \in \mathfrak{o}(CB), \quad \text{not all } 0.$$

Let $\xi^1, \ldots, \xi^n$ be points in the image of the ball of radius ρ by f, say $\xi^j = f(\zeta^j)$. Then there exists $z^0 \in \mathbf{C}^d$ such that

$$| f(z^0 u^j) - \xi^j | \leq \frac{1}{B} \quad \text{for} \quad j = 1, \ldots, n \quad \text{and} \quad | z^0 | \leq \frac{1}{\delta}$$

Furthermore, if ω^1 is a nonzero period and $\Lambda = \mathfrak{o}\omega^1$, the element z^0 can be chosen such that

$$\| z^0 u^j - \zeta^j \|_\Lambda \ll \frac{1}{B}.$$

Proof. Let

$$v^j = u^j / u^1; \quad j = 2, \ldots, n.$$

This makes sense because we know that no coordinate of u^1 is 0. By the preceding lemma, for any $w^2,..., w^n$ we can solve the inequalities

$$\left\| \eta \frac{u^j}{u^1} - \frac{w^j}{u^1} \right\|_{\mathfrak{o}} \ll \frac{1}{B}; \qquad \eta \in \mathfrak{o}, \quad \|\eta\| \ll \frac{1}{\delta}.$$

or in other words, for some $\gamma \in \mathfrak{o}$, after multiplying by ω^1,

$$\left\| \eta \frac{\omega^1 u^j}{u^1} - \frac{\omega^1 w^j}{u^1} \right\|_A \ll \frac{1}{B}; \qquad j = 2,..., n.$$

Let

$$z^0 = \eta \frac{\omega^1}{u^1} + \frac{\zeta^1}{u^1}.$$

Then

$$\| z^0 u^1 - \zeta^1 \|_A = 0.$$

On the other hand, for $j = 2,..., n$ we have:

$$\| z^0 u^j - \zeta^j \|_A = \left\| \eta \frac{\omega^1}{u^1} u^j - \frac{\omega^1 w^j}{u^1} + \frac{\omega^1 w^j}{u^1} + \frac{\zeta^1}{u^1} u^j - \zeta^j \right\|_A.$$

All we have to do is solve linearly for w^j in the equation

$$\frac{\omega^1 w^j}{u^1} + \frac{\zeta^1}{u^1} u^j - \zeta^j = 0,$$

to get the other inequalities

$$\| z^0 u^j - \zeta^j \|_A \ll \frac{1}{B}; \qquad j = 2,..., n.$$

The corresponding inequality when we apply f results from the fact that f satisfies a Lipschitz condition in the given neighborhood of 0. This proves our theorem.

Remark. For the applications, one may assume that u^1 is a period, in which case it is obvious that no component is equal to 0, and the argument in Theorem 6 simplifies slightly because ω^1 cancels u^1 in several places.

APPENDIX 2: INTERPOLATION

1. Polynomials with Many Zeros

In this first section, we recall the statement of a criterion of Masser, showing that a polynomial with too many zeros is identically zero.

Let P be a polynomial in d complex variables, of degree at most L in each variable. There exists a constant c depending only on d such that if P has a zero within distance at most $cL^{1/2}$ of each point of the ball of radius 1, then P is identically zero.

For the proof, see [24]. The result obviously applies to zeros in a ball of fixed radius ρ, with an appropriately smaller constant $c(\rho)$, by making the appropriate dilation.

2. Estimates for a Holomorphic Function

We want to show that an entire function that takes on very small values at well-distributed points in a ball of radius R also takes on small values at other points in this ball. In one variable, this is usually done by a variation of Cauchy's formula (Hermite interpolation formula). In several variables, when the function has many zeros, this is implied by the Jensen-Schwarz formula and has been used already in the theory of transcendental numbers [5, 6]. Siu has pointed out to me that this formula in several variables was already proved earlier by Stoll [29], Proposition 1.6 and Theorem 1.7. We shall recall here the relevant statement obtained by putting together the corollary of Proposition 3 and Corollary 1 of what is called the Schwarz lemma in Bombieri-Lang [6].

The notation throughout this section is self contained, independent of what precedes. Throughout this section we let Γ be a set of elements in $\mathbf{C}^d$, of cardinality S. We let σ be the minimum Euclidean distance between any two points of Γ. We let M be a positive integer.

LEMMA 1. *Let F be a holomorphic function on the ball of radius R in $\mathbf{C}^d$. Suppose that Γ is contained in the ball of radius $R_1 \leqslant R/4$, and that F has zeros of multiplicity M at each point of Γ. Then for any point w in the ball of radius R_1, we have the estimate*

$$\log |F(w)| \leqslant \log |F|_R - CMS \left(\frac{\sigma}{R_1}\right)^{2d-2} \log \left(\frac{R}{R_1}\right),$$

where the constant C depends only on d.

I am much indebted to Siu for pointing out that the analog of the above estimate when the function only takes a small value at the points can be carried out by constructing very simply an auxiliary polynomial having the prescribed values at the given points, and subtracting it from the function.

Given any point $\eta \in \Gamma$, we must find a polynomial which has a given Taylor expansion up to order M at that point and vanishes to order M at all other points in Γ. We can then take the sum of such polynomials to solve our problem. For simplicity, suppose that given the point is the origin. Let

$$Q(z) = \sum_{|m| \leq M} c(m) z_1^{m_1} \cdots z_d^{m_d}$$

be the given expansion, so that $D^{(m)}Q(0) = c(m)/m!$. Let

$$P(z) = \prod_{\gamma \neq 0} (\langle z, \gamma \rangle - \langle \gamma, \gamma \rangle)^M,$$

where the product is taken over all elements of Γ not equal to the given point, and thus $\neq 0$ with the present normalization. We have

$$(\langle z, \gamma \rangle - \langle \gamma, \gamma \rangle)^{-M} = \langle \gamma, \gamma \rangle^{-M} \left(1 - \frac{\langle z, \gamma \rangle}{\langle \gamma, \gamma \rangle}\right)^{-M} (-1)^{-M},$$

which has a binomial series expansion at the origin, using

$$(1-t)^{-M} = \sum \binom{-M}{\nu} t^\nu.$$

Note that $\langle \gamma, \gamma \rangle^{-M} \leq \tau^{-2M}$ and $\binom{-M}{\nu} \leq 2^{2M}$. Let $t = \langle z, \gamma \rangle / \langle \gamma, \gamma \rangle$. Viewing t^ν as a polynomial in z, we get an estimate for its coefficients, i.e.

$$|t^\nu| \leq C^\nu \sigma^{-\nu} |\Gamma|^\nu,$$

where $|\Gamma|$ is the maximum of the absolute values of elements in Γ.

Let $[P^{-1}]_M$ denote the Taylor polynomial of order M for P^{-1} in the above expansion, and let

$$F_0^* = Q[P^{-1}]_M P.$$

Then F_0^* is a polynomial having the following properties.

(i) It has the given expansion Q at the origin to order M.

(i′) It vanishes to order M at all other points of Γ.

(ii) It has total degree $\leqslant 3MS$.

(iii) Its coefficients are bounded by

$$\sigma^{-3MS} C^{MS} |\Gamma|^{2MS} \max |c(m)|,$$

for some constant C depending only on d.

These properties are obvious from the construction and the estimates already mentioned. The estimate of (iv) also holds if instead of normalizing at the origin we make a translation and obtain a corresponding polynomial for any point of Γ. Thus, we have the following lemma, which is Siu's interpolation.

LEMMA 2. *Let F be holomorphic on the ball of radius R in $\mathbf{C}^d$, and suppose that Γ is contained in the ball of radius $R/4$. There exists a polynomial F^* having the following properties:*

(i) $D^{(m)} F(\gamma) = D^{(m)} F^*(\gamma)$ *for all* $\gamma \in \Gamma$ *and* $|m| \leqslant M$.

(ii) F^* *has degree* $\leqslant 3MS$.

(iii) *Its coefficients are bounded by*

$$\sigma^{-3MS} C^{MS} R^{3MS} \max_{(m),\gamma} |D^{(m)} F(\gamma)|$$

for some constant C depending only on d.

Applying Lemma 1 to $F - F^*$, and observing that a monomial of total degree $\leqslant 3MS$ on the ball of radius R is bounded by R^{3MS}, we obtain an estimate for F.

LEMMA 3. *Let F be holomorphic on the ball of radius R in $\mathbf{C}^d$. Suppose that Γ is contained in the ball of radius R_1, and let w be a point in the ball of radius R_1. Then*

$$|F(w)| \leqslant \frac{|F|_R}{\exp\left[CMS \left(\frac{\sigma}{R_1}\right)^{2d-2} \log\left(\frac{R}{R_1}\right)\right]}$$
$$+ \left(\frac{C'R}{\sigma}\right)^{5MS} \max_{\substack{|m| \leqslant M \\ \gamma \in \Gamma}} |D^{(m)} F(\gamma)|,$$

where C, C' depend only on d.

An estimate for derivatives of F is obtained from the estimate of Lemma 3 by repeated application of Cauchy's formula with respect to each one of the variables $z_1, \ldots, z_d$, i.e.

$$D^{(r)}F(w) = \frac{r!}{(2\pi i)^{|r|}} \oint \cdots \oint \frac{F(z)\, dz_1 \cdots dz_d}{(z_1 - w_1)^{r_1+1} \cdots (z_d - w_d)^{r_d+1}}.$$

APPENDIX 3: LOCAL ESTIMATES FOR MEROMORPHIC MAPS

Even though in this paper we could carry out the proofs by using points that stayed uniformly away from the divisor of poles of the affine coordinates of a projective embedding, it is clear that there will be other applications when this is not possible. Hence, for the convenience of future work, we make here some remarks describing what tools one can use to deal with such more complicated cases.

First, one has the basic estimate describing how a function blows up near its poles.

LEMMA 1. (Lojasiewicz inequality). *Let f be meromorphic in the neighborhood of a point, i.e., a quotient of two holomorphic functions, $f = g/h$. Let Z be the divisor of h in a compact neighborhood W of the point Then there exist constants c, such that for all points $z \in W$, z lying at distance $\geq r$ from Z we have*

$$|f(z)| \geq \frac{C}{r^c}.$$

Proof. The proof is a simple consequence of the Weierstrass preparation theorem, which essentially reduces the question to the polynomial case. Lojasiewicz actually proves the result in the real case, which is somewhat harder. Cf. his notes from IHES, which as far as I know have not had more formal publication.

If $f = (f_1, \ldots, f_{d'})$ are the affine coordinate functions giving the embedding of an abelian variety, then the compactness of the torus allow us to apply the local lemma to the whole torus. Let $(f)_\infty$ be the union of the polar divisors of the functions f_i. For z outside this polar divisor, the differential $df(z) = J(z)$ is defined, as a linear map from $\mathbf{C}^d$ to $\mathbf{C}^d$. Let $\Delta(z)$ be its determinant. We note that

$$J^{-1}(z)$$

is defined for z outside $(\Delta)_0$, and we can find a divisor Z [perhaps somewhat bigger than $(\Delta)_0$] such that f is invertible on the complement of Z. We call such a divisor Z sufficiently large with respect to f.

LEMMA 2. *Let Z be sufficiently large with respect to $f = (f_1, ..., f_d)$. Then there exists C', c' depending only on f, Z such that for all points z lying at distance $\geq r$ from Z we have*

$$|df(z)| \geq \frac{C'}{r^{c'}}.$$

Proof. Any two norms on a finite dimensional vector space are equivalent. We can take the sup norm of the coefficients on the space of matrices. Then we can apply Lemma 1 to the components of J^{-1}, which have poles in $(\Delta)_0$, to see that

$$|J^{-1}(z)| \ll \frac{1}{r^{c'}}.$$

Multiplying both sides by $|J(z)|$ proves the lemma.

Whenever the differential df at a point is nonsingular, we know by the inverse mapping theorem that f is invertible. In the applications, it is necessary to know precisely what kind of uniformity is available in such instances, and one uses the following calculus lemma.

LEMMA 3. *Let f be a holomorphic map, invertible in the closed ball $B(r)$ of radius r centered at the origin 0. Assume that there exists a constant K such that*

$$|f'(0)^{-1}| \leq K.$$

Let $0 < s < 1$. If

$$|f'(z) - f'(w)| \leq \frac{s}{K}$$

for all $z, w \in B(r)$, then

$$f(B(r)) \supset B((1-s)r) K^{-1}.$$

Proof. Consider $f'(0)^{-1} \circ f$. Then

$$|f'(0)^{-1} f'(z) - f'(0)^{-1} f'(w)| \leq s.$$

By calculus, cf. for instance my *Real Analysis* Lemma of Chapter VI, Section 1 we conclude,

$$f'(0)^{-1} fB(r) \supset B((1-s)r),$$

whence

$$f(B(r)) \supset f'(0) B((1-s)r).$$

The lemma follows at once.

Finally, when dealing with something like Theorem 6 of Appendix 1, i.e., the generalization of Masser's theorem in the higher dimensional case, the point z^0 which we found there may lie on the divisor of poles and it becomes necessary to move away slightly from z^0, without losing some essential estimates. This is done by a slightly deeper lemma than those already mentioned.

LEMMA 4. *Let Z be an analytic divisor in a closed ball of $\mathbf{C}^d$, of finite radius. There exists numbers C and r_0 with*

$$0 < r_0 < 1$$

having the following property. Given $z^0 \in Z$ and $0 < r < r_0$, there exists a point z^1 such that

$$|z^0 - z^1| \leq Cr \quad \text{and} \quad \operatorname{dist}(z^1, Z) \geq r.$$

Proof. This lemma needs more extensive tools, either the resolution of singularities, or stratification of the divisor Z into nonsingular pieces behaving reasonably well as they approach the singular points. Such stratifications are also dealt with in the Lojasiewicz notes. Probably the best reference will be to a forthcoming paper of Hironaka, "On the Lojasiewicz inequality," to appear some time within the next year. The lemma as stated here is an easy consequence of the tools developed in that paper. The geometric idea behind it is that even if we have a bad singularity at z^0, we can move in a direction somewhere in the middle of singular branches of the divisor so that the distances of the point z^1 to our original z^0 and to the divisor Z are of the same order of magnitude.

In applications, e.g., Masser's inductive procedure applied to Abelian varieties, the point z^0 may be very close to the divisor Z, even though not on it. Lemma 4 can be applied by replacing z^0 with a point of Z at closest distance to z^0, so that z^1 is then reasonably far away from both z^0 and Z, as described in Lemma 4. For instance, suppose we are trying to deal with a situation as in Theorem 4, §3, and wish to use Theorem 6 of Appendix 1, §3. Then we take

$$q = CM^{d+2}$$

with some appropriate constant C. We pick z^1 such that

$$|z^1 - z^0| \ll \frac{1}{q} \quad \text{and} \quad \text{dist}(z^1 u^n, Z) \geq \frac{2}{q}.$$

Then

$$|z^1 u^j - z^0 u^j| \ll \frac{1}{M^{d+2}}$$

and

$$|f(z^1 u^j) - f(z^0 u^j)| \ll \frac{1}{M^{d+2}}; \quad \text{for } j = 0,\ldots, n-1,$$

whence

$$|f(z^1 u^j) - \xi^j| \ll \frac{1}{M^{d+2}}; \quad \text{for } j = 0,\ldots, n-1.$$

In other words, we have not destroyed the good effects of Theorem 6 with respect to $j = 0,\ldots, n-1$.

We can then consider the ball $B(z^1 u^n, 1/q^b)$ where b will be select appropriately large, compared with the exponents c, c' of Lemmas 1 and 2. For $z, w \in B(z^1 u^n, 1/q^b)$ we find

$$|f'(u)^{-1} \circ f'(z) - f'(u)^{-1} \circ f'(w)| \ll q^c |f'(z) - f'(w)|$$
$$\ll q^c q^{c'} |z - w|$$
$$\ll \frac{1}{q^{b-2c}} \leq \frac{1}{2}.$$

By the calculus lemma, it follows that the image of $B(z^1 u^n, 1/q^b)$ under f contains

$$B\left(f(z^1 u^n), \frac{1}{q^{b+c}}\right).$$

In this manner, Masser's lemma guaranteeing that a polynomial is equal to zero if it has enough zeros that are sufficiently close together can again be applied in a context similar to that of Theorem 4, but when the closeness of z^0 to the divisor of poles has to be taken into account.

References

1. A. ALTMAN, The size function on abelian varieties, *Trans. Amer. Math. Soc.* **164** (1972), 153–161.
2. A. BAKER, Linear forms in the logarithms of algebraic numbers, *Mathematika* **13** (1966), 204–216.

3. A. BAKER, On the periods of the Weierstrass $\wp$ function, *Symp. Math.* **IV** (1970), 155-174.
4. A. BAKER, An estimate for the $\wp$-function at an algebraic point, *Amer. J. Math.* **XVII** (1970), 619-622.
5. E. BOMBIERI, Algebraic values of meromorphic maps, *Invent. Math.* **10** (1970), 267-287.
6. E. BOMBIERI AND S. LANG, Analytic subgroups of group varieties, *Invent. Math.* **11** (1970), 1-14.
7. J. COATES, An application of the division theory of elliptic functions to diophantine approximation, *Invent. Math.* **11** (1970), 167-182.
8. J. COATES, Linear forms in the periods of the exponential and elliptic functions, *Invent. Math.* **12** (1971), 290-299.
9. H. DAVENPORT, Analytic methods for diophantine equations and diophantine inequalities, Lecture Notes, Univ. of Michigan, 1962.
10. N. FELDMAN, An elliptic analogue of an inequality of Gelfond, *Trans. Moscow Math. Soc.* **18** (1968), 71-84.
11. S. LANG, Diophantine Geometry, Interscience, New York, 1962.
12. S. LANG, Transcendental points on group varieties, *Topology* **1** (1962), 313-318.
13. S. LANG, Algebraic values of meromorphic functions, *Topology* **3** (1962), 183-191.
14. S. LANG, Algebraic values of meromorphic functions II, *Topology* **5** (1966), 363-370.
15. S. LANG, Diophantine approximations on toruses, *Amer. J. Math.* **86** (1964), 521-533.
16. S. LANG, Introduction to Transcendental Numbers, Addison Wesley, Reading, Mass., 1966.
17. S. LANG, Les formes bilinéaires de Néron et Tate, Séminaire Bourbaki, 1963-1954, No. 274.
18. S. LANG, Introduction to Diophantine Approximation, Addison Wesley, Reading, Mass., 1966.
19. S. LANG, Abelian Varieties, Interscience, New York, 1958.
20. S. LANG, Introduction to Algebraic and Abelian Functions, Addison Wesley, Reading, Mass., 1972.
21. S. LANG, Integral points on Curves, *Pub. Math. IHES*, Paris, 1960.
22. D. MASSER, Transcendence properties of elliptic functions, to appear.
23. D. MASSER, Algebraic points of an elliptic function, to appear.
24. D. MASSER, Polynomials in many variables, to appear.
25. A. NÉRON, Quasi-fonctions et hauteurs sur les variétés abéliennes, *Ann. Math.* **82** (1965), 249-331.
26. W. SCHMIDT, Lectures on diophantine approximation, Univ. of Colorado, 1970.
27. T. SCHNEIDER, Zur theorie der Abelschen Funktionen und Integrale, *J. Reine Angew. Math.* (1941), 110-128.
28. G. SHIMURA, Complex multiplication of abelian varieties, *Publ. Math. Soc. Japan*, 1961.
29. W. STOLL, Normal families of non-negative divisors, *Math. Zeitschr.* **84** (1964), 154-218.

Printed by the St Catherine Press Ltd., Tempelhof 37, Bruges, Belgium.

DIVISION POINTS OF ELLIPTIC CURVES AND ABELIAN FUNCTIONS OVER NUMBER FIELDS.

By Serge Lang.

Let A be an elliptic curve defined over a number field K. Let p be a prime number. Let A_{p^n} be the group of points of order p^n. If f, g are two positive functions of positive integers, we write $f \ll g$ if there exists a constant C such that $f(n) \leq Cg(n)$ for all n. We write $f \gg \ll g$ if $f \ll g$ and $g \ll f$, and we then say that f, g have the same order of increase. If A does not have complex multiplication, a theorem of Serre asserts that the degree

$$[K(A_{p^n}) : K]$$

has the order of increase p^{4n}, or equivalently that the Galois group of $K(A_{p^n})$ over K is of bounded index in $GL_2(\mathbf{Z}/p^n\mathbf{Z})$ for arbitrary n. Serre's arguments rely heavily on local p-theory, the Hodge structures of Tate, the formal group p-adically, etc. [7].

It occurred to me that a totally different approach could be taken to the problem. Let $A^{(p)}$ be the union of all groups A_{p^n}. Then $\text{Gal}(K(A^{(p)})/K)$ is closed in $GL_2(\mathbf{Z}_p)$, and is a Lie subgroup. Therefore it is *a priori* clear that the degrees above have order of increase p^n, or p^{2n}, or p^{3n}, or p^{4n}. Suppose that the order of increase was only like p^{3n} or less. View the elliptic curve as parametrized by the Weierstrass function. Then the values of $\wp$ taken at the division points ω/p^n are subject to the corresponding restriction. We have here an analogy with the situation arising in the theory of transcendental numbers, where one wants to prove that the values of certain functions cannot be algebraic, at certain points. Under the assumption that the values are algebraic, and lie in a fixed number field K, one has a technique which plays analytic estimates against arithmetic estimates to arrive at a contradiction. Of course, here, the number field is not fixed. However, as a first step in showing that there is something to the analogy, I shall prove by the above technique:

$$[K(A_{p^n}) : K] \text{ cannot be } \ll p^n.$$

Received November 10, 1972.

Copyright © 1975 by Johns Hopkins University Press.

Of course, this is much weaker than Serre's theorem. If one could prove an order of increase strictly greater than p^{2n}, then fairly simple arguments (bottom of p. IV-11 in Serre's book [7]), bypassing almost all of [7], would yield the final result. However, even the above theorem is not completely trivial by Serre's techniques. The proof involves the Tate parametrization p-adically, in the case of bad reduction, and a ramification analysis in the case of supersingular reduction, which uses the local formal group. On the other hand, the arguments to be given here are short, and are at the most basic level of the theory, namely the addition formula of the Weierstrass function (not even its differential equation). Instead of being p-adic, they take place at infinity. They also apply to abelian functions, and we discuss this below.

Even if a refinement of the method were to lead to the full Serre p^{4n}, this would not supercede Serre's theory, which, as mentioned above, gives results concerning the ramification properties of the fields under consideration. In addition, Serre's methods give very good estimates to determine when, say, the Galois group of $\mathbf{Q}(A_p)$ over $\mathbf{Q}$ is $GL_2(\mathbf{Z}/p\mathbf{Z})$ in case $K = \mathbf{Q}$. All the constants in the proof below are explicitly computable, but I have not checked how well they allow for computations in practice. Finally, Serre also proves that for almost all p, $\text{Gal}\,(K(A_p)/K)$ is $GL_2(\mathbf{Z}/p\mathbf{Z})$. The difficulty in extending the present method to this case seems to be similar to the difficulty mentioned at the end in Remark 2, involving the relations between the function of one and several variables.

On the other hand, the method obviously leads to a possibility of proving the isogeny theorem. In its original form, the conjecture states that if two elliptic curves A, B over a number field have Galois isomorphic p-adic representations, then they are isogenous. Equivalently, if they have no complex multiplication, this can be expressed in terms of field extensions, namely: If $K(A^{(p)}) = K(B^{(p)})$ then A and B are isogenous.

Here again, such a condition can be viewed as a restriction on the values of the Weierstrass functions parametrizing A and B, say $\wp_A$ and $\wp_B$. Let $[\omega_1, \omega_2]$ and $[\omega_3, \omega_4]$ be the periods of $\wp_A$ and $\wp_B$ respectively. Then hopefully considering the five functions

$$e^{2\pi i z}, \quad \wp_A(\omega_1 z), \quad \wp_A(\omega_2 z), \quad \wp_B(\omega_3 z), \quad \wp_B(\omega_4 z)$$

and applying refinements of the present techniques may lead to a proof of the isogeny theorem (which so far, Serre's techniques have not succeeded in reaching either). In order to prove the isogeny theorem, it suffices to prove the inequality

$$[K(A_{p^n}, B_{p^n}) : K] \gg p^{n(4+\varepsilon)}$$

This comes from the fact that the intersection of the fields

$$K(A_{p^n}) \cap K(B_{p^n}),$$

which is Galois over the field of all p-power roots of unity, must be finite, because $SL_2(\mathbf{Z}_p)$ does not contain a closed non-trivial subgroup of infinite index.

Proof of the theorem. We shall prove a slightly stronger version of the result stated previously. For the convenience of the reader, we give the formulation on elliptic curves, and only afterwards make comments concerning the applicability to abelian varieties.

THEOREM. *Let $\zeta_{p^n} = e^{2\pi i/p^n} = e(u_0/p^n)$ where $u_0 = 2\pi i$. Let f be a Weierstrass $\wp$-function associated with an elliptic curve A defined over the number field K. Let u_1 be a fundamental period of f. Then we cannot have*

$$[K(e(u_0/p^n), f(u_1/p^n)) : K] \ll p^n. \qquad (*)$$

Before starting the proof proper, we recall an auxiliary fact. A simple lemma of Siegel (cf. any book on transcendental numbers, e.g. [5], Chapter I) states that a system of q linear equations in N variable, with coefficients in the algebraic integers $\mathfrak{o}_K$, such that all the conjugates are bounded by C, has a non-trivial solution in $\mathfrak{o}_K$, bounded by a constant times

$$(NC)^{q/(N-q)}.$$

We shall deal with linear equations whose coefficients will be monomials in the division points $f(u_1 r/p^n)$ and roots of unity, and hence we need estimates for these. Since the function f has a pole of order 2 at the lattice points and no other poles, it follows trivially that for any integer r with $1 \leq r < p^n$ we have

$$|f(u_1 r/p^n)| \ll p^{2n}.$$

A similar estimate applies to any conjugate of $f(u_1 r/p^n)$.

Recall that a denominator for an algebraic number ξ is a positive integer a such that $a\xi$ is an algebraic integer. We need to estimate denominators for such division values, and it is known that such denominators are bounded. Explicit parametrizations of the elliptic curve p-adically as in Lutz [6] show this, giving explicit bounds (cf. also Cassels [3]). As Serre pointed out to me, one can also see this by a general argument. The exponential map on the elliptic curve on $\mathbf{C}_p$ (completion of the algebraic closure of $\mathbf{Q}_p$) converges in a neighborhood of the origin which does not contain any point of finite order. Hence the points of finite order are a discrete set in A_{C_p}, and since the poles of f are isolated, the boundedness follows at once.

We now assume the inequality (*), and derive a contradiction. Since $K(\zeta_{p^n})$ has degree $\gg p^n$, enlarging K be a finite extension if necessary, one may then assume without loss of generality that for all n,

$$[K(e(u_0/p^n), f(u_1/p^n)) : K] = K(\zeta_{p^n}).$$

Form the function

$$F(z) = \sum_{\lambda_0=1}^{L^2} \sum_{\lambda_1=1}^{L} \alpha_{\lambda_0 \lambda_1} e(u_0 z)^{\lambda_0} f(u_1 z)^{\lambda_1},$$

where $\alpha_{(\lambda)} = \alpha_{\lambda_0 \lambda_1}$ are coefficients to be chosen suitably in $K(\zeta_{p^m})$, so that

$$F(r/p^m) = 0, \quad 1 \leq r < p^m.$$

Here, m is a large positive integer, and we shall make L explicit in a moment. We view these conditions as linear equations in unknowns $\alpha_{(\lambda)}$ in $K(\zeta_{p^m})$. We have:

Number of variables $= L^3$

Number of equations $= p^m - 1$.

To apply Siegel's lemma, we make the number of variables approximately equal to the number of equations, say $L = [2p^{m/3}]$, so that L^3 is approximately $8p^m$.

The coefficients of the linear equations have absolute values $\ll p^{cmL}$ for some constant c, so do their conjugates, and a denominator for these conjugates.

Let $\zeta = \zeta_{p^m}$. For some integer $M \leq p^m$ (in fact, of that order of magnitude), the powers ζ^ν, $\nu = 0, \ldots, M$ form a basis for $K(\zeta)$ over K. The dual basis consists of the numbers $\zeta^\nu / g'(\zeta)$, where g is the irreducible equation for ζ over K, and

$$g'(\zeta) \text{ divides } p^m$$

since $g(X)$ divides $X^{p^m} - 1$. We express the monomials occurring as coefficients of the linear equations in terms of the above basis. If $\xi \in K(\zeta)$, and $\xi = \Sigma x_\nu \zeta^\nu$, with $x_\nu \in K$, then

$$x_\nu = \text{Tr}(\xi \zeta^\nu / g'(\zeta)).$$

Our system of linear equations in $K(\zeta)$ is therefore reduced to a system of linear equations in K, with L^3M variables, and $(p^m - 1)M$ equations. The expression in terms of the trace allows us to estimate easily the absolute values of the conjugates and denominators of the coefficients as $\ll p^{cmL}$ again. Clearing denominators, we can apply Siegel's lemma, and we see that there exists a solution of the system

$$F(r/p^m) = 0, \quad 1 \leq r < p^m,$$

with algebraic integers $\alpha_{(\lambda)}$ in K, not all 0, such that:

The maximum of the absolute values of the conjugates of the numbers $\alpha_{(\lambda)}$ satisfies

$$\|\alpha_{(\lambda)}\| \ll p^{cmL}. \tag{1}$$

Let n be the maximal integer such that

$$F(r/p^n) = 0, \quad 1 \leq r < p^n.$$

Such an integer exists because of the algebraic independence assumption, which implies that f is not identically zero. Then for some r_0 we have

$$F(r_0/p^{n+1}) \neq 0, \quad 1 \leq r_0 < p^{n+1}.$$

We express f as quotient of theta functions, $f = \theta/\psi$, relatively prime. Then $F(z)\psi(u_1 z)^L$ is an entire function. Since ψ is holomorphic, we have an estimate

$$\left| \frac{1}{\psi(u_1 r_0/p^{n+1})} \right| \ll p^{cnL}. \tag{2}$$

Observe that $F(z)$ has period 1. Let δ be a small number, say $\delta = 1/10$. For integers s satisfying $1 \leq s \leq R^{1-\delta}$, where $R = p^{n/2}$, we also have

$$F\left(s + \frac{r}{p^n}\right) = 0.$$

Therefore dividing F by the product of the linear factors

$$G(z) = \prod \left(z - \left(s + \frac{r}{p^n} \right) \right)$$

taken for $2 \leq s < R^{1-\delta}$ and $1 \leq r < p^n$ yields a function which is still holomorphic at the points $s + r/p^n$. We have

$$F(r_0/p^{n+1}) = \left[\frac{F(z)\psi(u_1 z)^L}{\psi(u_1 r_0/p^{n+1})^L} \frac{G(r_0/p^{n+1})}{G(z)} \right]_{\text{evaluated at } r_0/p^{n+1}}$$

We now use the maximum modulus principle on the circle of radius R. We have

$$|F(z)\psi(u_1 z)^L|_{|z|=R} \ll C^{R^2 L} C^{RL^2} p^{cnL}, \qquad (3)$$

taking into account the fact that the theta functions are of strict order 2. The G-quotient is estimated by

$$\left| \frac{G(r_0/p^{n+1})}{G(z)} \right|_{|z|=R} \ll \frac{1}{R^{\delta R^{1-\epsilon} p^n}}. \qquad (4)$$

Hence finally putting the above estimates together, yields

$$|F(r_0/p^{n+1})| \ll \frac{C^{R^2 L} C^{RL^2} p^{cnL}}{R^{\delta R^{1-\epsilon} p^n}}.$$

This upper bound has been obtained by an analytic estimate.

On the other hand, the algebraic number $F(r_0/p^{n+1})$ has degree $\ll p^n$, and we have seen that the absolute values of its conjugates, and a denominator, are $\ll p^{cnL}$. Taking the product over the conjugates, multiplied by a denominator yields an ordinary integer of absolute value ≥ 1. Consequently we get the opposite inequality

$$\frac{1}{p^{cnLp^n}} \ll |F(r_0/p^{n+1})|. \qquad (5)$$

Comparing our two inequalities yields

$$1 \ll \frac{C^{R^2 L} p^{cnLp^n}}{R^{\delta R^{1-\epsilon} p^n}}.$$

Since we choose $L \ll p^{n/3}$, $R = p^{n/2}$, we see that the numerator on the right grows at most like

$$C^{np^n + n/3},$$

whereas the denominator grows at least like

$$R^{\delta p^{n} + n(1-\delta)/2}$$

Letting m, and hence n, tend to infinity yields a contradiction, which proves our theorem.

Remarks.

Remark 1. The hypothesis that A has no complex multiplication was not used! If one makes this hypothesis, then one can use the two algebraically independent functions

$$\wp(\omega_1 z) \quad \text{and} \quad \wp(\omega_2 z),$$

and form the auxiliary function

$$\sum \alpha_{(\lambda)} e(\omega_0 z)^{\lambda_0} \wp(\omega_1 z)^{\lambda_1} \wp(\omega_2 z)^{\lambda_2}$$

with $\lambda_i \leq L$. We could let λ_0 range to L^2 because the ordinary exponential function is entire of strict order 1. The higher rate of growth of theta functions is one of the points where one encounters a difficulty, which even by adjusting the parameter L does not lead to an improvement of the estimate with p^{2n} instead of p^n appearing in estimate (5), when we took a product over conjugates. Forming the above auxiliary function would be a way of using the hypothesis, however.

Remark 2. The difficulty in showing that the desired degree cannot be $\ll p^{2n}$ is analogous to the difficulty encountered in the transcendence problem of values

$$e^{u_i v_j}, \quad i = 1, 2; \ j = 1, 2, 3;$$

as in Chapter II, [5]. Taking derivatives directly on the function $F(z)$ introduces power products of the transcendental periods. On the other hand, separating the variables and considering the function of several variables similar to that in Baker [2],

$$F^*(z_0, z_1, z_2) = \sum \alpha_{(\lambda)} e(\omega_0 z_0)^{\lambda_0} \wp_1(\omega_1 z_1)^{\lambda_1} \wp_2(\omega_2 z_2)^{\lambda_2},$$

it seems that at some point one must induce on the diagonal, otherwise I don't see how to use the hypothesis that the functions $\wp_i(\omega_i z)$ are algebraically independent, with the *same* variable z. The hope is that there is nevertheless a way to use the differential equation which will allow adjusting the parameters of the proof in such a way as to obtain the estimate with p^{2n} instead of p^n. Multiplicities of zeros would be one way to improve that rate of growth of the denominator in (4).

Remark 3. The method also applies to abelian functions. Let A be an abelian variety defined over a number field K, and let f be an abelian function in the function field $K(A)$. Let u_1 be a basic period, so that the function

$$z \mapsto f(u_1 z)$$

is meromorphic, not constant. Then all conjugates of $f(u_1 r/p^n)$ satisfy an estimate $\ll p^{cn}$. It seems to be a difficult problem to prove a similar estimate for denominators. As this question is quite different from the main considerations of this paper, I leave it aside for now. However, one can use only interpolation values r such that the estimate is satisfied, as in [5], Chapter II, §4 and then one gets:

Let K_n be the field obtained from K by adjoining the p^n-th roots of unity, and all points $f(u_1 r/p^n)$, with $1 \leq r < p^n$. Then we cannot have $[K_n : K] \ll p^n$.

Remark 4. For extensions of the theorem, one may consider abelian functions

$$f_1(u_1 z), \ldots, f_m(u_m z),$$

which are algebraically dependent. This occurs if, for instance, $u_1, \ldots, u_m$ are linearly independent over the rationals, and A has a trivial ring of endomorphisms, cf. [5], p. 41. One can also consider values u_i which are elliptic logarithms of algebraic points, i.e., points where the $\wp$-function takes on algebraic values, thus giving estimates on the degrees of fields obtained by the division values of such points. When the u_i are periods, however, Serre pointed out that for abelian varieties, the Galois group operation must preserve all "algebraic" structures, e.g. the skew-symmetric form into the roots of unity and others having to do with algebraic cycles, so that the Galois group of $K(A^{(p)})$ over K cannot be open in $GL_{2d}(\mathbf{Z}_p)$ when dim $A > 1$, even though the above functions $f_i(u_i z)$ are algebraically independent. This seems to be an analogous kind of degeneracy to the Riemann relations of a period matrix (cf. again [5], p. 41).

YALE UNIVERSITY

REFERENCES.

[1] A. Baker, "On the periods of the Weierstrass $\wp$ function, Proceedings of the Rome conference on number theory," 1969.
[2] A. Baker, "On the quasi-periods of the Weierstrass ζ-function," *Nach. Akad. Wiss. Gottingen II, Math. Phys. Klasse*, Nr. 16 (1969).
[3] J. W. Cassels, "Diophantine equations with special reference to elliptic curves," *Journal of the London Mathematical Society*, 41 (1962), pp. 193–291.
[4] J. Coates, "The transcendence of linear forms in ω_1, ω_2, η_1, η_2, $2\pi i$," *American Journal of Mathematics* (1971), pp. 385–397.
[5] S. Lang, Introduction to transcendental numbers, Addison Wesley, 1967.
[6] E. Lutz, "Sur l'equation $y^2 = x^3 - Ax - B$ sur les corps p-adiques," *J. reine angew. Math.*, 177 (1937), pp. 204–210.
[7] J. P. Serre, "Abelian l-adic representations and elliptic curves," W. A. Benjamin, 1968.

Units in the Modular Function Field. I

Dan Kubert* and Serge Lang*

Department of Mathematics Yale University, New Haven, Conn. 06520, USA

Let F_N be the modular function field of level N, i.e. the function field of the modular curve usually denoted by $X(N)$ over the field of N-th roots of unity over Q. We shall be interested in the units of the affine ring R_N which is the integral closure of $Z[j]$ in F_N, or $QR_N = R_N \otimes Q$. Such units have numerous applications, to be considered in a sequence of papers.

We shall first establish correspondences between the modular curves and the Fermat curves, by constructing units u, v in the integral closure of $Z[j]$ in the modular function field, satisfying

$$u + v = 1,$$

and then taking the n-th root of u and v. These units are closely associated with n-th division points on the generic curve with invariant j. From the case of level 2, one sees that a proof of the Mordell conjecture (finite number of rational points in any number field) for either type of curves (Fermat curve or modular curve) implies the conjecture for the other type.

The construction of u, v has implications both for the integral points and also the rational points on Fermat (resp. modular) curves, and both will be discussed. In particular, we reduce the proof that hyperelliptic or superelliptic curves have only a finite number of integral points, to the case of $x^3 + y^3 = 1$.

There are also implications for the problem of determining when a special elliptic curve over a number field k has a division point rational over k. Let

$$\pi : X(N) \to P^1$$

be the natural rational map of $X(N)$ onto the j-line. Assume that every subfield of F_N properly containing $Q(j)$ has genus ≥ 2. This condition is satisfied for example when N is a sufficiently large prime. Indeed, one knows that all subgroups of $PSL_2(Z/pZ)$ are either of Lie type, the largest being Borel, i.e. consisting of matrices

$$\begin{pmatrix} a & b \\ 0 & a^{-1} \end{pmatrix}$$

or the normalizers of a non-split Cartan subgroup, or small groups, namely A_4, S_4, A_5 (symmetric, alternating groups of the stated order). The genus of F_N and of the subfields fixed under the Borel subgroup is known, cf. for instance Shimura [Sh], and tends to infinity with N. Presumably so does the genus of the fixed field

---

* Supported by NSF grants.

of the Cartan subgroup, although we don't know a precise reference giving this genus. It is clear that the genus of the fields fixed under the small groups also goes to infinity with N, by the Hurwitz genus formula.

Note that we are not asserting that all subfields of F_N for N prime tending to infinity have genus ≥ 2. There may exist many such subfields of genus 1, but they do not contain $\mathbf{Q}(j)$.

Assume now the Mordell conjecture for subfields of F_N properly containing $\mathbf{Q}(j)$, and of genus ≥ 2.

Let k be a number field containing the N-th roots of unity. Assume that N is such that every subfield of kF_N properly containing $k(j)$ has genus ≥ 2. Then for all but a finite number of values $\bar{j}$ of j in k, we have

$$[k(\pi^{-1}(\bar{j})):k] = [kF_N:k(j)].$$

Proof. For each value $j = \bar{j}$ in k, the point of $X(N)$ lying above $\bar{j}$ has a decomposition group, which is a subgroup H of the Galois group G of $X(N)$ over $X(1) = \mathbf{P}^1$. Suppose that for infinitely many such values of $\bar{j}$ this decomposition group is not trivial, so that we can take it equal to the same subgroup H. Its fixed field is not $k(j)$, and has genus ≥ 2. It has a rational point in k corresponding to such values of $\bar{j}$, contradicting the Mordell conjecture.

The above statement is an irreducibility property, and will be called the **irreducibility conjecture** for the modular curves. The hypothesis concerning the genus of intermediate fields is satisfied, for instance, when N is a sufficiently large prime number. Thus the Mordell conjecture gives a uniform bound for any degeneracy of the group of p-points on elliptic curves defined over k, with variable invariant in k. On the other hand, Serre's theorem gives a uniform bound for a fixed invariant and variable primes. One expects a theorem containing both of these aspects, i.e. uniformity with respect to pairs $(p, \bar{j})$ for all primes p and $\bar{j}$ in k, for instance a statement such as this one:

For all but a finite number of pairs $(p, \bar{j})$ with $\bar{j} \in k$, the Galois group of the p-primary torsion of an elliptic curve A over k with invariant $\bar{j}$ is equal to $GL_2(\mathbf{Z}_p)$.

On the other hand, for the modular curves, we shall construct units u, v satisfying $u + v = 1$. For prime level p, we also construct such units whose divisor at infinity is a p-multiple. Specializing into a number field, and using a local regularity property at infinity, we shall see that for any point of the modular curves in a number field k (not at infinity), the algebraic numbers obtained by specializing these functions u, v in k have an ideal factorization which is a p-th power. More generally, we obtain a homomorphism from the group of divisors at infinity on the modular curve (generated by the so-called cusps) into the group of ideals (fractional) in the number field essentially by a simple pull back. This seems to be one of the algebraic-geometric mechanisms which lie behind the work of Demyanenko [D] and Kubert [K] in their studies of rational points on modular curves. It can be used as an alternative to arguments of Brylinski [Br] (who works on the Néron model, rather than with the Tate curve, as we do). It joins this train of thoughts with that of constructing units in the modular function fields [L 1], Chapter 18, § 6. Such results show how one can parametrize both units and ideal classes in number fields in an algebraic-geometric way.

Finally we study the relation of the group of units with the group of divisor classes of degree 0 at infinity generated by the cusps. Manin [M] and Drinfeld [Dr] have proved that this group is finite. Using units formed with the sigma function directly (essentially the Siegel functions) one can form the regulator of these units with their orders at infinity. Say for a prime power, the group $G = GL_2(\mathbf{Z}/p^n\mathbf{Z})$ operates on the points at infinity, and the isotropy of the standard cusp is the group of matrices

$$\begin{pmatrix} 1 & b \\ 0 & d \end{pmatrix}.$$

There is a direct decomposition $G = CG_\infty$, where C is the Cartan group of units in the unramified extension of degree 2 over $\mathbf{Q}_p$, mod p^n. We can then express the regulator by using the Frobenius determinant formula as a product of linear terms. We are indebted to Tate for his help in showing how to compute these linear terms (character sums). Up to factors which are Gauss sums, they are generalized Bernoulli numbers $B_{2,\chi}$ of Leopoldt, whence not equal to zero. This gives an explicit bound for the order of the divisor class group at infinity, and an independent proof of the Manin-Drinfeld theorem.

We are publishing the first two papers of the series simultaneously but separately because they are logically independent of each other.

They clearly form only the beginning of the study of units in the modular function field. The present introduction serves both parts. New introductions will be written for subsequent papers of this series, dealing with units and divisor classes at infinity on the modular curves.

Some of the program to be developed is already clear. First, we must analyse the projective limit of the representation of the group of automorphisms of the modular function field (of all levels, and especially prime power level) on the units and divisor classes. This is a 2-dimensional generic analogue for GL_2 (and its non-split unramified Cartan) of the 1-dimensional theories (with respect to $\mathbf{Z}_p$) centering around the Iwasawa theory, the Kubota-Leopoldt zeta functions, the formulation by Mazur of the appropriate p-adic measures, the values of zeta functions at negative integers, Hecke operators. Note that the "Stickelberger element" which annihilates ideal classes in cyclotomic fields is related to and has analogies with certain "Hecke operators" which annihilate the divisor classes at infinity on the modular curves.

Second, having the description of the generic situation, we must specialize to number fields, to the case of non-integral j-invariants. One must then prove an irreducibility theorem. We know Serre's results for the non-degeneracy of the representation of the Galois groups of division points on the Deuring-Tate module associated with division points. A corresponding theorem must be proved for the specialization of the divisor classes at infinity into the ideal classes. The generic Stickelberger element specializes to one in number fields of division points. Problems of uniformity will arise with respect to the three variables, j, p, p^n ($n \to \infty$). Several threads which were developed separately in the meantime come together at this point, and the diophantine role of the units, as for instance in the first paper, should be important later for the irreducibility proofs.

In writing the first two papers, we have kept in mind the need for their ultimate applications, and possible interaction. We therefore try to write them and subsequent ones so that logically independent parts can be read in a self contained manner, but also such that cross references can easily be established and understood.

I. Diophantine Applications

Contents

§ 1. Unramified correspondences 70
§ 2. Construction of units such that $u+v=1$ 77
§ 3. Construction of units and divisor classes for prime and prime power level 82
§ 4. Specializations of divisors and functions at infinity 84
§ 5. Specialization to complex multiplication 91

For basic definitions of the modular function field F_N, we refer the reader to Shimura [Sh] or Lang [L1]. The modular function field is that of the modular curve over the field of N-th roots of unity, and therefore by definition contains ζ_N. We let throughout R_N denote the integral closure of $\mathbf{Z}[j]$ in F_N, and R_N^∞ the integral closure of $\mathbf{Z}[1/j]$ in F_N.

§ 1. Unramified Correspondences

The Fermat-Thue-Siegel (Fermat for short) curve has played a basic role in the theory of integral points, see for instance Siegel [S], Lang in similar connections [L3], and following Gelfond [G], Baker [B], Baker-Coates [B-C], Coates [Co]. In these papers, it arises through the Siegel identity

$$\frac{x_3-x_1}{x_2-x_1}\frac{t-x_2}{t-x_3}+\frac{x_2-x_3}{x_2-x_1}\frac{t-x_1}{t-x_3}=1. \tag{1}$$

In the present paper, we shall use the simpler identity

$$\frac{x_3-x_1}{x_2-x_1}+\frac{x_2-x_3}{x_2-x_1}=1, \tag{2}$$

which is for instance satisfied by the λ-function in the theory of elliptic functions. We discuss this in detail at the end of this section. We put

$$u=\frac{x_3-x_1}{x_2-x_1} \quad \text{and} \quad v=\frac{x_2-x_3}{x_2-x_1},$$

so that

$$u+v=1.$$

Let $x^n=u$ and $y^n=v$. Then x, y satisfy the Fermat equation

$$x^n+y^n=1,$$

defining the Fermat curve Φ.

Units in the Modular Function Field. I

Let R be a finitely generated ring of transcendence degree 1 over $\mathbf{Z}$ and let K be its quotient field. Let $V = \operatorname{spec} R$, and let k be the algebraic closure of $\mathbf{Q}$ in K. We may view V as an affine variety, defined over k, and $K = k(V)$ as the function field of V over k. We shall also assume that R is integrally closed. If x_1, x_2, x_3 above are elements of K, then

$$k(u) = k(v) = k(u, v)$$

is a subfield of K, defining a rational map of the curve V into the projective line $\mathbf{P}^1$, that is the u-line. Let

$$W = \operatorname{spec} R[x, y].$$

Then we have a natural map

$$f: W \to V,$$

and W gives a correspondence between V and the Fermat curve,

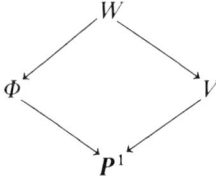

This correspondence is especially important in cases when we have additional information of an arithmetic nature about the functions u and v. For instance, let us assume known that the Fermat curve for some $n \geq 3$ has only a finite number of points in any subring of a number field finitely generated over $\mathbf{Z}$ (a special case of Siegel's theorem, or rather its generalization by Mahler-Lang, considerably easier to prove than the general case).

If there exist units u, v in R such that $u + v = 1$, then the affine curve V has only a finite number of points in any subring $\mathfrak{o}$ of k finitely generated over $\mathbf{Z}$.

We prove this from the corresponding property of the Fermat curve as follows. We take an integer $n \geq 3$. For any homomorphism

$$\varphi: z \mapsto \bar{z}$$

of R into $\mathfrak{o}$, that is, any point of V in $\mathfrak{o}$, the images $\bar{u}, \bar{v}$ are units in $\mathfrak{o}$. Consequently for any extension of φ to a homomorphism of $R[x, y]$, the images $\bar{x}, \bar{y}$ are units in the integral closure of $\mathfrak{o}$. Let $E = k(\bar{x}, \bar{y})$. Then the degree

$$[k(\bar{x}, \bar{y}) : k]$$

is bounded by the generic degree $[W : V]$. Furthermore, E is unramified over k except at the primes dividing n, and the primes $\mathfrak{p}$ of k at which the elements of $\mathfrak{o}$ are not $\mathfrak{p}$-integral. This is a finite set of primes. We use an elementary lemma of algebraic number theory (for a proof, see [L 1], Chapter 17, § 1).

Lemma. *Let k be a number field, let S be a finite set of primes in k, and let d be a positive integer. There is only a finite number of extensions of k of degree $\leq d$, unramified outside S.*

We conclude that the points $(\bar{x}, \bar{y})$ can lie only in a finite number of extensions of k. Therefore, if V has infinitely many points in $\mathfrak{o}$, then the Fermat curve has infinitely many points in the integral closure of $\mathfrak{o}$ in one of these extensions, a contradiction.

In view of the preceding result, a ring with units u, v such that $u + v = 1$ should have a special name, say a **Fermat ring**.

One essential feature in the above arguments was the bound on the ramification obtained by extracting n-th roots. In the context of rational points (rather than integral points), we can also use Weil's result [W], proved jointly with Chevalley for the case of curves, [C-W]:

Theorem on Unramified Extensions. *Let $f: W \to V$ be an unramified covering of a projective non-singular variety V by a variety W, defined over a number field k. There exists a positive integer c having the following property. For any point x of V in the algebraic closure of k, the relative discriminant of $k(f^{-1}(x))$ over $k(x)$ divides c.*

In particular, for any rational point x of V in k, the extension $k(f^{-1}(x))$ ramifies only in a fixed finite set of primes of k, and its degree is bounded. Thus the lemma applies, to show that there is only a finite number of such extensions, and we conclude:

Corollary 1. *If V has infinitely many rational points in k, then W has infinitely many rational points in some finite extension of k.*

Let us say that V is **Mordellic** if it has only a finite number of rational points in any number field.

Corollary 2. *V is Mordellic if and only if W is Mordellic.*

Remark 1. In the Chevalley-Weil Theorem, if $\mathfrak{p}$ is a prime of k such that the unramified covering $f: W \to V$ has non-degenerate reduction mod $\mathfrak{p}$, to an unramified covering

$$f_\mathfrak{p}: W_\mathfrak{p} \to V_\mathfrak{p},$$

then for any point $x \in V_k$ the extension $k(f^{-1}(x))$ is unramified above $\mathfrak{p}$. Thus the set of primes dividing c can be determined explicitly if one knows the covering explicitly, and so can the finite number of extensions unramified outside the given finite set.

Remark 2. We recall that in any finitely generated multiplicative group of complex numbers, there exists only a finite number of elements u, v such that $u + v = 1$, cf. [L2]. Thus the existence of such units in a natural way is always a remarkable event.

Example 1. Elliptic Curves and Superelliptic Curves

Chabauty [Ch] proved Siegel's theorem (finiteness of integral points) for elliptic curves by this kind of method. To do this, he constructs a correspondence between an elliptic curve A and the Fermat curve, such that the correspondence is unramified over A except at infinity. The proof is easy, and can be phrased in terms of

Units in the Modular Function Field. I

"units" as follows. Let P, Q be two distinct points of order exactly two on A. The covering "multiplication by 2" of A onto itself is unramified, and the inverse image of the origin 0 by this covering consists of the points of order 2 on A. Let A be defined over a number field k over which these points are rational. Let $R = k[A]$ be the affine coordinate ring of A over k consisting of all functions having poles only at 0. (Over the complex numbers, it is the ring generated by the Weierstrass functions.) Let R_2 be the integral closure of R in the function field of the covering. Let u, v be functions in R_2 such that their divisors are of the form

$$(u) = 2(P) - 2(O) \quad \text{and} \quad (v) = 2(Q) - 2(O).$$

(Explicitly, say $u = \wp - e_1$ in classical notation.)

Then u, v are units in R_2. The linear system $L(2(O))$ has dimension 2, and contains the constants, as well as the functions u and v. Consequently there exist constants $a, b \neq 0$ such that

$$u + av = b.$$

Replacing u and v by constant multiples, we may assume without loss of generality that

$$u + v = 1.$$

We are now in the general situation discussed previously, since to deal with integral points, we may replace the finitely generated algebra over k by a finitely generated ring over $\mathbf{Z}$ containing the functions u, v. The finiteness of integral points on the covering now follows from the existence of these units, and the finiteness on the base elliptic curve A follows from the theorem on unramified extensions recalled above.

Observe that the reduction of the diophantine problem to the Fermat curve $x^3 + y^3 = 1$ is effective. Thus this method of proof is effective if the Fermat curve can be handled in an effective way.

Another argument is used by Gelfond [G], Baker and Coates [B-C], who deal with Siegel's identity (1) instead of (2), and make all arguments explicit. The Chabauty paper has not been referred to in recent times, perhaps because no one was able to understand the rest of the paper. Chabauty claims to prove Siegel's theorem for general curves by a similar method. Indeed, he claims to prove that given any projective non-singular curve V, there exists a rational map $V \to \mathbf{P}^1$ on the projective line, and an elliptic curve $A \to \mathbf{P}^1$ such that the pull back of A to V is unramified of degree 2 over V. Unfortunately, this second step is invalid. If it were valid, the Jacobian of the unramified covering would contain an elliptic curve. Over the complex numbers, generically if W/V is an unramified covering of degree 2, then the quotient of the Jacobian varieties J_W/J_V is simple, and thus cannot contain an elliptic curve. If W is the maximal unramified abelian covering of V of exponent 2, then the Jacobian variety of W is in general isogenous to a product of J_V with such simple factors, so that Chabauty's assertion cannot hold. We are indebted to Mumford and Clemens for giving us the references to the literature [Cl], [Mu], [Re], [Ro]. The quotients of Jacobians have been studied classically under the name "Prym varieties".

The question then arises whether for some abelian (or even non-abelian) unramified covering W of V the factor J_W/J_V contains an elliptic curve. This would suffice for diophantine applications as above. We know of no evidence suggesting an affirmative answer in general, and the case of 2-coverings suggests a negative answer.

On the other hand, it is true for a lot of the curves which have been studied classically, e.g. hyperelliptic curves. More generally, consider the **superelliptic** curve V defined by the equation

$$y^n = f(x) = (x - a_1)\ldots(x - a_r).$$

For instance, $y^n - x^r = 1$ is of this type. We assume for simplicity that the numbers a_i are distinct. Suppose that n, r are distinct prime numbers, and $n \geq 3$. Then the curve V' defined by

$$y^n = (x - a_1)(x - a_2)$$

is ramified only at a_1, a_2 and infinity, of the same order as V. Hence the pull back $V' \otimes V$ is unramified over V. Furthermore, V' is hyperelliptic since x is quadratic over the function field in y. The same idea as above in the hyperelliptic case (when $n=2$, $r \geq 4$ we use four roots instead of two) yields an elliptic curve A such that $A \otimes V'$ is unramified over V'.

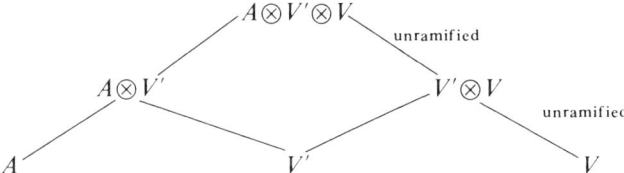

The notion of unramified correspondence is probably sufficiently useful and widespread to have a notion. If V, V' are two curves (always assume complete non-singular), let us introduce the partial ordering V **dominates** V',

$$V' \prec V$$

if there exists an unramified covering W of V which is also a possibly ramified covering of V', so that we have the picture

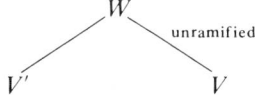

We can summarize our previous discussion by the statement:

If V is hyperelliptic, or superelliptic as above, then V dominates an elliptic curve.

For such curves, we then have the integral diophantine statement:

Integral Domination Theorem. *Let V be a curve which dominates an elliptic curve, over a number field k. Let V_0 be an affine subset of V. Then V_0 has only a finite number of points in any finitely generated subring of k, if $x^3 + y^3 = 1$ does.*

185

Units in the Modular Function Field. I

Proof. Let

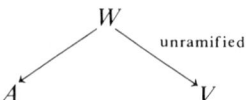

be the domination diagram. Let W_0 be the inverse image of V_0 in W, so that W_0 is affine. Let A_0 be the image of W_0 in A, so that A_0 is also affine. We may take the origin of A to be a point at infinity with respect to A_0. Then, as we have seen by the Chabauty argument, there is an unramified covering

$$A' \to A$$

(multiplication by 2) such that we can find units u, v in the affine ring A_0' satisfying $u+v=1$. As we have already seen, this reduces the integral diophantine statement to the Fermat curve, and concludes the proof. Observe that the reduction is effective.

Even though the method of unramified correspondences fails for general curves, another question arises, whether the desired type of correspondence exists with the Fermat curve, independently of any elliptic curve. In other words, given a projective non-singular curve V and an affine open subset V_0, does there exist an unramified covering W of V, and units u, v in the corresponding affine coordinate ring of W, such that $u+v=1$? Whenever this is the case, we get a correspondence

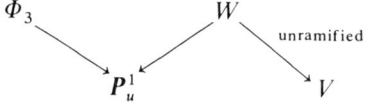

which reduces the study of integral points on V to integral points on Φ_3. By the results of the next example and the existence of appropriate units on the modular curve proved in § 2, we see that the Fermat curve Φ_n has such a correspondence with Φ_3.

Example 2. Modular Curves

Let F_N be the modular function field of level N over the rational numbers, containing the N-th roots of unity. (For the definition, cf. Shimura [Sh] and Lang [L 1].) For the modular curves (models of the modular function fields), we shall construct units u, v in § 2 satisfying $u+v=1$. Therefore we get an independent proof of the following fact.

Let $Y = \mathrm{spec} R_N$, where R_N is the integral closure of $\mathbf{Z}[j]$ in the modular function field F_N, for $N \geq 3$. Then Y has only a finite number of points in $\mathfrak{o}$.

This theorem is used in some proofs of Shafarevic's theorem concerning the finiteness of isomorphism classes of elliptic curves over a number field having good reduction at all but a finite number of primes (cf. [L 1], Chapter 17, § 1).

Our construction joins a train of thought which appeared in the work of Demyanenko [D] and Kubert [K] in their studies of rational points on modular curves, with that of constructing units in the modular function fields [L1], Chapter 18, § 6.

The simplest case of level 2 has especially interesting features. Let $k = \mathbf{Q}(e^{2\pi i/n})$ and again let

$$u = x^n, \quad v = y^n,$$

so that $k(u) = k(u, v)$ is the fixed field of the function field $k(x, y)$ under the group of automorphisms

$$(x, y) \mapsto (\zeta x, \zeta' y),$$

where ζ, ζ' are n-th roots of unity. We now put

$$u = \lambda = \frac{e_2 - e_3}{e_1 - e_3},$$

where λ is the generator of the modular function field of level 2 in the theory of elliptic functions, and e_i ($i = 1, 2, 3$) has the usual meaning, cf. [L1], Chapter 18, § 6. Let

$$\mathbf{P}_u^1 = \mathbf{P}_\lambda^1$$

be the projective u-line (i.e. λ-line), so that we have rational maps

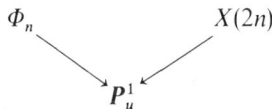

from the Fermat curve and the modular curve $X(2n)$ onto this line. Note that the points $u = 0, 1, \infty$ lie above $j = \infty$. It is obvious that the Fermat curve is ramified of order n above $u = 0, 1, \infty$, and that this is its only ramification over the projective line. It is known from the ramification theory of elliptic functions that $X(2n)$ has exactly the same ramification properties. Therefore we conclude that

the pull back of the Fermat curve Φ_n over $X(2n)$ is unramified over each one of these curves.

We believe that, except for a few low values of n, the fundamental groups of such pullbacks are non-congruence subgroups. The group belonging to $u^{1/n}$ has been considered classically, cf. [K-F], p. 658. Even though Klein and Fricke seem to use the phrase "congruence subgroup" in a sense different from the current one, it seems nevertheless that they prove that the group of $\lambda^{1/n}$ is not a congruence subgroup in the modern sense. We shall return to this question in a later paper of this series.

The Fermat conjecture here appears in a natural context ("the only rational points over $\mathbf{Q}$ should be at cusps"). Furthermore, the fact the pull backs above are unramified shows that:

The Fermat curve Φ_n is Mordellic if and only if the modular curve $X(2n)$ is Mordellic.

The proof consists of the lemma already mentioned together with the theorem on unramified extensions.

Thus we see that the diophantine properties of the Fermat curve and the modular curve affect each other. It is however not clear if ultimately the Mordell conjecture can be proved by knowing enough about the rational points of modular curves, or if conversely, diophantine properties of the modular curve can be reduced to those of the Fermat curve as in Demyanenko and Kubert. Note that an effective solution of the Mordell conjecture implies that the Fermat conjecture over Q can then be verified for each n, but a proof for all n simultaneously involves additional difficulties, having to do with uniform irreducibility.

§ 2. Construction of units such that $u+v=1$

Let F_N be the modular function field of level N over Q. (Cf. Shimura [Sh] and Lang [L 1].) By assumption, F_N contains the N-th roots of unity. We let

$$\zeta = \zeta_N = e^{2\pi i/N}$$

be the primitive N-th root of unity. We let R_N be the integral closure of $Z[j]$ in the modular function field F_N, and we let QR_N be the integral closure of $Q[j]$ in F_N. Let $q = q_\tau = e^{2\pi i \tau}$.

We shall be interested in constructing units in R_N, or $R_N[1/N]$. An element of R_N has a $q^{1/N}$-expansion

$$\sum \xi_\nu q^{\nu/N},$$

with ξ_ν in the ring of algebraic integers of $Q(\zeta)$. Its conjugates over $Q(j)$ have similar expansions. Such an element is a unit in R_N if and only if each conjugate has no zero or pole on the upper half plane, and the lower term of the q-expansion is equal to a unit in $Q(\zeta)$ times a (fractional) power of q. Indeed, by the q-expansion principle, the elementary symmetric functions of such an element are elements of $Z[j]$, and similarly for the inverse of such an element. We shall construct units by verifying for specific functions that their q-expansion satisfy the above condition.

Our first construction of units uses differences of Weierstrass functions, fitting the Eq. (2) of § 1. Later in II, we construct more generally units from the sigma function.

Let f_0 be the normalized Weber function

$$f_0(z;\tau) = f_0(z;[\tau,1]) = -2^7 3^5 \frac{g_2 g_3}{\Delta} \wp(z;\tau),$$

so that f_0 is homogeneous of degree 0, and has a q-expansion

$$f_0 = PS_z(q)\left[1 + \frac{12 q_z}{(1-q_z)^2} + 12 \sum_{m,n=1}^{\infty} n q^{mn}(q_z^n + q_z^{-n} - 2)\right] \tag{2.1}$$

where:

$$q = q_\tau = e^{2\pi i \tau}$$

$$q_z = e^{2\pi i z} .$$

$PS_z(q)$ is a power series in q with integer coefficients, and starting with q.

We fix a positive integer $N \geq 3$ and for any element

$$a = (a_1, a_2) \in \mathbf{Q}^2/\mathbf{Z}^2$$

with precise denominator N we let

$$f_a(\tau) = f_0(a_1\tau + a_2; \tau) = f_0(a_1\tau + a_2; [\tau, 1]) .$$

Thus $f_a(\tau)$ for generic τ is a normalized x-coordinate on "the" generic elliptic curve having invariant $j(\tau)$.

Theorem 1. *Let $a, b, c, d \in \mathbf{Q}^2/\mathbf{Z}^2$ have precise denominator N. Assume that*

$$a \not\equiv \pm b \quad \text{and} \quad c \not\equiv \pm d \quad (\mathrm{mod}\, \mathbf{Z}) .$$

Then

$$u = \frac{f_a - f_b}{f_c - f_d}$$

is a unit in QR_N.

Proof. The following assertions are obvious from [L1], Chapter 6, §§ 1–3. The conjugates of u over $Q(j)$ are of the same form, namely

$$\frac{f_{a'} - f_{b'}}{f_{c'} - f_{d'}} .$$

Since both u and u^{-1} are holomorphic and nowhere zero on the upper half plane, the elementary symmetric functions of u and its conjugates over $Q(j)$ are polynomials in j with rational coefficients, so that u and u^{-1} are integral over $Q[j]$, whence u is a unit in QR_N. (Easily verified to be units in $R_N[1/N]$.)

We now want to see that in certain cases, u is even a unit in R_N. For this, it is convenient to switch notation to free some letters. Thus we let (r, s) be a pair of integers such that

$$(r, s, N) = 1 ,$$

and we let

$$f_{r,s}(\tau) = f_0\left(\frac{r\tau + s}{N}; \tau\right) .$$

This value depends only on the residue classes of $r, s \bmod N$, and so we can pick $0 \leq r, s \leq N-1$.

Theorem 2. *Let a, b, c, d be integers prime to N, and assume that $a \pm b$ and $c \pm d$ are also prime to N. Then the element*

$$u = \frac{f_{a(r,s)} - f_{b(r,s)}}{f_{c(r,s)} - f_{d(r,s)}}$$

Units in the Modular Function Field. I

is a unit in R_N. Its q-expansion has coefficients in $\mathbf{Z}[\zeta]$, and begins with a unit in $\mathbf{Z}[\zeta]$ times a (fractional) power of q.

Proof. It suffices to prove this last property for every conjugate of u over $\mathbf{Q}(j)$. On the other hand, it follows at once from [L 1], Chapter 6, §§ 1–3 that the conjugates of u over $\mathbf{Q}(j)$ are of the same form as u, namely

$$\frac{f_{a(r',s')} - f_{b(r',s')}}{f_{c(r',s')} - f_{d(r',s')}},$$

with the same a, b, c, d and another pair (r', s') primitive mod N. Thus we shall investigate the lowest coefficient of each difference

$$f_{ar,as} - f_{br,bs}.$$

Note that $f_{ar,as}$ depends only on the residue classes of ar and as mod N. We put

$$z = \frac{ar}{N}\tau + \frac{as}{N}$$

in the q-expansion for f_0. It is clear that the coefficients of powers of $q^{1/N}$ are elements of $\mathbf{Z}[\zeta]$. We may also interpret ar/N as the smallest positive fraction representing its residue class mod $\mathbf{Z}$, and similarly when dealing with b, c, d instead of a.

Observe that in the quotient expressing u, the power series $PS_Z(q)$ of (3) cancels out. Taking differences in the numerator and denominator also cancels out the constant 1 as well as the sum

$$\sum_{m,n=1}^{\infty} -2nq^{mn}.$$

The factor of 12 than also disappears in taking the quotient. Thus in forming the element u, the only part of the function $f_{a(r,s)}$ which will ultimately give a contribution is

$$f^*_{ar,as} = \sum_{n=1}^{\infty} nq_z^n + \sum_{m,n=1}^{\infty} nq_\tau^{mn}(q_z^n + q_z^{-n}),$$

where

$$z = \frac{ar}{N}\tau + \frac{as}{N},$$

and ar/N, as/N are interpreted as the smallest positive rational numbers representing their residue classes mod $\mathbf{Z}$.

We now distinguish cases. For the first two, we assume $r \not\equiv 0 \pmod{N}$.

Case 1. $2r \equiv 0 \pmod{N}$. This is equivalent to $N = 2r$, having selected r to be an integer between 0 and $N-1$. Then N is even, and a, b, c, d are odd. The residue classes

$$\frac{ar}{N}, \frac{br}{N}, \frac{cr}{N}, \frac{dr}{N}$$

are therefore all congruent to $\frac{1}{2}$ (mod $\mathbf{Z}$). For $f^*_{a(r,s)}$ we therefore have the q-expansion

$$\sum_{n=1}^{\infty} nq^{n/2}\zeta^{nas} + \sum_{m,n=1}^{\infty} nq^{mn}(q^{n/2}\zeta^{nas} + q^{-n/2}\zeta^{-nas}).$$

Combining terms, this can be rewritten

$$\sum_{n=1}^{\infty} \sum_{m=0}^{\infty} nq^{n/2}q^{mn}(\zeta^{nas} + \zeta^{-nas}).$$

Therefore in the present case, we obtain

$$f^*_{a(r,s)} - f^*_{b(r,s)} = \sum_{n=1}^{\infty} \sum_{m=0}^{\infty} nq^{n/2}q^{mn}(\zeta^{nas} - \zeta^{nbs} + \zeta^{-nas} - \zeta^{-nbs}). \tag{2.2}$$

Put

$$t = \zeta^s.$$

We have the identity

$$t^{na} - t^{nb} + t^{-na} - t^{-nb} = t^{na}(1 - t^{-n(b+a)})(1 - t^{n(b-a)}). \tag{2.3}$$

The leading term of (2.2) is obtained when $n=1$ and $m=0$, and is therefore equal to

$$q^{1/2}(\zeta^{as} - \zeta^{bs} + \zeta^{-as} - \zeta^{-bs}). \tag{2.4}$$

It has the factorization (2.3) with $n=1$. Every coefficient in the power series of (2.2) is divisible by this first term in $\mathbf{Z}[\zeta]$. We may therefore factor it out. Taking the quotient

$$u = \frac{f^*_{a(r,s)} - f^*_{b(r,s)}}{f^*_{c(r,s)} - f^*_{d(r,s)}}$$

and recalling that all elements $1 - \zeta^v$, with v prime to N, are associates in $\mathbf{Z}[\zeta]$, i.e. their quotients are units, we conclude:

In the present case, the q-series for u has coefficients in $\mathbf{Z}[\zeta]$, and its lowest term is equal to a unit in $\mathbf{Z}[\zeta]$.

Case 2. $2r \not\equiv 0 \pmod{N}$. We use the fact that $(1-q_z)^2$ is invertible in $\mathbf{Z}[\zeta][[q^{1/N}]]$. Consequently the leading term of

$$f^*_{a(r,s)} - f^*_{b(r,s)}$$

is

$$q^{ar/N}\zeta^{as} + q^{1-ar/N}\zeta^{-as} - q^{br/N}\zeta^{bs} - q^{1-br/N}\zeta^{-bs}. \tag{2.5}$$

Since $2r \not\equiv 0 \pmod{N}$, the four terms in this last expression necessarily have different powers of q, and the coefficient of each power of q is a root of unity. Hence the lowest term is a power of q whose coefficient is a root of unity. Thus the numerator of u is a power series of the desired type, and the denominator also by the same argument. More precisely, we can describe the lowest term.

Let e be the smallest positive residue class mod $\mathbf{Z}$ *of*

$$\frac{ar}{N}, 1 - \frac{ar}{N}, \frac{br}{N}, 1 - \frac{br}{N}.$$

191

Let e' be the smallest positive residue class mod $\mathbf{Z}$ of

$$\frac{cr}{N},\ 1-\frac{cr}{N},\ \frac{dr}{N},\ 1-\frac{dr}{N}.$$

Then the lowest term of the q-series for u in the present case is

$$q^{e-e'}$$

times a root of unity.

If $r \not\equiv 0 \pmod{N}$, then either $2r \equiv 0 \pmod{N}$ or $2r \not\equiv 0 \pmod{N}$. Hence there remains one case to consider.

Case 3. $r \equiv 0 \pmod{N}$. Then s is prime to N, and

$$f^*_{0,as} - f^*_{0,bs} = \frac{\zeta^{as}}{(1-\zeta^{as})^2} - \frac{\zeta^{bs}}{(1-\zeta^{bs})^2} + \text{higher degree term}$$

where the higher degree terms have coefficients in $\mathbf{Z}[\zeta]$. Put

$$t = \zeta^s.$$

Then the constant term in the present case is

$$Q(a,b) = t^b \frac{(t^{a-b}-1)(t^{a+b}-1)}{(1-t^a)^2 (1-t^b)^2}, \tag{2.6}$$

and therefore u has an expression

$$u = \frac{Q(a,b) + \ldots}{Q(c,d) + \ldots}$$

where the other terms in the numerator and denominator besides $Q(a,b)$ and $Q(c,d)$ are power series in $q^{1/N}$ with coefficients in $\mathbf{Z}[\zeta]$. The algebraic numbers

$$1-t^a,\ 1-t^b,\ 1-t^c,\ 1-t^d$$

are all associate, i.e. their quotients are units. Multiplying both numerator and denominator by

$$(1-t^a)(1-t^b)$$

does not change u, and shows that u is a quotient of elements in

$$\mathbf{Z}[\zeta]((q^{1/N})),$$

whose leading coefficient is a unit in $\mathbf{Z}[\zeta]$. Therefore:

In the present case, the q-series for u begins with a unit in $\mathbf{Z}[\zeta]$.

This concludes the proof of Theorem 2.

Remark 1. By expressing the differences of Weierstrass functions by means of the sigma function, one can also give a proof of Theorem 2 by proving that certain quotients of sigma functions are units in R_N. We shall do this in the second part.

Remark 2. Some of Demyanenko's and Brylinski's results can also be proved by using units. Indeed, instead of looking functions on the elliptic curve one looks at the corresponding modular function. For instance let N be odd, let (r, s) be primitive mod N, and let (p, q) be primitive mod 2. Let

$$\psi_{r,s} = \frac{\wp'\left(\frac{r\tau+s}{N}; \tau\right)}{\wp\left(\frac{r\tau+s}{N}; \tau\right) - \wp\left(\frac{p\tau+q}{2}; \tau\right)}.$$

Then $u = \psi_{r,s}/\psi_{2r,2s}$ and $1-u$ are units in R_N, as is immediately verified by the same method. One can write $\psi_{r,s}$ in an abbreviate way as, say, $\wp'/(\wp - e_1)$ evaluated at a point of odd order on the elliptic curve. What is involved here is a commutative diagram, which on one hand goes through the generic elliptic curve over the modular curve, and then to the special fiber over a number field; on the other hand, restricts at once to the modular curves and then to the number field. The latter avoids entirely the Néron models, using instead the simple computations of orders of q-expansions. It is therefore much easier to handle.

§ 3. Units and Divisor Classes for Prime and Prime Power Level

In this section we carry out a similar construction for prime level, but such that the units obtained have a divisor at infinity which is the N-th multiple of a divisor, thus giving rise to a divisor class of period N on $X(N)$, when $N = p$ is a prime.

On a generic elliptic curve A with invariant $j(\tau)$ we consider pairs (C, P) consisting of a cyclic subgroup C of order N and a point P of order N, also of order N mod C. Given two such pairs, (C, P) and (C', P'), there exists $\alpha \in GL(A_N)$ such that

$$\alpha C = C' \quad \text{and} \quad \alpha P = P'.$$

We let h be the Weber function satisfying the following condition. If $A_C \approx C/[\tau, 1]$ and P is the point on A_C represented by $\frac{r\tau+s}{N}$ in C, then

$$h_A(P) = f_0\left(\frac{r\tau+s}{N}; [\tau, 1]\right).$$

We may form the function

$$h_{A/C}(P_C),$$

where P_C is the residue class of P in A/C, which we abbreviate by $h_{A/C}(P)$. If C corresponds to the lattice $L \supset [\tau, 1]$ such that

$$L/[\tau, 1]$$

is cyclic of order N, then

$$h_{A/C}(P) = f_0\left(\frac{r\tau+s}{N}; L\right).$$

Such lattices L are easily classified, and a generating point P also. They are of two types:

Case 1. $L=[\tau, 1/N]$, P represented by $r\tau/N$, with $(r, N)=1$.

Case 2. $L=\left[\dfrac{\tau+v}{N}, 1\right]$, P represented by s/N, with $(s, N)=1$, and $0\leq v\leq N-1$.

Let a be an integer prime to N. Then the conjugates of

$$h_{A/C}(aP)$$

over $\mathbf{Q}(j)$ are of the same form, namely

$$h_{A/C'}(aP'),$$

if the pair (C', P') is conjugate to (C, P). In other words, for any automorphism σ of F_N over $\mathbf{Q}(j)$ we have

$$(h_{A/C}(aP))^\sigma = h_{A/\sigma C}(aP^\sigma),$$

because we can take A defined over $\mathbf{Q}(j)$, fixed by σ.

Theorem 3. *Let a, b, c, d be integers prime to the prime number N, such that $a\pm b$ and $c\pm d$ are prime to N. Let r be prime to N. Let $f = f_0$ be the Weber function. Let*

$$u = \frac{f(ar\tau; N\tau) - f(br\tau; N\tau)}{f(cr\tau; N\tau) - f(dr\tau; N\tau)}.$$

Then u is a unit in R_N, and its conjugates over $\mathbf{Q}(j)$ are the elements

$$u_{C,P} = \frac{h_{A/C}(aP) - h_{A/C}(bP)}{h_{A/C}(cP) - h_{A/C}(dP)},$$

for all pairs (C, P) as above. In case 1, the expansion of $u_{C,P}$ at infinity has no fractional powers of q. In case 2, it begins with a unit in $\mathbf{Z}[\zeta]$. The divisor of u at infinity on $X(N)$ is the N-th multiple of a divisor.

Proof. Let $\tau' = N\tau$. Then the same argument as in § 2 shows that u is integral over $\mathbf{Z}[j(\tau')] = \mathbf{Z}[j(N\tau)]$. Since $j(N\tau)$ is integral over $\mathbf{Z}[j(\tau)]$, it follows that u is integral over $\mathbf{Z}[j]$, whence u is a unit in R_N, because u and u^{-1} have the same form.

As to the orders at infinity: In case 1, it is clear from the q-expansion of f_0 recalled in (2.1) that no fractional power of q occurs in the q-expansion of u. In case 2, one sees at once by a computation completely similar to that of Theorem 2, case 3, that the q-expansion of the corresponding conjugate of u begins with a unit. In particular in this case, the order of this conjugate at the infinite prime is 0.

Since $X(N)$ is ramified of order N over $X(1)$ at infinity, it follows that the order of u at every infinite prime on $X(N)$ is divisible by N, thereby proving Theorem 3.

In an analogous fashion one can also construct units in the prime power case. We recall that $\Gamma_1(N)$ is the subgroup of $SL_2(\mathbf{Z})$ consisting of matrices

$$\begin{pmatrix} 1 & b \\ 0 & 1 \end{pmatrix}$$

and that $X_1(N)$ is the complete curve having $\Gamma_1(N)\backslash\mathfrak{H}$ as affine open subset. We let $R_1(N)$ be the integral closure of $\mathbf{Z}[j]$ in the function field (over $\mathbf{Q}$) of $X_1(N)$.

Theorem 4. *Let $a, b, c \in \mathbf{Q}^2/\mathbf{Z}^2$ have denominators which are a power of a prime p, and such that $a, b, c, a \pm b, a \pm c, b \pm c$ have exact period p^{n-k} with $0 < k < n$. Let $N = p^n$ and let*

$$u = \frac{f_a - f_b}{f_a - f_c}.$$

Then u is a unit in $R_1(N)$, and the divisor of u on $X_1(N)$ is a p^k-multiple of some divisor on $X_1(N)$.

Proof. Similar to all the preceding proofs, and omitted.

Remark. Over the rationals, viewing $X(N)$ as defined over $\mathbf{Q}$, one sees that by using units whose divisors at infinity are N-th multiples, then the correspondence with the Fermat curve shows that if the modular curve has infinitely many rational points, so does the Fermat curve. Indeed, since the group of units is trivial, when extracting the N-th roots of an N-th power, we still find a rational number if N is an odd prime. See Theorem 7 of the next section.

§ 4. Specializations of Divisors and Functions at Infinity

In this section we analyse what happens to a function in the modular function field when specialized at a point of the modular curve where the j-invariant is not integral.

We return to an arbitrary positive integer N. We consider the group of units R_N^* in R_N. We let R be the integral closure of $\mathbf{Z}[j]$ in F_N, and R^∞ the integral closure of $\mathbf{Z}[1/j]$ in F_N. We let *in this section* $V(N)$ be the scheme whose function field is F_N, given by

$$V(N) = \mathrm{spec}(R) \cup \mathrm{spec}(R^\infty).$$

We let

$$\mathscr{D}^\infty$$

be the free abelian group on $V(N)$ generated by the minimal prime ideals in R^∞ lying above the ideal $(1/j)$ in $\mathbf{Z}[1/j]$. We call $\mathscr{D}^\infty$ the group of **divisors at infinity** on $V(N)$.

Remark. An element f of F_N has a $V(N)$-divisor in $\mathscr{D}^\infty$ if and only if f is a unit in R_N.

Proof. Suppose that the $V(N)$-divisor of f is at infinity. The ring R, which is Noetherian integrally closed, is equal to the intersection of the local (discrete valuation) rings

$$R = \bigcap_\mathfrak{p} R_\mathfrak{p}$$

taken over all minimal prime ideals $\mathfrak{p}$ in R. By assumption, f is a unit in each $R_\mathfrak{p}$, whence f lies in R. The same argument applied to $1/f$ shows that f is a unit in R. Conversely, if f is a unit in R, then its $V(N)$-divisor carries only minimal primes in R^∞ which do not contain any prime number [otherwise they would be represented also on spec(R)]. Hence these minimal primes lie over $(1/j)$ in $\mathbf{Z}[1/j]$ and are at infinity, thereby proving the remark.

The Tate curve over a complete local ring, rather than discrete valuation ring, was first considered in [L 4], cf. [L 1], Chapter XV. The question of regularity of the modular schemes is studied deeply in Deligne-Rapoport [De-Ra]. For the convenience of the reader, we describe the regularity at infinity, where we can give a very short and simple proof just from what is done in [L1], Chapter XV.

Theorem 5. *Let R^∞ be the integral closure of $\mathbf{Z}[1/j]$ in the modular function field F_N, for any positive integer N. Let M be a maximal ideal in R^∞ containing a prime number p and $1/j$. Then the local ring R_M^∞ is regular. In fact, let $\mathfrak{o}$ be the ring of algebraic integers in the cyclotomic field $\mathbf{Q}(\zeta_N)$ and let $\mathfrak{p} = M \cap \mathfrak{o}$. Let $\wedge$ denote completion. Then there is a natural isomorphism*

$$\hat{R}_M^\infty \cong \hat{\mathfrak{o}}_\mathfrak{p}[[q^{1/N}]].$$

Proof. The completion $\hat{R}_M^\infty$ is integrally closed by commutative algebra (e.g. EGA, Chapter IV, 7.8.3), and contains

$$\hat{\mathfrak{o}}_\mathfrak{p}[[1/j]] = \hat{\mathfrak{o}}_\mathfrak{p}[[q]].$$

The field of modular functions F_N can be identified with the field of "normalized" x-coordinates of N-th division points on the generic elliptic curve A with invariant j, as in Shimura [Sh] or [L1], p. 67. We let h_A be the normalized Weber function. Let K be the quotient field of $\hat{\mathfrak{o}}_\mathfrak{p}[[q]]$. Let B be the Tate curve as in [L1], Chapter XV, defined over K, also having invariant j. Then there exists an isomorphism

$$\lambda : A \to B$$

over some finite extension of K, uniquely determined up to ± 1 since End(A) = $\mathbf{Z}$. We have

$$h_B(\lambda a) = h_A(a), \quad \text{all} \quad a \in A_N.$$

Furthermore, $K(B_N) = K(q^{1/N})$. The integral closure of $\hat{\mathfrak{o}}_\mathfrak{p}[[q]]$ in this field of N-th division points $K(B_N)$ is therefore obviously equal to the power series ring

$$\hat{\mathfrak{o}}_\mathfrak{p}[[q^{1/N}]]$$

(which is integrally closed, and integral over $\hat{\mathfrak{o}}_\mathfrak{p}[[q]]$).

We conclude that the isomorphism λ induces an embedding of $\hat{R}_M^\infty$ into $\hat{\mathfrak{o}}_\mathfrak{p}[[q^{1/N}]]$. The only subfields of $K(q^{1/N})$ containing K are of type $K(q^{1/d})$ with $d|N$ by Kummer theory. Since the modular function field F_N is ramified of order N at infinity, it follows that $\hat{R}_M^\infty$ must be equal to $\hat{\mathfrak{o}}_\mathfrak{p}[[q^{1/N}]]$, thus proving the last assertion of the theorem.

In particular, a local ring is regular if and only if its completion is regular, so the first assertion also follows.

Let $V(N)^\infty$ be the subset of spec R^∞ consisting of those prime ideals containing $1/j$. We view the maximal ideal M in Theorem 5 as the maximal ideal of a closed point on $V(N)^\infty$, and we have the inclusion

$$\operatorname{spec} R_M^\infty \to V(N)^\infty.$$

A prime divisor on $V(N)^\infty$ passing through M is identified with a minimal prime of R_M^∞ containing $(1/j)$.

Corollary. *Given a point M as in the theorem, there exists a unique prime divisor on $V(N)^\infty$ passing through M, i.e. M contains a unique minimal prime containing $1/j$.*

Proof. It suffices to prove the assertion for the completion

$$\operatorname{spec} R_M^\infty = \operatorname{spec} \hat{\mathfrak{o}}_\mathfrak{p}[[q^{1/N}]].$$

In this case, it is clear that the ideal generated by $q^{1/N}$ is the unique minimal prime ideal of $\hat{\mathfrak{o}}_\mathfrak{p}[[q^{1/N}]]$ containing q (or $1/j$).

Just as in algebraic number theory, or the abstract theory of extensions of discrete valuations to an extension field, we can describe the (geometric) prime divisors at infinity also in the following manner. Let $Q_\zeta = Q(\zeta_N)$ be the cyclotomic field of N-th roots of unity over Q. The rational function field $Q_\zeta(j)$ is embedded canonically as a subfield of $Q_\zeta((1/j)) = Q_\zeta((q))$. The primes at infinity of F_N are in bijection with the extensions of this embedding to embeddings of F_N in $Q_\zeta((q^{1/N}))$, regarding two such extended embeddings as equivalent if they differ by an automorphism of $Q_\zeta((q^{1/N}))$ over $Q_\zeta((q))$. Such automorphisms are obviously of Kummer type, multiplying $q^{1/N}$ by an N-th root of unity. The (equivalence class) of embeddings inducing the same prime P will be called a P-**embedding**.

The closed point M on $V(N)^\infty$ will often be called an **arithmetic point**. If P is a prime divisor at infinity passing through M, then the ideal

$$PR_M^\infty$$

is principal because the local ring R_M^∞ is regular, and is generated by a prime element t which is called a local parameter at P. In geometric language, P is defined locally in a neighborhood of M by the equation

$$t = 0.$$

The maximal ideal M is generated by two elements,

$$M = (\pi, t),$$

and the point M is locally the intersection of the two hypersurfaces

$$\pi = 0 \quad \text{and} \quad t = 0,$$

which intersect transversally at M. A P-embedding of F_N in the power series field $Q_\zeta((q^{1/N}))$ determines a power series expansion for t, called the q-**expansion at P**,

$$t = c_1 q^{1/N} + \ldots.$$

The coefficients c_i lie in $Q(\zeta_N)$, $i = 1, 2, \ldots$.

Theorem 6. *Given a prime P at infinity, there exists a local parameter t in F_N whose q-expansion at P has coefficients c_i in $Z[\zeta_N]$, i.e. algebraic integers, and such that c_1 is a unit in $Z[\zeta_N]$.*

Proof. The modular function field F_N has a "standard" prime at infinity, determined by the complex q-expansions, and denoted by P_∞. Given any other prime P at infinity, there exists an element $\alpha \in SL_2(Z)$ such that $P = \alpha P_\infty$. In other words, two primes at infinity are conjugate under the action of the Galois group of F_N over F_1. It suffices to prove that there exists a function $f \in F_N$ whose q-expansion at P_∞ has the desired form,

$$f_{P_\infty}(q) = \sum c_i q^{i/N},$$

with c_i integral, and c_1 equal to a unit, for then we can take

$$t = f^{\alpha^{-1}}$$

as the desired parameter at P.

The construction of f is then obvious. The function

$$u = \frac{f_{1,1} - f_{1,0}}{f_{0,1} - f_{1,0}}$$

is immediately seen to have a q-expansion of the form

$$\zeta^{-1}(\zeta - 1)q^{1/N} + \text{higher terms},$$

and the coefficients of the higher terms are all integral, divisible by differences $\zeta^n - 1$ or $\zeta^{-n} - 1$, which in turn are divisible by $\zeta - 1$. Thus the function

$$f = (\zeta - 1)^{-1} u$$

has the desired properties.

Remark. For level $N = 2$, this is the significance of the factor 2 which appears in the classical formula, reproduced as $E_{3,1}^{1/4}$ of [L1], Chapter 18, § 4, p. 251.

Let $k \supset \mathbf{Q}(\zeta_N)$. A point of $V(N)$ into a number field k, not at infinity, is identified with a homomorphism

$$\varphi: R = R_N \to k.$$

At a given prime $\mathfrak{p}$ in k we have the local ring $\mathfrak{o}_\mathfrak{p}$, whence a map

$$\mathrm{spec}(\mathfrak{o}_\mathfrak{p}) \to X(N).$$

By pull back we can define a homomorphism from the divisors on $X(N)$ into the group of divisors of $\mathrm{spec}(\mathfrak{o}_\mathfrak{p})$, essentially the cyclic group generated by $\mathfrak{p}$. We describe this pull back completely in the following elementary manner.

For w in the local ring of φ, denote $\varphi(w)$ by $\bar{w}$. Let P be a prime divisor at infinity. Let $\mathfrak{p}$ be a prime ideal in k. If $\bar{j}$ is integral at $\mathfrak{p}$, we let

$$m(\varphi, \mathfrak{p}, P) = 0.$$

If $\bar{j}$ is not integral at $\mathfrak{p}$, we consider the composite homomorphism

$$R^\infty \to \bar{R}^\infty \to \bar{R}^\infty \bmod \mathfrak{p},$$

and let M be its kernel. Then by Theorem 5,

$$PR_M^\infty = (t)$$

is a principal ideal, whose generator t is determined up to a unit in R_M^∞. Thus the order

$$m(\varphi, \mathfrak{p}, P) = \mathrm{ord}_\mathfrak{p} \bar{t}$$

is well defined, and we let

$$(\varphi P)(\mathfrak{p}) = \bar{P}(\mathfrak{p}) = (\bar{t})_\mathfrak{p} = \mathfrak{p}^{m(\varphi, \mathfrak{p}, P)}.$$

We may view $m(\varphi, \mathfrak{p}, P)$ as the multiplicity of intersection of the geometric prime divisor P and the arithmetic prime divisor

$$\mathrm{spec}\,\mathfrak{o}_\mathfrak{p} \to X(N)$$

at M.

Finally, to each prime divisor P at infinity we associate the ideal

$$\varphi P = \bar{P} = \prod_\mathfrak{p} (\varphi P)(\mathfrak{p}),$$

and extend this to divisors $D = \sum v_P P$ by linearity, so that we obtain a fractional ideal

$$\varphi D = \bar{D} = \prod_P \varphi(P)^{v_P}.$$

The association

$$D \mapsto \varphi D = \bar{D}$$

is a homomorphism of $\mathscr{D}^\infty$ into the group of fractional ideals in k.

We let $\mathscr{C}^\infty$ be the factor group of $\mathscr{D}^\infty$ by the subgroup of divisors of units in R_N. Then the preceding association gives rise to a homomorphism of the divisor classes $\mathscr{C}^\infty$ into the group of ideal classes in k.

Theorem 7. *If a unit u in R_N has a divisor at infinity which is the m-th multiple of some divisor, then for any point φ as above, the ideal factorization of $\bar{u} = \varphi u$ is an m-th power.*

The proof is obvious, from the preceding discussion, but the result is useful, as in papers of Kubert and Brylinski, where m-th roots of such values are then extracted, and the fields generated by them analysed.

The theorem is a formulation in terms of commutative algebra of the following simple geometric idea. We view u as a function on the "variety" $X(N)$ viewed as scheme over $\mathbf{Z}$. If V is a variety, and W is a subvariety, and u is a function on V, then under appropriate conditions of completeness and non-singularity, the divisors of u on V, and the induced function

$$\bar{u} = u|W$$

on W are related by the relation

$$(\bar{u})_W = (u)_V \cdot W,$$

where the dot is the intersection product. In particular, if $(u)_V$ is the N-th multiple of a divisor on V, then $(\bar{u})_W$ is the N-th multiple of a divisor on W. The corollary (which obviously holds in a general context) formalizes this argument in the arithmetic case, when W is the "subvariety" associated with the ring of integers in a number field.

In the special case of the divisor of a function in the corollary, we can also describe the global factorization. Our point of $X(N)$ in k not at infinity is represented by a homomorphism

$$R \to \bar{R}$$

of R in k. Let $\mathfrak{p}$ be a prime ideal of k. If $\bar{j}$ is integral at $\mathfrak{p}$, then $\bar{u}$ lies in $\bar{R}$ and is a unit at $\mathfrak{p}$. Suppose that $\bar{j}$ is not integral at $\mathfrak{p}$. We compose $R^\infty \to \bar{R}^\infty$ with reduction mod $\mathfrak{p}$, to obtain a homomorphism of R^∞ into a finite field, whose kernel is a maximal ideal M. We have seen that the local ring R_M^∞ is regular. We may view M as a point on $S = \text{spec}(R^\infty)$. The irreducible elements of the local ring define the prime divisors on S passing through M. Since u is a unit in R, it has a factorization

$$u = \pi_1^{m_1} \ldots \pi_n^{m_n} w, \quad (\pi_i \text{ irreducible in } R_M^\infty),$$

where w is a unit in R_M^∞, and by Theorem 3, each exponent m_i is divisible by N. Specializing, we obtain

$$\bar{u} = \bar{\pi}_1^{m_1} \ldots \bar{\pi}_n^{m_n} \bar{w},$$

where $\bar{w}$ is a unit at $\mathfrak{p}$.

If $(\bar{u}) = \mathfrak{a}^N$ for a fractional ideal $\mathfrak{a}$, it would be quite important to determine the period of $\mathfrak{a}$ in specific cases.

If the function u has a q-expansion whose first coefficient is a unit, and such that all other coefficients are $\mathfrak{p}$-integral, then the order of the specialized value $\bar{u}$ at $\mathfrak{p}$ can also be described in terms of the order of the q-expansion, as follows. Let A be a complete discrete valuation ring with maximal ideal $\mathfrak{m}$. Given an element $\bar{t} \in \mathfrak{m}$, and the power series ring $A[[t]]$ in one variable t, there exists a unique continuous homomorphism

$$A[[t]] \to A$$

such that $t \mapsto \bar{t}$. This will be applied when $t = q^{1/N}$.

In addition, let $k = \mathbf{Q}(\zeta_N)$ be the cyclotomic field, let $\mathfrak{o} = \mathfrak{o}_k$ be its ring of integers. Then any homomorphism

$$\varphi : R^\infty \to A$$

induces a homomorphism

$$\mathfrak{o} \to A.$$

Let us assume that $A/\mathfrak{m}$ has characteristic p.

Let $\mathfrak{p}$ be the kernel in $\mathfrak{o}$ of the composite homomorphism

$$\mathfrak{o} \to A \to A/\mathfrak{m}.$$

Then we get an induced continuous homomorphism of the completion

$$\hat{\mathfrak{o}}_\mathfrak{p} \to A.$$

Viewing $q^{1/N} = t$ as a variable, there is a corresponding homomorphism of power series rings

$$\hat{\mathfrak{o}}_\mathfrak{p}[[q^{1/N}]] \to A[[q^{1/N}]].$$

Let $\bar{q}^{1/N}$ be an element of $\mathfrak{m}$. We see that there exists a unique continuous homomorphism

$$\hat{\mathfrak{o}}_{\mathfrak{p}}[[q^{1/N}]] \to A$$

which is equal to φ on $\mathfrak{o}_{\mathfrak{p}}$, and maps $q^{1/N}$ on $\bar{q}^{1/N}$.

The above notation will be preserved in the next theorem.

Theorem 8. *Let A be a complete discrete valuation ring with maximal ideal $\mathfrak{m}$ such that $A/\mathfrak{m}$ has characteristic p. Let*

$$\varphi: R^\infty \to A$$

be a homomorphism such that $1/\bar{j} = \varphi(1/j)$ lies in $\mathfrak{m}$. Let M be the kernel of the composite homomorphism

$$R^\infty \to A \to A/\mathfrak{m},$$

and R_M^∞ the corresponding local ring. Let $\mathfrak{p} = M \cap \mathfrak{o}$. Then we have a commutative diagram

$$\begin{array}{ccc} R_M^\infty & \xrightarrow{\varphi} & A \\ \downarrow & & \uparrow \\ \hat{R}_M^\infty & \xrightarrow{\approx} & \hat{\mathfrak{o}}_{\mathfrak{p}}[[q^{1/N}]] \end{array}.$$

Proof. The homomorphism $R_M^\infty \to A$ is continuous, since the maximal ideal M is the inverse image of $\mathfrak{m}$ in R^∞. Therefore it extends uniquely to a continuous homomorphism of $\hat{R}_M^\infty$ into A. The isomorphism (identification) of Theorem 5 determines a special value $\bar{q}^{1/N}$ in A, so that Theorem 8 is obvious from the preceding discussion.

Corollary. *Let in addition u be in $\hat{R}_M^\infty$, admitting a power series expansion determined by the bottom isomorphism*

$$u = \sum_{n \geq r} a_n q^{n/N}$$

with a_r a unit in $\hat{\mathfrak{o}}_{\mathfrak{p}}$ and $a_n \in \hat{\mathfrak{o}}_{\mathfrak{p}}$ for all n. Then

$$\operatorname{ord}_{\mathfrak{m}} \bar{u} = \frac{r}{N} \operatorname{ord}_{\mathfrak{m}} \bar{q}.$$

Proof. Obvious.

Theorem 9. *Let $\varphi: R = R_N \to k$ be a point of $V(N)$ in a number field k, not at infinity. Let $\mathfrak{p}$ be a prime of k dividing the denominator of $\varphi j = \bar{j}$. There exists only one prime at infinity P on $V(N)^\infty$ such that $m(\varphi, \mathfrak{p}, P) \neq 0$, and for that prime P we have*

$$m(\varphi, \mathfrak{p}, P) = -\frac{1}{N} \operatorname{ord}_{\mathfrak{p}} \varphi j.$$

Proof. The corollary of Theorem 5 already tells us that there is only one prime at infinity passing through M, so for all but one prime P the ideal PR_M^∞ is the unit ideal, and the multiplicity of intersection is 0. For the special prime P,

Units in the Modular Function Field. I

we know from Theorem 6 that there is a local parameter having its corresponding q-expansion equal to

$$t = \sum c_i q^{i/N} = c_1 q^{1/N} + \dots,$$

where c_1 is a unit in $\mathbf{Z}[\zeta_N]$ and all c_i are integral. The point φ is induced by an evaluation of the power series corresponding to

$$q^{1/N} \mapsto \bar{q}^{1/N}.$$

The order at $\mathfrak{p}$ of the special value $\varphi t = \bar{t}$ is therefore the order at $\mathfrak{p}$ of $\bar{q}^{1/N}$. But $\operatorname{ord}\bar{q} = -\operatorname{ord}\bar{j}$, so our theorem is proved.

§ 5. Specializations to Complex Multiplication

When we specialize j to an algebraic number which is integral, then of course the group of divisor classes at infinity disappears, but there is still something interesting happening to the units. Here we give merely a useful formulation of the reciprocity law governing such units in the case of complex multiplication.

Let k be imaginary quadratic and let $\mathfrak{o} = \mathfrak{o}_k$ be the ring of algebraic integers in k. Let $\mathfrak{f}$ be an ideal of $\mathfrak{o}$, $\mathfrak{f} \neq \mathfrak{o}$. (By *ideal* we always mean contained in $\mathfrak{o}$. Otherwise, we use the name *fractional ideal*.) We let $G_\mathfrak{f}$ be the ray class group with conductor $\mathfrak{f}$. We first represent ray classes as in Hasse [H].

We say that a pair $(t, \mathfrak{b})$ consisting of an element $t \in k^*$ and a fractional ideal $\mathfrak{b}$ prime to $\mathfrak{f}$ **represents a class** $R \in G_\mathfrak{f}$ if

$$t\mathfrak{f}\mathfrak{b}^{-1} = \mathfrak{c}$$

is an ideal in R, prime to $\mathfrak{f}$. Thus $(t) = \mathfrak{b}\mathfrak{c}/\mathfrak{f}$. Any class can be represented by such a pair, for given an ideal $\mathfrak{c} \in R$, we let $\mathfrak{b}$ be prime to $\mathfrak{f}$ such that $\mathfrak{b}\mathfrak{a}\mathfrak{f}^{-1} = (t)$ is principal. Then $(t, \mathfrak{b})$ represents the ray class of $\mathfrak{c}$.

Lemma 1. (i) *The pairs $(t, \mathfrak{b})$ and $(t', \mathfrak{b}')$ represent the same class if and only if there exists $\xi \in k^*$ such that*

$$\mathfrak{b}' = \xi \mathfrak{b} \quad \text{and} \quad t'/\xi t \equiv 1 \,(\text{mod}^* \mathfrak{f}).$$

(ii) *Under these circumstances, we have*

$$t' \equiv \xi t \,(\text{mod}\, \mathfrak{b}').$$

Proof. Suppose $(t, \mathfrak{b})$ and $(t', \mathfrak{b}')$ represent the same class, so that

$$t\mathfrak{f}\mathfrak{b}^{-1} = t'\mathfrak{f}\mathfrak{b}'^{-1}\alpha,$$

with some $\alpha \equiv 1 \,(\text{mod}^* \mathfrak{f})$. The element $\xi = t'\alpha/t$ achieves what we want. The converse of (i) is clear. This proves (i).

As for (ii),

$$t' - \xi t \in \mathfrak{b}' \Leftrightarrow \frac{t'}{\xi t} - 1 \in \frac{\mathfrak{b}'}{\xi t}.$$

Let $(t) = \mathfrak{c}\mathfrak{b}\mathfrak{f}^{-1}$ and $t' = \mathfrak{c}'\mathfrak{b}'\mathfrak{f}^{-1}$. Then $(t'/\xi t) = \mathfrak{c}'/\mathfrak{c}$. Hence

$$\mathfrak{b}'/\xi t = \mathfrak{f}/\mathfrak{c},$$

and our congruence condition is equivalent with

$$\frac{t'}{\xi t} - 1 \in \mathfrak{f}/\mathfrak{c}.$$

This is verified locally at each prime. If $\mathfrak{p}|\mathfrak{f}$, then

$$\mathrm{ord}_\mathfrak{p}\left(\frac{t'}{\xi t} - 1\right) \geq \mathrm{ord}_\mathfrak{p}\mathfrak{f}.$$

On the other hand, $t'/\xi t$ has at most $\mathfrak{c}$ as polar divisor. The congruence of (ii) now follows at once.

Let N be an integer > 1. An **invariant of level** N is defined to be a function of pairs $f(z, L)$, where L is a lattice in C, $z \in \frac{1}{N}L$, $z \notin L$, satisfying the following two conditions.

F1. f is homogeneous of degree 0, that is

$$f(\lambda z, \lambda L) = f(z, L), \quad \text{all} \quad \lambda \in C^*.$$

In particular, taking $\lambda = -1$ shows that $f(-z, L) = f(z, L)$, i.e. f is even in its first variable.

F2. f depends only on the residue class of $z \bmod L$.

The next lemma relates such invariants and ray classes.

Lemma 2. *Assume that* $\mathfrak{f}^{-1} \subset \frac{1}{N}\mathfrak{o}$. *Then*

$$f(t, \mathfrak{b})$$

is independent of a pair $(t, \mathfrak{b})$ *representing a given ray class* $\bmod \mathfrak{f}$.

Proof. If $(t', \mathfrak{b}')$ represents the same ray class, then by Lemma 1 and properties **F1, F2** we obtain

$$f(t', \mathfrak{b}') = f(\xi t, \mathfrak{b}') = f(t, \mathfrak{b}).$$

We may therefore define the **invariant**

$$f(R) = f(t, \mathfrak{b})$$

for any ray class R represented by $(t, \mathfrak{b})$.

Let r, s be integers not both $\equiv 0 \pmod{N}$. Let $L = [z_1, z_2]$ be a lattice, with

$$\tau = z_1/z_2$$

in the upper half plane. We define

$$f_{r,s}(\tau) = f_{r,s}\begin{pmatrix} z_1 \\ z_2 \end{pmatrix}$$

to be

$$f\left(\frac{rz_1+sz_2}{N}, L\right) = f\left(\frac{r\tau+s}{N}, [\tau, 1]\right).$$

The functions $f_{r,s}$ then satisfy the following condition.

F3. *If $\gamma \in SL_2(\mathbf{Z})$ then $f_{r,s}(\gamma\tau) = f_{(r,s)\gamma}(\tau)$.*

We shall say that f is of **Fricke type** if the functions $f_{r,s}$ lie in the modular function field F_N, and satisfy the additional condition

F4. $\sigma_d f_{r,s} = f_{r,ds}$.

where d is a positive integer, $(d, N) = 1$, and σ_d is the automorphism of F_N induced by the automorphism of $\mathbf{Q}(\zeta_N)((q^{1/N}))$ such that

$\zeta_N \mapsto \zeta_N^d$ and $q^{1/N}$ is fixed by σ_d.

If f if of Fricke type, then for any $g \in \Pi\, GL_2(\mathbf{Z}_p)$ we have

F5. $f_{r,s}^{\sigma(g)} = f_{(r,s)g}$.

(The notation is the same as in [L1], Chapters VI and VII.)

Proof. The effect of $\sigma(g)$ is determined by the congruence class of $g \bmod N$. Write

$$g \equiv \begin{pmatrix} 1 & 0 \\ 0 & d \end{pmatrix} \beta \pmod{N},$$

with some $\beta \in SL_2(\mathbf{Z})$. Then

$$\sigma(g) = \sigma_d \sigma(\beta).$$

Hence using **F4** and **F3** we find

$$f_{r,s}^{\sigma(g)}(\tau) = f_{r,s}^{\sigma_d\sigma(\beta)}(\tau) = f_{r,ds}(\beta\tau) = f_{(r,s)g}(\tau),$$

as desired.

Using the Shimura reciprocity law, we shall now give a proof for the reciprocity law of invariants, axiomatizing in the present context the arguments given for the similar result concerning the Siegel-Ramachandra invariants in [L1], Chapter 19, Theorem 3. It can then be applied to other cases, e.g. values of units as constructed in § 2, or to the Weber functions (actually the original case as in Hasse).

Theorem. *Let f be an invariant of level N, and of Fricke type. Assume $N \in \mathfrak{f}$. Let $R, S \in G_\mathfrak{f}$, and let (S, k) be the Artin automorphism on the ray class field $k_\mathfrak{f}$ of k with conductor $\mathfrak{f}$. Then $f(R)$ lies in $k_\mathfrak{f}$, and*

$$f(R)^{(S,k)} = f(RS).$$

Proof. Let $\mathfrak{a}$ be an ideal in S, $\mathfrak{c}$ an ideal in R, both prime to $\mathfrak{f}$, and

$$\mathfrak{tfb}^{-1} = \mathfrak{c}.$$

Then

$$\mathfrak{tfb}^{-1}\mathfrak{a} = \mathfrak{ca},$$

and $(t, \mathfrak{b}\mathfrak{a}^{-1})$ represents RS. Let $\mathfrak{b}=[z_1, z_2]$. Let $\alpha \in GL_2^+(\mathbf{Q})$ be such that

$$\alpha^{-1}\begin{pmatrix} z_1 \\ z_2 \end{pmatrix} \text{ is a basis of } \mathfrak{a}^{-1}\mathfrak{b}.$$

Then $\det \alpha = N\mathfrak{a}^{-1}$. Let s be an ideal of k such that

$$s_p = 1 \quad \text{for all} \quad p|N\mathfrak{a} = d,$$
$$s_p \mathfrak{o}_p = \mathfrak{a}_p^{-1} \quad \text{if} \quad p \nmid N\mathfrak{a}.$$

Let

$$t = \frac{r_1 z_1 + r_2 z_2}{N}.$$

By Shimura's reciprocity law [Sh] and [L1], Chapter XI, §1 we get

$$f(R)^{(s,k)} = f_{r_1, r_2}^{\sigma(q(s))}(z_1/z_2).$$

There exists $g_p \in GL_2(\mathbf{Z}_p)$ such that

$$q_p(s_p) = g_p \alpha, \quad g = (\ldots, g_p, \ldots).$$

Then by **F5**, we see that

$$f(R)^{(s,k)} = f_{r_1, r_2}^{\sigma(g)\sigma(\alpha)}(z) = f_{(r_1, r_2)g}(\alpha z).$$

We then have

$$f(RS) = f(t, \mathfrak{a}^{-1}\mathfrak{b}) = f\left(t, \alpha\begin{pmatrix} z_1 \\ z_2 \end{pmatrix}\right) = f_{(r_1, r_2)\alpha^{-1}}(\alpha z).$$

This proves that

$$f(R)^{(s,k)} = f(RS).$$

It shows that the effect of (s, k) depends only on the ray class of S, and hence that $f(R)$ lies in the ray class field with conductor $\mathfrak{f}$. It also proves the desired formula.

As pointed out by Robert [Rob], the Siegel-Ramachandra invariants can be constructed directly by pairs $(t, \mathfrak{b})$. His argument holds axiomatically in the present context, as follows.

Theorem. *Let f be a function satisfying **F1** and **F2**. Let $\mathfrak{a}$ be an ideal in a ray class R mod $\mathfrak{f}$, let $\mathfrak{d}$ be the different of $k/\mathbf{Q}$, and let*

$$\mathfrak{a}\mathfrak{d}^{-1}\mathfrak{f}^{-1} = \mathfrak{c} = [z_1, z_2], \quad z_1/z_2 \in \mathfrak{H}.$$

Then

$$f((\mathrm{tr}z_2)z_1 - (\mathrm{tr}z_1)z_2, \mathfrak{c}) = f(\bar{R}),$$

where $\bar{R}$ is the complex conjugate of R.

Proof. We need a lemma.

Lemma. *Let $\mathfrak{c} = [z_1, z_2]$ be a fractional ideal with $\mathrm{Im}(z_1/z_2) > 0$. Let $D = D(\mathfrak{o}_k)$ be the discriminant. Then*

$$(\mathrm{tr}z_2)z_1 - (\mathrm{tr}z_1)z_2 = \sqrt{D}\,N\mathfrak{c}.$$

205

Proof. If we replace (z_1, z_2) by $(\lambda z_1, \lambda z_2)$ with $\lambda \in k^*$, then both sides change by $\lambda\bar\lambda = N\lambda$. Hence it suffices to prove the lemma for $\mathfrak{c}=[z,1]$, i.e. $z_2=1$, and $z = x + y\sqrt{D}$, $y>0$. The left hand side of our formula is equal to

$$2z - 2x = 2y\sqrt{D}.$$

Hence we have only to show that

$$2y = N\mathfrak{c}.$$

But

$$D(\mathfrak{c}) = N\mathfrak{c}^2 D(\mathfrak{o}) = N\mathfrak{c}^2 \cdot D,$$

and

$$D(\mathfrak{c}) = \begin{vmatrix} 1 & z \\ 1 & \bar z \end{vmatrix}^2 = (2y)^2 D.$$

Since $y>0$, our lemma is proved.

For the theorem, observe that $\mathfrak{d} = \mathfrak{o}\sqrt{D}$. Then

$$\mathfrak{a}\mathfrak{d}^{-1}\mathfrak{f}^{-1} = [z_1, z_2] = \mathfrak{a}\mathfrak{f}^{-1}\mathfrak{D}^{-1} = \frac{t}{\sqrt{D}}\mathfrak{b}^{-1} = \frac{t}{\sqrt{D}}\frac{\bar{\mathfrak{b}}}{N\mathfrak{b}}.$$

We find:

$$f\left((\mathrm{tr}\, z_2)z_1 - (\mathrm{tr}\, z_1)z_2, \begin{pmatrix} z_1 \\ z_2 \end{pmatrix}\right) = f\left(\sqrt{D}\frac{Nt}{|D|N\mathfrak{b}}, \frac{t}{\sqrt{D}}\frac{\bar{\mathfrak{b}}}{N\mathfrak{b}}\right)$$
$$= f(-Nt, t\bar{\mathfrak{b}})$$
$$= f(-\bar t, \bar{\mathfrak{b}})$$
$$= f(\bar R),$$

as was to be shown.

References

[B]   Baker, A.: Contributions to the theory of diophantine equations. Phil. Trans. Royal Soc. London Series A, Math. and Physical Sciences No. **1139**, Vol. 263 (1968), pp. 173—208

[B-C]   Baker, A., Coates, J.: Integer points on curves of genus 1. Proc. Camb. Phil. Soc. **67** (1970), pp. 595—602

[Br]   Brylinski, J.: Torsion sur les courbes élliptiques (to appear)

[Ch 1]   Chabauty, C.: Sur le théorème fondamental de la théorie des points entiers et pseudo-entiers des courbes algébriques. C. R. Acad. Sci. Paris No. **217** (1943), pp. 336—338

[Ch 2]   Chabauty, C.: Démonstration de quelques lemmes de rehaussement. C. R. Acad. Sci. Paris No. **217** (1943), pp. 413—415

[C-W]   Chevalley, C., Weil, A.: Un théorème d'arithmétique sur les courbes algébriques. C. R. Acad. Sci. Paris (1930), pp. 570—572

[C]   Clemens, H.: On Prym varieties (to appear)

[Co]   Coates, J.: An effective analogue of a theorem of Thue. Acta Arithmetica, three papers: **I**, Vol. 15 (1969), pp. 279—305; **II**, Vol. 16 (1970), pp. 399—412; **III**, Vol. 16 (1970), pp. 425—435

[De-Ra] Deligne, P., Rapoport, M.: Les schémas de modules des courbes élliptiques. Modular functions of one variable II. Springer Lecture Notes No. **349**, pp. 143—316

[Dem 1] Demjanenko, V. A.: Torsion of elliptic curves. Izv. Akad. Nauk SSSR, Ser. Mat. Tom **35** (1971), No. 2, AMS translation, pp. 289—318

[Dem 2] Demjanenko, V. A.: On the uniform boundedness of the torsion of elliptic curves over algebraic number fields. Izv. Akad. Nauk SSSR, Ser. Mat. Tom **36** (1972), No. 3, AMS translation, pp. 477—490
[EGA] Grothendieck, A.: Eléments de géometrie algébrique. Pub. IHES, Chapter **IV**, 7.8.3, 7.8.6
[Dr] Drinfeld, V. G.: Two theorems on modular curves. Functional analysis and its applications, Vol. **7**, No. 2, translated from the Russian, April-June 1973, pp. 155—156
[F] Fricke, R.: Elliptische Funktionen und ihre Anwendungen, Vol. 1, pp. 450—451; Leipzig: Teubner Verlag 1930
[G] Gelfond, A. O.: Transcendental and algebraic numbers. Moscow, 1952; translated, Dover press, 1960
[H] Hasse, H.: Neue Begründung der komplexen Multiplikation, I and II. J. Reine Angew. Math. **157** (1927), pp. 115—139 and **165** (1931), pp. 64—88
[He 1] Hellegouarch, Y.: Une propriété arithmétique des points exceptionnels rationnels d'ordre pair d'une cubique de genre 1, C. R. Acad. Sci. Paris **260** (1965), pp. 5989—5992
[He 2] Hellegouarch, Y.: Applications d'une propriété arithmétique des points exceptionnels d'ordre pair d'une cubique de genre 1. C. R. Acad. Sci. Paris 260 (1965), pp. 6256—6258
[He 3] Hellegouarch, Y.: Étude des points d'ordre fini des variétés abéliennes de dimension un definies sur un anneau principal, J. Reine angew. Math. **244** (1970), pp. 20—36
[He 4] Hellegouarch, Y.: Points d'ordre fini sur les courbes élliptiques. C. R. Acad. Sci. Paris Ser. A-B **273** (1971), pp. 540—543
[Iw] Iwasawa, K.: Lectures on p-adic L-functions. Annals of Math. Studies No. 74
[K] Klein, F.: Über die elliptischen Normalkurven der n-ten Ordnung und die zugehörigen Modulfunktionen der n-ten Stufe. Leipziger Abh. Bd. 13 (1885), p. 339
[K-F] Klein, F., Fricke, R.: Vorlesungen über die Theorie der elliptischen Modulfunktionen, Vol. 2. Johnson Reprint Corporation, NY, and Teubner Verlag, Stuttgart 1966 (from the 1890 edition)
[K] Kubert, D.: Universal bounds on the torsion of elliptic curves. J. London Math. Soc. (to appear)
[L 1] Lang, S.: Elliptic functions. Addison Wesley, Reading (1974)
[L 2] Lang, S.: Integral points on curves. Pub. IHES (1960)
[L 3] Lang, S.: Diophantine geometry. New York: Interscience 1962
[L 4] Lang, S.: Isogenous generic elliptic curves. Am. J. Math. 1972
[Le] Leopoldt, H.: Eine Verallgemeinung der Bernoullischen Zahlen. Abh. Math. Sem. Hamburg (1958), pp. 131—140
[Ma] Manin, J.: Parabolic points and zeta-functions of modular curves. Izv. Akad. Nauk SSSR, Ser. Mat. Tom **36** (1972), No. 1, AMS translation, pp. 19—64
[Mu] Mumford, D.: Prym varieties (to appear)
[N] Newman, M.: Construction and application of a class of modular functions. Proc. London Math. Soc. (3) **7** (1957), pp. 334—350
[Ra] Ramachandra, K.: Some applications of Kronecker's limit formula. Ann. of Math. **80** (1964), pp. 104—148
[Re] Recillas, S.: A relation between curves of genus three and curves of genus 4. Ph. D. dissertation, Brandeis University 1970 (see especially, pp. 107—108, and Theorem 2 of Chapter IV et sequences)
[Rob] Robert, G.: Unités élliptiques. Bull. Soc. Math. France, Memoire No. 36 (1973)
[Ro] Roth, P.: Über Beziehungen zwischen algebraischen Gebilden von Geschlechtern drei und vier. Monatshefte **22** (1911)
[Sh] Shimura, G.: Introduction to the arithmetic theory of automorphic functions. Iwanami Shoten and Princeton University Press 1971
[Si] Siegel, C. L.: Über einige Anwendungen diophantischer Approximationen. Abh. Preuss. Akad. Wiss. Phys. Math. Kl. (1929), pp. 41—69
[We] Weil, A.: Arithmétique et géometrie sur les variétés algébriques. Act. Sci. et Ind. No. **206**. Paris: Hermann 1935

Received July 4, 1975

Units in the Modular Function Field

II. A full Set of Units

Dan Kubert* and Serge Lang*

Department of Mathematics, Yale University, New Haven, Conn. 06520, USA

Contents

1. Construction of Units with the Sigma Function 175
2. The Regulator at Infinity 179
3. Rank of the Units over $Z[j]$ 186

This is the second in a series of papers about units in the modular function field. The main introduction was given in the first paper, and we follow the same terminology. We prove a "unit theorem" just as in number theory, where the number of independent units is equal to the number of primes at infinity minus 1. In the present case of the function field of a curve, units have to be taken modulo constants, of course. Ultimately, a more refined version should also take into account these constants, in order to relate the specializations of these units with those in number fields, e. g. those of complex multiplication.

Whereas the classical theory (with respect to Z_p^*) expresses class numbers as products of $B_{1,\chi}$, in the present instance we get a regulator which can be factored into an expression involving $B_{2,\chi}$. This paper is devoted to establishing the link between the function theory and the abstract algebraic theory of certain character sums on a Cartan group, giving the factorization of the regulator. In [KL 2], we treat these character sums independently of the function theory. In the next paper of the series, we treat separately the corresponding elements in the group algebra, giving rise to the analogue of the Stickelberger element which annihilates the divisor classes at infinity, and leads into the p-adic distributions in the sense of Mazur, and Iwasawa.

§ 1. Construction of Units with the Sigma Function

In this section we construct modular forms of weight 1 on $\Gamma(2N^2)$, weight $2N$ on $\Gamma(N)$, whose divisors are at infinity, and whose quotients are units in R_N or $R_N[1/N]$. We first recall some terminology.

A **form of degree** k is a function

$$h\begin{pmatrix}\omega_1\\ \omega_2\end{pmatrix}, \quad \mathrm{Im}(\omega_1/\omega_2)>0,$$

---

* Supported by NSF grants.

satisfying the homogeneity property

MF 1. $h\left(\lambda \begin{pmatrix} \omega_1 \\ \omega_2 \end{pmatrix}\right) = \lambda^k h\begin{pmatrix} \omega_1 \\ \omega_2 \end{pmatrix}, \quad \lambda \in \mathbf{C}^*.$

Let Γ be, say, a subgroup of $SL_2(\mathbf{Z})$. We say that a form as above is **modular with respect to** Γ if it satisfies the additional property

MF 2. $h\left(\gamma\begin{pmatrix} \omega_1 \\ \omega_2 \end{pmatrix}\right) = h\begin{pmatrix} \omega_1 \\ \omega_2 \end{pmatrix}, \quad \gamma \in \Gamma.$

One should also introduce conditions of meromorphy or holomorphy. For our purposes, we need a special case, and we say that the **form is at infinity** if it also satisfies the third condition:

MF 3. *The function $h(\tau, 1)$ for τ in the upper half plane is holomorphic and has no zero. For each $\gamma \in SL_2(\mathbf{Z})$, the function*

$$h\left(\gamma\begin{pmatrix} \tau \\ 1 \end{pmatrix}\right)$$

is meromorphic at infinity, i.e. has a power series expansion in terms of a fractional power of q.

If X is the completed curve of $\Gamma \backslash \mathfrak{H}$, then we also say that such a modular form is **on X**. We say the modular form is **defined over a field** K if the coefficients of its q-expansion lie in K. In our applications, $K = \mathbf{Q}(\zeta)$ for some root of unity ζ.

The condition at infinity guarantees that we can define the order of a form at each point at infinity (cusp), and thus the order vector.

Let η_1, η_2 be the corresponding quasi-periods of the Weierstrass zeta function associated with the period lattice $[\omega_1, \omega_2]$. Let N be an integer >1, let r, s be integers not both congruent to 0 mod N. We define the **Klein forms**

$$\mathfrak{k}_{r,s}(\omega_1, \omega_2) = \mathfrak{k}\left(\frac{r}{N}, \frac{s}{N}; \begin{pmatrix} \omega_1 \\ \omega_2 \end{pmatrix}\right)$$

$$= \exp\left(-\frac{r\eta_1 + s\eta_2}{N} \frac{r\omega_1 + s\omega_2}{N}\right) \sigma\left(\frac{r\omega_1 + s\omega_2}{N}; \omega_1, \omega_2\right).$$

If the integer N is fixed throughout a discussion, we index the Klein form $\mathfrak{k}$ just with (r, s). When we deal with relations between Klein forms of various levels, however, it is necessary to keep the N in the notation. Other possibilities which may also be convenient are

$$\mathfrak{k}\left(\frac{r\omega_1 + s\omega_2}{N}; \begin{pmatrix} \omega_1 \\ \omega_2 \end{pmatrix}\right).$$

Since the sigma function is homogeneous of degree 1, and the quasi periods are homogeneous of degree -1, it follows that the Klein form is homogeneous of degree 1 in $[\omega_1, \omega_2]$, that is

$$\mathfrak{k}_{r,s}(\lambda\omega_1, \lambda\omega_2) = \lambda \mathfrak{k}_{r,s}(\omega_1, \omega_2), \quad \lambda \in \mathbf{C}^*.$$

Klein introduced this particular normalization of the sigma function because it is easier than theta functions to handle with respect to transformations in $SL_2(\mathbf{Z})$. For the convenience of the reader we reproduct some of the basic properties of the Klein forms as in Fricke [F], pp. 450, 451, 452, or Klein-Fricke [K-F], p. 22. They give an easier approach to the Siegel functions and the construction of units than through the fundamental theta function [L 1], as in the work of Siegel [Si], Ramachandra [Ra], and Robert [Ro].

The Klein forms satisfy the following properties.

K 1. If $\alpha = \begin{pmatrix} a & b \\ c & d \end{pmatrix}$ is in $SL_2(\mathbf{Z})$, then

$$\mathfrak{k}_{r,s}\left(\alpha\begin{pmatrix}\omega_1\\\omega_2\end{pmatrix}\right) = \mathfrak{k}_{(r,s)\alpha}\begin{pmatrix}\omega_1\\\omega_a\end{pmatrix}.$$

We recall the standard transformation property of the sigma function with respect to the periods, namely

$$\sigma(z+m_1\omega_1+m_2\omega_2) = (-1)^{m_1m_2+m_1+m_2} e^{(m_1\eta_1+m_2\eta_2)(z+\frac{1}{2}(m_1\omega_1+m_2\omega_2))}\sigma(z).$$

From this we obtain at once for arbitrary integers a, b:

K 2. $\mathfrak{k}_{r+aN,s+bN} = (-1)^{ab+a+b} e^{-2\pi i(as-br)/2N} \mathfrak{k}_{r,s}.$

(Of course, one uses the Legendre relation.)

Now suppose that

$$\alpha = \begin{pmatrix} a & b \\ c & d \end{pmatrix} \equiv \begin{pmatrix} 1 & 0 \\ 0 & 1 \end{pmatrix} \pmod{N},$$

and $\alpha \in SL_2(\mathbf{Z})$, so $\alpha \in \Gamma(N)$. Then

$$a-1 \equiv d-1 \equiv c \equiv b \equiv 0 \pmod{N}.$$

We write

$$ar+cs = r + \left(\frac{a-1}{N}r + \frac{c}{N}s\right)N$$

$$br+ds = s + \left(\frac{b}{N}r + \frac{d-1}{N}s\right)N.$$

If $\alpha \in \Gamma(N)$ we therefore find

K 3. $\mathfrak{k}_{r,s}\left(\alpha\begin{pmatrix}\omega_1\\\omega_2\end{pmatrix}\right) = \varepsilon(\alpha)\mathfrak{k}_{r,s}$

where $\varepsilon(\alpha)$ is a $(2N)$-th root of unity, namely

$$\varepsilon(\alpha) = -(-1)^{(\frac{a-1}{N}r+\frac{c}{N}s+1)(\frac{b}{N}r+\frac{d-1}{N}s+1)} e^{2\pi i(br^2+(d-a)rs-cs^2)/2N^2}.$$

It follows that $\mathfrak{k}_{r,s}^{2N}$ is invariant under $\Gamma(N)$, and satisfies the first two conditions of a modular form, of weight $2N$.

Let $\tau = \omega_1/\omega_2$ and $q = q_\tau = e^{2\pi i \tau}$. Let $\zeta = e^{2\pi i/N}$. From the q-expansion for the sigma function found in any book on elliptic functions (for instance [L]), we immediately get the q-product expansion for $\mathfrak{f}_{r,s}(\tau) = \mathfrak{f}_{r,s}(\tau, 1)$.

K 4. $\mathfrak{f}_{r,s}(\tau) \prod_{n=1}^{\infty} (1-q^n)^2 (2\pi i) =$
$$-e^{2\pi i s(r-N)/2N^2} q^{r(r-N)/2N^2} (1 - \zeta^s q^{r/N}) \prod_{n=1}^{\infty} (1 - \zeta^s q^{n+r/N})(1 - \zeta^{-s} q^{n-r/N}).$$

We recall that
$$\Delta^{1/12} = (2\pi i) q^{1/12} \prod_{n=1}^{\infty} (1 - q^n)^2.$$

Multiplying the expression in **K 4** by $q^{1/12}$ yields the product for

$\mathfrak{f}_{r,s} \Delta^{1/12}$,

which we write for reference.

K 5. $\mathfrak{f}_{r,s}(\tau) \Delta(\tau)^{1/12} =$
$$-q^{1/12} e^{2\pi i s(r-N)/2N^2} q^{r(r-N)/2N^2} (1 - \zeta^s q^{r/N}) \prod_{n=1}^{\infty} (1 - \zeta^s q^{n+r/N})(1 - \zeta^{-s} q^{n-r/N}).$$

To get $\mathfrak{f}_{r,s}\begin{pmatrix}\omega_1\\\omega_2\end{pmatrix} \Delta(\tau)^{1/12}$ one merely multiplies the above expression by ω_2.

Let
$$g_{r,s}(\tau) = (\mathfrak{f}_{r,s}(\tau) \Delta(\tau)^{1/12})^{12N}$$

If $L = [z_1, z_2]$ is a lattice in C with $\mathrm{Im}(z_1/z_2) > 0$, and $t \in \frac{1}{N}L$, we write
$$t \equiv \frac{rz_1 + sz_2}{N} \bmod L,$$

and define
$$g(t, L) = g_{r,s}(z_1/z_2).$$

Then g is an invariant of level N. It has the q-expansion

K 6. $g_{r,s}(\tau) =$
$$\zeta^{6rs} q^{N + 6r^2/N - 6r} (1 - \zeta^s q^{r/N})^{12N} \prod_{n=1}^{\infty} [(1 - \zeta^s q^{n+r/N})(1 - \zeta^{-s} q^{n-r/N})]^{12N}.$$

From this is clear that g is of Fricke type, i. e. that the operation $\zeta \mapsto \zeta^d$ induces $g_{r,s} \mapsto g_{r,ds}$.

Furthermore, the function $g_{s,-r}$ is the Siegel function in the terminology of [L], Chapter 19, § 2, i.e. it is the function used by Siegel in [Si]. Therefore we call again the functions $g_{r,s}$ the **Siegel functions**, and we call g the **Siegel invariant**.

It is clear that the functions $g_{r,s}$ have no zero or pole on the upper half plane, and the coefficients in the q-expansion are algebraic integers in $Q(\zeta)$. Therefore $g_{r,s}$ is a unit in QR_N.

Theorem 1.1. *Let (r, s) be primitive* mod N.

(i) *If N is composite, i.e. has at least two prime factors, then the functions $g_{r,s}$ are units in R_N.*

Units in the Modular Function Field II. 179

(ii) *If $N = p^N$ is a prime power, then $g_{r,s}$ is a unit in $R_N[1/p]$.*
(iii) *For any positive integer N and any integer a prime to N, the function*

$$g_{ar,as}/g_{r,s}$$

is a unit in R_N.

Proof. This is immediate from the q-expansion taking $0 \leq r, s \leq N$. Observe that if $r \not\equiv 0 \pmod{N}$ then the q-expansion of $g_{r,s}$ begins with a root of unity times a pure power of q. Hence the only leading coefficients which are not units in the cyclotomic field occur when $r \equiv 0 \pmod{N}$, in which case they are powers of

$$1 - \zeta^s,$$

and have the obvious divisibility property according as N is a prime power or not. When we take a quotient $g_{ar,as}/g_{r,s}$, we insure that the leading coefficient is a unit in every case, thus taking care of (iii).

As a corollary, we obtain again the fact that the functions constructed in [K-L], §2 and §3 are units, by using the formula expressing the Weierstrass function in terms of the sigma function,

$$\wp(z) - \wp(w) = \frac{\sigma(z+w)\sigma(z-w)}{\sigma(z)^2 \sigma(w)^2}.$$

This shows how the functions of §2 and §3 are quotients of Klein functions. The transcendental exponents involving $\eta_1, \eta_2, \omega_1, \omega_2$ cancel out, and we can apply (iii) of Theorem 1.1 to conclude that powers of the elements of [K-L], §2, §3 are products of Siegel functions, in fact products of quotients of type (iii), whence units in R_N.

Remark. Various authors have considered functions in the modular function field of what is usually denoted by $X_0(N)$, which have their divisors concentrated at cusps, e.g. Newman [N] and Ogg [O]. These functions were obtained as quotients of the discriminant function $\Delta(m\tau)/\Delta(\tau)$. Working with the Klein forms and Siegel functions is easier, goes farther, and especially one can profit from the fact that the full modular function field F_N is Galois over $Q(j)$, so that one can use Galois and Hilbert theory of primes in a Galois extension, as described in the next section.

§ 2. The Regulator of Units at Infinity

The group $GL_2(\mathbf{Z}/N\mathbf{Z})$ is represented as the group of automorphisms of the modular function field F_N over $Q(j)$, and the kernel of the representation is ± 1. We have a standard prime P_∞ at infinity, induced by the natural embedding of F_N in the complex power series field

$$Q(\zeta_N)((q^{1/N})), \quad q = e^{2\pi i \tau}$$

by means of the q-expansion. We let G_∞ be the isotropy group of the prime at infinity. We contend that G_∞ is the group of matrices

$$\begin{pmatrix} 1 & b \\ 0 & d \end{pmatrix}$$

in $GL_2(Z/NZ)$. The matrices

$$\begin{pmatrix} 1 & b \\ 0 & 1 \end{pmatrix}$$

have the effect of multiplying $q^{1/N}$ by a root of unity, and consequently do not change the prime. The matrices

$$\begin{pmatrix} 1 & 0 \\ 0 & d \end{pmatrix}$$

leave $q^{1/N}$ fixed, and send $\zeta \mapsto \zeta^d$, so that they do not change the prime either. Conversely, any matrix which leaves fixed the orders at P_∞ of all the Fricke functions, or the Siegel functions, must be of the above type.

Observe that $GL_2(Z/NZ)$ acts on $(Z/NZ)^2$ on the left, if elements of $(Z/NZ)^2$ are viewed as column vectors. Let

$$e_1 = \begin{pmatrix} 1 \\ 0 \end{pmatrix}.$$

Then

$$\begin{pmatrix} a & b \\ c & d \end{pmatrix} e_1 = \begin{pmatrix} a \\ c \end{pmatrix}$$

is the first column. Given a primitive pair $\begin{pmatrix} a \\ c \end{pmatrix}$ mod N, we can extend it to a matrix in $GL_2(Z/NZ)$. Hence

$$GL_2(Z/NZ)/G_\infty$$

is represented by primitive column vectors mod N, and G_∞ is the isotropy group of e_1.

We can split G_∞ in $GL_2(Z/NZ)$ by a commutative subgroup. Everything breaks up according to prime powers, so we consider that case. At first we work p-adically.

Let p be a prime number. Let $G_p = GL_2(Z_p)$ and let $G_{\infty,p}$ consist of the matrices

$$\begin{pmatrix} 1 & b \\ 0 & d \end{pmatrix}, \quad b \in Z_p, \quad d \in Z_p^*.$$

Let C_p be the group of units in the unramified extension of degree 2 over Q_p (i.e. the Witt vectors in F_{p^2}), viewed as embedded in G_p by the regular representation over Z_p. Sometimes C_p is called the **"non-split Cartan subgroup"**. Then the map

$$C_p \times G_{\infty,p} \to GL_2(Z_p)$$

given by the product is a bijection, so we have a unique decomposition

$$G_p = C_p G_{\infty,p}.$$

Proof. For simplicity we omit the index p. The elements $\notin Q_p$ of C have characteristic roots which are quadratic irrational over Q_p (their reductions mod p

are even irrational over the prime field F_p). So $C \cap G_\infty = \{1\}$. We now prove that C acts transitively on the primitive vectors in Z_p^2. This suffices to prove our assertion.

Let R be the ring or integral elements of the unramified extension of degree 2 over Q_p. Then R is a free module over Z_p (for instance, a basis may be taken to be $\{1, u\}$, where u is a unit whose residue class mod p generates F_{p^2} over F_p). The maximal ideal of R is precisely pR, and the units R^* of R consist therefore of the primitive elements of R, i.e. those which can be written as linear combination of basis elements with coefficients in Z_p not all of which are divisible by p. Since the units form a group, it is clear that C acts transitively on the primitive vectors, as desired.

Of course, the argument works generally to give a decomposition of $GL_n(Z_p)$.

We may now reduce mod p^n. Let $C(p^n)$ denote the reduction of C mod p^n, and similarly for $G_\infty(p^n)$. Then we have a unique decomposition

$$GL_2(Z/p^nZ) = C(p^n)G_\infty(p^n).$$

We leave the verification to the reader.

We return to composite N,

$$N = \prod_p p^{n(p)}.$$

Let

$$C(N) = \prod_{p|N} C_p(p^{n(p)}).$$

Let $G(N) = GL_2(Z/NZ)$, and

$$G_\infty(N) = \prod G_{\infty,p}(p^{n(p)}).$$

We have

$$G(N) = C(N)G_\infty(N).$$

We note that the Galois group of the modular function field F_N over $Q(j)$ is $G(N)/\pm 1$, which we denote by $G(N)^\pm$, with a similar notation for $C(N)^\pm$. Then we have the unique decomposition

$$G(N)^\pm = C(N)^\pm G_\infty(N).$$

Suppose given for each primitive pair (r, s) mod N a modular form $h_{r,s}$ of some fixed weight, even in the sense that

$$h_{r,s} = h_{-r,-s},$$

and on which $GL_2(Z/NZ)$ operates by matrix multiplication on the indices, as with the Klein functions,

$$h_{r,s}^\alpha = h_{(r,s)\alpha}.$$

We also assume that their Fourier coefficients (coefficients of the $q^{1/N}$) lie in $Q(\zeta_N)$, and that the matrix

$$\begin{pmatrix} 1 & 0 \\ 0 & d \end{pmatrix}$$

operates by sending $\zeta \mapsto \zeta^d$ and leaving $q^{1/N}$ fixed. Finally, we assume that the divisor of each $h_{r,s}$ on $X(N)$ has only components at infinity, i.e. at cusps.

In view of the previous discussion, a **primitive** vector corresponds to a unique element $\alpha \in C(N)$, and we may write

$$h_{r,s} = h_\alpha.$$

The primes at infinity on F_N (i.e. lying above $j \to \infty$) correspond to the orbit of $C(N)^{\pm}$. The order vector

$$h \mapsto (\ldots, \mathrm{ord}_P h, \ldots)$$

indexed by the points P at infinity can therefore be also described as

$$h \mapsto (\ldots, \mathrm{ord}_\infty h^\beta, \ldots),$$

for β ranging over $C(N)^{\pm}$. The order ord_∞ is the order with respect to the standard point P_∞ as before, given by the embedding in the power series in $q^{1/N}$.

We shall say that the system of forms $\{h_\alpha\}$ is **full** if the rank of their order vectors is equal to the number of points at infinity on F_N, i.e. equal to

$$v_\infty(N) = |C(N)^{\pm}| = (G : \pm G_\infty).$$

The **regulator** of the system of forms h_α is defined to be the absolute value of the determinant

$$\det \mathrm{ord}_\infty h_\alpha^\beta = \det \mathrm{ord}_\infty h_{\alpha\beta},$$

with $\alpha, \beta \in C(N)^{\pm}$. By the Frobenius determinant formula, cf. for instance [L 1], Chapter 21, § 2 we have

$$\det \mathrm{ord}_\infty h_{\alpha\beta} = \prod_\chi \sum_\alpha \chi(\alpha) \mathrm{ord}_\infty h_\alpha,$$

where χ ranges over the character group of $C(N)^{\pm}$, and α ranges over $C(N)^{\pm}$. Therefore we obtain:

Theorem 2.1. *The system of forms $\{h_\alpha\}$ is full if and only if for each character χ of $C(N)^{\pm}$ we have*

$$\sum_\alpha \chi(\alpha) \mathrm{ord}_\infty h_\alpha \neq 0.$$

This is the case if and only if the lattice spanned by their order vectors is of finite index in the free abelian group generated by the points at infinity. The regulator is equal to this index.

Let h_0 be any one of the forms h_α. Then h_α/h_0 is a modular function in F_N, and is a unit in QR_N. Again by the Frobenius determinant formula, we get:

Theorem 2.2. *If the forms $\{h_\alpha\}$ are full, then the functions h_α/h_0 are full in the group of divisors of degree 0 at infinity.*

Let $J^\infty(N)$ denote the group of divisor classes of degree 0 generated by the points at infinity on $X(N)$, i.e. the primes at infinity of F_N. If the forms $\{h_\alpha\}$ are full, then we see that $J^\infty(N)$ is finite, and its order divides the regulator of the functions $\{h_\alpha/h_0\}$.

If
$$\alpha = \begin{pmatrix} a & b \\ c & d \end{pmatrix},$$

then we let

$$T_N = T: \mathrm{Mat}_2(\mathbf{Z}/N\mathbf{Z}) \to \mathbf{Z}/N\mathbf{Z}$$

be defined by

$$T\alpha = a.$$

Then T is linear and surjective.

If $\mathfrak{t}_\alpha = \mathfrak{t}_{r,s}$ is the Klein form, then $h_\alpha = \mathfrak{t}_\alpha^{2N}$ is invariant under $\Gamma(N)$ by **K 3** of § 5, and

$$\mathrm{ord}_\infty \mathfrak{t}_\alpha^2 = f(T\alpha).$$

where

$$f(x) = f_N(x) = \left\langle \frac{x}{N} \right\rangle \left(\left\langle \frac{x}{N} \right\rangle - 1 \right), \quad x \in \mathbf{Z}/N\mathbf{Z},$$

and $\langle t \rangle$ is the smallest real number ≥ 0 in the residue class of t mod $\mathbf{Z}$. Therefore

$$\sum \chi(\alpha) \, \mathrm{ord}_\infty \mathfrak{t}_\alpha^2 = \sum \chi(\alpha) f(T\alpha),$$

where the sum is taken over $\alpha \in C(N)^\pm$.

Theorem 2.3. (i) *If N is a prime power, then the primitive Klein forms $\mathfrak{t}_\alpha^{2N}$ are full.*

(ii) *If N is composite, then the primitive Klein forms are not full.*

Proof. The prime power case will be proved in [KL 2]. From the q-expansions, we know that in the prime power case, the Siegel functions are not units in R_N, only in $R_N[1/N]$. The opposite holds in the composite case. It follows at once that the primitive Klein forms cannot be full in the composite case.

In the prime case, or prime power case, we tried two methods to prove the fullness. The first method consisted in decomposing the character sum of Theorem 2.1 in orbits under the action of $(\mathbf{Z}/p^n\mathbf{Z})^*$. One could then prove quite easily and shortly that the character sum does not vanish, whenever the character induced by χ on $(\mathbf{Z}/p^n\mathbf{Z})^*$ is not trivial. On the other hand, when this character was trivial there seemed to be some difficulty.

Tate pointed out to us that by expressing the function f in terms of its Fourier coefficients, a simple manipulation decomposes the character sum into essentially standard Gauss sums, and a sum involving f on $\mathbf{Z}/p^n\mathbf{Z}$. He also pointed out to us that this last type of sum is precisely equal to the generalized Bernoulli numbers $B_{2,\chi}$ of Leopoldt, which are values of Dirichlet L-series at -1. This approach works in all cases. It also generalizes considerably, and hence a general treatment of this sort of algebra is given in a separate paper [KL 2]. The reader is advised to carry out the proof first in the case of prime level, where only the easier parts of the prime power case play a role.

The second method attempted to relate the multiplicative independence of the Klein-Siegel functions modulo constants with the independence results of Ramachandra [Ra] concerning units in ray class fields obtain from complex multiplication. Presumably, the independence of units proved by Ramachandra should be stronger than the generic case, and should imply it by a specialization argument, but this method remains to be worked out.

The next sections are devoted to the analysis of the sum

$$\sum \chi(\alpha) f(T\alpha).$$

Since f is an even function, i.e. $f(x) = f(-x)$, and for Theorem 2.3 we use only even characters, we only increase the generality by analysing the sum with arbitrary characters.

In the composite case when the primitive Klein forms are not full, it becomes necessary to take a power product of Klein forms, not necessarily primitive, in order to obtain a full set of forms, and we make here the appropriate definitions and general comments.

Let $\mathfrak{K}^{2N}$ be the group of products

$$\prod_{r,s} \mathfrak{k}_{r,s}^{2Ne(r,s)}$$

taken over all pairs (r, s) with $0 \leq r, s \leq N-1$, and not both $r, s \equiv 0 \pmod{N}$. We call such pairs **non-trivial** pairs. The exponents $e(r, s)$ are taken to be integers. We call $\mathfrak{K}^{2N}$ the **Klein group**. This is a temporary definition, useful for our present purposes of proving fullness. In a subsequent paper, we shall give the more general definition of the Klein group of level N, i.e. the group of elements expressible as power products of Klein forms, and modular of level N.

We consider the order mapping

$$h \mapsto (\ldots, \mathrm{ord}\, h^\alpha, \ldots), \qquad h \in \mathfrak{K},$$

with $\alpha \in C(N)^\pm$ as before. We say that the group $\mathfrak{K}^{2N}$ is **full** if the rank of the order vectors is $v_\infty(N)$.

Theorem 2.4. *The group $\mathfrak{K}^{2N}$ of Klein forms is full.*

As mentioned in the introduction of our first paper, this gives another proof for the Manin-Drinfeld theorem that the points at infinity are of finite order in the Jacobian of $X(N)$ (cf. [Dr], [Ma].)

Remark 1. It is unusual and always significant for points on a curve to be of finite order in the Jacobian, in view of the Mumford-Manin conjecture that there is only a finite number of such points. For a discussion of this, and a reduction of the proof of the conjecture to a problem concerning the Galois group of division points, cf. [L 2].

Remark 2. We regard it as a strong possibility that the units in QR_N which can be expressed as power products of the Klein forms constitute the full group of units (modulo constants). This would require different arguments to prove.

The fullness of the Klein group will be proved by the following lemmas, completed by [KL 2].

Lemma 1. *Let K be some commutative monoid. Let C be a finite abelian group of automorphisms of K, of order v. Let*

$$\lambda : K \to R^v$$

be a homomorphism. Let

$$L : K \to R^v = R^C$$

be the map $L(h) = (\ldots, \lambda(h^\alpha), \ldots)_{\alpha \in C}$. The following conditions are equivalent.
  (i) $L(\mathfrak{K})$ *has rank* v.
  (ii) *There exists* $h \in \mathfrak{K}$ *such that* $\det \lambda(h^{\alpha\beta}) \neq 0$.
  (iii) *For every character χ of C there exists $h \in \mathfrak{K}$ such that*

$$\sum_{\alpha \in C} \chi(\alpha) \lambda(h^\alpha) \neq 0.$$

Proof. Assume (i). Then the vector $(\ldots, \chi(\alpha), \ldots)_{\alpha \in C}$ cannot be orthogonal to $L(\mathfrak{K})$ whence (iii) follows.

It is obvious that (ii) implies (i).

Finally, assume (iii). We prove (ii). For each character ψ we can find an element $h_\psi \in \mathfrak{K}$ such that

$$\sum_\alpha \chi(\alpha) \lambda(h_\psi^\alpha) \neq 0.$$

Let

$$h = \prod h_\psi^{e(\psi)}$$

where the exponents $e(\psi)$ are integers ≥ 0 still to be determined. Then by the Frobenius determinant formula,

$$\det \lambda(h^{\alpha\beta}) = \prod_\chi \sum_\alpha \chi(\alpha) \lambda(h^\alpha).$$

Thus we have to pick the $e(\psi)$ such that for *every* character χ we have

$$\sum_\psi e(\psi) \left[\sum_\alpha \chi(\alpha) \lambda(h_\psi^\alpha) \right] \neq 0.$$

This amounts to finding a vector $E = (\ldots, e(\psi), \ldots)$ which is not perpendicular to a finite number of non-zero vectors, and can be done with sufficiently general positive integers, thereby proving (ii).

Lemma 2. *To prove the fullness of the Klein group for level N, it suffices to prove that*

$$\sum_{\alpha \in C(c)^\pm} \chi(\alpha) f_c(T_c(\alpha)) \neq 0$$

for every character χ with conductor c, with $c \neq 1$, and also that for the trivial character,

$$\sum_{\alpha \in C(N)^\pm} f_N(T_N(\alpha)) \neq 0.$$

(Since f_N is strictly negative, this last condition is obviously satisfied.)

Proof. Let

$$h_{r,s} = \mathfrak{k}_{r,s}^{2N}.$$

Let $d = N/c$. The vector

$(dr, ds)\alpha \pmod{N}$

depends only on the residue class of α mod c, i.e. on the image of α in $C(c)$. Therefore we obtain

$$2\sum_{\alpha \in C(N)^\pm} \chi(\alpha) \operatorname{ord} h^\alpha_{dr,ds} = \sum_{\alpha \in C(N)} \chi(\alpha) \operatorname{ord} h_{(dr,ds)\alpha}$$
$$= \frac{\phi(N)}{\phi(c)} \sum_{\alpha \in C(c)} \chi(\alpha) \operatorname{ord} h_{(dr,ds)\alpha}.$$

The character induced by χ on $C(c)$ is still even. This proves our lemma.

Theorem 2.5. *If χ is a primitive even character then*

$$\sum \chi(\alpha) f_N(T_N \alpha)) \neq 0.$$

This theorem will be proved, and the sum evaluated in the next paper dealing with the general theory of Bernoulli distributions, and not involving any function theory. See Theorem 5.3 of [KL 2]. In the present paper, we described how the function theoretic problems can be reduced to purely group theoretic ones.

In the prime power case, we shall see the character sum

$$\sum \chi(\alpha) f_N(T_N(\alpha))$$

for non-trivial even character χ is equal to the analogous sum taken for the corresponding primitive character, which is proved to be non-zero (Theorems 2.4 and 5.4 of [KL 2]). Therefore we get:

Theorem 2.6. *Assume that $N = p^n$ is a prime power. Then the determinant*

$\det f_N(T_N(\alpha\gamma)) \neq 0$.

This is the statement which proves the fullness of the primitive Klein forms (Theorem 2.3 (1)).

§ 3. Rank of the Units over $Z[j]$

We have already observed that the primitive Klein forms are not full for composite N, because of special divisibility properties occuring in the prime power case, and not the composite case. In fact, the rank of the groups of the two different types of units can be determined explicitly, as follows.

Let S_N be the group generated by the Siegel functions of level dividing N. Let $\bar{S}_N$ be this group modulo constants. Let

$$Z_N = \left(\frac{1}{N} \mathbf{Z}^2 / \mathbf{Z}^2\right) \Big/ \pm 1,$$

and let Z_N^* be the subset of Z_N consisting of elements whose representatives in $\frac{1}{N} \mathbf{Z}^2 / \mathbf{Z}^2$ are primitive of level N. Let S_N^* be the set of Siegel functions or primitive level N. By Theorem 2.4, the group $\bar{S}_N$ is free of order $|Z_N^*| - 1$.

A more refined result is proved in Kubert [K]. By an admissible divisor E of N we mean a product

$$\prod p^{n(p)}$$

over some of the primes dividing N, but with the same exponent $n(p)$ that this prime p divides N. Kubert proves that there exists a free basis for $\bar{S}_N$ consisting of elements from the sets S_E^* for all admissible divisors E of N. Let

$$\{h_{E,i}\}$$

be such a free basis, where the elements $h_{E,i}$ lie in S_E^*. (For simplicity of notation, we omit the bar over $h_{E,i}$.) Moreover, if E is a prime power p^n, $\{h_{E,i}\}$ consists of $\{g_{r,s}\} - \{g_{1,0}\}$.

If $E = p^{n(p)}$ is a prime power, then $(\mathbf{Z}/p^{n(p)}\mathbf{Z})^*$ acts naturally on $\mathbf{Z}_{p^{n(p)}}^*$ by multiplication. We let σ_p be the number of orbits.

We let $\bar{U}$ denote the subgroup of $\bar{S}_N$ consisting of those elements which can be represented by a unit in R_N (the integral closure of $\mathbf{Z}[j]$ in the modular function field of level N).

Theorem 3.1. *The group* $\bar{U}$ *has rank*

$$|Z_N^*| - 1 - \sum_{p|N}(\sigma_p - 1).$$

Proof. We want to know when an element

$$h = c \prod_{E,i} h_{E,i}^{m(E,i)}$$

is a unit. If E is composite, then $h_{E,i}$ is a unit by Theorem 1.1, so we need look only at the prime power case, when the above product is taken over those E which are prime powers. We let

$$h_E = \prod_i h_{E,i}^{m(E,i)},$$

so that

$$h = c \prod_{E \text{ prime power}} h_E.$$

Lemma. *Such an element* h *is in* $\bar{U}$ *if and only if for all admissible prime power* E *the element* h_E *is in* $\bar{U}$.

Proof. It is obvious that if each h_E is in $\bar{U}$ then h is in $\bar{U}$. Conversely, suppose that h is in $\bar{U}$, so that there exists a constant c such that ch is a unit in R_N, and this constant may be selected to be an algebraic number, actually lying in the field of N-th roots of unity. It is clear from the q-expansion that h is a unit in $R_N[1/N]$. Using a prime ideal factorization of c, and going to a larger field where these ideals become principal we can write

$$c = \prod_{p|N} c_p$$

where each c_p is an algebraic number which is a unit at all primes not dividing p. Then

$$c_p h_{p^{n(p)}}$$

is necessarily a unit over $\mathbf{Z}[j]$. This reduces the question of whether h is a unit to that of its p-components, i.e. to the prime power case.

Therefore let $N=p^n$, and we write $Z(N)=\mathbf{Z}/N\mathbf{Z}$, $Z(N)^*=(\mathbf{Z}/N\mathbf{Z})^*$. We write a Siegel function $g(r)=g_{r_1,r_2}$ in the form

$$g(r)=c(r)g^*(r),$$

where $c(r)$ is in the field of N-th roots of unity, and $g^*(r)$ has a q-expansion starting with a fractional power of q having coefficient 1, and such that all coefficients lie in $\mathbf{Z}[\zeta]$. In fact, we have:

$$c(r,s)=\begin{cases}\zeta^{rs} & \text{if } r\ne 0\\ (1-\zeta^s)^{12N} & \text{if } r=0.\end{cases}$$

From this it is clear that the order at p of the constant $c(r)$ has the following properties:

It is constant over each orbit of $Z(p^n)^*$, and there is precisely one orbit for which it is not zero, namely the orbit of $(0,1)$.

Let $\{S\}$ be the set of orbits of Z_{p^n} under $Z(p^n)^*$, and let s_1 be the orbit of $(1,0)$. A power product

$$h=\prod_{s\in S-s_1}\prod_{r\in s}g(r)^{m(r)}\prod_{r\in s_1-(1,0)}g(r)^{m(r)}$$

has the leading coefficient

$$c(h)=c\prod_S\prod_{r\in s}c(r)^{m(r)}.$$

Furthermore, h is a unit over $\mathbf{Z}[j]$ if and only if it is a unit at every divisor extending a minimal prime in $\mathbf{Z}[j]$. These are of two types: Geometric, and arithmetic involving reduction modulo some prime number. Of these, obviously only those involving reduction mod p are relevant, and we see that h is a unit of and only if the leading coefficient of every conjugate of h over $\mathbf{Q}(j)$ is a unit in $\mathbf{Z}[\zeta]$. In view of the properties of the constant $c(r)$, this is equivalent with the system of linear equations:

$$\operatorname{ord}_p(1-\zeta)^{12N}+\sum_{r\in s}m(r)=-\operatorname{ord}_p c \quad \text{for} \quad s\in S-s_1$$
$$\operatorname{ord}_p(1-\zeta)^{12N}+\sum_{r\in s_1-(1,0)}m(r)=-\operatorname{ord}_p c.$$

Given $\operatorname{ord}_p c$, these conditions are clearly independent, so the number of linearly independent conditions is exactly equal to σ_p-1 as we let $\operatorname{ord}_p c$ vary.

Therefore, in the prime power case, the rank of $\bar U$ is

$$|Z^*_{p^n}|-1-(\sigma_p-1)=|Z^*_{p^n}|-\sigma_p.$$

Furthermore, $\bar S_{p^n}/\bar U_{p^n}$ has rank σ_p-1. Therefore $\bar S_N/\bar U$ has rank

$$\sum(\sigma_p-1).$$

Since $\bar S_N$ has rank $|Z^*_N|-1$, it follows that $\bar U_N$ has rank

$$|Z^*_N|-1-\sum(\sigma_p-1),$$

as was to be shown.

Theorem 3.2. *Let $\bar{U}$ be the group of units in S_N over $\mathbf{Z}[j]$, modulo roots of unity. Then $\bar{U}$ has rank*

$$|Z_N^*| - 1 - \sum_{p|N}(\sigma_p - 1).$$

Proof. We first reduce to the prime power case. Then $\{g_{r,s}\}$ (the primitive Siegel functions) will give a free basis for $\bar{S}_{p^n}$ (we don't have to excise $g_{1,0}$), and the final condition reads

$$\sum_{r \in s} m(r) = 0, \quad \text{all} \quad s \in S.$$

So we get σ_p independent conditions, and $\bar{S}_{p^n}/\bar{U}_{p^n}$ has order σ_p. Therefore if N is arbitrary, $\bar{S}_N/\bar{U}_N$ has rank $\sum \sigma_p$. Now $\bar{S}_N$ has rank

$$|Z_N^*| - 1 + \sum_{p|N} 1,$$

so $\bar{U}$ has the desired rank.

References

[Dr] Drinfeld, V.G.: Two theorems on modular curves, Functional Analysis and its applications, Vol. 7 No. 2, translated from the Russian, April-June 1973, pp. 155—156

[F] Fricke, R.: Elliptische Funktionen und ihre Anwendungen, Vol. One, Teubner Verlag, Leipzig (1930) pp. 450—451

[Iw] Iwasawa, K.: Lectures on p-adic L-functions, Annals of Math. Studies No. 74

[K-F] Klein, F., Fricke, R.: Vorlesungen über die Theorie der elliptischen Modulfunktionen, Vol. Two, Johnson Reprint Corporation, New York, and Teubner Verlag, Stuttgart (1966), from the 1890 edition

[K] Kubert, D.: A system of free generators for the universal even ordinary distribution on Q^k/Z^k, to appear

[KL 1] Kubert, D., Lang, S.: Units in the modular function field I, Diophantine applications, this volume

[KL 2] Kubert, D., Lang, S.: Distributions on toroidal groups, to appear

[L 1] Lang, S.: Elliptic Functions, Addison Wesley, Reading, Mass. 1974

[L 2] Lang, S.: Division points on curves, Ann. mat. pura ed appl. IV, Tomo LXX (1965) pp. 229—234

[Le] Leopoldt, H.: Eine Verallgemeinung der Bernoullischen Zahlen, Abh. Math. Sem. Hamburg (1958) pp. 131—140

[Ma] Manin, J.: Parabolic points and zeta functions of modular curves, Izv. Akad. Nauk SSSR, Ser. Mat. Tom 36 (1972) No. 1, AMS translation pp. 19064

[N] Newman, M.: Construction and application of a class of modular functions, Proc. London Math. Soc. (3) (1957) pp. 334—350

[O] Ogg, A.: Rational points on certain elliptic modular curves, AMS conference, St. Louis, 1972, pp. 221—231

[Ra] Ramachandra, K.: Some applications of Kronecker's limit formula, Ann. of Math. 80 (1964) pp. 104—148

[Ro] Robert, G.: Unités élliptiques, Bull. Soc. Math. France, Mémoire No. 36 (1973)

[Sh] Shimura, G.: Introduction to the arithmetic theory of automorphic functions, Iwanami Shoten and Princeton University Press, 1971

[Si] Siegel, C.L.: Lectures on advanced analytic number theory, Tata Institute Lecture Notes, 1961

(Received July 4, 1975)

Units in the Modular Function Field

III. Distribution Relations

Dan Kubert* and Serge Lang*

Department of Mathematics, Yale University, New Haven, Conn. 06520, USA

Contents

1. Definition of Distributions 273
2. The Stickelberger Element 275
3. The Robert Relations 278
4. The Siegel Functions as a Universal Distribution 282

In this part, we give several relations among units and divisor classes at infinity on the modular curves $X(N)$, following the pattern of the Mazur distributions [Ma 1], [Ma 2], [M-SwD]. Similar relations are also those satisfied by Stickelberger elements in number fields, in cyclotomic towers, or with respect to certain projective limits of class field towers, about which there is developing a fairly substantial literature, for instance McKenzie [McK], or [L 2], Chapter IV, § 4; Coates and Sinnott [Co-S], Siegel [Si 1], [Si 2] Rideout [R], Serre [Se], etc., following the work of Iwasawa [I].

We first recall the basic relation of Bernoulli numbers (for B_2) on the projective system $\{Z/NZ\}$. We then discuss the analogous relations on Z^2/NZ^2, and on the Cartan group $C(N)$ used in [KL II] in connection with cusps and units in the modular function field. This gives rise to the analogue of the Stickelberger element, which annihilates the divisor classes at infinity. Finally, we point out that relations among units due to Klein, Ramachandra and Robert may be expressed in terms of Mazur distributions.

§ 1. Definition of Distributions

We follow Mazur. Let $G = \lim G/G_n$ be a compact group, which is the projective limit of finite factor groups G/G_n, where G_n is an open subgroup. Let $T_n = G/G_n$. Let $\{\varphi_n\}$ be a family of functions of T_n into some abelian group W. We say that this family is **consistent**, or defines a **distribution** on G, if it satisfies the following condition:

DIST. Let $x \in T_n$, and let $\pi: T_{n+1} \to T_n$ be the canonical map. Then

$$\sum_{\pi y = x} \varphi_{n+1}(y) = \varphi_n(x).$$

---

* Supported by NSF grants.

The reason for the word "distribution" is as follows. Suppose that W is the additive group of a ring. Let f be a function on G/G_n for some n. Then the value

$$\sum_{x \in T_n} f(x)\varphi_n(x)$$

is independent of the choice of n for which f is defined mod G_n, and can be defined as the integral

$$\int_G f \, d\varphi.$$

Remark. We have indexed the projective system $\{G/G_n\}$ by the sequence of positive integers. This is for convenience. Often in the applications it is indexed by positive integers ordered by divisibility.

The condition **DIST** can be expressed in terms of group rings. Suppose given for each n an element $\theta_n \in W[G/G_n]$. The canonical homomorphism

$$\pi : G/G_{n+1} \to G/G_n$$

induces an algebra homomorphism

$$W[G/G_{n+1}] \to W[G/G_n].$$

Each element θ_n can be written in the form

$$\theta_n = \sum \varphi_n(x) x,$$

for some function $\varphi_n : G/G_n \to W$. The condition that for all n,

$$\pi \theta_{n+1} = \theta_n$$

is obviously equivalent to the condition **DIST** on the family of functions $\{\varphi_n\}$.

Example. The simplest example of a projective system of groups is that of $\mathbf{Z}/N\mathbf{Z}$. There is a useful commutative diagram, using $\frac{1}{N}\mathbf{Z}/\mathbf{Z}$ instead, ordering integers by divisibility. Suppose $M|N$. Let $D = N/M$.

$$\begin{array}{ccc} \frac{1}{N}\mathbf{Z}/\mathbf{Z} & \xrightarrow{N} & \mathbf{Z}/N\mathbf{Z} \\ {\scriptstyle D}\downarrow & & \downarrow{\scriptstyle \text{can}} \\ \frac{1}{M}\mathbf{Z}/\mathbf{Z} & \xrightarrow{M} & \mathbf{Z}/M\mathbf{Z} \end{array}$$

In the diagram, the horizontal maps are isomorphisms. The right vertical map is the canonical map, and the left vertical map is multiplication by D. The same diagram of course holds with $\mathbf{Z}$ replaced by $\mathbf{Z}^2 = \mathbf{Z} \times \mathbf{Z}$, or with more general groups with congruence conditions mod N and mod M. One of our applications will be when the group is the Cartan group $C(N)$ as in [KL II], § 2. This is carried our in the next section.

Example. Here we give the example of the Bernoulli distribution, in the special case relevant to the study of units. We consider the polynomial

$$Q(t) = B_2(t) = t^2 - t + \tfrac{1}{6}.$$

Units in the Modular Function Field III. 275

We note that $Q(t)$ is the unique polynomial of degree 2 with leading coefficient 1 such that

$$Q(t) = Q(1-t)$$

and

$$\int_0^1 Q(t)dt = 0.$$

Lemma 1. *We have the identity*

$$N \sum_{x=0}^{N-1} Q\left(\frac{t+x}{N}\right) = Q(t).$$

Proof. It is obvious that the left hand side is of degree 2, has leading coefficient 1, and has the same values at t and $1-t$. The integral is also obviously equal to 0.

For the next lemma, if t is a real number, we let $\langle t \rangle$ be the smallest real number ≥ 0 in the residue class of $t \bmod \mathbf{Z}$.

Lemma 2. *We have*

$$N \sum_{x=0}^{N-1} Q\left(\left\langle t + \frac{x}{N} \right\rangle\right) = Q(\langle Nt \rangle).$$

Proof. The family of elements of $\mathbf{R}/\mathbf{Z}$ given by

$$t + \frac{x}{N} \pmod{\mathbf{Z}}, \quad x = 0, \ldots, N-1$$

coincides with the family of elements whose N-th multiples are equal to $Nt \pmod{\mathbf{Z}}$. Positive representatives for these elements mod $\mathbf{Z}$ are given by

$$\frac{\langle Nt \rangle + x}{N}, \quad x = 0, \ldots, N-1.$$

Therefore Lemma 1 implies the relation of Lemma 2.

For $x \in \mathbf{Z}/N\mathbf{Z}$, we define

$$Q_N(x) = Q\left(\left\langle \frac{x}{N} \right\rangle\right).$$

Then Lemma 2 shows that

$$\{NQ_N\}$$

is a distribution on the system $\{\mathbf{Z}/N\mathbf{Z}\}$.

§ 2. The Stickelberger Element

On $\mathbf{Z}^2/N\mathbf{Z}^2$ we define the function

$$v_N(a,c) = Q_N(a).$$

Theorem 2.1. *The family of functions $\{v_N\}$ defines a distribution on the projective system $\{\mathbf{Z}^2/N\mathbf{Z}^2\}$.*

Proof. In order to prove the compatibility relation, it suffices to do it for two integers

$$M \quad \text{and} \quad N = pM$$

where p is a prime number. The inverse image of the element

$$\left(\frac{a}{M}, \frac{c}{M}\right) \in \frac{1}{M} \mathbf{Z}^2/\mathbf{Z}^2$$

in $\frac{1}{N} \mathbf{Z}^2/\mathbf{Z}^2$ consists of

$$\left(\frac{a}{pM} + \frac{v}{p}, \frac{c}{pM} + \frac{\lambda}{p}\right), \quad v, \lambda = 0, \ldots, p-1.$$

Thus

$$\sum_{\lambda=0}^{p-1} \sum_{v=0}^{p-1} Q\left(\left\langle \frac{a}{pM} + \frac{v}{p} \right\rangle\right) = \sum_{\lambda=0}^{p-1} \sum_{v=0}^{p-1} Q\left(\left\langle \frac{a/M + v}{p} \right\rangle\right)$$

$$= Q\left(\left\langle \frac{a}{M} \right\rangle\right).$$

This proves the theorem.

Let $C(N)$ be the Cartan subgroup as defined in [KL II], § 2. (For prime power $p^n = N$, it is the group of units in the quadratic unramified extension of $\mathbf{Z}_p$, mod p^n. In general, it is defined by multiplicativity.)

If

$$\alpha = \begin{pmatrix} a & b \\ c & d \end{pmatrix}$$

is in $\mathrm{Mat}_2(\mathbf{Z}/N\mathbf{Z})$, we let as before

$$T\alpha = a \in \mathbf{Z}/N\mathbf{Z}.$$

We define

$$v_N(\alpha) = Q_N(T\alpha).$$

An element α of $C(N)$ is determined by its first column, and conversely, and such a column is characterized as a primitive vector mod N.

We have a commutative diagram:

$$\begin{array}{ccccc}
\left(\frac{1}{N}\mathbf{Z}^2/\mathbf{Z}^2\right)^* & \longleftarrow & (\mathbf{Z}^2/N\mathbf{Z}^2)^* & \longleftarrow & C(N) \\
{\scriptstyle N/M} \downarrow & & \downarrow {\scriptstyle r_M} & & \downarrow {\scriptstyle r_M} \\
\left(\frac{1}{M}\mathbf{Z}^2/\mathbf{Z}^2\right)^* & \longleftarrow & (\mathbf{Z}^2/M\mathbf{Z}^2)^* & \longleftarrow & C(M)
\end{array}$$

226

Units in the Modular Function Field III.

We denote reduction mod M by r_M. The map on the left is multiplication by N/M. The horizontal arrows are bijections. The star indicates the subset of primitive elements. Factoring each group by ± 1 would also yield a similar commutative diagram.

Theorem 2.2. *Let $M|N$, and suppose that every prime p dividing N also divides M. Then the compatibility relation holds for $\bar{\alpha} \in C(M)$:*

$$\sum_{r_M \alpha = \bar{\alpha}} v_N(\alpha) = v_M(\bar{\alpha}).$$

In particular, the family of functions $\{v_{p^n}\}$ forms a distribution on the projective system

$$\{C(p^n)\}.$$

Proof. In the argument of Theorem 2.1, let us suppose that p also divides M. Then the elements of the inverse image

$$\left(\frac{a}{pM} + \frac{v}{p}, \frac{c}{pM} + \frac{\lambda}{p}\right), \quad v, \lambda = 0, \ldots, p-1$$

are primitive if (a, c) is primitive mod N. Therefore these elements can be viewed as the first columns of elements in $C(N)$, whence the desired compatibility relation follows.

We let the **Stickelberger element** of level N be

ST 1. $\theta(N) = \sum_{\gamma \in C(N)} v_N(\gamma) \gamma$.

This is an element of the group ring $Q[C(N)]$. The preceding theorem can then be expressed in the form

$$r_M \theta(N) = \theta(M),$$

whenever M and N are divisible by the same primes and M divides N.

Application to Units

We assume that the reader is acquainted with the Siegel functions, and the fact that we can index them by elements of the Cartan subgroup (mod ± 1), as in [KL II], § 2. If g is a modular function in the modular function field F_N of level N, i.e. the function field of the curve $X(N)$ over $Q(\zeta_N)$, and g is a unit in QR_N (where R_N is the integral closure of $Z[j]$ in F_N), then the divisor of g is concentrated at infinity. We define the order mapping

$$v : (QR_N)^* \to Q[C(N)]$$

by

$$v(g) = \sum_{\gamma \in C(N)} (\text{ord } g^\gamma) \gamma, \quad \text{ord} = \text{ord}_{q^{1/N}},$$

We write g^γ instead of $g^{\sigma(\gamma)}$ for simplicity, in the light of the operation of $GL_2(Z/NZ)$ on F_N over $Q(j)$. Cf. [L 1], Chapter VI, § 3.

If $g = g_\alpha$ is a Siegel function, then we have the relation

ST 2. $N^2 \theta(N)\alpha = v(g_\alpha),$

which is immediate from the definitions. Thus the Stickelberger element converts the "cusp" α into the divisor of the function g_α. In particular, multiplied by a suitable integer, it annihilates the divisor classes at infinity.

In [KL II] we shifted back and forth between the order functions of the Siegel functions and the Klein forms. Here again, we must take this into account for the next theorem. We let

$$Q^0(X) = X^2 - X, \quad Q_N^0(x) = Q^0\left(\left\langle \frac{x}{N} \right\rangle\right)$$

$$f_N(\alpha) = Q_N^0(T\alpha).$$

We let $C(N)^\pm = C(N)/\pm 1$. Since the function f_N is even, it is now more appropriate to look at the group algebra $Q[C(N)^\pm]$. We put

$$\theta^0(N) = \sum_\gamma f_N(\gamma)\gamma,$$

and call this the **augmented Stickelberger element** (of level N). The sum is taken over γ in $C(N)^\pm$.

Theorem 2.3. *If $N = p^n$ is a prime power, then $\theta^0(N)$ is invertible in the group algebra $Q[C(N)^\pm]$.*

Proof. We know from the fullness of the Klein forms [KL II], [KL D], that the determinant of the order vectors is non-zero. This means that the elements

$$u_\alpha = \sum_\gamma f_N(\gamma)\alpha\gamma, \quad \alpha \in C(N)^\pm,$$

are linearly independent over Q. Therefore a linear combination of these elements with rational coefficients is equal to 1, and this immediately shows that there exists an element of the group ring $Q[C(N)^\pm]$ which is an inverse for the augmented Stickelberger element. In other words, the augmented Stickelberger element is a unit in the group ring.

The group $C(N)^\pm$ induces a group of automorphisms of the modular curve $X(N)$, and therefore a group of automorphisms of its Jacobian variety $J(N)$. We denote its image in $\text{Aut } J(N)$ by $C_J(N)$. We get a corresponding representation

$$Q[C(N)^\pm] \to \text{End}_Q J(N) = Q \otimes \text{End } J(N),$$

and we let $\theta_J(N)$ be the image of $\theta(N)$ under this representation. Similarly, we let

$$\theta_J^0(N)$$

be the image of $\theta^0(N)$ under the representation. We obtain:

Corollary. *Some integral multiple of $\theta_J^0(N)$ is an isogeny.*

Proof. A unit in an algebra remains a unit under a representation.

§3. The Robert Relations

Strictly speaking the relations between the Siegel functions given in this section were proved in increasing degree of generality by Klein [K-F], p. 68, 275, also Part V, Chapter 2 throughout; Ramachandra [Ra], and Robert [Ro]. We shall

Units in the Modular Function Field III.   279

prove them by a method which apparently goes back to Klein, who only obtained partial results.

We need some notation. Let L be a lattice in $\mathbf{C}$ and let t be a complex number such that $Nt \in L$ but $t \notin L$. Let

$$L = [\omega_1, \omega_2], \quad \text{with} \quad \text{Im}\,\omega_1/\omega_2 > 0.$$

Write $t = (r\omega_1 + s\omega_2)/N$. We define

$$g_N(t, L) = \mathfrak{k}\left(\frac{r}{N}, \frac{s}{N}; \begin{pmatrix}\omega_1\\\omega_2\end{pmatrix}\right)^{12N} \Delta(L)^N.$$

Thus g_N is the Siegel function in a convenient notation.

Theorem 3.1. *Let $L \supset L'$ be two lattices. Let*

$$n = (L : L').$$

Let N be an integer > 1, and $t \in \frac{1}{N}L'$, but $t \notin L'$. Let

$$\{t_1, \ldots, t_n\}, \quad t_1 = 0,$$

be coset representatives of L' in L. Finally let

$$m = \text{scm}(N, n')$$

be the smallest common multiple of N and n', where n' is the smallest positive integer such that $n'L \subset L'$. Then

$$\prod_{i=1}^{n} g_m(t + t_i, L') = g_m(t, L).$$

Proof. The proof is based on translating the relation into a relation about modular functions. In order to prove that two modular functions of level m are equal it suffices to prove:

(i) They have the same orders of zeros and poles.
(ii) At the standard prime at infinity, the leading coefficient of their q-expansions coincide.

Indeed, from (i) we conclude that their quotient is a constant, and this constant must be equal to 1 by (ii). In our applications, we deal with modular functions having zeros or poles only at the points at infinity, so we need only check the orders of zeros and poles of the conjugates of the given modular functions over $\mathbf{C}(j)$ at the standard prime at infinity to verify (i). In fact, we may do so for a fixed power of the two modular functions, since for such a power, the orders get multiplied by the same integer.

The integer m in the theorem is chosen to be least such that each point $t + t_i$ has period m with respect to L'. It is a function of the index $(L : L')$ and of the period of t. If the assertion of the theorem is proved for a triple (L, L', t) and m, then it follows when m is replaced by any multiple, from the definition of the invariant g_m. Let

$$L \supset L_1 \supset L'$$

229

for some sublattice L_1 such that $(L_1 : L') = p$ is prime. Let

$$u_1, \ldots, u_{n'}$$

be coset representatives of L_1 in L (with $u_1 = 0$), and let

$$t_1, \ldots, t_p$$

be coset representatives of L' in L_1. Then

$$t_k + u_j \quad (1 \leq k \leq p, 1 \leq j \leq n')$$

are coset representatives of L' in L. By induction on the number of prime factors of $(L : L')$, we see that if we have proved our relation for prime index, then

$$\prod_{i=1}^{n} g_m(t + t_i, L') = \prod_{k=1}^{p} \prod_{j=1}^{n'} g_m(t + t_k + u_j; L')$$
$$= \prod_{k=1}^{p} g_m(t + t_i; L_1)$$
$$= g_m(t, L).$$

This reduces to proving the relation for the integer m when

$$(L : L') = p,$$

which we assume.

In this case, we can take $L = [\tau, 1]$, and L' is any one of the lattices

$$[p\tau, 1], \quad [\tau + b, p], \quad 0 \leq b \leq p - 1.$$

Coset representatives of L/L' are given by:

$e\tau$, with $0 \leq e \leq p - 1$, if $L' = [p\tau, 1]$

e, with $0 \leq e \leq p - 1$, if $L' = [\tau + b, p]$.

The point t can then be written

$$t = \frac{rp\tau + s}{N} \quad \text{if} \quad L' = [p\tau, 1],$$

$$t = \frac{r(\tau + b) + sp}{N} \quad \text{if} \quad L' = [\tau + b, p].$$

Lemma. *For any r, s not both $\equiv 0 \pmod{N}$, the following pairs of modular functions have the same order at infinity, i.e. in their q-expansions:*

(1) $\prod_{e=0}^{p-1} g_m\left(\frac{r(\tau + b) + sp}{N} + e; [\tau + b, p]\right)$ and $g_m\left(\frac{r(\tau + b) + sp}{N}; [\tau, 1]\right)$

(2) $\prod_{e=0}^{p-1} g_m\left(\frac{rp\tau + s}{N} + e\tau; [p\tau, 1]\right)$ and $g_m\left(\frac{rp\tau + s}{N}; [\tau, 1]\right)$.

Proof. In the first case, we put

$$\tau' = \frac{\tau + b}{p}.$$

Then $q_{\tau'} = q_{\tau/p} \zeta_p^{rb}$, and $\text{ord}_{q_\tau} = \frac{1}{p} \cdot \text{ord}_{q_{\tau'}}$. From the basic q-product for the Siegel functions, we see that each term on the left hand side has order at q equal to $1/p$ times the order at q of the right hand side. This proves case (1).

The second case depends on a different phenomenon. This time, we let

$$\tau' = p\tau,$$

so that $\text{ord}_{q_{\tau'}} = \frac{1}{p} \cdot \text{ord}_{q_\tau}$. The order at q of the left hand side is equal to the sum

$$p \sum_{e=0}^{p-1} 6mQ\left(\left\langle \frac{pr/N + e}{p} \right\rangle\right),$$

which, by the basic relation of the Bernoulli polynomial, is equal to

$$6mQ\left(\left\langle \frac{rp}{N} \right\rangle\right).$$

This is precisely equal to the order of the right hand side, and our lemma is proved.

The conjugates of a modular function $f(\tau)$ over $C(j)$ are given by the action of $SL_2(\mathbf{Z})$, and are of the form $f(\gamma\tau)$ for $\gamma \in SL_2(\mathbf{Z})$. Under the action of γ, the point t changes, but one sees immediately that the lemma suffices to prove that the two functions occurring say in part (2) of the lemma have the same orders at all points at infinity. This takes care of (i).

For (ii) we proceed in an analogous manner. In this case, we have only to look at one conjugate of the modular functions under consideration, say corresponding to the sublattice $L' = [\tau, p]$. From the q-product **K6** for the Siegel functions, we see that the leading coefficient of the q-expansion of the left hand side is given by the following values:

(a) $r \equiv 0 \pmod{N}$:

$$(1 - e^{2\pi i ps/N})^{12m} \prod_{k=0}^{p-1} e^{2\pi i 6 \frac{r}{N}(\frac{s}{N} + \frac{k}{p})} = (1 - e^{2\pi i ps/N})^{12m} e^{2\pi i 6 rsp/N^2}.$$

(b) $r \not\equiv 0 \pmod{N}$: $e^{2\pi i 6 rsp/N^2}$.

The same q-product for $g_m\left(\frac{r\tau + ps}{N}; [\tau, 1]\right)$ shows that the leading coefficient of the function on the right hand side has the same value, and our theorem is proved.

We reformulate some of Theorem 3.1 in a more suggestive notation.

Let r, s be two integers, not both $\equiv 0 \pmod{N}$. Let us define

$$g\left(\frac{r}{N}, \frac{s}{N}\right)$$

to be the function on the upper half plane whose value at τ is

$$g\left(\frac{r}{N}, \frac{s}{N}; \tau\right) = \mathfrak{k}\left(\frac{r\tau + s}{N}; [\tau, 1]\right) \eta(\tau)^2,$$

where $\eta(\tau)$ is the Dedekind eta function, i.e. the standard 12-th root of $\Delta(\tau)$. We view g as a mapping from $\frac{1}{N}\mathbf{Z}^2/\mathbf{Z}^2$ into the multiplicative group of geometric units in the modular function field, modulo constants. We may define $g(0)$ any way we want, say we define $g(0)$ to be an element multiplicatively independent of all other values of g. The product

$$\prod g(t)$$

taken over all $t \in \frac{1}{N}\mathbf{Z}^2/\mathbf{Z}^2$, and $t \neq 0$ is a unit in the field of rational functions of j, and hence is constant. The Robert relations of Theorem 3.1 then show:

Corollary. *The mapping g is a distribution, on $\left\{\frac{1}{N}\mathbf{Z}^2/\mathbf{Z}^2\right\}$.*

In the next section, we shall axiomatize properties of this distribution, which will be called the **Siegel distribution**.

Remark. The statement and proof of Theorem 3.1 is much more elaborate than what is needed just to prove that g is a distribution. The corollary can be proved in just a few lines, because the sublattice involved are much simpler.

§ 4. The Siegel Functions as a Universal Distribution

Let

$$Z_N = \left(\frac{1}{N}\mathbf{Z}^2/\mathbf{Z}^2\right)$$

The abelian groups Z_N form a projective system. If $M|N$, the connecting map $Z_N \to Z_M$ is multiplication by N/M. They also form an injective system, i.e. $Z_M \subset Z_N$, this inclusion being induced by the inclusion

$$\frac{1}{M}\mathbf{Z}^2/\mathbf{Z}^2 \subset \frac{1}{N}\mathbf{Z}^2/\mathbf{Z}^2 \subset \mathbf{Q}^2/\mathbf{Z}^2 .$$

Thus the connecting map may be written in the form

$$\frac{N}{M}\delta ,$$

where δ is the identity mapping on Z_N.

A distribution $\{g_N\}$ on $\left\{\frac{1}{N}\mathbf{Z}^2/\mathbf{Z}^2\right\}$ is said to be **even** if

$$g_N(t) = g_N(-t), \quad \text{all } t \in \frac{1}{N}\mathbf{Z}^2/\mathbf{Z}^2 .$$

Such a distribution induces in an obvious way a distribution on the projective system $\{Z_N\}$, again denoted by g.

We say that a distribution $\{g_N\}$ on the system $\{Z_N\}$ is **ordinary** if there exists an even function g on $\mathbf{Q}^2/\mathbf{Z}^2$ into some abelian group A such that g_N is the restriction of g to Z_N. Of course, we could also define an ordinary distribution on the system $\left\{\frac{1}{N} \mathbf{Z}^2/\mathbf{Z}^2\right\}$ in a like manner, and then ordinary distributions on $\{Z_N\}$ are precisely the distributions induced by even ordinary distributions on $\left\{\frac{1}{N} \mathbf{Z}^2/\mathbf{Z}^2\right\}$.

We let Z_N^* denote the subset of Z_N consisting of primitive elements. The order of Z_N^* is the same as that of $C(N)$.

Let

A_N = group generated by $g(Z_N)$.

Let us assume that our distribution g is normalized so that $g(0)$ is independent of all other values $g(r)$, $r \neq 0$. Then it can be shown (cf. [KL D], Theorem 3.2) that the rank of A_N is at most equal to the rank of the Siegel distribution. The Siegel invariant g gives an example of an ordinary distribution which achieves the maximal possible rank.

It is easy to see that distributions on $\{Z_N\}$ (ordinary) form a category, the morphisms being given by homomorphisms of their image group. The usual formal procedure shows that one can construct a universal such distribution, and the maximality of the rank of the Siegel units shows that the Siegel distribution is universal.

If we let

$A = \bigcup A_N$

be the image group of g, and if $\varphi: A \to A'$ is a homomorphism, then $\varphi \circ g$ is again a distribution (ordinary, normalized).

Example 1. In the case of the Siegel distribution, we can take φ to be the order function at the standard point at infinity, and we recover the distribution of Theorem 2.1.

Example 2. Let $f \in (QR_N)^*$ be a unit in the modular function field F_N of level N. Then the divisor of f is concentrated at the points at infinity, and f has no zero or pole on the upper half plane $\mathfrak{H}$. The differential df/f is of the third kind on the modular curve $X(N)$ (the completion of $Y(N) = \Gamma(N) \backslash \mathfrak{H}$). Let us fix a determination of $\log f$ on $\mathfrak{H}$.

For each $\gamma \in \Gamma(N)$ there exists an integer $\langle f, \gamma \rangle \in \mathbf{Z}$ such that for all $z \in \mathfrak{H}$ we have

$(\log f)(\gamma z) = \log f(z) + \langle f, \gamma \rangle 2\pi i$.

Indeed, the symbol $\langle f, \gamma \rangle$ is equal to the integral

$\langle f, \gamma \rangle = \int_z^{\gamma z} \frac{df}{f}$,

which may also be viewed as an integral on $X(N)$ along a cycle, lifted to the upper half plane. The value is independent of z, for by the change of variables formula, for any $w \in \mathfrak{H}$ we have

$$\int_z^w \frac{df}{f} = \int_{\gamma z}^{\gamma w} \frac{df}{f},$$

since f is assumed invariant under $\Gamma(N)$. It is clear that the period is an integral multiple of $2\pi i$.

Again the change of variables formula shows that the symbol

$$\langle f, \gamma \rangle$$

is bimultiplicative in f and γ. In particular, on the right, it factors through the abelianized group $\Gamma(N)^{ab}$, which is a finite abelian group. Also for each γ we obtain a homomorphism

$$f \mapsto \langle f, \gamma \rangle$$

which may be denoted by $\langle \gamma \rangle$.

If $\gamma \in \mathrm{SL}_2(\mathbf{Z})$, we pick some power γ^m such that γ^m lies in $\Gamma(N)$, and define

$$\langle f, \gamma \rangle = \frac{1}{m} \langle f, \gamma^m \rangle.$$

It is immediately verified that this is well defined, and defines a homomorphism on the group generated by the Siegel functions of all levels (modulo constants). In this manner, we obtain a distribution of the type considered above, on the projective system $\{Z_N\}$.

We do not go much more deeply now into the properties of the symbol, other than for the following remark.

If $\langle f, \gamma \rangle = 0$ for some fixed f and all γ, then f is constant.

Proof. Fix some point $z_0 \in \mathfrak{H}$. The integral

$$g(z) = \int_{z_0}^z \frac{df}{f}$$

defines a holomorphic function on $\Gamma(N) \backslash \mathfrak{H} = Y(N)$. Integrating df/f around small circles around each point at infinity yields 0. Hence g cannot have a singularity at such a point, and hence g is a holomorphic function on $X(N)$, whence constant. This implies that $df = 0$, i.e. f is constant.

We shall return to the study of the symbol in a later paper, in connection with Kummer theory.

References

[Co-Si] Coates, J., Sinnott, W.: An analogue of Stickelberger's Theorem for the higher K-groups, Inv. Math. 24 (1974) pp. 149—161
[I] Iwasawa, K.: Lectures on p-adic L-functions, Annals of Math. Studies No. 74
[K-F] Klein, F., Fricke, R.: Vorlesungen über die Theorie der elliptischen Modulfunktionen, Vol. II, Johnson reprint corporation, New York, 1966 (from the 1892 edition)
[KL II] Kubert, D., Lang, S.: Units in the modular function field II. A full set of units, Math. Ann. 218 (1975) pp. 175—189

[KL D] Kubert, D. and Lang, S.: Distributions on toroidal groups, to appear, Math. Zeitschrift.
[L 1]   Lang, S.: Elliptic Functions, Addison Wesley, Reading, 1973
[L 2]   Lang, S.: Algebraic Number Theory, Addison Wesley, Reading, 1970
[Ma 1] Mazur, B.: Courbes élliptiques et symboles modulaires, Seminaire Bourbaki No. 414, June 1972
[Ma 2] Mazur, B.: Bourbaki report on p-adic distributions and p-adic L-series, 1973
[Ma-SD] Mazur, B., Swinnerton-Dyer, H. P. F.: Arithmetic of Weil curves, Inv. Math. 25 (1974) pp. 1—61
[McK]   McKenzie, R.: Class group relations in cyclotomic fields, Am. J. Math. 74 (1952) pp. 759—763
[Ra]   Ramachandra, K.: Some applications of Kronecker's limit formula, Ann. of Math. 80 (1964) pp. 104—148
[Ro]   Robert, G.: Unités élliptiques, Mémoire No. 36, Bull. Soc. Math. France 1973
[R]   Ridout, D.: A generalization of Stickelberger's theorem, PhD thesis, McGill University, Montreal, 1970
[Se]   Serre, J. P.: Formes modulaires et fonctions zeta p-adiques, in Modular functions of one variable III, Springer Lecture notes in Mathematics 350, pp. 191—268
[Si 1] Siegel, C. L.: Bernoullische Polynome und quadratischer Zahlkörper, Göttingen Nachr. 2 (1968) pp. 7—38
[Si 2] Siegel, C. L.: Über die Fourierschen Koeffizienten von Modulformen, Göttingen Nachr. 3 (1970) pp. 15—56

(Received July 4, 1975)

Diophantine Approximation on Abelian Varieties with Complex Multiplication

John Coates (Cambridge)* and Serge Lang (New Haven)**

Let A be an abelian variety of dimension d defined over a number field K. Throughout this note we assume that A has complex multiplication in the sense that $k = \text{End}(A) \otimes \mathbf{Q}$ is a totally imaginary field which is a quadratic extension of totally real field, and that $[k:\mathbf{Q}] = 2d$, see [10]. Chapter 2. We can represent the complex points $A_\mathbf{C}$ as a quotient

$$\exp: \mathbf{C}^d \to A_\mathbf{C}$$

by means of a projective embedding with theta functions $(\theta_0, \ldots, \theta_{d'})$, whose quotients $f_i = \theta_i/\theta_0$ are abelian functions. After a linear transformation of $\mathbf{C}^d$ if necessary, we assume that the differential of the exponential map at the origin is algebraic, defined over K. It is well known that the representation of k in $\mathbf{C}^d$ is isomorphic to the direct sum $\sum \tau_i$, where τ_i runs through a set of representatives of the pairs of complex conjugate embeddings of k in $\mathbf{C}$. Assuming K is sufficiently large (and in particular, contains k and all its conjugates), we can therefore make a linear change of variables in $\mathbf{C}^d$ with coefficients in K so that k acts on the new variables $(z_1, \ldots, z_d)$ by

$$\gamma(z_1, \ldots, z_d) = (\gamma_1 z_1, \ldots, \gamma_d z_d),$$

where $\gamma_i = \tau_i(\gamma)$. As in [5] we call this a strong normalization of the exponential map. When A has dimension 1, the exponential map associated with the Weierstrass function is automatically strongly normalized.

Let $|\ |$ be the sup norm on $\mathbf{C}^d$. If $\alpha = (\alpha_i)$ and $\beta = (\beta_i)$ are in $\mathbf{C}^d$, we define $\alpha\beta$ to be $(\alpha_i \beta_i)$. We say that $u \in \mathbf{C}^d$ is an algebraic point of the exponential map if $\exp(u)$ is algebraic.

Theorem. *Let n be an integer ≥ 1, and let $u^0, \ldots, u^n$ be algebraic points for the strongly normalized exponential map, which are linearly independent over k. Let $\kappa > 4 + 2d + 8d(2d-2) + 8d^2(n+1)$. Let d_0 be a given integer. Let $\alpha^1, \ldots, \alpha^n$ range over all vectors in $\mathbf{C}^d$ whose coordinates are algebraic of degree $\leq d_0$. Then there*

---
* Supported by NSF grant and Sloan Foundation
** Supported by NSF grant

exists a positive number C such that, if H is the maximum of the heights of $\alpha^1, \ldots, \alpha^n$, we have the inequality

$$|\alpha^1 u^1 + \cdots + \alpha^n u^n - u^0| > C e^{-\tau(H)},$$

where $\tau(H) = (\log H)^\kappa$, and C is independent of H.

Lower bounds for linear forms in logarithms of exponential maps are one of the main objectives of the theory of diophantine approximations. Baker and Feldman [1] have given such bounds for ordinary logarithms, and for special cases especially involving the periods of elliptic functions.

Masser, in his important thesis [7], obtained an inequality of the above type in the case of elliptic curves with complex multiplication, but with the function $\tau(H) = H^\varepsilon$. Lang [5] proved the corresponding transcendence result on abelian varieties with complex multiplication. Then Masser [8] obtained the measure of transcendence with the function $\tau(H) = H^\varepsilon$ on abelian varieties, with complex multiplication.

In an unpublished manuscript [3], it was shown how the corresponding transcendence result could be obtained on elliptic curves with complex multiplication provided one had an appropriate theorem for the degree of the field of division of algebraic points, see also [2]. In this note we point out that by returning to those ideas, it is possible to give a much simplified proof for the theorem, obtaining even the better result with our function $\tau(H)$ as a power of $\log H$. We shall use the following theorem of Ribet [9] (Bashmakov [11] for elliptic curves). If q is a positive integer, we let A_q be the group of q-torsion points on A.

Ribet's Theorem. *Let $P^1, \ldots, P^n$ be points of A_K, linearly independent over End A. For all but a finite number of primes q, the Galois group of division points*

$$K\left(A_q, \frac{1}{q}P^1, \ldots, \frac{1}{q}P^n\right) \quad \text{over} \quad K(A_q)$$

is isomorphic to the translation group by A_q^n,

$$\left(\frac{1}{q}P^1, \ldots, \frac{1}{q}P^n\right) \mapsto \left(\frac{1}{q}P^1 + a^1, \ldots, \frac{1}{q}P^n + a^n\right)$$

with $a^j \in A_q$.

The proof of the theorem depends on Ribet's theorem, and a Main Lemma, which is a summary of what one can deduce from the falsehood of the theorem by the usual techniques of the theory of transcendental numbers (construction of an auxiliary function and extrapolation) with the right choice of parameters. As in [5] we assume that the projective embedding of the complex torus has been changed at first by a projective linear transformation over K which is sufficiently general, and in particular is such that the abelian functions $f_1, \ldots, f_d$ give an analytic isomorphism of the ball $\mathbf{D}_{3\rho}(0)$ of radius 3ρ, centered at the origin, for some fixed small number ρ. We let $f = (f_1, \ldots, f_d)$.

Main Lemma. *Given the algebraic points $u^0, u^1, \ldots, u^n$ and a positive integer d_0. Let $\alpha^1, \ldots, \alpha^n$ be elements of $\mathbf{C}^d$ with algebraic components of degrees $\leq d_0$. Suppose*

that for H sufficiently large,
$$|\alpha^1 u^1 + \cdots + \alpha^n u^n - u^0| < e^{-\tau(H)}.$$

Let $M = M(H) = (\log H)^4$ and $L = (\log H)^b$ where $b = 4 - 1/(n+1)$. Then there exists a nonzero polynomial P in $(n+1)d$ variables, of degree $\leq L$ in each variable, with coefficients in K, such that

$$P\left(f\left(\frac{\gamma}{q}u^1\right), \ldots, f\left(\frac{\gamma}{q}u^n\right), f\left(\frac{\gamma}{q}u^0\right)\right) = 0$$

for all quotients γ/q satisfying:

q is a positive integer $\leq M$;

$\gamma \in \mathrm{End}(A)$ and $\max|\gamma_i| \leq M$;

$(\gamma/q)u^j \in \mathbf{D}_{2\rho}(0)$ modulo the period lattice.

This main lemma is the same as that in [5], except that $\tau(H)$ is now the new function, and the bounds for q and γ are changed accordingly. Also for simplicity we assume that the components of the α^j lie in K. The proof of the main lemma is identical with the proof given in [5], §4 through §10, except for the following changes. In the light of our new $\tau(H)$, inequality (7) must be replaced by $MS^{2d} \prec \tau(H)$, for $S = NM^{d(n+1)+2d-2}$. This value is now taken as the maximal value to which one applies the extrapolation on integral multiplies. In other words, in Lemma 9.1, we let

$$v \leq [d(n+1) + 2d - 2]/\delta.$$

In Lemma 10.1 when we apply Lemma 9.1, this means that we take the value for S given above again. Furthermore, we definitely take $b = 1$, so that $q \leq M$, and $\max|\gamma_i| \leq M$. At the end of the proof, this means that we take $T = M$. No other changes are needed. We shall combine the main lemma with the following theorem of Masser [7].

Masser's Theorem. *Let P be a polynomial in r complex variables, of degree at most L in each variable. Let B be the closed ball in $\mathbf{C}^r$ with radius $\delta > 0$ and any center. There exists a constant $c = c(r, \delta)$ such that if P has a zero within distance cL^{-1} of each point of B, then P is identically zero.*

We now give the proof of our theorem, say when some u^ℓ is a period. We shall show at the end either that one can assume this to be the case without loss of generality, or that one deals with a case which is even easier to handle. We let q be the largest prime $\leq M$, and we pick an element γ as in the main lemma with $(\gamma, q) = 1$. Note that the points $P^j = \exp(u^j)$ $(j \neq \ell)$ are linearly independent over $\mathrm{End}(A)$. Let t^j $(j \neq \ell)$ be arbitrary q-division points of the period lattice with $|t^j| \leq \rho$. By Ribet's theorem, there exists an element of the Galois group of the field of division points which, when applied to the polynomial equation of the main lemma, shows that the polynomial

$$P\left(X^0, \ldots, f\left(\frac{\gamma}{q}u^\ell\right), \ldots, X^n\right)$$

vanishes at $X^j = f\left(\dfrac{\gamma}{q} u^j + t^j\right)$. Plainly the points t^j come within a distance $\ll 1/q$ of each point of the ball of radius ρ. Since f is an analytic isomorphism, and hence preserves the order of magnitude of the distance between points, we see that the above polynomial vanishes at enough points to contradict Masser's theorem, unless it is identically zero, because $q \gg L^{1+\theta}$ for some $\theta > 0$. This means that all the coefficients of the above polynomial, which are themselves polynomials in X^ℓ, evaluated at $f\left(\dfrac{\gamma}{q} u^\ell\right)$, must be zero. Let $G(X^\ell)$ be any such polynomial. Again Masser's theorem will show that $G(X^\ell)$ is identically zero provided the points $\dfrac{\gamma}{q} u^\ell$ lie within distance $\ll 1/q$ of any point of the ball of radius ρ for γ satisfying the conditions of the main lemma. They do for the following reason. Note that $\dfrac{1}{q} u^j \in \mathbf{D}_\rho(0)$ for $j = 1, \ldots, n$. Let $\eta \in \mathrm{End}(A)$ be such that $\|\eta u^j\| \leq \rho q$. Let $Q_1, \ldots, Q_{r_1}$ be the prime ideals in k lying above q, which divide $\eta + 1$. We may assume $q \geq 2d + 1$. Then $\eta + 2$ is not divisible by these primes. Let $Q_{r_1+1}, \ldots, Q_{r_2}$ divide $\eta + 2$. Then $\eta + 3$ is not divisible by Q_i for $1 \leq i \leq r_2$. Continuing in this way, and using the fact that there are at most $2d$ primes of k lying above q we obtain some element $\eta + 1 + v$ with $v \leq 2d$, which is prime to q. Let $\gamma = \eta + 1 + v$. Such points $\dfrac{\gamma}{q} u^\ell$ have the desired property. Thus finally, the polynomial P is identically zero, a contradiction which proves the theorem in this case.

If the points $u^0, \ldots, u^n$ are linearly independent from the periods over k, then the same argument goes through without mentioning the point u^ℓ. If they are linearly dependent with the periods over k, then we may use a linear relation to reduce this case to the situation when some u^j is a period.

Remarks. In the case of elliptic curves, it is not necessary to use Masser's theorem. One simply notes that the polynomial equation in the main lemma implies that there exists an integer j with $0 \leq j \leq n$ such that the degree of $f(u^j/q)$ over $K(f(u^0/q), \ldots, f(u^{j-1}/q))$ if $j > 0$, and over K if $j = 0$, is smaller than it ought to be. This contradicts Bashmakov's theorem, or the fact that the degree of $K(A_q)$ over K is $\gg q^2$ (which is immediate from the theory of complex multiplication).

It is easy to reduce the constant κ in various ways, e.g. by using the stronger version of Masser's theorem, with $L^{-1/2}$ replacing L^{-1}. However, the measure of approximation obtained in this paper sets things up for abelian logarithms (with complex multiplication) where Baker was in his first work on ordinary logarithms. It is then a problem to determine to what extent Feldman's ideas for constructing the auxiliary function with binomial coefficients can be used to obtain the improvement with $\tau(H) = C \log H$, where C is a sufficiently large constant. We met specific difficulties when we attempted to do so.

Since Ribet's theorem is effective, being in the complex multiplication case, the constant C is, in theory, effectively computable.

For elliptic curves, the constant would also be computable in practice. The significance of this for diophantine equations is explained in [6].

Bashmakov's theorem has also been proved for elliptic curves without complex multiplication. What is missing to get a proof of the approximation theorem in this case is the analogue of the Main Lemma. At this time no one knows how to make the extrapolation procedure succeed, because the abelian (Weierstrass) functions have a growth of order 2, and one does not see how to construct enough zeros of the approximating auxiliary function to match this.

The analogue of Ribet's theorem for abelian varieties of dimension >1 has not yet been proved.

Finally, one can give some explicit dependence of the approximating function on the heights of the points $P_j = \exp u^j$. Let

$$U = \max \log H(P_j).$$

Let

$$\tau(H, U) = (\log H)^\kappa U^{\kappa'}$$

where κ' is greater than an expression similar to the lower bound for κ. Then one has the inequality

$$|\beta^1 u^1 + \cdots + \beta^n u^n - u^0| > C^{-\tau(H, U)},$$

where C depends only on n, d, $[K:\mathbf{Q}]$. The proof is the same, except that in the main lemma, one has to use $M = (\log H)^4 U^t$ with an appropriate value of t.

References

1. Baker, A.: Transcendental numbers. Cambridge: Cambridge University Press 1975
2. Coates, J.: An application of the division theory of elliptic functions to diophantine approximation. Inventiones math. **11**, 167–182 (1970)
3. Coates, J.: On the analogue of Baker's theorem for elliptic integrals. Unpublished 1972
4. Lang, S.: Diophantine Geometry. New York: Interscience 1962
5. Lang, S.: Diophantine approximation on abelian varieties with complex multiplication. Advances in Math. (1975)
6. Lang, S.: Diophantine approximations on toruses. Am. J. Math. **86**, 521–533 (1964)
7. Masser, D.: Elliptic functions and transcendence. Lecture Notes in Math. **437**. Berlin-Heidelberg-New York: Springer 1975
8. Masser, D.: Linear forms in algebraic points of abelian functions I. Math. Proc. Camb. Phil. Soc. **77**, 499–513 (1975)
9. Ribet, K.: Division points on abelian varieties. To appear, Compositions Math.
10. Shimura, G., Taniyama, Y.: Complex multiplication of abelian varieties. Pub. Math. Soc. Japan, 1961
11. Bashmakov, M.: Un théoreme de finitude sur la cohomologie des courbes élliptiques, C. R. Acad. Sci. Paris **270**. Série A, 999–1001 (1970)

Received September 20, 1975

John Coates
Pure Mathematics Department
Mill Lane
Cambridge, England

Serge Lang
Department of Mathematics
Yale University
New Haven, Conn. 06520, USA

Distributions on Toroidal Groups

Dan Kubert* and Serge Lang*

Department of Mathematics, Yale University
Box 2155, Yale Station, New Haven, Connecticut 06520, USA

The study of units in the modular function field [KL II] and [KL III] gave rise to certain purely algebraic theorems, which are of independent interest, and which will be developed in a self contained manner in the present paper. The results are organized in two parts.

The first part gives the results relating to "distribution" relations, arising in a projective system with a family of functions such that the sum of the values in the fiber lying above a given point is equal to the value at that point. By mixing Bernoulli numbers and the Cartan group used in [KL II], we get such distributions.

To determine the rank of the values of the distribution, one needs to evaluate certain character sums, in which the generalized Bernoulli numbers $B_{k,\psi}$ of Leopoldt occur as factors. These are not zero precisely when k has the same parity as ψ, i.e. $\psi(-1)=(-1)^k$. The second part deals with such sums.

The need for such sums when $k=2$, to give the fullness of the group of units in the modular function field developed into this rather general theory. Originally, Tate had helped us evaluate the sums with $B_{2,\psi}$, by suggesting using the Fourier expansion, and the same thing works in general.

Contents

Part One. Distribution relations 34

1. The unramified Cartan group 34
2. Distributions on $\mathbf{Q}^k/\mathbf{Z}^k$ 35
3. The enlarged distribution on $\mathbf{Q}^k/\mathbf{Z}^k$ 40

Part Two. Evaluation of character sums 43

4. Evaluation of Gauss sums for prime powers 43
5. The composite case 45

---

* Supported by NSF grants.

Part One. Distribution Relations

§1. The Unramified Cartan Group

Let p be a prime number, and let k be an integer ≥ 1. Let $\mathfrak{o}_p$ be the ring of p-adic integers in the unramified extension of degree k over $\mathbf{Q}_p$. Then $\mathfrak{o}_p$ is a free module of rank k over $\mathbf{Z}_p$. The group of units $\mathfrak{o}_p^*$ will be called the **Cartan group** at p. We assume that a basis for $\mathfrak{o}_p$ over $\mathbf{Z}_p$ has been chosen. Then $\mathfrak{o}_p^*$ is represented as a group of matrices, denoted by C_p, and also called the **Cartan group** at p, contained in $GL_k(\mathbf{Z}_p)$.

The elements of $\mathfrak{o}_p^*$ are obviously characterized as those linear combinations of basis elements such that not all coefficients in $\mathbf{Z}_p$ are divisible by p. We call elements of $\mathfrak{o}_p^*$ **primitive elements**.

We let $C_p(p^n)$ be the reduction mod p^n of C_p, and similarly for $\mathfrak{o}_p(p^n)$. If

$$N = \prod p^{n(p)}$$

is composite, we let

$$C(N) = \prod_{p|N} C_p(p^{n(p)}).$$

Each $C_p(p^{n(p)})$ is represented as a subgroup of $GL_k(\mathbf{Z}/p^{n(p)}\mathbf{Z})$. By the Chinese remainder theorem, we obtain a representation of $C(N)$ as a subgroup of $GL_k(\mathbf{Z}/N\mathbf{Z})$.

We denote an element $\alpha \in C(N)$ by the matrix

$$\alpha = \begin{pmatrix} a_1 & * & \cdots & * \\ a_2 & * & \cdots & * \\ \vdots & \vdots & & \vdots \\ a_k & * & \cdots & * \end{pmatrix}.$$

Let $\vec{e}$ be the transpose of the unit vector $(1, 0, \ldots, 0)$. Then the map

$$\alpha \mapsto \alpha\, e^1$$

gives a bijection of $C(N)$ with $(\mathbf{Z}^k/N\mathbf{Z}^k)^*$, i.e. the primitive elements of $\mathbf{Z}^k/N\mathbf{Z}^k$, and similarly for multiplication on the right of a unit row vector. We let T be the projection on the first coordinate, so that

$$T: \mathrm{Mat}_k(\mathbf{Z}/N\mathbf{Z}) \to \mathbf{Z}/N\mathbf{Z} \quad \text{and} \quad T: \mathbf{Z}^k/N\mathbf{Z}^k \to \mathbf{Z}/N\mathbf{Z}$$

is such that $T((a_{ij})) = a_{11}$ and $T(a_1, \ldots, a_k) = a_1$.

In particular, if α is the above matrix,

$$T(\alpha) = a_1.$$

Then T is linear, and surjective.

We let

$$R = R(N) = \prod \mathfrak{o}_p(p^{n(p)}),$$

Distributions on Toroidal Groups

so that R is an algebra over $\mathbf{Z}/N\mathbf{Z}$. Then $R^* = C(N)$, and T may be viewed as a functional on R.

We abbreviate $\mathbf{Z}^k(N) = \mathbf{Z}^k/N\mathbf{Z}^k$. The primitive elements are denoted by

$(\mathbf{Z}^k/N\mathbf{Z}^k)^* = \mathbf{Z}^k(N)^*$.

§2. Distributions on $\mathbf{Q}^k/\mathbf{Z}^k$

Let $\{G_N\}$ be a projective system of abelian groups, indexed by the positive integers, ordered by divisibility. Let

$$\mathbf{r}_M^N = \mathbf{r}_M: G_N \to G_M$$

be the connecting homomorphism if $M|N$, assumed surjective. Suppose we are given a family of functions

$g_N: G_N \to$ some abelian group.

We say that this family is a **distribution** if it satisfies the condition

$$\sum_{\mathbf{r}_M t = s} g_N(t) = g_M(s)$$

for every $s \in G_M$, and with the sum taken over all $t \in G_N$ which project on s under the connecting map.

A function

$g: \mathbf{Q}^k/\mathbf{Z}^k \to$ some abelian group

will be called an **ordinary distribution,** or **distribution** for short, if for every positive integer D and $r \in \mathbf{Q}^k/\mathbf{Z}^k$ we have

$$\sum_{Dt=r} g(t) = g(r).$$

To verify that such a function is a distribution, it suffices to do so when D is a prime number.

If $M|N$ we have a commutative diagram

$$\begin{array}{ccc} \dfrac{1}{N}\mathbf{Z}^k/\mathbf{Z}^k & \xrightarrow{N} & \mathbf{Z}^k/N\mathbf{Z}^k \\ {\scriptstyle N/M}\Big\downarrow & & \Big\downarrow{\scriptstyle \mathbf{r}_M} \\ \dfrac{1}{M}\mathbf{Z}^k/\mathbf{Z}^k & \xrightarrow{M} & \mathbf{Z}^k/M\mathbf{Z}^k \end{array}$$

where the horizontal maps are multiplication by N, M respectively, the map on the left is multiplication by N/M, and the map on the right is reduction mod M. Thus a distribution on $\mathbf{Q}^k/\mathbf{Z}^k$ also defines a projective system of functions on the groups $\mathbf{Z}^k/N\mathbf{Z}^k$, satisfying the obvious compatibility relation.

Example. We use the definition of Bernoulli polynomials given by

$$\frac{t\,e^{Xt}}{e^t - 1} = \sum B_n(X) \frac{t^n}{n!}.$$

If $x \in \mathbf{Z}/N\mathbf{Z}$, then x/N lies in $\frac{1}{N}\mathbf{Z}/\mathbf{Z}$. For $y \in \mathbf{R}/\mathbf{Z}$ we let

$$\langle y \rangle$$

be the smallest real number ≥ 0 in the residue class of $y \bmod \mathbf{Z}$.

We have fixed our integer k. We define, for $x \in \mathbf{Z}/N\mathbf{Z}$,

$$Q_N(x) = B_k\left(\left\langle \frac{x}{N} \right\rangle\right).$$

The basic fact about B_k is that the family of functions

$$\{N^{k-1} Q_N\}$$

defines a distribution on the projective system

$$\{\mathbf{Z}/N\mathbf{Z}\}.$$

This means that the following relation is satisfied for $y \in \mathbf{Q}/\mathbf{Z}$:

$$N^{k-1} \sum_{x=0}^{N-1} B_k\left(\left\langle y + \frac{x}{N} \right\rangle\right) = B_k(\langle Ny \rangle).$$

The reader can find proofs in Iwasawa [Iw], who however takes a different normalization for Bernoulli polynomials, so some adjustment has to be made. Eventually, the reader will find a treatment in [L].

Except for certain statements that some expressions are $\neq 0$, all the formalism of the sequel depends only on the distribution relation. Consequently, we assume only unless otherwise specified that

$$B_k : \mathbf{Q}/\mathbf{Z} \to F$$

is a map into some field F of characteristic zero, such that

$$\{N^{k-1} Q_N\}$$

is a distribution. We use the same letter as for the Bernoulli distribution because it is suggestive. We use the normalization involving the power N^{k-1} because such powers come in a natural way in the subsequent formalism. The function Q_N is defined by the same formula as above.

By extension of scalars, one may assume that F contains whatever irrationalities are needed (e.g. roots of unity) in the course of the investigation. One may, in applications, then take for F the rational field $\mathbf{Q}$, or a cyclotomic field, or the complex numbers, or a p-adic field, etc.

By going to a higher dimensional space $\mathbf{Q}^k/\mathbf{Z}^k$, we shall obtain a distribution where the extra power N^{k-1} disappears. Conversely, if we started with some distribution Q on $\mathbf{Q}/\mathbf{Z}$ and followed the same procedure on this higher dimensional space, we would obtain a distribution by dividing with N^{k-1}. For the moment, only the first type has come in applications. It is clear that the whole formalism can be generalized in various ways, partly by taking into account this duality, partly by going over to number fields rather than the rationals. We have struck a compromise, giving some generality, but keeping the paper easily readable for the reader merely interested in seeing the complete proof for the independence of units constructed in [KL II].

If ψ is a Dirichlet character with conductor c, then for any N divisible by c we define

$$B_{k,\psi} = N^{k-1} \sum_{w=0}^{N-1} \psi(w) B_k\left(\frac{w}{N}\right).$$

This is independent of the value of N, and can also be written

$$\int \psi \, dE_k, \quad \text{where} \quad E_k = N^{k-1} B_k.$$

When B_k is the Bernoulli distribution, then $B_{k,\psi}$ is the generalized Bernoulli number of Leopoldt.

This is of course an important case, so far the most important, for the known applications. Unfortunately, Iwasawa [Iw], p. 10 (following Leopoldt) takes the "wrong" definition of Bernoulli polynomials, although he ends up with the "right" values of Bernoulli numbers for $n > 1$, namely those numbers having the property that

$$L(1-k, \psi) = -\frac{B_{k,\psi}}{k}.$$

The proof is recalled in Iwasawa p. 11, together with the fact that if the parity of ψ and k is the same, then $B_{k,\psi} \neq 0$ (this comes from the value of the L-series as above). Conversely, if ψ and k have opposite parity, then $B_{k,\psi} = 0$.

On $\mathbf{Z}^k/N\mathbf{Z}^k$ we define the function

$$v_N(a_1, \ldots, a_k) = Q_N(a_1).$$

Theorem 2.1. *The family of functions $\{v_N\}$ defines a distribution on the projective system $\{\mathbf{Z}^k/N\mathbf{Z}^k\}$.*

Proof. In order to prove the compatibility relation, it suffices to do it for two integers

$$M \quad \text{and} \quad N = pM,$$

where p is a prime number. The inverse image of an element

$$\frac{a}{M} \in \frac{1}{M} \mathbf{Z}^k/\mathbf{Z}^k$$

in $\frac{1}{N} \mathbf{Z}^k/\mathbf{Z}^k$ consists of

$$\frac{a}{pM} + \frac{\lambda}{p}, \quad \lambda \in \mathbf{Z}^k/p\mathbf{Z}^k.$$

We sum independently over $\lambda_1 = 0, \ldots, p-1$ and $\lambda_i = 0, \ldots, p-1$ for $i > 1$. The values v_N of an element are independent of its last $k-1$ coordinates, so that the second sum introduces a factor p^{k-1}. Taking into account the basic property of the Bernoulli distribution proves the theorem.

We let

$$r_M \colon \mathbf{Z}^k/N\mathbf{Z}^k \to \mathbf{Z}^k/M\mathbf{Z}^k$$

be reduction mod M, and also use r_M when applied to other objects reduced mod M, e.g. $r_M \colon C(N) \to C(M)$.

The next result gives the more complicated relation similar to that of Theorem 2.1 when restrictions of primitivity are made.

Theorem 2.2. (i) *Let $M \mid N$, and assume that every prime dividing M also divides N. Let $\bar{a} \in Z^k(M)^*$ be a primitive element. Then*

$$\sum_{\substack{a \in Z^k(N)^* \\ r_M a = \bar{a}}} v_N(a) = v_M(\bar{a}).$$

(ii) *On the other hand, let $N = p^n M$, where $p \nmid M$. Again let $\bar{a} \in \mathbf{Z}^k(M)^*$ be a primitive element. Then*

$$\sum_{\substack{a \in Z^k(N)^* \\ r_M a = \bar{a}}} v_N(a) = v_M(\bar{a}) - v_M(p^{-1} \bar{a}).$$

Proof. In the first case, any element a in $\mathbf{Z}^k(N)$ lying above $\bar{a}$ is primitive, and so the preceding theorem applies. We suppose next that we deal with case (ii).

The non-primitive elements $a \in \mathbf{Z}^k(N)$ lying above $\bar{a}$ are in bijection with the elements $b \in \mathbf{Z}^k(N/p)$ satisfying the conditions

$$a = pb \quad \text{and} \quad b \equiv p^{-1} \bar{a} \pmod{M},$$

under the map

$$b \mapsto pb,$$

which sends $\mathbf{Z}/(N/p)\mathbf{Z}$ into $p\mathbf{Z}/N\mathbf{Z} \subset \mathbf{Z}/N\mathbf{Z}$. Therefore, the sum over primitive elements lying above a given $\bar{a}$ can be expressed as a difference

$$\sum_{a \text{ prim}} v_N(a) = \sum_{\text{all } a} v_N(a) - \sum_b B_k\left(\left\langle \frac{b_1}{N/p} \right\rangle\right)$$

$$= v_M(\bar{a}) - v_M(p^{-1}\bar{a})$$

by the distribution relation, as was to be shown.

Next, we interpret the last two theorems on the Cartan group. We define for an element $\alpha \in C(N)$,

$$v_N(\alpha) = Q_N(T\alpha) = B_k\left(\left\langle \frac{T\alpha}{N} \right\rangle\right).$$

Distributions on Toroidal Groups

Let $e_1 = (1, 0, \ldots, 0)$ be the first unit vector in $\mathbf{Z}^k/N\mathbf{Z}^k$. Then the map

$$\alpha \mapsto e_1 \alpha$$

is a bijection of $C(N)$ with the set of primitive elements of $\mathbf{Z}^k/N\mathbf{Z}^k$. Therefore

$$v_N(\alpha) = v_N(e_1 \alpha)$$

can be written in terms of the first row vector of α.

Theorem 2.3. (i) *Let $M|N$, and suppose that every prime p dividing N also divides M. Then the compatibility relation holds for $\bar{\alpha} \in C(M)$:*

$$\sum_{\mathbf{r}_M \alpha = \bar{\alpha}} v_N(\alpha) = v_M(\bar{\alpha}),$$

where $\mathbf{r}_M$ means reduction mod M. In particular, the family of functions $\{v_{p^n}\}$ forms a distribution on the projective system

$$\{C(p^n)\}.$$

(ii) *Suppose on the other hand that $N = p^n M$ and $p \nmid M$. Then*

$$\sum_{\substack{\alpha \in C(N) \\ \mathbf{r}_M \alpha = \bar{\alpha}}} v_N(\alpha) = v_M(\alpha) - v_M(p^{-1} \bar{\alpha}).$$

Proof. This is a reformulation of Theorem 2.2, given the bijection between $C(N)$ and primitive vectors in $\mathbf{Z}^k(N)^*$.

We define the **Stickelberger element** associated with the distribution, to be

$$\theta(N) = \sum_{\alpha \in C(N)} v_N(\alpha) \alpha.$$

Theorem 2.1 can be expressed in the form

$$\mathbf{r}_M \theta(N) = \theta(M).$$

whenever M and N are divisible by the same primes, and $M|N$.

We have the orthogonal idempotent associated with a character ψ of $C(N)$, namely

$$e_\psi = \frac{1}{|C(N)|} \sum_{\alpha \in C(N)} \bar\psi(\alpha) \alpha,$$

so that the Stickelberger element has the expression

$$\theta(N) = \sum_\psi S(\psi, v_N) e_\psi$$

where the Fourier coefficient is given by

$$S(\psi, v_N) = \sum_{\alpha \in C(N)} v_N(\alpha) \psi(\alpha).$$

We let ψ_Z be the restriction of ψ to $(\mathbf{Z}/N\mathbf{Z})^*$.

Theorem 2.4. *Let ψ be a character of $C(N)$, $\psi \neq 1$. (i) Let $M = \operatorname{cond} \psi$ be the conductor of ψ, and assume that every prime p dividing N also divides M. Let ψ_M be the primitive character on $C(M)$ through which ψ factors. Then*

$$S(\psi, v_N) = S(\psi_M, v_M).$$

(ii) *Let p be a prime, $p \mid N$, and $p \nmid \operatorname{cond} \psi$. Write $N = p^n M$, with $p \nmid M$. Let ψ_M be the character on $C(M)$ which gives the factorization of ψ through $C(M)$. Then*

$$S(\psi, v_N) = (1 - \psi(p)) S(\psi_M, v_M).$$

Proof. The compatibility relation of Theorem 2.2 gives the first relation.

As for (ii), we have for $M \neq 1$:

$$S(\psi, v_N) = \sum_{\bar{\alpha} \in C(M)} \psi_M(\bar{\alpha}) \sum_{\substack{\alpha \in C(N) \\ r_M \alpha = \bar{\alpha}}} v_N(\alpha).$$

We use Theorem 2.3(ii) giving the value of the inner sum at level M. We make the change of variables $\bar{\alpha} \mapsto p\bar{\alpha}$ to get

$$S(\psi, v_N) = S(\psi_M, v_M) - \psi(p) S(\psi_M, v_M),$$

which proves (ii).

Corollary. (i) *In case (i) of the theorem, the sum $S(\psi, v_N)$ is $\neq 0$ if and only if $B_{k, \psi_z} \neq 0$. If B_k is the Bernoulli distribution, this is the case if and only if ψ has the same parity as k.*

(ii) *Assume that B_k is the Bernoulli distribution. In case (ii), the sum is 0 if ψ has different parity from k, or if there exists a prime $p \mid N$, $p \nmid \operatorname{cond}(\psi)$, and $\psi(p) = 1$. The sum is $\neq 0$ otherwise.*

Proof. This depends on the explicit expressions which will be obtained in Theorem 5.3, where the number B_{k, ψ_z} occurs as a factor of the sum $S(\psi, v_N)$, and all other factors are $\neq 0$.

Remark. Theorem 2.4 gives a reduction for the evaluation of character sums $S(\psi, v_N)$ to the case when ψ is primitive. Theorems 5.1 and 5.2 then give a decomposition of these sums when ψ is primitive, into non-zero factors (ordinary Gauss sums) and the number B_{k, ψ_z}. This cumulative process therefore gives a complete formal description of the sums.

§3. The Enlarged Distribution on $\mathbf{Q}^k/\mathbf{Z}^k$

In this section we define on $\mathbf{Q}^k/\mathbf{Z}^k$ the analogue of the distribution which we called the Siegel distribution in [KL III].

If

$$r = (r_1, \ldots, r_k)$$

is a vector, we put

$$Tr = r_1.$$

Distributions on Toroidal Groups

Suppose that $r \in \mathbf{Q}^k/\mathbf{Z}^k$. Let N be a common denominator for the components of r. Define

$$g_N(r) = \frac{1}{|C(N)|} \sum_{\alpha \in C(N)} B_k(\langle T(r\alpha) \rangle) \alpha$$

as an element of the group space $F[C(N)]$. Note that r is a row vector, and multiplication by the $k \times k$ matrix α on the right makes sense.

It is clear that if M is a common denominator for the elements of r, and M divides N, then the reduction mod M of $g_N(r)$ is equal to $g_M(r)$. Thus we define

$$g(r) = \lim g_N(r)$$

in the projective limit of the group spaces $F[C(N)]$, ordered by divisibility, with connecting homomorphisms equal to the reduction maps.

Theorem 3.1. *The function* $g: \mathbf{Q}^k/\mathbf{Z}^k \to \lim F[C(N)]$ *is a distribution.*

Proof. This amounts to proving that

$$\sum_x g(r+x) = g(pr),$$

if the sum is taken over $x \in \frac{1}{p} \mathbf{Z}^k/\mathbf{Z}^k$. Note that the map

$$x \mapsto x\alpha$$

permutes the elements of $\frac{1}{p} \mathbf{Z}^k/\mathbf{Z}^k$, for any given α, because α is primitive, and therefore has an inverse. We find:

$$\sum_x B_k(\langle T((r+x)\alpha) \rangle) = \sum_y B_k(\langle T(r\alpha + y) \rangle),$$

where y ranges also over $\frac{1}{p} \mathbf{Z}^k/\mathbf{Z}^k$. Because of the projection on the first coordinate, we may separate the sum over y into a sum over the first coordinate, and over the others which give a factor of p^{k-1}. The desired relation then follows from the fundamental relation of Bernoulli polynomials, which define a distribution on $\mathbf{Q}/\mathbf{Z}$. This proves the theorem.

Let A_N be the vector space generated by the values $g(r)$, with

$$r \in \frac{1}{N} \mathbf{Z}^k/\mathbf{Z}^k.$$

We observe that $g(0)$ is a constant multiple of the augmentation element, that is

$$g(0) = \frac{B_k(0)}{|C(N)|} \sum_{\alpha \in C(N)} \alpha.$$

We let $\hat{C}_k(N) = \hat{C}_{B_k}(N)$ be the set of characters ψ on $C(N)$ such that $B_{k, \psi_Z} \ne 0$, where ψ_Z is the restriction of ψ to $\mathbf{Z}/N\mathbf{Z}$.

Theorem 3.2. *The dimension of A_N is equal to the cardinality of $\hat{C}_k(N)$.*

249

Proof. Let C be a finite abelian group, and let f be a function on C. In the application, $C = C(N)$ and f will be described later as a linear combination of elements $g(r)$ for suitable r. Consider the elements

$$\theta_\beta(f) = \theta_\beta = \sum_\alpha f(\alpha\beta)\alpha.$$

If e_ψ is the idempotent associated with a character ψ of C, then θ_β may also be written as a linear combination

$$\theta_\beta = \sum_\psi \bar\psi(\beta) S(\psi, f) e_\psi,$$

where

$$S(\psi, f) = \sum_{\alpha \in C} f(\alpha)\psi(\alpha).$$

From the linear independence of characters, we obtain:

Lemma 1. *The dimension of the space generated by the elements $\theta_\beta(f)$ is equal to the number of characters ψ such that*

$$S(\psi, f) \neq 0.$$

Lemma 2. *Let ψ be in $\hat{C}_k(N)$. Then there exists an element $r \in \frac{1}{N}\mathbf{Z}^k/\mathbf{Z}^k$ such that*

$$\sum_{\alpha \in C(N)} \psi(\alpha) B_k(\langle T(r\alpha)\rangle) \neq 0.$$

Proof. Let N' be the conductor of ψ. We take simply

$$r = \left(\frac{1}{N'}, 0, \ldots, 0\right).$$

Then the sum in the lemma differs by a non-zero constant factor from the sum

$$\sum_{\alpha \in C(N')} \psi(\alpha) B_k(\langle T\alpha\rangle) = S_{R(N')}(\psi, v_{N'}).$$

This last sum is $\neq 0$ by Theorem 5.3.

We come to the proof of Theorem 3.2 proper. We let ψ range over the elements of $\hat{C}_k(N)$. We view elements of the group space as functions on $C(N)$, i.e.

$$h = \sum h(\alpha)\alpha.$$

By Lemma 2, for any $\psi \in \hat{C}_k(N)$ (trivial or not!) we can find an element h_ψ in A_N such that

$$\sum_\alpha h_\psi(\alpha)\psi(\alpha) \neq 0.$$

Let

$$h = \sum m(\psi) h_\psi$$

Distributions on Toroidal Groups

where the sum is taken over $\psi \in \hat{C}_k(N)$. We contend that we can select positive integers $m(\psi)$ such that

$$S(\chi, h) \neq 0$$

for all characters $\chi \in \hat{C}_k(N)$. Indeed, we have

$$S(\chi, h) = \sum_\alpha \chi(\alpha) h(\alpha)$$
$$= \sum_\psi \sum_\alpha m(\psi) h_\psi(\alpha) \chi(\alpha)$$
$$= \sum_\psi m(\psi) \left(\sum_\alpha h_\psi(\alpha) \chi(\alpha)\right).$$

This amounts to finding a vector $(\ldots, m(\psi), \ldots)$ which is not perpendicular to a finite number of non-zero vectors, and can trivially be done with positive integral coordinates.

Using Lemma 1, it is clear that the dimension of the space generated by the elements $\theta_\beta(h)$ is at least equal to the cardinality of $\hat{C}_k(N)$. The opposite inequality is obvious, and the theorem is proved.

Corollary. *Let B_k be the Bernoulli distribution. If $N \neq 2$ then the dimension of A_N is equal to $\frac{1}{2}|C(N)|$. If $N = 2$, then it is equal to $|C(N)|$ if k is even, and 0 if k is odd.*

Proof. It is equal to the number of even characters of $C(N)$, by the fact already recalled concerning the non-vanishing of Bernoulli numbers $B_{k,\psi}$. If $N \neq 2$ then $-1 \neq 1$ in the Cartan group, and the assertion is obvious. If $N = 2$, then $-1 = 1$ in $C(N)$, and the second assertion is equally obvious.

Part Two. Evaluation of Character Sums

§4. Evaluation of Gauss Sums for Prime Powers

Let A be a finite commutative ring such that $p^n A = 0$ for some prime number p. Let

$$\mu: A \to \mathbf{C}^*$$

be a character (homomorphism of the additive group of A into roots of unity), and let

$$\chi: A^* \to \mathbf{C}^*$$

be a multiplicative character. For any function g on A define the sum

$$S(\chi, g) = \sum_{u \in A^*} \chi(u) g(u).$$

If $g = \mu$ then $S(\chi, \mu)$ has the form of a usual Gauss sum.

The **conductor of** χ is by definition the smallest power p^m such that χ is trivial on $1 + p^m A$. Let $c(\chi)$ be the conductor. Let

$$r(\chi) = n - m(\chi), \quad \text{if } c(\chi) = p^{m(\chi)}.$$

Then χ is called **trivial** (resp. **primitive** if and only if $r(\chi) = n$ (resp. $r(\chi) = 0$).

Lemma 1. *Assume:*

(i) *χ is non-trivial.*

(ii) *μ is non-trivial on $p^{n-1} A$.*

Let $0 \leq r$. If $r \neq r(\chi)$ then

$$S(\chi, \mu \circ p^r) = \sum_{u \in A^*} \chi(u) \mu(p^r u) = 0.$$

Proof. If $r \geq n$, we get $\sum \chi(u) = 0$. Let $r(\chi) < r < n$. We use the coset decomposition

$$A^* = \bigcup u_i (1 + p^{n-r} A).$$

Then

$$S = \sum_i \sum_{x \in p^{n-r}A} \chi(u_i) \chi(1+x) \mu(p^r u_i)$$

$$= \sum_i \chi(u_i) \mu(p^r u_i) \sum_{x \in p^{n-r}A} \chi(1+x).$$

Since χ is non-trivial on $1 + p^{n-r} A$, this sum is equal to 0.

Next assume that $r < r(\chi)$. We then use the coset decomposition

$$A^* = \bigcup u_i (1 + p^{n-r(\chi)} A).$$

Then our sum is equal to

$$\sum_i \sum_{x \in p^{n-r(\chi)}A} \chi(u_i) \mu(p^r u_i + p^r x)$$

$$= \sum_i \chi(u_i) \mu(p^r u_i) \sum_{x \in p^{n-r(\chi)}A} \mu(p^r x).$$

Then $\mu \circ p^r$ is non-trivial on $p^{n-r(\chi)} A$, whence the sum is again equal to 0.

Note. Instead of a ring over $\mathbf{Z}/p^n \mathbf{Z}$ we could also work over a number field. In the applications, p remains a prime ideal, so we use notation with p instead of a parameter π to fit in later. The formalism of Gauss sums is essentially standard, and we reproduce here a few lemmas only for the convenience of the reader.

In Lemma 1 we proved that the Gauss sum is 0 except possibly in one case. In the next lemma, we give its non-zero absolute value in that case, and a special ring A. The result would also be valid more generally with π instead of p.

Lemma 2. *Let $\mathfrak{o}$ be the ring of integers in an unramified extension of $\mathbf{Q}_p$. Let $\mathfrak{p} = p \mathfrak{o}$ and let $A = \mathfrak{o}/\mathfrak{p}^n$. Let χ, μ be as in Lemma 1, so that $r(\chi) \neq n$. Then*

$$|S(\chi, \mu \circ p^{r(\chi)})|^2 = \mathbf{N}\mathfrak{p}^{n+r(\chi)}.$$

Proof. By the case $r > r(\chi)$ of Lemma 1, if $z \in A$, $z \notin A^*$, then

$$S(\chi, \mu \circ z\, p^{r(\chi)}) = 0.$$

Therefore, denoting $|A^*| = \phi(\mathfrak{p}^n)$ we get

$$\sum_z |S(\chi, \mu \circ z\, p^{r(\chi)})|^2 = \phi(\mathfrak{p}^n) |S(\chi, \mu \circ p^{r(\chi)})|^2.$$

Distributions on Toroidal Groups   45

On the other hand, expanding out the sum yields

$$\sum_u \sum_v \sum_z \chi(u)\,\bar\chi(v)\,\mu(z\,p^{r(\chi)}(u-v)).$$

If u, v is a pair such that $u \equiv v \pmod{p^{n-r(\chi)}}$ then $\chi(u) = \chi(v)$, and the sum over z has $\mathbf{N}\mathfrak{p}^n$ terms in it. If $u \not\equiv v \pmod{p^{n-r(\chi)}}$, then the sum over z is equal to 0 by condition (ii) of Lemma 1. The value of Lemma 2 then follows at once.

Lemma 3. *Let $\mathfrak{o}$ be the ring of integers in an unramified extension of $\mathbf{Q}_p$. Let $\mathfrak{p} = p\,\mathfrak{o}$ and let $A = \mathfrak{o}/\mathfrak{p}^n$. Let μ be a character of A, non-trivial on $\mathfrak{p}^{n-1}A$. Let $0 \leq r < n$. Then*

$$\sum_{u\in A^*} \mu(p^r u) = \begin{cases} 0 & \text{if } r \neq n-1 \\ -\mathbf{N}\mathfrak{p}^{n-1} & \text{if } r = n-1. \end{cases}$$

Proof. First we sum over the units in $\mathfrak{o}/\mathfrak{p}^{n-r}$, multiplying this sum by $\mathbf{N}\mathfrak{p}^r$. Put $t = n - r$, and $\mu_r = \mu \circ p^r$. Then we write the sum over the units as the sum over all elements of $\mathfrak{o}/\mathfrak{p}^t$, minus the sum over all elements in the subgroup $p\,\mathfrak{o}/\mathfrak{p}^t$. The value of the lemma falls out at once.

Remark. Since in the applications we want to work with arbitrary fields of characteristic 0, which might turn out to be algebraic over p-adic fields, we then have to take the characters μ and χ to have values in such fields. We really don't need the precise absolute value, only whether certain sums are equal to zero or not, and this property is then independent of whether we took values in $\mathbf{C}$ or not.

§5. The Composite Case

Let N be an integer ≥ 2 and let R be a finite algebra over $\mathbf{Z}/N\mathbf{Z} = Z$. Let

$$L\colon R \to \mathbf{Z}/N\mathbf{Z} = Z$$

be linear.

Let χ be a character of R^*, and let χ_Z be its restriction to Z. Elements of R^* are denoted by α.

Let λ be an additive character on Z which makes Z self dual under the map $(z, z') \to \lambda(z\,z')$. If λ is complex valued, we can take $\lambda(z) = e^{2\pi i z/N}$. Otherwise, λ, χ have to have values in the original field F.

Let f be a function on Z and $\hat f$ its Fourier transform, which we normalize here conveniently so that

$$f(x) = \sum_y \hat f(y)\,\lambda(x\,y)$$

$$\hat f(y) = \frac{1}{N} \sum_x f(x)\,\bar\lambda(x\,y).$$

Then we have a sum decomposition for

$$S_R(\chi, f \circ L) = \sum_\alpha \chi(\alpha)\,f(L\alpha),$$

253

namely

$$S_R(\chi, f \circ L) = \sum_{d|N} \frac{\phi(N/d)}{\phi(N)} S_Z(\bar{\chi}_Z, \hat{f} \circ d) S_R(\chi, \lambda \circ L \circ d) \tag{1}$$

$$S_R(\bar{\chi}, \hat{f} \circ L) = \frac{1}{N} \sum_{e|N} \frac{\phi(N/e)}{\phi(N)} S_Z(\chi_Z, f^{\,-} \circ e) S_R(\bar{\chi}, \lambda \circ L \circ e). \tag{2}$$

As usual $f^{\,-}$ is the function such that $f^{\,-}(x) = f(-x)$. This is obtained by writing for each divisor d of N and each y such that $(y, N) = d$ the factorization

$$y = dw \quad \text{with } w \in (\mathbf{Z}/(N/d)\,\mathbf{Z})^*.$$

For each w there are $\phi(N)/\phi(N/d)$ elements $u \in (\mathbf{Z}/N\mathbf{Z})^*$ such that

$$u \equiv w \pmod{N/d}.$$

Hence

$$S_R(\chi, f \circ L) = \sum_{d|N} \frac{\phi(N/d)}{\phi(N)} \sum_\alpha \sum_u \chi(\alpha)\, \hat{f}(du)\, \lambda(L\,du\,\alpha).$$

Letting $\alpha \to \alpha u^{-1}$ yields the desired expression for our sum.

Write

$$N = \prod p^{n(p)}.$$

Then R splits as a direct product

$$R = \prod R_p,$$

where each R_p is a finite algebra over $\mathbf{Z}/p^{n(p)}\,\mathbf{Z} = Z_p$. The characters χ, λ and L split into products of their p-components χ_p, λ_p, L_p. It is trivial to prove the product decomposition

$$S_R(\chi, \lambda \circ L) = \prod_p S_{R_p}(\chi_p, \lambda_p \circ L_p). \tag{3}$$

We apply all of this to the case when

$$R = R(N) = \prod \mathfrak{o}_p/p^{n(p)}\,\mathfrak{o}_p,$$

and $\mathfrak{o}_p$ is the ring of integers in the unramified extension of $\mathbf{Q}_p$. Thus

$$R^* = C(N) = \prod C(p^{n(p)}) = \prod R_p^*$$

is the composite Cartan group. Write

$$N = cd,$$

where c is the conductor of χ. Then $c = \prod c_p$, where c_p is the conductor of χ_p for each $p|N$.

Note that by the change of variables $u \to -u$ we get

$$S_Z(\chi_Z, f^{\,-}) = \chi(-1) S_Z(\chi_Z, f).$$

Distributions on Toroidal Groups   47

The evaluation of the sums $S_{R(N)}(\chi, v_N)$ will be according to cases, where χ_{Z_p} is trivial or not trivial. However, we always assume that χ itself is not trivial. Recall that

$$v_N = Q_N \circ T,$$

where T is the projection on the first coordinate.

The next lemma will be used in all resulting cases.

Lemma 1. *Assume that χ_p is non-trivial for all $p|N$. Write $N = c(\chi)d$, where $c(\chi)$ is the conductor of χ. Then*

$$S_R(\chi, Q_N \circ T) = \frac{\phi(N/d)}{\phi(N)} S_Z(\bar{\chi}_Z, \hat{Q}_N \circ d) S_R(\chi, \lambda \circ T \circ d),$$

and

$$S_R(\chi, \lambda \circ T \circ d) \neq 0.$$

Proof. Formula (1) decomposes $S_R(\chi, f \circ T)$ into a sum, and by (3), each term of the sum decomposes into a product according to the prime factorization of N. By Lemma 1 of the prime power case §4, we conclude that there is only one possibly non-zero term in the sum, corresponding to the values $r(\chi_p)$ for all p, and this term is precisely as stated. The non-vanishing of the sum comes from its prime power decomposition, together with Lemma 2, §4 of the prime power case.

We are therefore led to studying the sum

$$S_Z(\bar{\chi}_Z, \hat{Q}_N \circ d) = S_N(\bar{\chi}_N, \hat{Q}_N \circ d), \quad d = d(\chi),$$

where $\chi_N = \chi_Z$ is the character restricted to $Z(N) = \mathbf{Z}/N\mathbf{Z}$. We shall first consider the case when all χ_{Z_p} are non-trivial. We then show that to each p such that χ_{Z_p} is trivial, we can split off a p-factor of the desired type.

Case 1. χ_{Z_p} *is non-trivial for all $p|N$.*

Theorem 5.1. *Let $N = c(\chi)d$. Assume that χ_{Z_p} is non-trivial for all $p|N$. Let c be the conductor of χ_Z, and let*

$$N = cde.$$

Then

$$S_R(\chi, Q_N \circ T) = \frac{1}{N(cd)^{k-1}} \frac{\phi(N/d)}{\phi(N)} \chi(-1) B_{k, \chi_Z} S_Z(\bar{\chi}_Z, \lambda \circ de) S_R(\chi, \lambda \circ T \circ d).$$

Proof. We consider the expression

$$S_R(\chi, Q_N \circ T) = \frac{\phi(N/d)}{\phi(N)} S_Z(\chi_Z, \hat{Q}_N \circ d) S_R(\chi, \lambda \circ T \circ d)$$

from Lemma 1. By (2), the prime power decomposition of (3), and the prime power case, Lemma 1 in §4, we see that there is only one term in the sum decomposition for $S_Z(\bar{\chi}_Z, \hat{Q}_N \circ d)$, and that is

$$S_Z(\bar{\chi}_Z, \hat{Q}_N \circ d) = \frac{1}{N} \frac{\phi(N/e)}{\phi(N)} S_Z(\chi_Z, Q_N^- \circ e) S_Z(\bar{\chi}_Z, \lambda \circ de).$$

We then have

$$S_Z(\chi_Z, Q_N \circ e) = \sum_{u \bmod N} \chi_Z(u) Q_N(e\, u)$$

$$= \frac{\phi(N)}{\phi(c\, d)} \sum_{w \bmod cd} \chi_Z(w) B_k\left(\left\langle \frac{e\, w}{N} \right\rangle\right)$$

$$= \frac{\phi(N)}{\phi(c\, d)} \sum_{w \bmod cd} \chi_Z(w) B_k\left(\left\langle \frac{w}{c\, d} \right\rangle\right)$$

$$= \frac{\phi(N)}{\phi(c\, d)} \frac{1}{(c\, d)^{k-1}} B_{k,\chi_Z}.$$

This proves the theorem.

Case 2. χ_{Z_p} is trivial for some $p|N$.

We shall compute this case when N is composite only when the conductor of χ is N. We shall see that a p-factor splits off, reducing the computation to a lower level not divisible by p.

Theorem 5.2. *Assume that the conductor of χ is N, and that χ_{Z_p} is trivial for some $p|N$. Write*

$$N = p^n M, \quad \text{where } p \nmid M.$$

Then

$$S_{R(N)}(\chi, Q_N \circ T_N) = \left(1 - \frac{\bar{\chi}_M(p)}{p^2}\right) S_{R_p}(\chi_p, \lambda_p \circ T_p) S_{R(M)}(\chi_{R(M)}, Q_M \circ T_M).$$

If $N = p^n$ is a prime power, then $R(M) = 0$ and

$$S_{R(M)}(\chi_{R(M)}, Q_M \circ T_M) = B_k(0).$$

Proof. We have

$$S_R(\chi, Q_N \circ T) = S_Z(\bar{\chi}_Z, \hat{Q}_N) S_Z(\chi_Z, \lambda \circ T).$$

Lemma 2. *If χ_{Z_p} is trivial, then*

$$S_Z(\bar{\chi}_Z, \hat{Q}_N) = \left(1 - \frac{\bar{\chi}_M(p)}{p^2}\right) S_M(\chi_M, \hat{Q}_M).$$

Proof. Divisors of N are of the form

$$p^r d, \quad \text{with } d|M.$$

Distributions on Toroidal Groups

By (2), (3) combined with Lemma 3, §4 (concerning prime powers), we see that there will be only two terms in the sum expression for $S_Z(\chi_Z, \hat{Q}_N)$ which need not vanish. Therefore we obtain:

Lemma 3. *If N is the conductor of χ, and χ_{Z_p} is trivial, then*

$$S_Z(\bar{\chi}_Z, \hat{Q}_N)$$
$$= \frac{1}{N}\chi(-1)\left[\sum_{d|M} \frac{p-1}{\phi(p^n)} \frac{\phi(M/d)}{\phi(M)} S_Z(\chi_Z, Q_N \circ p^{n-1}d) S_Z(\bar{\chi}_Z, \lambda \circ p^{n-1}d)\right.$$
$$\left. + \sum_{d|M} \frac{\phi(M/d)}{\phi(p^n)\phi(M)} S_Z(\chi_Z, Q_N \circ p^n d) S_Z(\bar{\chi}_Z, \lambda \circ p^n d)\right].$$

For this rest of this section, we let d be a divisor of M.

The second sum over $d|M$ is the easiest to treat. Each function appearing in it factors through $\mathbf{Z}/M\mathbf{Z}$. Therefore this second sum is equal to

$$\sum_{d|M} \frac{\phi(M/d)}{\phi(M)} \phi(p^n) S_M(\chi_M, Q_M \circ d) S_M(\bar{\chi}_M, \lambda_M \circ d). \tag{4}$$

In the first sum, the product decomposition according to relatively prime factors and the evaluation of the prime power sum in Lemma 3, §4, for a trivial character χ_{Z_p}, yields

$$\frac{p-1}{\phi(p^n)} S_N(\bar{\chi}_N, \lambda_N \circ p^{n-1}d) = \bar{\chi}_M(p) S_M(\bar{\chi}_M, \lambda_M \circ d). \tag{5}$$

Also in the first sum, we evaluate the term with Q_N. For each $u \in Z(N)^*$ we write the partial fraction decomposition

$$\frac{u}{N} \equiv \frac{u'}{p^n} + \frac{u''}{M} \pmod{\mathbf{Z}},$$

where u' is determined mod p^n and u'' is determined mod M. Then

$$u \equiv p^n u'' \pmod{M} \quad \text{and} \quad u \equiv M u' \pmod{p^n}.$$

So $(Mu', p^n u'')$ are the components (mod p^n, mod M) of u. Therefore

$$\chi(u) = \chi_M(p^n u'') \chi_{Z_p}(Mu') = \chi_M(p^n) \chi_M(u'') \chi_{Z_p}(M) \chi_{Z_p}(u')$$
$$= \chi_M(p^n) \chi_M(u'')$$

since we assumed that χ_{Z_p} is trivial.

Lemma 4. *Assume that χ_{Z_p} is trivial. Then*

$$S_N(\chi_N, Q_N \circ p^{n-1}d) = \chi_M(p) p^{n-1}\left(\frac{\bar{\chi}_M(p)}{p} - 1\right) S_M(\bar{\chi}_M, Q_M \circ d).$$

257

Proof. By the partial fraction decomposition

$$\frac{u}{pM} \equiv \frac{u'}{p} + \frac{u''}{M} \pmod{\mathbf{Z}}$$

we find that

$$S_N(\chi_N, Q_N \circ p^{n-1} d) = \sum_{u \bmod N} \chi(u) Q\left(\left\langle \frac{du}{pM} \right\rangle\right)$$

$$= \sum_u \chi_M(p\, u'') Q\left(\left\langle \frac{du'}{p} + \frac{du''}{M} \right\rangle\right)$$

$$= \chi_M(p) \sum_u \chi_M(u'') Q\left(\left\langle \frac{du'}{p} + \frac{du''}{M} \right\rangle\right)$$

$$= \chi_M(p) \sum_{u''} \chi_M(u'')\, p^{n-1} \sum_{u'=1}^{p-1} Q\left(\left\langle \frac{u'}{p} + \frac{du''}{M} \right\rangle\right)$$

(by Lemma 3, §3)

$$= \chi_M(p) \sum_{u''} \chi_M(u'')\, p^{n-1} \left[\frac{1}{p} Q\left(\left\langle \frac{p\, du''}{M} \right\rangle\right) - Q\left(\left\langle \frac{du''}{M} \right\rangle\right)\right]$$

$(u'' \mapsto p^{-1} u'')$

$$= \chi_M(p)\, p^{n-1} \sum_{u''} \left(\frac{\bar{\chi}_M(p)}{p} - 1\right) \chi_M(u'') Q\left(\left\langle \frac{du''}{M} \right\rangle\right).$$

This proves the lemma.

In the sum expression of Lemma 3 for $S_Z(\chi_Z, \hat{Q}_N)$ we may now substitute the values found in (4), (5) and Lemma 4. Combining them together and using (2) yields the proof of Lemma 2.

We return to the original expression

$$S_R(\chi, Q_N \circ T_N) = S_Z(\bar{\chi}_Z, \hat{Q}_N) S_N(\chi_N, \lambda_N \circ T_N)$$

$$= S_Z(\bar{\chi}_Z, \hat{Q}_N) S_{R_p}(\chi_{R_p}, \lambda_{p^n} \circ T_{p^n}) S_{R(M)}(\chi_{R(M)}, \lambda_M \circ T_M)$$

in which we substitute the value of Lemma 2. The terms involving M combine to yield

$$S_M(\chi_M, \hat{Q}_M) S_{R(M)}(\chi_{R(M)}, \lambda_M \circ T_M) = S_{R(M)}(\chi_{R(M)}, Q_M \circ T_M),$$

in other words, we get back the basic sum at level M.

We may put together the last two theorems to get a general expression in the most important case, when the conductor of χ is N.

Theorem 5.3. *Assume that the conductor of χ is N. Write*

$$N = DM,$$

where $(D, M) = 1$, D is divisible by those primes p such that χ_{Z_p} is trivial, and M is divisible by those primes p such that χ_{Z_p} is non-trivial. Let

$$M = ce, \quad \text{where } c = c(\chi_{Z(M)}).$$

Then

$$S_R(\chi, Q_N \circ T_N) = \prod_{p|D} \left(1 - \frac{\bar{\chi}_M(p)}{p^2}\right) S_R(\chi, \lambda_N \circ T_N) \frac{1}{M c^{k-1}} \chi(-1) B_{k,\chi z} S_{Z(M)}(\bar{\chi}_{Z(M)}, \lambda_M \circ e).$$

We give the prime power case explicitly as a special case.

Theorem 5.4. *Let $N = p^n$ be a prime power. Assume that χ is primitive, but χ_Z is trivial. Then*

$$S_R(\chi, Q_N \circ T_N) = B_k(0) \left(1 - \frac{1}{p^2}\right) S_R(\chi, \lambda_N \circ T_N).$$

When χ is not primitive, one can use the distribution relations of Theorem 2.4 to reduce the study of the sum to the primitive case.

Bibliography

[IW] Iwasawa, K.: Lectures on *p*-adic *L*-functions. Annals of Math. Studies No. 74. Princeton: Princeton University Paris 1972

[KL II] Kubert, D., Lang, S.: Units in the modular function field. II: A full set of units. Math. Ann. 175-189 (1975)

[KL III] Kubert, D., Lang, S.: Units in the modular function field. III: Distribution relations. Math. Ann. 273-285 (1975)

[LE] Leopoldt, H.: Eine Verallgemeinerung der Bernoullischen Zahlen. Abh. math. Sem. Univ. Hamburg **22**, 131-140 (1958)

[L] Lang, S.: Lectures on modular forms. To appear

Received October 13, 1975

Modular Functions in one Variable
Lecture Notes in Mathematics 601
(Bonn Conference) 1976, pp. 247-275

UNITS IN THE MODULAR FUNCTION FIELD

by

Dan Kubert* and Serge Lang*

1.	Units and divisor classes at infinity	248
2.	Construction of units, Klein forms	250
3.	Representation of the cusps	258
4.	Analysis of the regulator	259
5.	Distribution relations	260
6.	Characterization of the units	262
7.	Weierstrass units	264
8.	Applications to diophantine analysis	265
Appendix.	The Weierstrass eta function	270
Bibliography		274

*Supported by NSF grants

We shall mostly summarize what is in the list of papers [KL] of the bibliography, but we also include some new results. We only sketch the ideas of some of the proofs, referring to the original papers for a complete treatment.

§1. UNITS AND DIVISOR CLASSES AT INFINITY

Let H be the upper half plane. Let $H^* = H \cup \underset{\sim}{Q} \cup \infty$. Let $\Gamma(N)$ be the subgroup of $SL_2(\underset{\sim}{Z}) = \Gamma(1)$ consisting of the matrices $\equiv 1 \bmod N$. There is an affine curve $Y(N)$ defined over $\underset{\sim}{Q}(\mu_N)$ such that $Y(N)_C \approx \Gamma(N)\backslash H$, and its complete curve $X(N)$ is such that $X(N)_C \approx \Gamma(N)\backslash H^*$, where the isomorphism is complex analytic. The cusps, or points at infinity, are by definition $\underset{\sim}{Q}, \infty$, or their images in $X(N)$, so the complement of $Y(N)$ in $X(N)$.

We let R_N be the integral closure of $\underset{\sim}{Z}[j]$ in the function field F_N of $X(N)$ over $\underset{\sim}{Q}(\mu_N)$, and $\underset{\sim}{Q}R_N$ the integral closure of $Q[j]$. Elements of $(\underset{\sim}{Q}R_N)^*$ will be called <u>units</u>, and elements of R_N^* will be called <u>units over</u> Z. The rank of $(\underset{\sim}{Q}R_N^*)/$constants is obviously bounded by No. of cusps $- 1$.

THEOREM 1.1. *The rank of the units mod constants is equal to the number of cusps - 1.*

Viewing $X(N)$ as embedded in its Jacobian, this theorem is equivalent to the fact that the cusps are of finite order in the Jacobian. Let x_1, x_2 be two cusps. We denote by $\{x_1, x_2\}$ the functional on the space of differentials of first kind given by

$$\{x_1, x_2\}: \omega \mapsto \int_{x_1}^{x_2} \omega \; .$$

A priori, $\{x_1, x_2\}$ lies in $H^1(X(N), R)$. Manin and Drinfeld have shown that $\{x_1, x_2\}$ in fact lies in $H^1(X(N), Q)$. This is a third equivalent statement to the above. They use Hecke operators. Their method suggests generalizations to higher dimensional bounded symmetric domains. We shall leave this method aside in the present exposition, but shall discuss the manner in which explicit units may be constructed to prove the theorem.

Units may be characterized as elements of the function field all of whose zeros and poles are at the cusps. Let f be a unit, and let

$$f = \sum a_n q^{n/N}$$

be its q-expansion, which may have a finite number of terms with $n < 0$. Let $a_m \neq 0$ be the lowest non-zero coefficient. We also write $a_m = a_m(f)$. The following criterion will be used constantly.

   If the coefficients $a_n(f \circ \gamma)$ *are algebraic integers, and the lowest coefficient* $a_m(f \circ \gamma)$ *is a unit in* $Q(\mu_N)$ *for all* $\gamma \in SL_2(Z)$, *then* f *is a unit over* Z.

§2. CONSTRUCTION OF UNITS, KLEIN FORMS

Modular functions whose only zeros and poles are at the cusps have been constructed previously as in Newman [Ne], Ogg [O], and also [L 1], Chapter 18, §6. Here we follow [KL II]. Let L be a lattice, $L = [w_1, w_2]$. The Weierstrass zeta function satisfies

$$\zeta(z+w, L) = \zeta(z, L) + \eta(w, L),$$

where $\eta(z, L)$ is R-linear in z. In fact, it can be shown that

$$\eta(z, L) = s_2(L) z + \frac{\pi}{NL} \bar{z},$$

where NL is the area of a fundamental domain, and $s_2(L)$ is the limit $\lim_{s \to 0} \sum_{w \neq 0} 1/w^2 |w|^{2s}$. Cf. appendix. We won't need this. We define the **Klein forms** (homogeneous of degree 1 in z, L) by

$$k(z, L) = e^{-\eta(z, L) z/2} \sigma(z, L).$$

Let $\psi(w, L) = 1$ if $w/2 \in L$ and -1 otherwise. Then

$$k(z+w, L) = k(z, L) \, e^{-\frac{\pi i}{NL} \operatorname{Im}(z\bar{w})} \psi(w, L).$$

We shall abbreviate $W = \begin{pmatrix} w_1 \\ w_2 \end{pmatrix}$ and $W_\tau = \begin{pmatrix} \tau \\ 1 \end{pmatrix}$. We write

$$z = a_1 w_1 + a_2 w_2 = a \cdot W$$

where $a = (a_1, a_2)$ is a pair of real numbers, not both in $\underline{Z}$. We then also put $\underline{k}_a(W) = \underline{k}(z,L)$. For $\lambda \in C^*$ we have

$$\underline{k}_a(\lambda W) = \lambda \underline{k}_a(W).$$

If $b = (b_1, b_2) \in \underline{Z}^2$ then

$$\underline{k}_{a+b}(W) = \varepsilon_o(a,b) \underline{k}_a(W)$$

where

$$\varepsilon_o(a,b) = (-1)^{b_1 b_2 + b_1 + b_2} e^{-2\pi i (b_1 a_2 - b_2 a_1)/2}$$

has absolute value 1. As usual we put $\underline{k}_a(\tau) = \underline{k}_a(W_\tau)$. Then we define the <u>Siegel function</u> (homogeneous of degree 0)

$$\underline{k}_a(\tau) \Delta^{1/12}(\tau) = g_a(\tau)$$

$$= - q_\tau^{\frac{1}{2} B_2(a_1)} e^{2\pi i a_2 (a_1 - 1)/2} (1-q_z) \prod_{n=1}^{\infty} (1-q_\tau^n q_z)(1-q_\tau^n / q_z)$$

where $q_\tau = e^{2\pi i \tau}$ and $q_z = e^{2\pi i z}$, and $B_2(X) = X^2 - X + \frac{1}{6}$.

We note that $z \mapsto \log|g(z)|$ is the local component of the Néron-Tate height function at infinity, cf. [Ner], last page.

We shall be concerned with $(a_1, a_2) = (r_1/N, r_2/N)$, $r_1, r_2 \in \underline{Z}$. We may take $0 \leq a_1, a_2 \leq 1$, and it is convenient to write $\langle t \rangle$ for the smallest real number ≥ 0 in the residue class of $t \mod \underline{Z}$. The expression for $\varepsilon_o(a,b)$ above shows that $\varepsilon_o(a,b)^{2N} = 1$. Also if $\alpha = \begin{pmatrix} a & b \\ c & d \end{pmatrix} \in SL_2(\underline{Z})$,

$$\underline{k}_a(\alpha W) = \underline{k}_{a\alpha}(W) = \epsilon(\alpha)\underline{k}_a(W),$$

this last equality holding for $\alpha \in \Gamma(N)$ with $\epsilon(\alpha) =$

$$-(-1)^{(\frac{a-1}{N}a_1+\frac{c}{N}a_2+1)(\frac{b}{N}a_1+\frac{d-1}{N}a_2+1)} e^{2\pi i(ba_1^2+(d-a)a_1a_2-ca_2^2)/2N^2}.$$

Therefore $\epsilon(\alpha)^{2N} = 1$, $\epsilon(\alpha) = 1$ if $\alpha \in \Gamma(2N^2)$, $\underline{k}_a$ is a modular form on $\Gamma(2N^2)$, and $\underline{k}_a^{2N}$ is on $\Gamma(N)$.

THEOREM 2.1. If $a = r/N$, write $\underline{k}_r$ instead of $\underline{k}_a$. A product

$$\prod_r \underline{k}_r^{m(r)}$$

(with a finite family $m(r)$ of integers $\neq 0$) has level N if and only if

$$\sum m(r)r_1^2 \equiv \sum m(r)r_2^2 \equiv \sum m(r)r_1r_2 \equiv 0 \bmod N$$

if N is odd, and if N is even, then the first two congruences should be $\equiv 0 \bmod 2N$.

THEOREM 2.2. Let $a \in \frac{1}{N}\underline{Z}^2/\underline{Z}^2$ be primitive.
1) If N is composite, then g_a is a unit over $\underline{Z}$.
2) If $N = p^n$ is a prime power, then g_a is a unit in $R_N[1/p]$.
3) If $c \in \underline{Z}$, $c \neq 0$ is prime to N, then g_{ca}/g_a is a unit over $\underline{Z}$.

In particular, if we have a point $y \in Y(N)$ in a number

field K such that $j(y) \in \underline{o}_K$, then we get a homomorphism

$$R_N^* \longrightarrow R_N^*(y) \subset \underline{o}_K^*$$

thus parametrizing units in K. These may be called <u>modular units</u> in the field of N-th division points of an elliptic curve over K. Analogously, if $j(y)$ is not integral, one gets by pull back a homomorphism from the divisors at infinity on $X(N)$ into the group of fractional ideals, inducing a homomorphism from divisor classes to ideal classes, of which it would be very interesting to know significant examples when it is non-degenerate. Cf. [KL I], the last chapter of [L], Ramachandra [Ra] and Robert [Ro] for cases with complex multiplication. The Shimura reciprocity law can be used with good effect in this case to describe the behavior of the special units under the Galois group.

In [KL II], the rank of the units over $\underline{Z}$ is determined, and a criterion is given for a unit to be a unit over $\underline{Z}$ as follows. Let

$$g = \prod_a g_a^{m(a)}$$

be a finite product, with integer exponents $m(a)$, $a \in \underline{Q}^2$ and $a \notin \underline{Z}^2$. Write

$$g = g_{comp} \prod_p g^{(p)},$$

where g_{comp} is the partial product taken over those a

which have composite period with respect to $\underline{Z}^2$, and $g^{(p)}$ is the product taken over those a having p-power period for the prime p. We know that g_{comp} is a unit over $\underline{Z}$, and it is easy to see that

g <u>is a unit over</u> $\underline{Z}$ <u>if and only if</u> $g^{(p)}$ <u>is a unit over</u> $\underline{Z}$ <u>for each prime</u> p.

Now fix a prime p, and suppose

$$g = g^{(p)} = \prod g_a^{m(a)}$$

is expressed as a product where a has p-power period mod $\underline{Z}^2$. Using the distribution relation (see Theorem 5.1 below), we can achieve that g is equal to such a product in which the exact periods of elements a are all equal to a fixed power p^n, in which case we say that the product is normalized. Note that $\underline{Z}_p^*$ operates on $\underline{Q}_p^2/\underline{Z}_p^2$. We consider the orbits under this action.

THEOREM 2.3. <u>Let</u> $g = g^{(p)} = \prod g_a^{m(a)}$ <u>be a normalized product. Then</u> g <u>is a unit over</u> $\underline{Z}$ <u>if and only if for each orbit of</u> $\underline{Z}_p^*$ <u>we have</u>

$$\sum_{a \in orbit} m(a) = 0.$$

We could also formulate a similar criterion without dealing with normalized products.

Robert's construction of units in the complex multiplication case is seen to come from the following modular construction. Let $L \subset L'$ be lattices with $(L':L)$ odd. Define
$$\underline{k}(z, L'/L) = \underline{k}(z,L)^{(L':L)} / \underline{k}(z, L').$$
Then $\underline{k}(z, L'/L)$ is elliptic and periodic with respect to L (the anti-holomorphic term and $\psi(w, L)$ drop out!). We have
$$\underline{k}(z, L'/L) = \prod \frac{1}{\wp(z,L) - \wp(w,L)}$$
where the product is taken over $w \in (L'/L)/\pm 1$, and $w \neq 0$, because both sides have the same zeros and poles.

Let $\alpha \in \text{Mat}_2^+(\mathbb{Z})$ be an integral matrix with odd determinant, and let $N\alpha = \det \alpha$. Define
$$\varphi(z, \alpha, W) = \underline{k}(z, W)^{N\alpha} / \underline{k}(z, \frac{\alpha W}{N\alpha}).$$
In view of the periodicity property of $\underline{k}(z, L'/L)$, the above value depends only on the left coset of α with respect to $SL_2(\mathbb{Z})$, so we may assume that
$$\alpha = \begin{pmatrix} a & b \\ 0 & d \end{pmatrix}$$
is triangular. Let $A = \{\alpha_j, n_j\}_{j \in J}$ be a finite family, where α_j are matrices as above, and n_j are integers

such that $\sum_j n_j(N\underline{\alpha}_j - 1) = 0$. We let

$$\varphi(z,A,W) = \prod \varphi(z,\alpha_j,W)^{n_j}.$$

Then φ has weight 0 in (z,W), and is an even elliptic function. For $z = a \cdot W$ its q-expansion is easily determined from that of the Klein forms. We assume again that $a = (r_1/N, r_2/N)$ has denominator N, and similarly $b = (b_1, b_2)$. We let

$$\varphi(a \cdot W, \alpha, W) = \varphi_a(\alpha, W),$$

and similarly for $\varphi_a(A,W)$, replacing α by α_j and taking the product raising each term to the power n_j. <u>We assume that</u> $N\underline{\alpha}, N\underline{\alpha}_j$ <u>is prime to the denominator</u> N <u>of</u> a, b.

THEOREM 2.4. <u>The function</u>

$$\varphi_a(A,W)/\varphi_b(A,W)$$

<u>is a unit over</u> $\underline{Z}$.

Finally let $\delta(W) = \Delta^{1/12}(W)$. Let

$$\delta(\alpha,W) = \delta(W)^{N\underline{\alpha}}/\delta(\frac{\alpha W}{N\underline{\alpha}}),$$

$$\delta(A,W) = \prod \delta(\alpha_j, W)^{n_j}.$$

Then:

<u>The product</u> $\delta(A,W)\varphi_a(A,W)$ <u>is a unit over</u> $\underline{Z}$.

The proof is done by comparing lowest coefficients. Let $f = \sum a_n q^{n/N}$ be a power series in $q^{1/N}$, having possibly a finite number of polar terms. If g is another such power series, we write $f \sim g$ to mean that the coefficients of the lowest terms in f and g have a quotient which is a unit in a cyclotomic field. Then

$$\varphi(a_1\tau+a_2, \alpha, [\tau,1]) \sim \begin{cases} a & \text{if } a_1 \neq 0 \\ a \dfrac{\left(1 - q_1^{\langle a_2 \rangle}\right)^{N\alpha}}{1 - q_1^{\langle aa_2 \rangle}} & \text{if } a_1 = 0. \end{cases}$$

The theorem follows at once.

Specializing to complex multiplication, one gets Robert's units, using ideals $\underline{a}_j^{-1}$ instead of the matrices $\alpha_j/N\alpha_j$. If $L = [w_1, w_2]$, for any ideal $\underline{a}$ there is a matrix α such that $\underline{a}^{-1}L$ has the basis $\dfrac{\alpha W}{N\alpha}$. We can take α to be in triangular form by operating on the left with an element of $SL_2(Z)$. As a variation, if $\underline{a}_j = (\xi_j)$ is principal, then the product

$$\prod \xi_j^{-n_j} \underline{k}(z, \underline{a}_j^{-1}L/L)$$

is a unit in the ray class field with conductor $\underline{c}$, when L is an ideal, $z \in \underline{c}^{-1}L$, and all ideals $\underline{a}_j$ are odd, and prime to $\underline{c}$. Working with the Klein forms and the delta function separately gives a splitting of the group of units which may have arithmetic significance.

§3. REPRESENTATION OF THE CUSPS

On the modular function field F_N, the group $SL_2(\mathbb{Z}/N\mathbb{Z})$ operates in the usual way. Matrices $\sigma_d = \begin{pmatrix} 1 & 0 \\ 0 & d \end{pmatrix}$ with d prime to N operate by leaving $q^{1/N}$ fixed, and moving the coefficients of the q-expansion, $\sigma_d: \zeta_N \mapsto \zeta_N^d$. Then $GL_2(\mathbb{Z}/N\mathbb{Z})$ operates, and the isotropy group of ∞ is

$$G_\infty(N) = \left\{ \begin{pmatrix} 1 & b \\ 0 & d \end{pmatrix} \right\}.$$

There is a direct decomposition

$$GL_2(N) = C(N) G_\infty(N),$$

where $C(N)$ is described as follows. It is multiplicative over prime powers, $C(N) = \prod C(p^{n(p)})$ if $N = \prod p^{n(p)}$. The group C_p is taken to be the group of units $\mathfrak{o}_p^*$ in the unramified extension of degree 2 over $\mathbb{Z}_p$, under the regular representation. We let $C(p^n) = C_p \mod p^n$. Then $C(p^n)$ operates simply transitively on $(\mathbb{Z}^2/p^n\mathbb{Z}^2)^* = \mathbb{Z}^2(p^n)^*$, whence on $\frac{1}{p^n}\mathbb{Z}^2/\mathbb{Z}^2$ by

$$a \mapsto a\alpha.$$

Similarly, $C(N)$ operates simply transitively on $(\frac{1}{N}\mathbb{Z}^2/\mathbb{Z}^2)^*$.

If P_∞ is the standard place at infinity of F_N, then αP_∞ for $\alpha \in C(N)/\pm 1$ gives all the other places at infinity. By this device, the cusps can be analysed with a commutative group as in complex multiplication. See [KL II].

§4. ANALYSIS OF THE REGULATOR

Let $a \in \frac{1}{N}\mathbb{Z}^2/\mathbb{Z}^2$. Let g_a be the Siegel function. Then

$$\mathrm{ord}_\infty\, g_a = \tfrac{1}{2} B_2(\langle a_1 \rangle).$$

We may write the divisor of g_a in the group ring

$$(g_a) = \sum_\alpha \tfrac{1}{2} B_2(\langle T(a\alpha) \rangle)\alpha, \quad \text{with } \alpha \in C(N)/\pm 1.$$

We have put $T(a_1, a_2) = a_1$. If a is primitive, so $a = (1/N, 0)\beta$ with some β, then

$$(g_a) = (g_\beta) = \sum_\alpha \tfrac{1}{2} B_2(\langle T(\beta\alpha) \rangle)\alpha.$$

Let $f(\alpha\beta)$ denote the coefficient of α in this expression. Then we have the idempotent decomposition

$$(g_\beta) = \sum_\psi \bar{\psi}(\alpha) S(\psi, f) e_\psi,$$

where ψ ranges over the character group of $C(N)/\pm 1$, e_ψ is the orthogonal idempotent of ψ, and

$$S(\psi, f) = \sum \psi(\alpha) f(\alpha).$$

Let $\psi_\mathbb{Z}$ be the restriction of ψ to $(\mathbb{Z}/N\mathbb{Z})^*$. Then

$$S(\psi, f) = (\text{non-zero constant}) B_{2, \psi_\mathbb{Z}}$$

where

$$B_{2, \psi_\mathbb{Z}} = N \sum_{w \in (\mathbb{Z}/N\mathbb{Z})^*} \psi_\mathbb{Z}(w) B_2(\langle \tfrac{w}{N} \rangle),$$

provided that the conductor of ψ is N.

The factorization of the sum $S(\psi,f)$ is carried out in [KL D], following a suggestion of Tate, using Fourier expansions of the function f. The non-zero constant consists mostly of ordinary Gauss sums, and it is known that $B_{2,\psi_Z} \neq 0$ for even characters ψ_Z, so the regulator of the Siegel functions is not 0.

Along the same lines, one can construct a <u>Stickelberger element</u> in the group ring, annihilating the divisor classes at infinity [KL III]. This is a 2-dimensional analogue of the situation which has arisen in the work of Leopoldt, Iwasawa, etc.

§5. DISTRIBUTION RELATIONS

Let $G = \lim G/G_n$ be a compact group, which is the projective limit of finite factor groups $X_n = G/G_n$. Let $\{\varphi_n\}$ be a family of functions, where φ_n maps X_n into some abelian group. We say that this family is consistent, or defines a <u>distribution</u> on G, if it satisfies the following condition. Let $x \in X_n$, let $\pi: X_{n+1} \to X_n$ be the canonical map. Then

$$\sum_{\pi y = x} \varphi_{n+1}(y) = \varphi_n(x).$$

Suppose the values of φ_n are in a ring R. We let

$$\theta_n = \sum \varphi_n(x) x \in R[G/G_n].$$

The elements θ_n are compatible under the natural homomorphisms of the group rings. We are especially interested in distributions defined by a function φ on Q^2/Z^2 satisfying the condition

$$\sum_{Ds=r} \varphi(s) = \varphi(r)$$

for every positive integer D. Such distributions are called <u>ordinary</u>. If the condition is satisfied only for $r \neq 0$ we say the distribution is <u>punctured</u>.

THEOREM 5.1. <u>The map</u> $a \mapsto g_a$ <u>modulo roots of unity is an ordinary, punctured distribution.</u>

The relation in the case $r = 0$ has to be exceptional, since for instance g_0 is not defined. The product $\prod g_a$ for $Da = 0$, $a \neq 0$ in Q^2/Z^2 is a constant depending on D. Cf. Klein, Ramachandra [Ra] and Robert [Ro], the latter however not looking at things modularly, but through theta functions. The proofs of the relations are given from the q-expansions in [KL III].

In addition, Kubert [K 2] constructs a free basis for the units (of all level) modulo constants, showing that the distribution of Theorem 5.1 is a universal distribution (even) into abelian groups on which 2 is invertible. The combinatorics are relatively complicated, due in part to composite N, mostly to the ± 1 ambiguity. It is simple when N is a prime power.

If f is a unit, then df/f is a differential of third kind on the modular curve X(N). Let us fix a determination of log f on H. For each $\gamma \in \Gamma(N)$ there is an integer $\langle f,\gamma \rangle$ in $\mathbb{Z}$ such that for all $z \in H$ we have

$$(\log f)(\gamma z) = \log f(z) + \langle f,\gamma \rangle 2\pi i.$$

Each γ thus induces a homomorphism of the group of units modulo constants into $\mathbb{Z}$, and gives rise to a homomorphic image of the Siegel distribution.

Distributions involving the higher Bernoulli numbers are discussed in [KL D], independently of the context of units in the modular function field.

§6. CHARACTERIZATION OF THE UNITS

We ask whether the construction of units by means of the Klein forms and Siegel functions gives all units in the modular function field (of all levels). The answer is yes, up to possibly 2-torsion. Let $S(N)$ be the group generated by the Siegel functions g_a with $a \in \mathbb{Q}^2$ such that $Na \in \mathbb{Z}^2$, and the non-zero constants.

THEOREM 6.1. *Let* g *be a unit.*

(i) *If* $N = p^n$ *is a prime power, and* $g^m \in S(N)$ *for some integer* m > 0 *then* $g \in S(N)$.

(ii) **If** N **is composite and** ℓ **is an odd prime, such that** $g^\ell \in S(N)$, **then** $g \in S(N)$.

Hence the units modulo the union of the groups $S(N)$ **for all** N **form a 2-torsion group, equal to** 1 **if** N **is a prime power.**

The proof is in [KL IV]. It is based on a lemma of Shimura that the Fourier coefficients of a modular form have bounded denominators. Let us write

$$g = cq^\lambda g^*,$$

where c is a constant, and g^* is a power series starting with 1. The q-expansion of the Siegel functions shows that

$$g_a^* = 1 + \text{power series in } q^{1/N} \text{ with integral coeffs.}$$

Suppose

$$g^{*\ell} = \prod_a g_a^{*m(a)}$$

so that

$$g^* = \prod_a g_a^{*m(a)/\ell},$$

taking the root with the binomial series. Suppose first that N is a prime p. We look at the coefficient of $q^{1/p}$. In $g_a^{*m(a)/\ell}$ it is equal to

$$-\sum_{\nu=0}^{p-1} \frac{1}{\ell} m^*(\nu) \zeta^\nu$$

where ζ is a primitive p-th root of unity. We can choose a "good" basis for the cyclotomic integers in order to see that the coefficient has a denominator unless $g \in S(p)$. In general, when N is not prime, we look at the first non-constant coefficient and use induction.

The same method can be used to prove independence relations among the Siegel units, yielding rather easily when N is a prime power the fact that they have the proper rank as in Theorem 1.1, modulo constants. In this way one gets an independent proof that $B_{2,\chi} \neq 0$, without using the L-series. A similar method works to get the right rank when N is composite, but in a more complicated fashion.

§7. WEIERSTRASS UNITS

For $a \in \frac{1}{N}Z^2/Z^2$ let $\wp_a(\tau) = \wp(a_1\tau + a_2; [\tau,1])$.
Then

$$\frac{\wp_a - \wp_b}{\wp_c - \wp_d}$$

is a unit. Simple conditions can also be given on a, b, c, d to determine when such an expression is a unit over Z. Note that

$$\wp_a - \wp_b = \frac{\underline{k}_{a+b} \underline{k}_{a-b}}{\underline{k}_a^2 \underline{k}_b^2} .$$

Let $\underline{K}(N)$ be the group of forms expressible as products

$$\prod_a \underline{k}_a^{m(a)}$$

which are modular with respect to $\Gamma(N)$. Let $\underline{K}^+(N)$ be the subgroup of forms of even degree, i.e. such that $\sum m(a)$ is even. Let $W(N)$ be the subgroup generated by the Weierstrass forms, consisting of elements which are modular with respect to $\Gamma(N)$.

THEOREM 7.1. *We have* $W(N) = \underline{K}^+(N)$.

The proof in Kubert [K 1] is done by descent, and involves only combinatorial juggling with the quadratic relation and the parallelogram relations

$$(a+b) + (a-b) - 2(a) - 2(b).$$

§8. APPLICATIONS TO DIOPHANTINE ANALYSIS

The λ-function (generator of the modular function field of level 2) has the form of a Weierstrass unit, and so does $1-\lambda$. Put $u = \lambda$, $v = 1-\lambda$, so that $u+v = 1$. Let
$$u = x^N, \qquad v = y^N, \quad \text{so} \quad x^N + y^N = 1.$$
Fricke [F] already knew that this gives a parametrization of the Fermat curve by modular functions, which do not belong to a congruence subgroup. It gives a correspondence between $X(2N)$ and the Fermat curve $V(N)$, called standard.

In view of the identity

$$\frac{x_3-x_1}{x_2-x_1} + \frac{x_2-x_3}{x_2-x_1} = 1$$

we get infinitely many non-standard correspondences between the Fermat curve and modular curves. The standard one is especially good, however, because it is unramified above both $X(2N)$ and $V(N)$. From this and an old theorem of Chevalley-Weil (recalled in [KL I]), one gets:

> The Fermat curve has only a finite number of rational points in any number field if and only if the modular curve $X(2N)$ has this same property.

In [K 3], Kubert also proves, using the Weierstrass units:

> Let K be a number field, and ℓ a prime ≥ 5. There is a universal bound on the torsion of elliptic curves A over K such that ℓ does not divide the order of the Galois group $G(K(A_\ell)/K)$.

This follows the work of Demjanenko [D], which has not yet certifiably shown that the order of the torsion group over K is uniformly bounded, an outstanding conjecture.

The equation $u+v = 1$ has only a finite number of solutions with u, v in any finitely generated multiplicative group of complex numbers [L 3]. The Weierstrass units provide a significant example of such solutions.

The standard uniformization of the Fermat curve puts the Fermat conjecture in the light that the only rational points (over $\underset{\sim}{Q}$!) should be at the cusps (in line with the Ogg conjecture for the modular curves belonging to congruence subgroups), cf. [KL I].

The existence of units such that $u+v = 1$ also gives a direct reduction on the modular curves for the finiteness of integral points to the situation where one can apply Gelfond's idea to use a diophantine inequality for linear combinations of logarithms of algebraic numbers, proved by Baker in general.

Let F_N be the modular function field again. If we assume the Mordell conjecture for subfields of F_N properly containing $\underset{\sim}{Q}(j)$ and of genus ≥ 2, then we see that one gets the following irreducibility statement [KL I].

Let k be a number field containing the N-th roots of unity. Assume that N is such that every subfield of kF_N properly containing k(j) has genus ≥ 2 (satisfied if N is prime sufficiently large). Then for all but a finite number of values j_o of j in k, we have
$$[k(\pi^{-1}(j_o)) : k] = [kF_N : k(j)],$$
where $\pi: X(N) \longrightarrow X(1)$ is the natural map.

In fact, we conjecture following uniformity property as in [KL I], in connection with Serre's theorem.

<u>For</u> <u>all</u> <u>but</u> <u>a</u> <u>finite</u> <u>number</u> <u>of</u> $j_o \in k$, <u>and</u> <u>all</u> <u>but</u> <u>a</u> <u>finite</u> <u>number</u> <u>of</u> <u>primes</u> p, <u>the</u> <u>Galois</u> <u>group</u> <u>of</u> <u>the</u> p-<u>primary</u> <u>torsion</u> <u>of</u> <u>an</u> <u>elliptic</u> <u>curve</u> A <u>over</u> k <u>with</u> <u>invariant</u> j_o <u>is</u> <u>equal</u> <u>to</u> $GL_2(\mathbb{Z}_p)$.

Next we consider another aspect of torsion points.

The Manin-Mumford conjecture asserts that on a curve of genus ≥ 2 there are only a finite number of points of finite order in the Jacobian. The cusps on the modular curves provide significant examples of such points, according to the Manin-Drinfeld theorem (equivalent to Theorem 1.1). The question can be raised whether the cusps are also of finite order on curves which are quotients of non-congruence subgroups. This is true for the standard representation of the Fermat curve, as shown by Rohrlich [Roh], who determines completely the structure of the divisor class group generated by the cusps. On the other hand, Rohrlich has observed that the answer is negative in general. The argument goes as follows.

In [L 3], Lang reduces the Manin-Mumford conjecture to a Galois property of the field of torsion points on the Jacobian, namely that the index of the subgroup of the Galois group of the N-th torsion points over the given number field generated by the homotheties (that is, inducing multiplication by an integer prime to N on the N-th torsion points) should be bounded in $(Z/NZ)^*$. Recently Shimura has informed us that this property can be proved in the case of complex multiplication, and therefore:

> The Manin-Mumford conjecture is true in the case of complex multiplication.

In particular, it is true for the Fermat curve, which has complex multiplication.

By choosing infinitely many suitable non-standard correspondences of the Fermat curve with modular curves, i.e. representations as quotient of the upper half plane by non-congruence subgroups associated with units satisfying $u+v = 1$, Rohrlich shows that one would get infinitely many points on the curve of finite order in the Jacobian if the Manin-Drinfeld theorem were true in the non-congruence case, a contradiction.

APPENDIX

Because of its fundamental interest, we shall carry out here the analysis of the Weierstrass eta function $\eta(z,L)$ in detail. We recall first some facts about Eisenstein series.

By Kronecker's first limit formula or otherwise, we know that

$$E(\tau,s) = \sum \frac{y^s}{|m\tau+n|^{2s}} = \frac{\pi}{s-1} + O(1).$$

Let $L = [\omega_1, \omega_2]$, and let $\tau = \omega_1/\omega_2 \in H$. Let

$$E(L,s) = \sum_{\omega \neq 0} \frac{1}{|\omega|^{2s}}.$$

Then in a neighborhood of $s = 1$ we have

$$E(L,s) \sim \frac{\pi}{NL} \frac{1}{s-1},$$

where

$$NL = \frac{1}{2i}(\omega_1 \bar{\omega}_2 - \bar{\omega}_1 \omega_2)$$

is the area of the fundamental domain. The residue of $E(L,s)$ at $s = 1$ is therefore π/NL.

THEOREM A1. (i) <u>The function</u> $\sum \dfrac{1}{\omega^2 |\omega|^{2s}}$ <u>is holo-</u>

morphic at $s = 0$ (it is even an entire function).

(ii) The function defined for $z \notin L$ by

$$G(z,L,s) = \sum_{\omega \neq 0} \frac{\overline{z+\omega}}{|z+\omega|^{2s}} \quad \text{for} \quad \text{Re } s > 3/2$$

has an analytic continuation to Re $s > 1/2$, and is holomorphic at $s = 1$.

Proof. These are essentially well known, and can be proved by standard techniques using Poisson's summation formula. Cf. Siegel [Si], Theorem 3, p. 69.

In particular, we can define $s_2(L)$ as the value of the function in (i), at $s = 0$.

Following Birch-Swinnerton Dyer [B-SwD] we define

$$\zeta_s(z,L) = \frac{\overline{z}}{|z|^{2s}} + \sum_{\omega \neq 0} \left\{ \frac{\overline{z+\omega}}{|z+\omega|^{2s}} - \frac{\overline{\omega}}{|\omega|^{2s}} \left(1 - \frac{sz}{\omega} + \frac{\overline{z}}{\overline{\omega}}(1-s)\right) \right\}$$

$$= \frac{\overline{z}}{|z|^{2s}} + \sum_{\omega \neq 0} \left\{ \frac{\overline{z+\omega}}{|z+\omega|^{2s}} - \frac{\overline{\omega}}{|\omega|^{2s}} + \frac{sz}{|\omega|^{2s-2} \omega^2} \right.$$

$$\left. + \frac{\overline{z}}{|\omega|^{2s}} (s-1) \right\}$$

The series converges absolutely for Re s > 1/2 by the usual argument, and we have

$$\lim_{s \to 1} \zeta_s(z,L) = \zeta(z,L)$$

taking the limit for s real > 1. For s > 3/2 we can rearrange the terms in $\{\ \}$. Combining ω with $-\omega$ shows

$$\sum_{\omega \neq 0} \frac{\bar{\omega}}{|\omega|^{2s}} = 0.$$

Since $E(L,s)$ has only a simple pole at $s = 1$, it follows that

$$\sum{}' \frac{\bar{z}}{|\omega|^{2s}} (s-1) \sim \bar{z} \frac{\pi}{NL} \quad \text{for} \quad s \to 1.$$

Therefore the function

$$\sum_{\omega \neq 0} \left\{ \frac{\overline{z+\omega}}{|z+\omega|^{2s}} + \frac{s z}{|\omega|^{2s-2} \omega^2} \right\}$$

is holomorphic at $s = 1$. From the definition of $s_2(L)$ we then find:

$$\boxed{\zeta(z,L) = \frac{1}{z} + G(z,L,1) + s_2(L) z + \frac{\pi}{NL} \bar{z}.}$$

THEOREM A2. $\quad \eta(z,L) = s_2(L)z + \dfrac{\pi}{NL}\bar{z}$.

Proof. The expression

$$\frac{1}{z} + G(z,L,1) = \frac{\bar{z}}{|z|^{2s}} + G(z,L,s) \quad \text{at} \quad s=1$$

is periodic in L. Hence

$$\zeta(z+\omega,L) - \zeta(z,L) = s_2(L)\omega + \frac{\pi}{NL}\bar{\omega}.$$

$$= \eta(\omega,L).$$

The theorem follows by R-linearity. It may be viewed as a generalization of the Legendre relation, which is seen to result from the above by putting $z = \omega_1$ and $z = \omega_2$. The relation is known in the case of complex multiplication cf. Damerell, Acta Arith. XVII (1970) pp. 294 and 299, but we could find no reference for it in general.

Bibliography

[B-SwD]   B. BIRCH and H. SWINNERTON-DYER, Notes on elliptic curves II, J. reine angew. Math. 218 (1965) pp. 79-108

[De]   V. A. DEMJANENKO, On the uniform boundedness of the torsion of elliptic curves over algebraic number fields, Math. USSR Izvestija Vol. 6 (1972) No. 3 pp. 477-490

[Dr]   V. G. DRINFELD, Two theorems on modular curves, Functional analysis and its applications, Vol. 7 No. 2, AMS translation from the Russian, April-June 1973, pp. 155-156

[F]   R. FRICKE, Über die Substitutionsgruppen, welche zu den aus dem Legendre'schen Integralmodul $k^2(\omega)$ gezogenen Wurzeln gehören, Math. Ann. 28 (1887) pp. 99-118

[K 1]   D. KUBERT, Quadratic relations for generators of units in the modular function field, Math. Ann. 225 (1977) pp. 1-20

[K 2]   _____, A system of free generators for the universal even ordinary $Z_{(2)}$ distribution on $Q_\omega^{2k}/Z_\omega^{2k}$, Math. Ann. 224$^{(2)}$ (1976) pp. 21-31

[K 3]   _____, Universal bounds on the torsion of elliptic curves, J. London Math. Soc. to appear

[KL]   D. KUBERT and S. LANG, Units in the modular function field, Math. Ann.:
I, 1975, pp. 67-96
II, 1975, pp. 175-189
III, 1975, pp. 273-285
IV, 1977, pp. 223-242

[KL D]   ―――――――, Distributions on toroidal groups, Math. Z. 148 (1976) pp. 33-51

[L 1]   S. LANG, Elliptic Functions, Addison Wesley, 1973

[L 2]   ―――――, Division points on curves, Ann. Mat. pura et appl. IV, Tomo LXX (1965) pp. 229-234

[L 3]   ―――――, Integral points on curves, Pub. IHES No. 6 (1960) pp. 27-43

[Ma]   J. MANIN, Parabolic points and zeta functions of modular curves, Izv. Akad. Nauk SSSR, Ser. Mat. Tom 36 (1972) No. 1, AMS translation pp. 19-64

[Ner]   A. NÉRON, Quasi-fonctions et hauteurs sur les variétés abéliennes, Ann. Math. 82 (1965) pp. 249-331

[New]   M. NEWMAN, Construction and application of a class of modular functions, Proc. London Math. Soc. (3) (1957) pp. 334-350

[O]   A. OGG, Rational points on certain elliptic modular curves, AMS conference St. Louis, 1972, pp. 221-231

[Ra]   K. RAMACHANDRA, Some applications of Kronecker's limit formula, Ann. Math. 80 (1964) pp. 104-148

[Rob 1]   G. ROBERT, Unités elliptiques, Bull. Soc. Math. France, Mémoire No. 36 (1973)

[Rob 2]   ―――――――, Nombres de Hurwitz et unités elliptiques, to appear

[Roh]   D. ROHRLICH, Modular functions and the Fermat curve, to appear

[Si]   C. L. SIEGEL, Lectures on advanced analytic number theory, Tate Institute Notes, 1961, 1965

D. Kubert
Mathematics Department
Cornell University
Ithaca, N.Y. 14850

S. Lang
Departement of Mathematics
Yale University
New Haven, Conn. 06520

Lecture Notes in Mathematics

Edited by A. Dold and B. Eckmann

504

Serge Lang
Hale Trotter

Frobenius Distributions
in GL_2-Extensions

Distribution of Frobenius Automorphisms in
GL_2-Extensions of the Rational Numbers

Springer-Verlag
Berlin · Heidelberg · New York 1976

Authors

Serge Lang
Mathematics Department
Yale University
New Haven, Connecticut 06520
USA

Hale Freeman Trotter
Fine Hall
Princeton University
Princeton, New Jersey 08540
USA

Library of Congress Cataloging in Publication Data

Lang, Serge, 1927-
   Frobenius distributions in GL_2-extensions.

   (Lecture notes in mathematics ; 504)
   Bibliography: p.
   Includes index.
   1. Probabilistic number theory. 2. Galois theory.
3. Field extensions (Mathematics) 4. Numbers,
Rational. I. Trotter, Hale F., joint author. II. Title. III. Series: Lecture notes in mathematics
(Berlin) ; 504.
QA3.L28   no. 504   [QA241.7]   510'.8s   [512'.7]
                                                      75-45242

AMS Subject Classifications (1970): 10 K 99, 12 A 55, 12 A 75, 33 A 25

ISBN 3-540-07550-X Springer-Verlag Berlin · Heidelberg · New York
ISBN 0-387-07550-X Springer-Verlag New York · Heidelberg · Berlin

This work is subject to copyright. All rights are reserved, whether the whole or part of the material is concerned, specifically those of translation, reprinting, re-use of illustrations, broadcasting, reproduction by photo-copying machine or similar means, and storage in data banks.

Under § 54 of the German Copyright Law where copies are made for other than private use, a fee is payable to the publisher, the amount of the fee to be determined by agreement with the publisher.

© by Springer-Verlag Berlin · Heidelberg 1976
Printed in Germany
Offsetdruck: Julius Beltz, Hemsbach/Bergstr.

ACKNOWLEDGMENTS

Both Lang and Trotter are supported by NSF grants. The final draft of this monograph was written while Lang was at the Institute for Advanced Study, whose hospitality we appreciated.

We also thank Mrs. Helen Morris for a superb job of varityping the final copy.

INTRODUCTION

We are interested in a distribution problem for primes related to elliptic curves, but which can also be described solely in terms of the distribution of Frobenius elements in certain Galois extensions of the rationals. We therefore first describe this situation, and then indicate its connection with elliptic curves.

Let K be a Galois (infinite) extension of the rationals, with Galois group G. We suppose given a representation

$$\rho : G \longrightarrow \prod GL_2(Z_\ell)$$

which we assume gives an embedding of G onto an open subgroup of the product, taken over all primes ℓ. We let

$$\rho_\ell : G \longrightarrow GL_2(Z_\ell)$$

be the projection of ρ on the ℓ-th factor. We assume that there is an integer Δ such that if p is a prime and $p \nmid \Delta\ell$, then p is unramified in ρ_ℓ, or in other words, the inertia group at a prime of K above p is contained in the kernel of ρ_ℓ. Then the Frobenius class σ_p is well defined in the factor group

$$G_\ell = G/\text{Ker}\, \rho_\ell \,,$$

and $\rho_\ell(\sigma_p)$ has a characteristic polynomial which we assume of the form

$$X^2 - t_p X + p \,.$$

We assume that t_p is an integer independent of ℓ, and call t_p the trace of Frobenius. Finally we assume that the roots of the characteristic polynomial have absolute value $\sqrt{p}$, and are complex conjugates of each other. Let π_p be such a root.

Let t_0 be a given integer. Let k be a given imaginary quadratic field. We let $N_{t_0,\rho}(x)$ be the number of primes $p \leq x$ such that $t_p = t_0$. We let $N_{k,\rho}(x)$

be the number of primes $p \leq x$ such that $Q(\pi_p) = k$. If $k = k_D$ is the field

$$k_D = Q(\sqrt{D})$$

with discriminant D, we also write $N_D(x)$, $N_{D,\rho}(x)$ or $N_k(x)$ instead of $N_{k,\rho}(x)$.

We conjecture that there are constants $C(t_0,\rho)$ and $C(k,\rho) > 0$ such that we have the asymptotic relations

$$N_k(x) \sim C(k,\rho) \frac{\sqrt{x}}{\log x} \quad \text{and} \quad N_{t_0}(x) \sim C(t_0,\rho) \frac{\sqrt{x}}{\log x}.$$

The constants depend on k,ρ or t_0,ρ respectively. If $C(t_0,\rho) = 0$ then the asymptotic relation is to be interpreted to mean that $N_{t_0}(x)$ is bounded. If $C(t_0,\rho) \neq 0$, then the asymptotic relation has the usual meaning. We shall also see that $C(0,\rho) \neq 0$. Cf. Part I, §4, Remark 1.

Actually, define

$$\pi_{\frac{1}{2}}(x) = \sum_{p \leq x} \frac{1}{2\sqrt{p}}.$$

This is essentially

$$\int^x \frac{1}{2\sqrt{x}} d\pi(x) \sim \int^x \frac{1}{2\sqrt{x}} \frac{dx}{\log x},$$

which can be integrated by parts with $u = 1/\log x$ and $dv = \frac{dx}{2\sqrt{x}}$ to show that

$$\pi_{\frac{1}{2}}(x) \sim \frac{\sqrt{x}}{\log x}.$$

Both in the theoretical analysis and in the numerical computations, it is the asymptotic relations

$$N_k(x) \sim C(k,\rho) \pi_{\frac{1}{2}}(x) \quad \text{and} \quad N_{t_0}(x) \sim C(t_0,\rho) \pi_{\frac{1}{2}}(x)$$

which arise. Therefore it is more natural to deal with $\pi_{\frac{1}{2}}(x)$, rather than with the elementary form

$$\frac{\sqrt{x}}{\log x},$$

which converges asymptotically only slowly to $\pi_{\frac{1}{2}}(x)$.

Our arguments to make the conjecture plausible involve only the Galois representation ρ, the Tchebotarev and Hecke distribution theorems in finite Galois extensions, and a conjectured distribution function for the angles of Frobenius elements. One may view our study as a first attempt to formulate for certain infinite extensions distribution laws for Frobenius elements. On the other hand, the motivation also arises from the theory of elliptic curves as follows.

Let A be an elliptic curve over the rationals. Let $K = Q(A_{tor})$ be the field obtained by adjoining the coordinates of its torsion points. Then the Galois group admits a natural representation in $\prod GL_2(Z_\ell)$. We assume that A has no complex multiplication. We then know from Serre's work [S 2] that the representation is open in the product. It is also known that the other properties mentioned above are satisfied (especially that the roots of the characteristic polynomial are complex conjugates of each other, which is none other than the Riemann Hypothesis, Hasse's Theorem). When the representation arises from an elliptic curve, we then write also $N_{k,A}(x)$, etc., replacing ρ by A in the notation. We note that the constants $C(k, A)$ and $C(t_0, A)$ are obviously isogeny invariants of A. (Isogenies over the algebraic closure of Q are allowed.)

A prime p is called **supersingular** when $t_p = 0$. This is a standard interesting case in the theory of elliptic curves. There are numerous other characterizations of this case, which are however irrelevant for us here (cf. for instance Appendix 2 of [L 1]). Serre had already observed that the densities of supersingular primes, or those for which $Q(\pi_p) = k$, are zero ([S 1] for the supersingular case, private communication in the other). Mazur emphasized the importance of the case when $t_p = 1$ for the arithmetic of elliptic curves (see [Ma], Propositions 8.5 and 8.14) and called the prime p **anomalous** when $t_p = 1$. In the case of complex multiplication, if a prime is anomalous, then it lies in a quadratic progression, and the conjectured distribution of such primes can be reduced to a conjecture of Hardy-Littlewood, that it is of the form

$$C \frac{\sqrt{x}}{\log x}$$

for some constant C. The Galois group of a curve with complex multiplication is of course not a GL_2-group, and our situation is more complicated.

(Strictly speaking, one should define supersingular (resp. anomalous) by the condition $t_p \equiv 0$ (resp. $t_p \equiv 1$) mod p, but since $|t_p| < 2\sqrt{p}$, this amounts to the same thing for primes > 5, so the distinction is irrelevant for our purpose, which is to count primes asymptotically.)

The axiomatization of the distribution properties only in terms of the Galois group is important for eventual applications to representations arising from modular forms other than those associated with elliptic curves. One knows from the work of Swinnerton-Dyer [SwD] that they give rise to (essentially) GL_2-extensions of the rational numbers. (Cf. also Ribet [R].) The characteristic polynomial of a Frobenius element is of the form

$$X^2 - t_p X + p^{k-1} = 0 .$$

The analogue of our $\sqrt{p}$ is then $p^{(k-1)/2}$. This leads us to think that when $k \geq 4$, there is only a finite number of primes such that the Frobenius element belongs to the given quadratic field, or such that $t_p = 0$. This would be in line with the Lehmer conjecture that $\tau_p \neq 0$ for all p, where τ_p is the trace of Frobenius for the best known cusp form Δ from the theory of elliptic functions. For $k = 3$, one gets an intermediate asymptotic behavior, and for $k = 1$ one gets back to the oldest situation of actual densities, since the associated Galois group is finite. The case with $k > 2$ introduces enough perturbations in our arguments that we shall handle it elsewhere.

From a naive approach, one already suspects that the asymptotic behavior has something to do with $\pi_{\frac{1}{2}}(x)$. Indeed, the trace of Frobenius t_p must lie in the interval $|t_p| < 2\sqrt{p}$. Under equal probability that it hits any integer in this interval, this probability is $\frac{1}{4\sqrt{p}}$. Summing for $p \leq x$ yields the $\pi_{\frac{1}{2}}(x)$. In reality, the probabilities of hitting the different integers in the interval are not equal, but depend in a fairly complicated way on the Galois representation. In the imaginary quadratic case one wants the probability of coincidence of t_p with the trace of some integer of the field k with norm p. This probability involves an interaction between the field of division points and the maximal abelian extension of k, and becomes especially complicated when the intersection of these two fields is larger (as it may be, by a finite extension) than the field of all roots of unity over the rationals. The effect that this last complication can have on the probabilistic factor is one of the more interesting things we have encountered in the present study.

Tuskina [Tu] already conjectured an equivalent asymptotic formula for the distribution of supersingular primes, purely on the basis of empirical evidence (but without making any conjecture as to the value of the constant).

The computation of the constant makes it necessary to have an exact description of the Galois groups. This can be an arduous task. We obviously rely heavily on Serre [S 2], and also use ideas of Shimura [Sh], especially in determining the group of $X_0(11)$.

Our heuristic method is to consider probabilistic models in which we consider the sequence of traces of Frobenius $\{t_p\}$ to be a random sequence. We choose the simplest model for which almost all sequences have asymptotic properties consistent with the density theorems of Tchebotarev and Hecke, concerning the distribution of primes with given element of the Galois group, and in sectors of the plane, and also consistent with the Sato-Tate conjecture. We show that for this model, almost all sequences have an asymptotic behavior of the form mentioned (a constant times $\pi_{\frac{1}{2}}(x)$), and we compute this constant explicitly in terms of the Galois group. Our conjecture is that the sequence of Frobenius elements has this behavior. More precisely, say in the supersingular case, this amounts to saying that the probability that p is supersingular is asymptotic to $C(0,\rho) \cdot \frac{1}{2\sqrt{p}}$, and similarly in the other cases, using $C(k,\rho)$ instead of $C(0,\rho)$.

In the case of the quadratic field k with discriminant D, the constant is inversely proportional to $\sqrt{|D|}$, and can be expressed as a product of local factors depending on ℓ and D for almost all primes ℓ, as well as a factor depending on the special position of the Galois group in the product at a finite number of exceptional primes depending on ρ, and D. There is also a factor at infinity, derived from the Sato-Tate distribution.

The factors at finite primes can be expressed as integrals over ℓ-adic sets of certain functions which are Harish transforms. We develop ab ovo the theory of Harish transforms, which can be formulated completely naively in terms of the direct image of Haar measure under the trace-determinant map (i.e. the map which associates with each matrix the coefficients of its characteristic polynomial). The theory of this transform has independent interest, and is given in Part II, §7 and §8.

Our axiomatization involving only a GL_2-Galois extension of the rationals gives rise to various questions.

1. Are there such extensions (all of elliptic type, see §1 below) other than those arising from the division points of an elliptic curve?

This question does not seem to fit exactly in the general Langlands framework, since no assumption is made here about the associated zeta function of the GL_2-extension.

2. Cusp forms do not appear in the present work. Can the conjecture be even remotely approached for the case of elliptic curves by using explicit formulas for the coefficients of the associated cusp form, e.g. formulas as in Manin [Man]?

How do the congruence conditions and the finite part of the constant arising from the Galois representation translate into conditions on the coefficients of the associated cusp form? How can one describe only in terms of these coefficients the conditions which determine the "fixed trace progression," or the "given imaginary quadratic field progression"?

In some sense what we are about is to reconstruct the arithmetic of an elliptic curve without complex multiplication by piecing together the totality of elliptic curves with complex multiplication in a certain way. There should be something like a reciprocity law which bears to our conjectured asymptotic behavior a relationship analogous to that which the Artin reciprocity law bore to the Frobenius conjectured density properties, proved by Tchebotarev.

3. Again in the case of elliptic curves, can one give a condition on the analytic behavior of the associated Dirichlet series (zeta function) which implies our conjectured asymptotic property? In particular, is there significance to the partial Euler products taken over those p which are supersingular, or which correspond to a given imaginary quadratic field, and is there an L-series formalism attached to such products? The Hardy-Littlewood paper [H−L] is in two parts. The first shows how various Riemann Hypotheses imply distribution results. The second, including the conjecture on primes in quadratic progressions, limits itself to heuristic arguments. Therefore, even in that case, it would be interesting to see what analytic properties of zeta functions imply the conjectured asymptotic behavior.

4. Adapting to the present situation the classical view point of characterizing Galois extensions by those primes that split completely, it is reasonable to expect that two elliptic curves over the rationals are isogenous if they have the same set of supersingular primes, except possibly for a subset having an asymptotic order of magnitude strictly smaller than $\sqrt{x}/\log x$. Further comments are made on this in §4, when we have more precise definitions to discuss the matter technically.

For simplicity of expression, the conjecture may be weakened by requiring that the two curves have the same sets of supersingular primes, except for a finite number. In §4 we shall see that it may be strengthened by supposing merely that the common set of supersingular primes not be $O(\log \log x)$.

The elliptic curve A has a rational invariant j_A. For all non-exceptional primes p, we have

$Q(\pi_p) = k$ *if and only if* $j_A \equiv j(\mathfrak{o}) \pmod{\mathfrak{p}}$ *for some order* $\mathfrak{o}$
in k *and some prime* $\mathfrak{p}$ *over* p *in* $k(j(\mathfrak{o}))$.

According to a theorem of Deuring (cf. [L 1], Chapter 13, §4, Theorem 13), we can pick $\mathfrak{o}$ such that p splits completely in $k(j(\mathfrak{o}))$, and the above congruence condition has to be satisfied. It is standard (cf. [L 1], Chapter 8, §1, Corollary of Theorem 7) that there is only a finite number of imaginary quadratic orders $\mathfrak{o}$ such that $j(\mathfrak{o})$ lies in a given number field. However this approach through an increasing tower of orders and congruence conditions did not seem to lead towards a determination of the asymptotic behavior of the distribution of Frobenius elements belonging to the given quadratic field k.

Finally we observe that in the light of Yoshida's proof of the analogue of the Sato conjecture in the function field case [Y], it is possible that enough is known about the distribution of Frobenius elements in that case to be able to give a proof of the analogue of our conjecture. Of course, there is no question of having infinitely many supersingular values of j, which are necessarily finite, and in characteristic p one can only have imaginary quadratic fields as algebras of endomorphisms in which p splits completely, by Deuring's theorems. Cf. [L 1], Chapter 13 and 14. Except for these limitations, one expects a similar theory to hold. The approach through the congruence values $j(\mathfrak{o})$ as $\mathfrak{o}$ ranges over orders of k with conductor prime to p may in fact work in this case, in line with Ihara's ideas [I] which were used by Yoshida [Y].

There remains to say a few words about the logistics of this paper. In Part I, we discuss the fixed trace case In Part II, we treat the imaginary quadratic distribution. While the finite part of the constant stabilizes at finite level in Part I, it does not in Part II, and its theoretical analysis, as well as practical computation requires a more elaborate discussion. Finally, in Part III, we put together special computations dealing with the quadratic fields for which the GL_2-extension has an intersection with the maximal abelian extension of k which is strictly bigger than the field of all roots of unity. These cases are the most interesting.

For instance, the exceptionally large number of occurrences of $Q(\sqrt{-11})$ for $X_0(11)$ must be reflected in a correspondingly large prediction. (It occurs 88 times, when most other fields occur at most one-fourth this many times.) This requires a description of the field of 4-division points. Other cases require a similar description of the 3-division points. Since we felt that our computations should be checkable by anyone else interested to do so, we have included the full details in all cases.

In Part IV we present and discuss the numerical results for five curves and the first 5,000 primes. For one of the curves, $X_0(11)$, the calculation was pushed to include almost 190,000 primes. On the whole, the fit between actual and predicted values is good. We feel that the data are compatible with the conjecture. There are discrepancies, but they seem to lie within the range of reasonable statistical fluctuations.

PART I

SUPERSINGULAR AND FIXED TRACE DISTRIBUTION

PRELIMINARIES

1. The Galois representation in GL_2	17
2. Some notions of probability	20

THE DISTRIBUTION FOR FIXED TRACE

3. The probabilistic model	29
4. The asymptotic behavior	33

EXAMPLES

5. Serre curves, $M = 2q$, the general formula	41
6. Computations of Galois groups	46
7. The curve $y^2 = x^3 + 6x - 2$	51
8. The Shimura curve $X_0(11)$	55

PART II

IMAGINARY QUADRATIC DISTRIBUTION

Introduction 69

THE FIXED TRACE CASE

1. Fixed traces from the quadratic field 77
2. Computation of the constant for fixed trace 84

THE MODEL FOR THE MIXED CASE

3. The mixed Galois representations 91
4. The probabilistic model 104
5. The asymptotic behavior 108
6. The finite part of the constant as a quotient of integrals .. 112

COMPUTATIONS OF HARISH TRANSFORMS

7. Haar measure under the trace-determinant map on Mat_2. General formalism. 123
8. Relations with the trace-norm map on k 133
9. Computation of C_ℓ for almost all ℓ 141
10. The constant for Serre curves, $K \cap k_{ab} = Q_{ab}$ 143
11. The constant for $X_0(11)$ 149

PART III

SPECIAL COMPUTATIONS

Introduction — 157

GENERAL LEMMAS

1. Lemmas on commutator subgroups — 163
2. $G_2 = GL_2(Z_2)$ — 165
3. Cases when $K \cap k_{ab} = Q_{ab}$ — 174
4. $K \cap k_{ab}$ when $k = Q(\sqrt{-3})$ and $GL_2(Z_3)$ splits — 181
5. $K \cap k_{ab}$ in other cases — 185

$k = Q(\sqrt{-3})$

6. The action of $\mathcal{G}$ on $k(\Delta^{1/3})$ — 191
7. The constant for Serre fiberings, $k = Q(\sqrt{-3})$, $M = 2q$, q odd prime $\neq 3$, $\Delta = \pm q^n$ — 195
8. Computation of integrals — 201

$k = Q(i)$

9. The constant for Serre fiberings, q odd $\neq 3$ — 209

$k = Q(\sqrt{\Delta})$

10. The action of $\mathcal{G}$ on $k(A_2, \Delta^{1/4})$ when $k = Q(\sqrt{\Delta})$ — 215
11. The action of matrices on $k(A_4)$ — 218
12. Computation of integrals and the constant — 221

PART IV

NUMERICAL RESULTS

SUPERSINGULAR AND FIXED TRACE DISTRIBUTION

1. General discussion of results — 235
2. Tables
   Table I : Fixed trace distributions — 239
   Table II : Supersingular primes — 240
   Table III: Primes with $t_p = 1$ — 241
   Table IV: Traces of Frobenius — 242

IMAGINARY QUADRATIC DISTRIBUTION

3. General discussion of results — 249
4. Tables
   Table V : Imaginary quadratic distributions — 253
   Table VI : Primes associated with fields of small discriminant, for curves A and B — 258
   Table VII: Distribution of primes associated with small discriminants — 260

EXTENDED RESULTS FOR $X_0(11)$

5. Discussion and description of tables — 265
   Table VIII: Supersingular primes — 267
   Table IX : Imaginary quadratic distribution — 268
   Table X : Distribution of primes for fields with small discriminants — 269

Remarks on the Computations — 271

Bibliography — 273

PART I

SUPERSINGULAR AND FIXED TRACE DISTRIBUTION

PRELIMINARIES

§1. The Galois representation in GL_2

We fix a Galois (infinite) extension K of the rationals, with Galois group G, and a representation

$$\rho : G \longrightarrow \prod GL_2(Z_\ell)$$

which we assume gives an embedding of G onto an open subgroup of the product, taken over all primes ℓ. We let

$$\rho_\ell : G \longrightarrow GL_2(Z_\ell)$$

be the projection of ρ on the ℓ-th factor, and let G_ℓ be the factor group

$$G/\text{Ker}\, \rho_\ell \,.$$

We let K_ℓ be the fixed field of $\text{Ker}\,\rho_\ell$, so that G_ℓ is the Galois group of K_ℓ over Q. A Galois extension of the rationals with a representation of its Galois group as above will be called a GL_2-**extension**.

For each positive integer M, we may reduce $\prod GL_2(Z_\ell)$ mod M, and thereby obtain a factor representation

$$\rho_{(M)} : G \longrightarrow GL_2(Z/MZ) \,.$$

The factor group of G by $\text{Ker}\,\rho_{(M)}$ is denoted by G(M), and the fixed field of $\text{Ker}\,\rho_{(M)}$ by K(M). Then K(M) is a finite Galois extension with group G(M). We also call G(M) the **reduction of** G mod M.

We denote Z/MZ by Z(M), and use similar notation for other objects reduced mod M. For instance, $Z(M)^* = (Z/MZ)^*$ is the group of units in Z(M).

We denote by G_M the projection of $\rho(G)$ into the finite product

$$\prod_{\ell \mid M} GL_2(Z_\ell) \,,$$

and use ρ_M to denote the representation of G in this projection. We shall say that M splits ρ if we have an isomorphism

$$\rho(G) = \prod_{\ell \nmid M} GL_2(Z_\ell) \times G_M .$$

Note that G_M is open in the finite product.

On the other hand, let M be arbitrary, and let

$$r_M : \prod_{\ell \mid M} GL_2(Z_\ell) \longrightarrow GL_2(Z(M))$$

be the reduction map. We shall say that M is stable if the following condition is satisfied.

ST 1. $\qquad G_M = r_M^{-1}(G(M)) .$

It then follows at once that the next condition is also satisfied.

ST 2. *For every element* $\bar\sigma \in G(M)$, *we have*

$$r_M^{-1}(\bar\sigma) = \prod_{\ell \mid M} r_{\ell^{m(\ell)}}^{-1}(\bar\sigma(\ell^{m(\ell)}) ,$$

where $M = \prod \ell^{m(\ell)}$ *is the prime power decomposition of* M, *and* $\bar\sigma(\ell^{m(\ell)})$ *is the reduction of* $\bar\sigma$ *mod* $\ell^{m(\ell)}$.

We use the notation

$$G(M^\infty) = \lim_{n \to \infty} G(M^n) .$$

Then $G_M = G(M^\infty)$, and $G(M^\infty)$ is completely determined by $G(M)$ if M is stable. According to ST 2, G_M decomposes into a union of open sets, each of which is a product over $\ell \mid M$.

It will be natural to study G by picking an integer M which splits ρ and is also stable. Then any prime $\ell \nmid M$ is also stable, or as we also say, ρ is stable at level 1 for any prime $\ell \nmid M$.

As remarked already in the introduction, Serre has shown that the torsion points of an elliptic curve over the rationals without complex multiplication give rise to a GL_2-extension as above, for which there always exists some M which splits and stabilizes ρ.

We say that the extension K, or ρ, has **limited ramification** if there is a positive integer Δ having the following property. If p is a prime and $p \nmid \Delta \ell$, then p is unramified in K_ℓ, or as we also say, ρ_ℓ is **unramified** at p. Then the Frobenius class σ_p is well defined in G_ℓ, and $\rho_\ell(\sigma_p)$ has a characteristic polynomial which we assume of the form

$$\Phi_p(X) = X^2 - t_p X + p .$$

We say that ρ is **integral and consistent** if t_p is an integer independent of ℓ. We call t_p the **trace of Frobenius**.

For the rest of this paper, we assume that ρ has limited ramification, is integral, and consistent. We assume that the roots of the characteristic polynomial are complex conjugates of each other. We say that such ρ is of elliptic type.

We let $\mathfrak{H}$ be the upper half plane, and let

$$z_p = \text{root of } \Phi_p(X) \text{ in } \mathfrak{H} .$$

Thus z_p is the representative of the Frobenius class in the upper half plane, and is also sometimes called a **Frobenius element**.

We let N denote the absolute norm to $\mathbf{Q}$, so $Nz = z\bar{z}$. If z_p is a Frobenius element, then $Nz_p = p$.

Given an integer t, a prime p, such that $t^2 < 4p$, we let

$$z(t, p) = \text{root in } \mathfrak{H} \text{ of } X^2 - tX + p = 0 .$$

We let

$$\theta(z) = \text{angle of } z \in \mathfrak{H}, \qquad 0 < \theta(z) < \pi ,$$

We sometimes write $\theta(t, p)$ for $\theta(z(t, p))$.

This settles the basic notation concerning our GL_2-extension.

§2. Some notions of probability

The law of large numbers in probability will motivate part of our work, and we state some versions of it here for the convenience of the reader.

For each positive integer n suppose given a measured space X_n with positive measure μ_n such that the total measure of X_n is 1. We let

$$X = \prod X_n \quad \text{and} \quad \mu = \prod \mu_n$$

be the product space and product measure respectively. We view X as our probability space.

Theorem 2.1. *Suppose given a measurable subset S_n of X_n for each n. Assume that the limit exists,*

$$\lim_{n \to \infty} \mu_n(S_n) = L \ .$$

Then for almost all elements (sequences) $x = \{x_n\}$ in X, the density of n such that $x_n \in S_n$ exists and is equal to L. This means:

$$\lim_{N \to \infty} \frac{\#\{n \leq N, \ x_n \in S_n\}}{N} = L \ .$$

The above theorem has a simple intuitive content, but our main application requires a stronger version as follows.

Theorem 2.2. *Suppose given a measurable subset S_n of X_n for each n. Let $\{b_n\}$ be a sequence of positive real numbers tending monotonically to infinity. Assume that*

$$\sum \frac{1}{b_n^2} \mu_n(S_n) < \infty \ .$$

Then for almost all sequences x, we have

$$\#\{n \leq N, \ x_n \in S_n\} = \sum_{n=1}^{N} \mu_n(S_n) + o(b_N) \ .$$

The first theorem is obtained from this second by putting $b_n = n$. On the other hand, we obtain the following corollary.

Corollary. *Let p_n be the n-th prime. Let C be a positive number, and assume*

$$\mu_n(S_n) \sim C \frac{1}{2\sqrt{p_n}}.$$

Then for almost all sequences x,

$$\#\{n \leq N,\ x_n \in S_n\} \sim C \sum_{n \leq N} \frac{1}{2\sqrt{p_n}}.$$

Proof. Take

$$b_N = \sum_{n \leq N} \frac{1}{2\sqrt{p_n}}.$$

The corollary is stated precisely in a form which fits the applications, where $\pi_{\frac{1}{2}}(x)$ occurs.

In the applications, the indexing set for the measured spaces will be the prime numbers, so we write p instead of n. In the simplest case, i.e. in the study of the supersingular distribution, the measured space at p will essentially be the interval of integers t such that

$$-2\sqrt{p} < t < 2\sqrt{p},$$

and the measure will be a discrete measure designed to take into account the Tchebotarev-Sato-Tate distributions. In the imaginary quadratic case, it will be similar but more complicated, and we wait till the appropriate section to give the more precise description of the probability space. It is of course a major assumption that there exists a model of the above type for the distribution of Frobenius elements under consideration. However, in the theory of prime numbers, it seems to be the appropriate model for this and other distribution problems. For instance, we can use the same pattern to recover a conjecture of Hardy-Littlewood concerning primes in quadratic progression. This will fit in a natural way in the second part.

The situation is in fact a little more complicated, because we describe a probabilistic model as above for each given positive integer M, for the sequence of Frobenius elements in the Galois group at level M, i.e. in the representation

$\rho_{(M)}$ into $GL_2(Z/MZ)$. We then have to take a further limit over M (ordered by divisibility in the supersingular case, and somewhat more subtly in the imaginary quadratic case) in order to get the asymptotic behavior for the entire Galois group.

Appendix

For the convenience of the reader we include a proof of the probabilistic theorem. The first lemma, due to Kolmogoroff and formulated by him in probabilistic terms, will be a refinement of the fundamental lemma of integration theory, which asserts that given an L^1 (or L^2) Cauchy sequence, there exists a subsequence which converges absolutely almost everywhere. Here we give up on absolute convergence, but have conditions which make the full sequence converge pointwise almost everywhere.

Lemma. *For each n let h_n be a function on X_n, also viewed as function on X by projection on the n-th factor. Assume that*

$$\int h_n \, d\mu_n = 0.$$

Let

$$H_n(x) = \sum_{k=1}^{n} h_k(x)$$

be the partial sum. Assume that $\sum \|h_k\|_2^2$ converges. Then the limit

$$\lim_{n \to \infty} H_n(x)$$

exists for almost all $x \in X$.

Proof. We first note that the functions h_n are mutually orthogonal on X. The heart of the proof lies in the next statement.

Kolmogoroff's inequality. *Given ε, let*

$$Z = \left\{ x \in X, \max_{1 \leq k \leq n} H_k^2(x) \geq \varepsilon \right\}.$$

Then

$$\varepsilon \mu(Z) \leq \sum_{k=1}^{n} \|h_n\|_2^2.$$

Proof. Let

$$Y_k = \{x \in X, \ H_k^2(x) \geq \varepsilon \ \text{ and } \ H_i^2(x) < \varepsilon \ \text{ all } \ i < k\}.$$

In other words, Y_k is the set of points x such that $H_k^2(x)$ is the first partial sum at least equal to ε. Then the sets Y_k are disjoint, and we get the inequality

$$\varepsilon \sum_{k \leq n} \mu(Y_k) \leq \sum_{k \leq n} \int_{Y_k} H_k^2.$$

Write

$$H_k^2 = H_n^2 - 2H_k(H_n - H_k) - (H_n - H_k)^2.$$

The last term is negative, and we shall leave it out when we integrate. On the other hand, the middle term gives

$$\int_{Y_k} H_k(H_n - H_k) \, d\mu = 0.$$

This is due to the fact that H_k is effectively a function only of the first k variables, while $H_n - H_k$ is effectively a function only of the last $n-k$ variables. The integral splits into a product of integrals over the distinct variables, and is immediately seen to yield 0 as desired. Therefore we can replace H_k^2 by H_n^2 and then integrate over all of X, thereby giving as bound the L^2-norm squared of $\sum h_k$, which proves the asserted inequality.

We have assumed that $\sum h_k$ is in L^2, that is

$$\sum_{k=1}^{\infty} \|h_k\|_2^2 < \infty.$$

This means that for m_0 sufficiently large, and $n \geq m \geq m_0$ we get

$$\mu\{x \in X, \ \max (H_n - H_m)^2(x) \geq \varepsilon\} \leq \frac{1}{\varepsilon} \sum_{n=m}^{\infty} \|h_n\|_2^2 < \varepsilon.$$

Define

$$Z_i = \{x \in X, \ (H_n - H_m)^2(x) \geq 1/2^i \ \text{ if } \ m, n \geq m_0(i)\}.$$

Then Z_i has measure $\leq 1/2^i$ if we pick $m_0(i)$ sufficiently large. Let

$$W_n = Z_n \cup Z_{n+1} \cup \cdots$$

for large n, so that W_n has measure $\leq 1/2^{n-1}$. Then the partial sums $\sum h_k(x)$ converge for x not in W_n. Hence if we let W be the intersection

$$W = \cap W_n ,$$

then these partial sums converge for x not in W, and W has measure zero, thereby proving the lemma.

The next theorem is also due to Kolmogoroff, in that generality.

Theorem 2.3. *For each n let f_n be a function on X_n, and assume that*

$$\int f_n \, d\mu_n = 0 .$$

Let $\{b_n\}$ be a sequence of positive real numbers monotonically increasing to infinity. If

$$\sum \frac{1}{b_n^2} \|f_n\|_2^2 < \infty ,$$

then for almost all x the partial sums

$$F_n(x) = \sum_{k=1}^{n} f_k(x)$$

satisfy the estimate

$$F_n(x) = o(b_n) .$$

Proof. Let $h_n = f_n/b_n$ and apply the lemma, to the partial sums

$$H_n(x) = \sum_{k=1}^{n} h_k(x) = \sum_{k=1}^{n} f_k(x)/b_k .$$

The lemma says that these partial sums converge for almost all x. It is a trivial fact (proved by summation by parts) that if $\sum a_k$ is a convergent sequence then

$$\sum_{k=1}^{n} a_k b_k = o(b_n).$$

Applying this fact when $a_k = h_k(x)$ proves the theorem.

We have stated Theorem 2.3 under the normalization that the integral of the functions f_n is 0. This is of course not satisfied in general, but a translation reduces the general case to this special case. Indeed, suppose that ψ_n are functions such that

$$\int \psi_n \, d\mu_n = c_n$$

is a constant c_n. Define

$$f_n = \psi_n - c_n.$$

Then the integral of f_n is 0. In particular, suppose that ψ_n is the characteristic function of some subset S_n of X_n. Then

$$\|f_n\|_2^2 = \int (\psi_n - c_n)^2 \, d\mu_n = c_n - c_n^2.$$

Applying Theorem 2.3 to this situation yields Theorem 2.2, as desired.

THE DISTRIBUTION FOR FIXED TRACE

§3. The probabilistic model

We fix a positive integer M. For convenience, instead of an interval, we let the measured space associated with each prime p be the set of integers $\mathbb{Z}$. The measure μ_p on each fiber is assumed to be represented by a function

$$f_M(t, p) \geq 0, \qquad t \in \mathbb{Z},$$

with respect to counting measure. We also write

$$f_M(t, p) = \mathrm{pr}\{y_p = t\},$$

where y_p is the "random variable" in the language of the probabilists. The probability condition is that

$$\sum_t f_M(t, p) = 1,$$

and we assume in addition that

PR 1. $\qquad f_M(t, p) = 0 \quad if \quad |t| > 2\sqrt{p}$.

Let S be a congruence class mod M, say consisting of all those integers $\equiv t \pmod{M}$. Let $G(M)_t$ be defined by

$$G(M)_t = \{\sigma \in G(M),\ \mathrm{tr}\,\sigma \equiv t \bmod M\}.$$

By Tchebotarev, the density of primes p such that $t_p \equiv t \pmod{M}$ has a density equal to

$$\frac{|G(M)_t|}{|G(M)|}.$$

We define

$$F_M(t) = M\,\frac{|G(M)_t|}{|G(M)|},$$

so F_M depends only on the residue class of t mod M. The factor M is put there so that the average value of F_M is 1, i.e.

$$\sum_{t \bmod M} \frac{1}{M} F_M(t) = 1 .$$

For real y, define

$$\xi(y, p) = \frac{y}{2\sqrt{p}} ,$$

so that if $|\xi(y, p)| < 1$, then $\xi(y, p) = \cos \theta(y, p)$. We assume

ST. *There exists a continuous density function $\phi(\theta)$ on $[0, \pi]$ determining the distribution of angles $\theta(t_p, p)$.*

This means that the primes p such that $\theta(t_p, p)$ lies in a given interval $[\theta_1, \theta_2]$ have a density, namely

$$\int_{\theta_1}^{\theta_2} \phi(\theta) \, d\theta .$$

Changing variables, with $\xi = \cos \theta$ shows that the condition ST is equivalent with:

ST'. *There exists a continuous function*

$$g(\xi) = \frac{\phi(\theta)}{\sin \theta}$$

which is equal to 0 outside the interval $[-1, 1]$, such that the primes p for which $\xi(t_p, p)$ lies in an interval $[\xi_1, \xi_2]$ have a density, given by

$$\int_{\xi_1}^{\xi_2} g(\xi) \, d\xi .$$

We call g (or ϕ) the distribution function of Frobenius at infinity. Note that $g = g_\rho$ depends on ρ. When the representation ρ is that obtained from the torsion points of an elliptic curve without complex multiplication, then the Sato-Tate conjecture states that

$$\phi(\theta) = \frac{2}{\pi} \sin^2 \theta .$$

The function ϕ is of course normalized to have integral 1 over $[0, \pi]$. In this special case, we have

$$g(\xi) = \frac{2}{\pi} \sqrt{1 - \xi^2} .$$

Our main assumption concerning the probabilistic model is that the function f_M can be written in the form

PR 2. $$f_M(t, p) = c_p g(\xi(t, p)) F_M(t)$$

where c_p is a constant which is determined so that

$$\sum_t f_M(t, p) = 1 .$$

This is an assumption of independence for the behavior at infinity and over congruence classes. We shall now determine the asymptotic nature of c_p, and prove that

$$c_p \sim \frac{1}{2\sqrt{p}} .$$

The integral

$$1 = \int_{-1}^{1} g(\xi) \, d\xi$$

has approximating Riemann sums

$$\frac{1}{2\sqrt{p}} \sum_{-2\sqrt{p} < t < 2\sqrt{p}} g(\xi(t, p)), \qquad \text{for } p \to \infty .$$

For any integer t_0 we have the limit

(1) $$\lim_p \frac{1}{2\sqrt{p}} \sum_{t \equiv t_0 \bmod M} g(\xi(t, p)) = \frac{1}{M} .$$

Hence for given M we have the limit as $p \to \infty$,

(2) $$\lim_p \frac{1}{2\sqrt{p}} \sum_{t \equiv t_0(M)} g(\xi(t, p)) F_M(t_0) = \frac{F_M(t_0)}{M} .$$

Summing over the congruence classes t_0, we find that

$$\lim_p \frac{1}{2\sqrt{p}} \sum_t g(\xi(t, p)) F_M(t) = 1 .$$

Comparing with the definition and normalization of $f_M(t, p)$ proves that

$$c_p \sim \frac{1}{2\sqrt{p}},$$

as desired.

Going back to (2) with the knowledge that $c_p \sim \frac{1}{2\sqrt{p}}$ now shows:

Lemma 1.
$$\lim_p \mu_p(S) = \frac{1}{M} F_M(t_0)$$

where S is the congruence class of t_0 mod M.

The law of large numbers then proves that our probability assumption PR 2 implies that the density of p such that

$$y_p \equiv t_0 \bmod M$$

is precisely equal to

$$\frac{1}{M} F_M(t_0),$$

i.e. that our probability assumption is compatible with the Tchebotarev density property for the sequence of Frobenius elements to be viewed as a random sequence in our model.

Given a subinterval I of $[-1, 1]$, let $S_{I,p}$ be the set of integers y_p such that $\xi(y_p, p) \in I$. By definition,

$$\mu_p(S_{I,p}) = \sum_{\xi(t,p) \in I} f_M(t, p).$$

Lemma 2.
$$\lim_p \mu_p(S_{I,p}) = \int_I g(\xi) \, d\xi.$$

Proof. Entirely analogous to the proof of Lemma 1. We start with approximating Riemann sums for the integral

$$\int_I g(\xi) \, d\xi,$$

and use the analogue of (1) and (2) which have this number as a factor instead of the number 1. We leave the details to the reader.

Now the law of large numbers yields the compatibility with the Sato-Tate conjecture.

§4. The asymptotic behavior

Given the probabilistic model of the last section, and a random sequence $\{y_p\}$, and noting that

$$\xi(t_0, p) \longrightarrow \xi(0) = \phi(\pi/2) \qquad \text{as } p \longrightarrow \infty,$$

we get

$$\mathrm{pr}_M\{y_p = t_0\} = f_M(t_0, p) = c_p g(\xi(t_0, p)) F_M(t_0)$$

$$\sim \frac{1}{2\sqrt{p}} g(0) F_M(t_0)$$

$$\sim \frac{1}{2\sqrt{p}} \phi(\pi/2) F_M(t_0).$$

It is now reasonable to suppose that taking the limit for $M \to \infty$ (under the ordering by divisibility) yields the probability that $t_p = t_0$, or in other words, that the above value for large M gives an approximation of this probability. In fact, it will be proved in Theorem 4.2 that the limit

$$\lim_M F_M(t_0) = F(t_0)$$

exists, and is given by an absolutely convergent product. We then define

$$\boxed{C(t_0, \rho) = \phi(\pi/2) F(t_0)}$$

and our conjecture is that the number of primes $p \leq x$ such that $t_p = t_0$ satisfies the asymptotic property

$$\boxed{N_{t_0}(x) \sim C(t_0, \rho) \pi_{\frac{1}{2}}(x).}$$

In the case of elliptic curves, $\phi(\pi/2) = 2/\pi$.

The factor $F(t_0)$ arose from congruence conditions, and we now see how it decomposes into a product over primes.

Recall that

$$F_M(t) = M \frac{|G(M)_t|}{|G(M)|}.$$

We proceed to determine F_M as much as we can in a general context. First we have the multiplicativity.

Lemma 1. *Suppose that* $M = M_1 M_2$ *is a product of two relatively prime factors, and* $G(M) = G(M_1) \times G(M_2)$. *Then*

$$F_M(t_0) = F_{M_1}(t_0) F_{M_2}(t_0).$$

Proof. Obvious.

Next we study F_M as M becomes highly divisible. We first show that the limit $\lim_{n \to \infty} F_{M^n}(t)$ depends only on a stabilizing integer M_0, and then find an explicit value for the prime power factor for almost all ℓ. As a matter of notation, we write

$$F_{M^\infty}(t) = \lim_{n \to \infty} F_{M^n}(t)$$

whenever that limit exists. We identify $G(M)$ inside $GL_2(Z(M)) = GL_2(M)$.

Lemma 2. *Assume that* M_0 *is stable, that* $M_0 | M$, *and that* M_0 *is divisible by the same primes as* M. *If* $s \equiv t \pmod{M_0}$ *then*

$$|G(M)_s| = |G(M)_t|.$$

Furthermore,

$$F_M(t) = F_{M_0}(t).$$

Proof. There is a bijection between the two sets $G(M)_t$ and $G(M)_s$, given by

$$\sigma \longmapsto \sigma + \begin{pmatrix} s-t & 0 \\ 0 & 0 \end{pmatrix}$$

which makes the first assertion obvious. If $r: GL_2(M) \to GL_2(M_0)$ is the reduction map, then given $\bar\sigma \in GL_2(M_0)$ there are $(M/M_0)^4$ elements in the fiber $r^{-1}(\bar\sigma)$, and there are M/M_0 elements of Z/MZ lying above an element of $Z/M_0 Z$. Hence by the first part of the lemma, we find

$$(M/M_0)^4 |G(M_0)_t| = (M/M_0)|G(M)_t|,$$

whence

$$M|G(M)_t| = (M/M_0)^4 M_0 |G(M_0)_t|.$$

But

$$|G(M)| = (M/M_0)^4 |G(M_0)|.$$

Taking the quotient proves the lemma.

In particular, *if* M *is stable, then*

$$F_{M^n}(t) = F_M(t)$$

is constant, so the limit $F_{M^\infty}(t)$ *obviously exists and is equal to* $F_M(t)$.

Lemma 3. *If* $uI \in G(M)$ *for some integer* u (mod M), *then*

$$F_M(ut) = F_M(t).$$

Proof. The map $\sigma \mapsto u\sigma$ is a bijection of $G(M)_t$ and $G(M)_{ut}$.

Theorem 4.1. *Assume that* $G_\ell = GL_2(\mathbb{Z}_\ell)$. *Let* $r = r(\ell) = 1/\ell$. *Then*

$$F_{\ell^\infty}(0) = F_\ell(0) = \frac{1}{1-r^2}$$

$$F_{\ell^\infty}(1) = F_\ell(1) = \left(1 - \frac{r^3}{1-r-r^2}\right)^{-1}$$

Proof. By Lemma 3, and Lemma 2, we only have to consider $F_\ell(0)$ and $F_\ell(1)$, i.e. $t = 0$ or $t = 1$ at level 1.

Suppose first $t = 0$. Then $G(\ell)_0$ consists of the matrices

$$\begin{pmatrix} a & b \\ c & -a \end{pmatrix} \text{ such that } a^2 + bc \not\equiv 0 \bmod \ell.$$

If $a \equiv 0$, there are $(\ell-1)^2$ pairs b, c which give rise to possible matrices. the other hand, there are $\ell-1$ values for $a \neq 0$. Given b, there is a unique such that $bc \equiv -a^2$, so there are

$$\ell^2 - (\ell - 1)$$

pairs b, c such that $a^2 + bc \not\equiv 0 \pmod{\ell}$. Adding yields

$$|G(\ell)_0| = (\ell-1)^2 + (\ell-1)(\ell^2 - (\ell-1)) = \ell^2(\ell-1).$$

Furthermore, to get the case $t = 1$, we use the formula

$$\sum_{t \bmod \ell} |G(\ell)_t| = |G(\ell)_0| + (\ell-1)|G(\ell)_1|$$

and also

$$\sum_{t \bmod \ell} |G(\ell)_t| = |GL_2(F_\ell)| = \ell(\ell-1)(\ell^2-1).$$

We then solve for $|G(\ell)_1|$, to get

$$|G(\ell)_1| = \ell(\ell^2 - \ell - 1).$$

Thus

$$F_\ell(0) = \frac{\ell|G(\ell)_0|}{|GL_2(F_\ell)|} = \frac{\ell^2}{\ell^2 - 1} \quad \text{and} \quad F_\ell(1) = \frac{\ell(\ell^2-\ell-1)}{(\ell-1)(\ell^2-1)},$$

as desired.

We can put all these computations together. Using the multiplicativity lemma and the stabilizing Lemma 2, we find:

Theorem 4.2. *Assume that M splits and stabilizes ρ. Then the constant is given by*

$$C(t_0, \rho) = \phi(\pi/2) F_M(t_0) \prod_{\ell \nmid M} F_\ell(t_0).$$

In particular,

$$C(0, \rho) = \phi(\pi/2) \frac{\pi^2}{6} F_M(0) \prod_{\ell | M} (1 - 1/\ell^2).$$

Proof. The first assertion follows by putting the lemmas together. We then use the value found in Theorem 4.1,

$$F_\ell(0) = \frac{1}{1 - 1/\ell^2}.$$

Since $\zeta(2) = \pi^2/6$, the second formula follows at once.

In this expression, it is the factor $F_{M^\infty}(0)$, depending on the curve A, which distinguishes the special situation. This factor is entirely determined by the image of the representation

$$\rho_{(M)} : G(M) \longrightarrow GL_2(Z(M)) .$$

In the case of an elliptic curve, it follows from results of Serre (cf. the next section) that any M which splits and stabilizes ρ is necessarily even.

Remark 1. *If* $t = 0$ *then the constant* $C(0,\rho)$ *is not* 0. To see this, it suffices to observe that there is at least one element in the stable group G(M) with trace 0. Such an element is given by complex conjugation, for instance, so that $F_M(0) \neq 0$ for a stabilizing M.

On the other hand, for other values of $t_0 \neq 0$ it may happen that the constant $C(t_0,\rho)$ is equal to 0. A typical reason for this is the presence of rational points on the elliptic curve, which may impose congruence conditions on the traces of Frobenius.

Remark 2. Let A, B be two elliptic curves over the rationals, without complex multiplication. Then the Galois group $\mathcal{G}$ of $Q(A_{tor}, B_{tor})$ over the rationals is contained in the subgroup of the product of the Galois groups of $Q(A_{tor})$ and $Q(B_{tor})$ consisting of the elements which have the same effect on the field of all roots of unity, and Serre conjectured that it is of finite index (he also proved this when one of the j-invariants is not integral). Assuming that this is the case, and arguing heuristically in a manner analogous to that used for one curve, assuming also in addition that probability distributions on the two curves are essentially independent, one comes up with an estimated asymptotic behavior for the set of primes p which are supersingular for both A and B. The number of such primes $\leq x$ should in particular be

$$O\left(\sum_{p \leq x} \frac{1}{p}\right) = O(\log \log x) .$$

Therefore, the conjecture concerning a strengthening of the isogeny theorem is that if two elliptic curves over the rationals without complex multiplication have a common set of supersingular primes of asymptotic order of magnitude which is not $O(\log \log x)$, then the curves are isogenous.

For the curves we looked at, there are just two coincidences of supersingular primes among the first 5,000 primes. The prime 19 is supersingular for A and $X_0(11)$, and 2,411 is supersingular for C and D. The function $\log \log x$ grows so slowly that it would be hard to get statistically meaningful data.

Of course a similar conjecture can be made about primes whose Frobenius elements for two given curves both generate the same quadratic field.

EXAMPLES

41

§5. Serre curves, M = 2q, the general formula

Serre has shown that G is always contained in a subgroup of index 2 in $\prod GL_2(Z_\ell)$, determined by conditions on the quadratic symbol with respect to the field $Q(\sqrt{\Delta})$. For our purposes, we need only a special case.

Let q be an odd prime. The group $GL_2(Z_q)$ has a unique subgroup of index 2, denoted by E_q. It is the set of elements σ such that det σ is a square in Z_q^* (or equivalently, a square mod q). Since the determinant of the scalar matrices consists of the squares, we see that E_q is simply the product of the scalar matrices times $SL_2(Z_q)$.

For the prime 2, we have the reduction homomorphism

$$GL_2(Z_2) \longrightarrow GL_2(2) \approx S_3 \, ,$$

and we let E_2 be the subgroup of elements σ such that the reduction $\bar\sigma$ mod 2 is an even permutation.

We call E_q or E_2 the group of even elements, and let O_q, O_2 be their respective cosets, called the cosets of odd elements. We define Serre's subgroup S_{2q} of $GL_2(Z_2) \times GL_2(Z_q)$ to be

$$S_{2q} = (E_2 \times E_q) \cup (O_2 \times O_q) \, .$$

It is of index 2, and 2q stabilizes this subgroup. Cf. [S 2], 5.5, Proposition 22, where Serre shows that if the discriminant of the field $Q(\sqrt{\Delta})$ is $\pm q^n$, then the image of ρ is always contained in

$$S_{2q} \times \prod_{\ell \neq 2, q} GL_2(Z_\ell) \, ,$$

which we call Serre's subgroup of $\prod GL_2(Z_\ell)$.

If $M = 2q$ is stable, and $G(M) = S_{2q}(M)$, we also call $G(M)$ the Serre subgroup in $GL_2(Z(M))$.

As the supersingular case is especially important, the next two theorems give the value of $F_M(0)$. Complete tables are established after that for all values $F_M(t)$.

333

Theorem 5.1. *Let* $M = 2q$ *with* q *an odd prime. Assume*:

(i) M *is stable*.

(ii) $G(M)$ *is Serre's subgroup of index* 2 *in* $GL_2(Z(M))$.

Then

$$F_M(0) = \begin{cases} \dfrac{2q(2q^2 - 3q + 1)}{3(q-1)(q^2-1)} & \text{if } \left(\dfrac{-1}{q}\right) = 1 \\[2ex] \dfrac{2q(2q^2 - q - 1)}{3(q-1)(q^2-1)} & \text{if } \left(\dfrac{-1}{q}\right) = -1 \end{cases}$$

Remark. The two cases depending on the parity of $\left(\dfrac{-1}{q}\right)$ have values which differ only in a second order term, and tend to $\dfrac{4}{3}$ as $q \to \infty$.

Theorem 5.1 will be proved below. If we combine the values given for $F_{2q}(0)$ in this theorem with the general expression for the asymptotic supersingular distribution found in §4, we get the asymptotic supersingular distribution in the Serre case.

Theorem 5.2. *Let* $M = 2q$ *where* q *is an odd prime. Assume*:

(i) M *splits and stabilizes* ρ.

(ii) $G(M)$ *is Serre's subgroup of index* 2 *in* $GL_2(Z(M))$.

Then

$$C(0, \rho) = \phi\left(\frac{\pi}{2}\right) \frac{\pi^2}{6} F_{2q}(0)(1 - 1/4)(1 - 1/q^2) .$$

In particular, for the ordinary Sato-Tate distribution,

$$C(0, \rho) = \frac{\pi}{4} F_{2q}(0)(1 - 1/q^2) .$$

We shall now compute the values $F_{2q}(t)$ when q is an odd prime and $2q$ is stable. By definition,

$$F_{2q}(t) = \frac{2q |G(2q)_t|}{3q(q-1)(q^2-1)} ,$$

because $G(2q)$ is of index 2 in

$$GL_2(2) \times GL_2(q) ,$$

and $GL_2(2)$ has order 6, while $GL_2(q)$ has order $q(q-1)(q^2-1)$. Thus our task is to compute $G(2q)_t$ for various values of t. The result is given in the following table. Since it turns out that $|G(2q)_t|$ is divisible by q in all cases, the table gives the quotient by q, i.e. it gives the values of

$$\frac{1}{q}|G(2q)_t|.$$

		q|t	q∤t
$\left(\frac{-1}{q}\right) = 1$	2|t	$2q^2 - 3q + 1$	$2q^2 - 2q - 1$
	2∤t	$q^2 - 1$	$q^2 - q - 2$
$\left(\frac{-1}{q}\right) = -1$	2|t	$2q^2 - q - 1$	$2q^2 - 2q - 3$
	2∤t	$q^2 - 2q + 1$	$q^2 - q$

For example, if $\left(\frac{-1}{q}\right) = 1$, then the table shows that

$$|G(2q)_0| = q(2q^2 - 3q + 1)$$

and consequently $F_{2q}(0)$ has the value stated in Theorem 5.1. In general, we may write

$$F_{2q}(t) = \frac{2q^2 \cdot \text{table entry}}{3q(q-1)(q^2-1)}.$$

We now give the proof that the table is correct.

By definition, Serre's subgroup G is the group

$$G = [(E_2 \times E_q) \cup (O_2 \times O_q)] \times \prod_{\ell \nmid 2q} GL_2(Z_\ell).$$

Our notation is such that

$$E(q)_t = \{\sigma \in E(q), \text{ tr } \sigma \equiv t \bmod q\}.$$

335

We note that if $u \in F_q^*$ and $u \neq 0$, then the map

$$\sigma \longmapsto u\sigma$$

gives a bijection

$$E(q)_t \longrightarrow E(q)_{ut},$$

so the sets $E(q)_t$ all have the same cardinality for $t \neq 0$. Hence

$$|E(q)| = \frac{1}{2} q(q-1)(q^2-1) = |E(q)_0| + (q-1)|E(q)_1|.$$

We compute $|E(q)_0|$. We note that there is a direct product decomposition

$$E(q)_0 = (F_q^*/\pm 1) SL_2(q)_0,$$

where $F_q^*/\pm 1$ denotes a set of representatives of F_q^* mod ± 1. Thus it suffices to count the order of $SL_2(q)_0$. Let

$$\sigma = \begin{pmatrix} a & b \\ c & -a \end{pmatrix}$$

be an element of $SL_2(q)_0$, so $-bc = 1 + a^2$. We want to know how many such elements there are.

Case 1. $\left(\frac{-1}{q}\right) = 1$.

There are two values of a such that $1 + a^2 = 0$, and to each such value there are $2q - 1$ pairs (b, c) with $-bc = 0$. There are $q - 2$ values of a such that $1 + a^2 \neq 0$, and for each such a, there are $q - 1$ pairs (b, c) with $-bc = 1 + a^2$. This yields

$$|SL_2(q)_0| = 2(2q-1) + (q-1)(q-2) = q^2 + q$$

whence

$$|E(q)_0| = \frac{1}{2}(q^3 - q).$$

Case 2. $\left(\frac{-1}{q}\right) = -1$.

Then $1 + a^2 \neq 0$ for all a, whence

$$|SL_2(q)_0| = q^2 - q \qquad \text{and} \qquad |E(q)_0| = \frac{1}{2}(q^3 - 2q^2 + q).$$

Next, we have

$$GL_2(q)_0 = E(q)_0 \cup O(q)_0$$

from which we can compute the order of $O(q)_0$. Arithmetic then yields the following table of orders for $E(q)_t$ and $O(q)_t$.

	$t = 0$	$t \neq 0$
$\left(\frac{-1}{q}\right) = 1,\ E(q)_t$	$\frac{1}{2} q(q^2 - 1)$	$\frac{1}{2} q(q^2 - q - 2)$
$\left(\frac{-1}{q}\right) = -1,\ O(q)_t$		
$\left(\frac{-1}{q}\right) = 1,\ O(q)_t$	$\frac{1}{2} q(q^2 - 2q + 1)$	$\frac{1}{2} q(q^2 - q)$
$\left(\frac{-1}{q}\right) = -1,\ E(q)_t$		

The individual case for the prime 2 is done by inspection:

	$t = 0$	$t = 1$
$E(2)_t$	1	2
$O(2)_t$	3	0

The table for $|G(2q)_t|$ then follows trivially.

In order to apply these tables to concrete instances, we must determine explicitly that certain elliptic curves have the Serre group as Galois group of their division points, or groups related to this subgroup. The next sections are devoted to the proofs in two important instances which are fairly typical, and relatively complicated.

§6. Computations of Galois groups

In this section we recall some techniques for the computation of Galois groups.

Refinement lemmas

The first lemma was proved originally by Shimura [Sh] with some restrictive conditions, and was proved by Serre [S 3] in general. See also [L 1], Chapter 17, §4.

Lemma 1. *Let* H *be a closed subgroup of* $GL_2(Z_\ell)$ *whose projection mod* ℓ *contains* $SL_2(F_\ell)$. *Then* H *contains* $SL_2(Z_\ell)$ *if* $\ell \geq 5$.

We refer to this lemma as the refinement lemma. There are others of a slightly more abelian nature as follows. Cf. Shimura [Sh].

Let $M_\ell = Mat_2(Z_\ell)$, and since ℓ is fixed, write $M = M_\ell$. Let

$$V_n = I + \ell^n M_\ell$$

be the group of matrices $\equiv I \mod \ell^n$. The map

$$X \longmapsto I + \ell^n X$$

induces an isomorphism

$$M/\ell M \xrightarrow{\approx} V_n/V_{n+1} .$$

Consider the case when $\ell \geq 3$. The map raising to the ℓ-th power gives an isomorphism

$$V_n/V_{n+1} \approx V_{n+1}/V_{n+2} .$$

Let U_1 be a closed subgroup of V_1 and let $U_n = U_1 \cap V_n$. We have an embedding

$$U_1/U_2 \longrightarrow V_1/V_2 \approx M/\ell M .$$

We may refer to U_1/U_2 as the associated vector space to U_1 at level 1, or as the tangent space to U_1. It has dimension ≤ 4.

Lemma 2. *Let ℓ be odd, and let U_1 be a closed subgroup of V_1 whose associated vector space at level 1 has dimension 4. Then $U_1 = V_1$.*

Proof. From the isomorphism $U_1/U_2 \approx V_1/V_2$ raised to the ℓ-th power recursively, we get an isomorphism

$$U_n/U_{n+1} \approx V_n/V_{n+1} \, ,$$

and it then follows that $U_1/U_n = V_1/V_n$, whence taking the limit shows $U_1 = V_1$, as desired.

For the prime 2, as usual, one has to start the recursive process mod 8, so we have an isomorphism

$$V_2/V_3 \approx V_n/V_{n+1} \, , \qquad n \geq 2 \, .$$

From this we get the similar lemma:

Lemma 3. *Let $\ell = 2$. Let U_1 be a closed subgroup of V_1. If $U_2/U_3 = V_2/V_3$ then $U_2 = V_2$. If furthermore $U_1/U_3 = V_1/V_3$, in other words, if the reduction of U_1 mod 8 contains*

$$I + 2M_2 \;(\text{mod } 8) \, ,$$

then $U_1 = V_1$.

Proof. Clear.

Abelianization

We let G be our usual group with representation ρ. The field K of which G is the Galois group contains all the roots of unity, and this cyclotomic field is the maximal abelian extension of the rationals.

Let G′ denote the closure of the commutator subgroup, and $G^{ab} = G/G'$. Then G^{ab} is the Galois group of the cyclotomic field.

Lemma 4. *Assume that M is divisible by all the primes dividing Δ. Then*

$$G'_M = G_M \cap \prod_{\ell \mid M} SL_2(\mathbb{Z}_\ell) \, .$$

Proof. We observe that SL_2 is the kernel of the det map,

$$\det : \prod GL_2(Z_\ell) \longrightarrow \prod Z_\ell^* .$$

By Kronecker's theorem, the only abelian extensions of the rationals ramified only at primes dividing M are contained in the field of M^∞-th roots of unity, and the lemma follows at once.

Remark. If no assumption is made concerning the ramification, then one has only an inclusion

$$G'_M \subset G_M \cap \prod_{\ell \mid M} SL_2(Z_\ell) .$$

For an example of this which causes trouble, see the section on the Shimura curve.

A theorem of Serre

In this part we recall techniques of Serre showing that the Galois group of division points is nearly GL_2.

Let G be a subgroup of a product

$$G \subset (G_1 \times G_2) ,$$

and assume that the projections of G on the two factors are G_1 and G_2 respectively. Let

$$H_1 = pr_1 [G \cap (G_1 \times \{e_2\})] \quad \text{and} \quad H_2 = pr_2 [G \cap (\{e_1\} \times G_2)] .$$

Then there is an isomorphism

$$G_1/H_1 \approx G_2/H_2$$

whose graph is induced by G. We call this Goursat's lemma.

If L is a possibly infinite set of primes, and G is again the Galois group of the division points of an elliptic curve A over Q, we let G_L be the projection on the L-th factor. In other words, we have the representation

$$\rho_L : G \longrightarrow \prod_{\ell \in L} GL_2(Z_\ell) ,$$

and we denote its image by G_L.

We say that a prime occurs in a group if it divides the order of some solvable Jordan-Holder constituent for the group.

We let $M_\ell = \text{Mat}_2(Z_\ell)$.

Theorem 6.1. *Let M be divisible by 2, 3 and all primes dividing the discriminant Δ of A. Let L consist of the primes not dividing M. Assume:*

(i) $G(\ell) = GL_2(F_\ell)$ *for all* $\ell \in L$.

(ii) *If* $\ell \in L$, *then* ℓ *does not occur in* G_M.

Then M splits ρ, and
$$G_L = \prod_{\ell \in L} GL_2(Z_\ell) .$$

Proof. This is proved by Serre's arguments reproduced in [L 1], Chapter 17, §4 and §5. We sketch the proof. We have $G \subset G_M \times G_L$. We let H_M and H_L be the corresponding Goursat subgroups. The refinement lemma, together with the fact that
$$SL_2(F_\ell)/\pm 1$$
is simple, imply that the groups
$$(\cdots, 1, SL_2(Z_\ell), 1, \cdots) , \qquad \ell \in L ,$$
are contained in G_L and also in H_L. Since $I + \ell M_\ell$ is an ℓ-group, the hypothesis (ii) implies also that
$$(\cdots, 1, I + \ell M_\ell, 1, \cdots)$$
is contained in H_L. Using the determinant map which gives the effect on the field of roots of unity then shows that G_L is the full product of all $GL_2(Z_\ell)$, and G_L/H_L is abelian. It is the Galois group of a cyclotomic field which ramifies only in L, and whose intersection with the field of G_M/H_M must be Q, because only primes dividing M ramify in ρ_M. Hence H_L is all of G_L, and it follows that M splits ρ, thus proving the theorem.

The techniques involved in determining certain Galois groups involve all the above, and two main additional ones:

First the use of the Tate parametrization, which locally gives unipotent elements, cf. [L 1], Chapter 15.

Second, the use of Frobenius elements, whose traces are known from tables, and which we find ad hoc to satisfy certain congruence properties, ultimately showing that the Galois group is large. This is related to the method used by Shimura [Sh] when he investigated the Galois group of division points of $X_0(11)$.

§7. The curve $y^2 = x^3 + 6x - 2$

The above curve is considered by Serre as 5.9.2, p. 318, [S 2]. It has

$$\Delta = -2^6 3^5 \quad \text{and} \quad j = 2^9 3 .$$

Serre has shown that $G_\ell = GL_2(Z_\ell)$ for $\ell \neq 2, 3$, and also that

$$G(2) = GL_2(F_2) \quad \text{and} \quad G(3) = GL_2(F_3) .$$

Theorem 6.1 shows that 6 splits ρ. All we have to determine is

$$\rho_{2,3} : G \longrightarrow G_2 \times G_3 ,$$

and we shall prove:

Theorem 7.1. *The above curve is a Serre curve with* $q = 3$.

The proof is given in the following lemmas.

Lemma 1. $G(6) = S_6(6)$.

Proof. Compatibility on the field $Q(\sqrt{\Delta})$ requires that the group of torsion points is contained in the Serre group, so we have $G(6) \subset S_6(6)$. A machine computation checks that the points of order 2 have degree 6 over Q, and that the cubic field above the quadratic field $Q(\sqrt{\Delta})$ is not contained in the field of 3-torsion points. Indeed, the machine finds that the prime 67 splits completely in the field of x-coordinates of A_3 (A is the curve), but does not split completely in $Q(A_2)$. The equality $G(6) = S_6(6)$ follows because the orders of the two groups are equal.

The image $\rho_{2,3}(G)$ in $G_2 \times G_3$ gives a Goursat correspondence, and we let H_2, H_3 be the corresponding kernels, so that

$$G_2/H_2 \approx G_3/H_3 .$$

We must determine H_2 and H_3.

343

Lemma 2. $H_3 \supset I + 3M_3$ and $G_3 = GL_2(\mathbb{Z}_3)$.

Proof. For the prime 23, tables show that $t_{23} = 6$. The characteristic polynomial of Frobenius σ_{23} is

$$X^2 - 6X + 23 \equiv (X-1)(X+4) \pmod 9 .$$

Since $-1 \not\equiv 4 \pmod 3$ we can diagonalize $\rho_3(\sigma_{23})$ over $\mathbb{Z}_3$, and

$$\rho_3(\sigma_{23}) \equiv \begin{pmatrix} 1 & 0 \\ 0 & -4 \end{pmatrix} \pmod 9 .$$

Furthermore,

$$\rho_3(\sigma_{23}^2) \equiv I + 3 \begin{pmatrix} 0 & 0 \\ 0 & 5 \end{pmatrix} \pmod 9 .$$

But $\left(\frac{23}{3}\right) = -1$, and $\rho_2(\sigma_{23})$ is odd in G_2, so

$$\rho_2(\sigma_{23}^2) \equiv I \pmod 2 .$$

Taking 4-th powers, we conclude:

The sequence $\rho_2(\sigma_{23})^{4^n}$ converges to I in G_2.
A subsequence of $\rho_3(\sigma_{23})^{4^n}$ converges to an element τ in G_3, and

$$\tau \equiv I + 3 \begin{pmatrix} 0 & 0 \\ 0 & 5 \end{pmatrix} \pmod 9 .$$

Since $\rho_2(\tau) = I$, it follows that τ lies in H_3. We write

$$\tau = I + 3Y .$$

Since H_3 is normal in G_3, the elements

$$I + 3g Y g^{-1} \pmod 9$$

are also in H_3 for $g \in GL_2(3)$. Hence the associated vector space in $M_3/3M_3$ is 4-dimensional. This shows that H_3 contains $I + 3M_3$. Since Serre proved that $G(3) = GL_2(F_3)$, it follows also that $G_3 = GL_2(\mathbb{Z}_3)$, thereby proving the lemma.

Lemma 3. $H_2 \supset I + 2M_2$ and $G_2 = GL_2(\mathbb{Z}_2)$.

Proof. Since $\left(\frac{31}{3}\right) = 1$, it follows that the Frobenius element σ_{31} is even in $G(2)$. Tables show that $t_{31} = 8$, so

$$\sigma_{31} \in I + 2M_2 \ .$$

Write $\sigma_{31} = I + 2Y$. Then the characteristic equation for σ_{31} is

$$X^2 - 8X + 31 = 0 \ ,$$

whence the characteristic polynomial for Y is

$$Y^2 - 3Y + 6 \equiv (Y+2)(Y+3) \pmod{8} \ .$$

Since $2 \not\equiv 3 \pmod 2$ we can diagonalize Y over Z_2. Hence

$$\rho_2(\sigma_{31}) \equiv I + 2 \begin{pmatrix} -2 & 0 \\ 0 & -3 \end{pmatrix} \pmod 8 \ .$$

Furthermore, $\rho_3(\sigma_{31})$ is a root of

$$X^2 - 8X + 31 \equiv X^2 + X + 1 \pmod 3 \ .$$

Hence

$$\rho_3(\sigma_{31})^3 \equiv I \pmod 3 \ .$$

It follows that the sequence $(\sigma_{31}^3)^{9^n}$ converges to I in G_3. There is a subsequence which converges to an element τ in G_2, and τ must lie in H_2 since its projection by ρ_3 is trivial. We must have

$$\tau \equiv I + 2 \begin{pmatrix} -2 & 0 \\ 0 & -3 \end{pmatrix} \pmod 8 \ .$$

Conjugation by elements of $GL_2(2)$ shows that we get all elements of

$$I + 2M_2 \pmod 4$$

in $H_2 \pmod 4$. Hence

$$G(4) = GL_2(4) \ .$$

Once more, conjugation by $GL_2(4)$ shows that we get all elements of

$$I + 2M_2 \pmod 8$$

in H_2 (mod 8). The recursive procedure takes hold, and we have found that H_2 contains $I + 2M_2$. Using Serre's result that $G(2) = GL_2(F_2)$, we conclude that $G_2 = GL_2(Z_2)$, thereby proving the lemma.

The theorem is immediate from the lemmas.

Remark. The use of Frobenius elements in the above proof and in the next section is similar to that made by Shimura [Sh].

§8. The Shimura curve $X_0(11)$

The curve $X_0(11)$ is defined over the rationals by the equation

$$y^2 + y = x^3 - x^2 - 10x - 20.$$

It has $\Delta = -11^5$ and $j = -2^{12}\,31^3/11^5$. Cf. Shimura [Sh]. We let ρ be its Galois representation, and G as before the Galois group of all its division points. A number of facts are known about G, and we shall recall the proofs of some of them, as we determine ρ.

The crucial primes are 2, 5, 11. We begin by looking at 5. The equation satisfied by the x-coordinate of the points of order 5 (actually the equation for the points of order n, given recursively) can be found in Weber, among other places. In the present case, it is determined by machine computation to be

$$0 = (x-5)(x-16)(5x^2+5x-29)(x^4+15x^3+120x^2+200x+155)(x^4+x^3+11x^2+41x+101).$$

The linear factor shows that there is a rational point of order 5, whose coordinates are given by (5,5). Actually, one can even exhibit the coordinates of its integral multiples, namely
$$(5,5),\ (5,-6),\ (16,60),\ (16,-61).$$

In particular, $G(5)$ is contained in the upper triangular Borel subgroup. The fact that the above equation has no irreducible factor of degree ≥ 5 shows that the group $G(5)$ cannot contain an element of order 5. Therefore we conclude that

$$G(5) = \left\{ \begin{pmatrix} 1 & 0 \\ 0 & u \end{pmatrix},\ u \in F_5^* \right\}.$$

The next theorem determines the 5-adic group G_5.

Theorem 8.1. G_5 *is the inverse image in* $GL_2(\mathbb{Z}_5)$ *of* $G(5)$, *in other words 5 is stable.*

Proof. Let V be the group of matrices which are

$$\equiv \begin{pmatrix} 1 & 0 \\ 0 & u \end{pmatrix} \mod 5, \qquad \text{some } u \in Z_5^* \,.$$

We use the refinement Lemma 2. We let $U = G_5$ so $U_1 = U \cap V_1$ consists of those elements of U which are $\equiv I \pmod 5$. Furthermore, V_1 is the full group $I + 5M_5$. It will suffice to prove:

There exist four elements $a_i \, (i = 1, \cdots, 4)$ in G_5 such that

$$a_i \equiv I + 5X_i \pmod{25} \,,$$

and X_1, X_2, X_3, X_4 are linearly independent mod 5.

We exhibit such elements by means of Frobenius elements.

From tables, for $p = 13$ we get $t_{13} = 4$, so G_5 contains a matrix σ_{13} which is a root of

$$\Phi_{13}(X) = X^2 - 4X + 13 \,.$$

Note that

$$\Phi_{13}(X) \equiv (X - 6)(X + 2) \pmod{25} \,,$$

and $-6 \not\equiv 2 \pmod 5$. We can therefore diagonalize σ_{13} over Z_5, and we assume our coordinates so chosen that

$$\sigma_{13} = \begin{pmatrix} 6 + 25a & 0 \\ 0 & -2 + 25b \end{pmatrix} \,.$$

We then obtain

$$\sigma_{13}^4 \equiv I + 5 \begin{pmatrix} -1 & 0 \\ 0 & 3 \end{pmatrix} \pmod{25} \,.$$

In particular, σ_{13}^4 is not scalar mod 25.

Again from tables, for the prime $p = 653$ we have $t_{653} = -41$, and σ_{653} has characteristic polynomial

$$X^2 - 41X + 653 \equiv (X + 11)(X - 2) \pmod{25} \,.$$

Since $11 \not\equiv 2 \bmod 5$, it follows that σ_{653} is diagonalizable over Z_5, and since

$$11^4 \equiv (-2)^4 \equiv 16 \pmod{25}$$

we obtain

$$\sigma_{653}^4 \equiv 16I \equiv I + 5 \cdot 3I \pmod{25},$$

which is a scalar mod 25. Using the same basis as we picked to diagonalize σ_{13}, we see that σ_{653}^4 provides us with a diagonal element with respect to this basis mod 25. Therefore, from σ_{13}^4 and σ_{653}^4 we already obtain two linearly independent matrices in U_1/U_2, corresponding to the tangent space for the diagonal matrices.

To find a third element independent from the two preceding ones, we pick σ_{31} with $t_{31} = 7$. The characteristic polynomial is

$$X^2 - 7X + 31 \equiv (X-1)^2 \pmod 5.$$

From the fact that G(5) is a diagonal group, and $(\sigma_{31} - I)^2 \equiv 0 \pmod 5$, we may write σ_{31} in the form

$$\sigma_{31} = I + 5Z, \qquad Z \in \mathrm{Mat}_2(Z_5),$$

and we see that Z satisfies the characteristic equation

$$Z^2 - Z + 1 = 0,$$

which is irreducible mod 5. Consequently Z is not triangular with respect to any basis. Since we already have the diagonal elements, there exists a third element of the form

$$I + 5 \begin{pmatrix} 0 & x \\ y & 0 \end{pmatrix} \pmod{25} \qquad x, y \not\equiv 0 \pmod 5$$

linearly independent from the others mod 5.

On the other hand, the space of matrices Y (mod 5) such that $I + 5Y$ belongs to G (mod 25) is invariant under conjugation by G. If we conjugate by

$$\begin{pmatrix} 1 & 0 \\ 0 & u \end{pmatrix},$$

we see that matrices

$$\begin{pmatrix} 0 & u^{-1}x \\ uy & 0 \end{pmatrix}$$

349

are such elements Y. From this it is immediate that we can get a fourth linearly independent matrix Y (mod 5), thereby concluding the proof.

Next we list more facts to be used about $X_0(11)$.

It is isomorphic to the Tate curve over Q_{11}. One can verify this by using the criterion for multiplicative reduction, cf. Tate [T], §6, Theorem 5 (the tangents at the double point are rational over Q_{11}). The order of its j-invariant is -5.

The equation shows that the curve has good reduction for all primes except 11. Serre [S 2] has proved that

$$G(\ell) \approx GL_2(F_\ell) \text{ for all primes } \ell \neq 5 \ .$$

All of this shows that $X_0(11)$ satisfies the hypotheses of the next theorem. Observe already that by the refinement lemma, we can conclude that

$$G_\ell = GL_2(Z_\ell), \qquad \ell > 5 \ .$$

We want to see that the image of G in $GL_2(Z_2) \times GL_2(Z_{11})$ is Serre's subgroup. In his paper, Serre asserts a similar statement for several curves (5.5.6, 5.5.7, 5.5.8, see p. 311), and he kindly communicated a proof to us. We reproduce his arguments in the following theorem, for the convenience of the reader.

Theorem 8.2. *Let* A *be an elliptic curve over the rationals. Let* L *be a set of primes containing* 2 *and an odd prime* $q \geq 5$. *Assume:*
(i) A *has good reduction at all* $\ell \in L$, $\ell \neq q$.
(ii) A *is isomorphic to the Tate curve over* Q_q, *and*

$$\text{ord}_q j_A$$

is not divisible by 2 *or* 3.
(iii) $G(\ell) \approx GL_2(F_\ell)$ *for all* $\ell \in L$.
Then the image of

$$\rho_L : G \longrightarrow \prod_{\ell \in L} GL_2(Z_\ell)$$

is Serre's subgroup of index 2, *namely*

$$\rho_L(G) = S_{2q} \times \prod_{\ell \neq 2, q} GL_2(Z_\ell) \ .$$

Proof. By Theorem 6.1, we know that $2 \cdot 3 \cdot q$ splits ρ, and the $GL_2(Z_\ell)$ split off for $\ell \geq 5$, $\ell \neq q$. Without loss of generality, we may therefore assume that L consists of 2, q and possibly 3. For the sake of concreteness, let us assume that $3 \in L$.

Using the isomorphism with the Tate curve over Q_{11}, one sees that G_2 (resp. G_3) contains the matrix (unipotent)

$$\begin{pmatrix} 1 & 1 \\ 0 & 1 \end{pmatrix}.$$

By (iii) and the fact that the image of $GL_2(Z_\ell)$ under the determinant map is Z_ℓ^*, one sees easily that even for $\ell = 2$ or 3 one has

$$G_\ell = GL_2(Z_\ell).$$

We leave this easy part to the reader.

We now prove that the map

$$\rho_{2,3} : G \longrightarrow GL_2(Z_2) \times GL_2(Z_3)$$

is surjective. By Goursat's lemma, this amounts to determining the possible isomorphisms from a quotient of $GL_2(Z_2)$ with a quotient of $GL_2(Z_3)$. The argument concerning the unipotent matrix above can again be used to see that the image of $\rho_{2,3}$ contains

$$\begin{pmatrix} 1 & Z_2 \\ 0 & 1 \end{pmatrix} \times \{1\} \quad \text{and} \quad \{1\} \times \begin{pmatrix} 1 & Z_3 \\ 0 & 1 \end{pmatrix}.$$

Since this image is invariant by conjugation, it contains

$$SL_2(Z_2) \times \{1\} \quad \text{and} \quad \{1\} \times SL_2(Z_3).$$

Since

$$\det{}_{2,3} : G \longrightarrow Z_2^* \times Z_3^*$$

is surjective, it follows that the image of $\rho_{2,3}$ is the full product

$$GL_2(Z_2) \times GL_2(Z_3).$$

Finally, we determine the image of

$$\rho_{2,3,q} : G \longrightarrow [GL_2(Z_2) \times GL_2(Z_3)] \times GL_2(Z_q).$$

Since the order $q \geq 5$ does not occur in the orders of (a Jordan-Holder decomposition of) $GL_2(\mathbb{Z}_2) \times GL_2(\mathbb{Z}_3)$, and since $SL_2(\mathbb{F}_q)/\pm 1$ is simple, it follows that in Goursat's lemma, any possible quotient of $GL_2(\mathbb{Z}_q)$ must have order prime to q and must be abelian. One sees easily that:

$$GL_2(\mathbb{Z}_\ell)^{ab} = \mathbb{Z}_\ell^* \qquad \text{(via det)} \qquad \text{if } \ell \geq 3$$

$$GL_2(\mathbb{Z}_2)^{ab} = \{\pm 1\} \times \mathbb{Z}_2^* \qquad \text{(via } \varepsilon \times \text{det)} \qquad \text{if } \ell = 2,$$

where ε is the homomorphism

$$\varepsilon : GL_2(\mathbb{F}_2) = S_3 \longrightarrow \{\pm 1\}.$$

Let

$$\psi : G_2 \times G_3 \times G_q \longrightarrow G_2^{ab} \times G_3^{ab} \times G_q^{ab}$$

be the factor commutator group mapping. As we have seen, the group G must contain $(G_2 \times G_3)'$ and G_q'. Thus $\psi(G)$ is a subgroup

$$\psi(G) \subset \{\pm 1\} \times \mathbb{Z}_2^* \times \mathbb{Z}_3^* \times \mathbb{Z}_q^*,$$

and because the roots of unity are contained in the field of division points, we know that $\psi(G)$ must project onto

$$\mathbb{Z}_2^* \times \mathbb{Z}_3^* \times \mathbb{Z}_q^*.$$

On the other hand, $\psi(G)$ is necessarily contained in the subgroup consisting of elements

$$(\pm 1, u_2, u_3, u_q)$$

such that

$$\left(\frac{u_q}{q}\right) = \pm 1,$$

which has index 2. Therefore $\psi(G)$ is equal to this subgroup. This implies that

$$\rho_L(G) = S_{2q} \times GL_2(\mathbb{Z}_3),$$

thereby proving the theorem.

In the application, we let

$$L = \{2, 3, \ell \geq 7\}, \qquad q = 11.$$

Theorem 6.2 gives us the structure of G_L for the Shimura curve. We then have to see how it mixes with G_5, which amounts to determining the Goursat subgroups. We shall need additional notation for this, and the result is stated at the end, as Theorem 8.3.

We have

$$G \subset G_5 \times G_L, \qquad \text{and} \qquad G_L = S_{2q} \times \prod_{\substack{\ell \neq 2, q \\ \ell \in L}} GL_2(Z_\ell).$$

We let H_5 and H_L be the corresponding Goursat subgroups, giving an isomorphism

$$G_5/H_5 \approx G_L/H_L,$$

i.e. giving the identification of G_5 and G_L on the common subfield $K_5 \cap K_L$, which is a finite extension of Q. We shall determine explicitly what this subfield is, and what the above isomorphism is.

We have an exact sequence

$$1 \longrightarrow U_5 \longrightarrow G_5 \longrightarrow Z(4) \longrightarrow 1,$$

where $U_5 = 1 + 5M_5$. The map $H_5 \to Z(4)$ is surjective because $Q(\mu_5)$ is disjoint from K_L (the field of fifth roots of unity $Q(\mu_5)$ is ramified only at 5, and K_L is unramified at 5). Thus we conclude:

$$(G_5 : H_5) \text{ is a power of } 5.$$

Lemma 1. G_5/H_5 is abelian, and H_L contains:
(i) $SL_2(Z_\ell)$ for $\ell \neq 2, 5, 11 = q$.
(ii) $(1 + 2M_2) \times (1 + qM_q)$ as subgroup of $S_{2q} \subset H_L$.
Furthermore, $S_{2q}(2q)$ contains the product

$$E(2) \times SL_2(q),$$

where $E(2)$ is the group of even elements at 2.

Proof. The first two parts (i) and (ii) follow from the standard Jordan-Holder consideration, either because of the simplicity of $SL_2(F_\ell)/\pm 1$ for $\ell \geq 5$, or because primes dividing the orders of certain subgroups on the right do not occur in H_5, that is, are not equal to 5. The last statement follows similarly, noting that the even elements form a subgroup of order $3 \neq 5$, and that $SL_2(q)$ has a Jordan-Holder decomposition of a simple group and a group of order $2 \neq 5$. The prime 5 occurs only once in the part of G_L^{ab} ramified only at 11. It follows at once that G_5/H_5 has order 1 or 5. This proves the lemma.

Lemma 2. We have
$$H_L = H_{2q} \times \prod_{\ell \neq 2,5,11} GL_2(Z_\ell)$$
and $(S_{2q}:H_{2q}) = 1$ or 5.

Proof. The group G_5/H_5 is the Galois group of a cyclic extension of the rationals which is ramified only at 11 or possibly 5, and of degree 1 or 5. On the other hand, G_L/H_L is the Galois group of an abelian extension which is unramified at 5. Hence the common subfield $K_5 \cap K_L$ must be the 5-part of the field of 11-th roots of unity,
$$Q(\mu_{11}),$$
or it must be trivial. This proves the lemma, while also providing additional information.

We must determine $K_5 \cap K_{22}$, or in other words $H_{2q} = H_{2 \cdot 11}$. For this we need to make some mappings more explicit.

We know that G_5 is the group of matrices
$$a = \begin{pmatrix} 1+5a & 5b \\ 5c & u \end{pmatrix}$$
where $a, b, c \in Z_5$ and $u \in Z_5^*$. There is a homomorphism
$$\psi : a \longmapsto a(a) = a \pmod 5$$
whose kernel contains the commutator subgroup G_5', and consists of the matrices
$$\begin{pmatrix} 1+25a & 5b \\ 5c & u \end{pmatrix}.$$

The homomorphism ψ cannot factor through

$$\det_5 : G_5 \longrightarrow Z_5^*,$$

because, for instance, we have $\psi(\sigma) \neq 0$ and $\det \sigma = 1$ if σ is the matrix

$$\begin{pmatrix} 6 & 0 \\ 0 & 1/6 \end{pmatrix}.$$

The map $\psi \circ \mathrm{pr} : G_{2 \cdot 11} \times G_5 \to Z(5)$ which factors through the projection on G_5 is abelian. Let $M = 2 \cdot 11 \cdot 5$. Let

$$\rho_M : G \longrightarrow GL_2(Z_2) \times GL_2(Z_{11}) \times GL_2(Z_5)$$

be the partial representation on those three factors. Then $\psi \circ \mathrm{pr} \circ \rho_M$ is abelian on G, and hence factors through G'_M, which is the intersection of G with SL_2. This means that $\psi \circ \mathrm{pr} \circ \rho_M$ factors through the determinant map on G. In other words, the following diagram commutes on G, with some homomorphism $\lambda_{22,5}$.

$$(1 \times H_5) \subset G \xrightarrow{\rho_M} G_{22} \times G_5 \xrightarrow{\psi \circ \mathrm{pr}} Z(5)$$

$$\det \downarrow \qquad \nearrow \lambda_{22,5}$$

$$Z_{22}^* \times Z_5^*$$

In symbols, this means

$$\psi \circ \mathrm{pr} \circ \rho_M = \lambda \circ \det \circ \rho_M.$$

We apply this to the subgroup $\{1\} \times H_5$ which is contained in G, and conclude:

Lemma 3. *We have $H_5 \neq G_5$ and $H_5 \subset \mathrm{Ker}(\psi - \lambda \circ \det)$, where $\lambda = \lambda_5$ is the restriction of $\lambda_{22,5}$ to the second factor. The map*

$$\psi - \lambda \circ \det : G_5/H_5 \longrightarrow Z(5)$$

is an isomorphism.

Since, as we have already remarked, G_5/H_5 is abelian of order 5, it now follows that G_L/H_L is abelian of order 5, and factors through the projection on

G_{11} and the determinant map to give a natural isomorphism

$$\phi_{11} \circ \mathrm{pr}_{11} : G_L/H_L \approx Z(11)^*/\pm 1 ,$$

as Galois group of the 5-part of the cyclotomic field $Q(\mu_{11})$. Thus our task is to find explicitly the isomorphism

$$\phi_5 : G_5/H_5 \longrightarrow Z(11)^*/\pm 1 .$$

The group on the right is cyclic, and we can select ± 2 as a generator (primitive root). We can now state the main theorem tying up all the lemmas together.

Theorem 8.3. *The homomorphism* λ *is trivial. The isomorphism* $G_5/H_5 \to Z(11)^*/\pm 1$ *is given by*

$$\phi_5 : \sigma \longmapsto (\pm 2)^{\psi(\sigma)} .$$

The number $M = 2 \cdot 25 \cdot 11$ *splits and stabilizes* ρ. *The group* G_{110} *consists of the elements* $(\sigma_5, \sigma_2, \sigma_{11})$ *such that*

$$\phi_5(\sigma_5) = \phi_{11}(\sigma_{11}) \qquad \text{and} \qquad (\sigma_2, \sigma_{11}) \in S_{2 \cdot 11} .$$

Proof. We have seen that the map

$$\psi_1 = \psi - \lambda \circ \det : G_5/H_5 \longrightarrow Z(5)$$

is an isomorphism, and identifying $Z(5)$ with its action on the 11-th roots of unity, we may write

$$\psi_1(\sigma) = \det \rho_{11}(\sigma)/\pm 1 .$$

Note that λ is defined on $Z_5^*(25)$, because $\mathrm{Im}\,\lambda$ is cyclic and the group $1 + 5Z_5$ is cyclic. There is a unique element $w \in Z(5)$ such that

$$\lambda \circ \det(\sigma) = w \frac{(\det \sigma)^4 - 1}{5} .$$

We shall prove that $w = 0$. Let

$$\sigma = \begin{pmatrix} 1 + 5a & 5b \\ 5c & u \end{pmatrix} , \qquad \psi(\sigma) = a .$$

Then $\det \sigma = u(1+5a) \pmod{25}$, $\operatorname{tr} \sigma = 1+u+5a$, and

$$(1-u)a = \frac{\operatorname{tr} \sigma - \det \sigma - 1}{5} \pmod 5.$$

If $\det \sigma \not\equiv 1 \pmod 5$ then

$$\psi(\sigma) = a = \frac{\operatorname{tr} \sigma - \det \sigma - 1}{5(1-u)} = \frac{\operatorname{tr} \sigma - \det \sigma - 1}{5(1-\det \sigma)} \pmod 5.$$

Thus for $p \not\equiv 1 \pmod 5$ we can calculate $\psi(\sigma_p)$ from $t_p = \operatorname{tr} \sigma_p$ and $p = \det \sigma_p$, and in particular obtain the following table.

$p = \det \sigma_p$	t_p	$\psi_1(\sigma_p) = \det \sigma_p$ in $Z(11)^*/\pm 1$	$\psi(\sigma_p)$	$\dfrac{(\det \sigma_p)^4 - 1}{5}$
2	-2	± 2	1	-2
13	4	± 2	1	2

The table shows that $\psi_1(\sigma_2) = \psi_1(\sigma_{13})$ and $\psi(\sigma_2) = \psi(\sigma_{13})$, and therefore $\lambda \circ \det (\sigma_2) = \lambda \circ \det (\sigma_{13})$. Since the entries in the last column of the table are not congruent (mod 5), w must be 0. Since $\psi_1 = \pm 2$ when $\psi = 1$, the isomorphism ϕ_5 must be as given in the statement of the theorem.

Having determined the Galois group, we saw that

$$M = 2 \cdot 11 \cdot 5^2$$

stabilizes and splits ρ. We are then in a position to determine the constant at the bad primes.

Theorem 8.4. *Let* $q = 11$. *For the curve* $X_0(11)$, *we have*

$$F_M(t) = F_{2q}(t) F_{25}(t).$$

Proof. It suffices to prove that if t_1 is the residue class of t mod $M_1 = 2 \cdot 11$ and t_2 is the residue class of t mod $M_2 = 25$, then

$$|G(22)_{t_1}| \, |G(25)_{t_2}| = 5 \, |G(22 \cdot 25)_t| \, .$$

Let

$$\phi_1 : G(22) \longrightarrow Z(5) \quad \text{and} \quad \phi_2 : G(25) \longrightarrow Z(5)$$

be the two homomorphisms such that $G(M) = \text{Ker } \lambda$, where

$$\lambda = \phi_1 \otimes 1 - 1 \otimes \phi_2 \, .$$

We have a correspondence

$$G(M)_t \longrightarrow G(22)_{t_1} \times G(25)_{t_2}$$

which associates

$$\sigma \longmapsto \left(\sigma_1, \sigma_2 + \begin{pmatrix} 5a & 0 \\ 0 & -5a \end{pmatrix} \right),$$

with $a \in Z(5)$. It is clear that this association establishes a one-to-five correspondence which makes the theorem obvious.

Remark. The value $F_{2q}(t)$ is exactly the same as that found in §5 for the Serre curves, in the tables.

There remains to determine $F_{25}(t)$.

Theorem 8.5. *We have:*

$$F_{25}(t) = \begin{cases} 5/4 & \text{if } t \not\equiv 1 \bmod 5 \\ 0 & \text{if } t \equiv 1 \bmod 5 \, . \end{cases}$$

Proof. The value 0 is clear. For the others, note that we have for $t, s \not\equiv 1 \pmod 5$:

$$|G(25)_t| = |G(25)_s| \, ,$$

arising from the bijection

$$\sigma \longmapsto \sigma + \begin{pmatrix} 0 & 0 \\ 0 & s-t \end{pmatrix}.$$

The desired values then follow at once from the definitions.

PART II

IMAGINARY QUADRATIC DISTRIBUTION

We let k be an imaginary quadratic field, with discriminant D, so $k = \mathbf{Q}(\sqrt{D})$. We let w be the number of roots of unity in k, and we let h be the class number. We let $\mathfrak{o}$ be the integers of k.

Let A be an elliptic curve over the rationals. For each prime p where A has good reduction, we have a Frobenius endomorphism π_p, and we want to describe conjecturally a probabilistic model for which the sequence of traces t_p such that $\mathbf{Q}(\pi_p) = k$ forms a random sequence. We work entirely in the Galois theory setting of division points, and thus axiomatize the situation. The matter is briefly reviewed in §3. We shall see in §5 that the probability that $\mathbf{Q}(\pi_p) = k$ is conjecturally asymptotic to

$$C(k, A) \frac{1}{2\sqrt{p}},$$

for some constant $C(k, A) > 0$. The conjecture implies that the number of primes $p \leq x$ for which $\mathbf{Q}(\pi_p) = k$ is asymptotic to

$$C(k, A) \pi_{\frac{1}{2}}(x),$$

where

$$\pi_{\frac{1}{2}}(x) = \sum_{p \leq x}' \frac{1}{2\sqrt{p}} \sim \frac{\sqrt{x}}{\log x}.$$

We let K be the field of division points. We go through similar steps as for the fixed trace distribution discussed in Part I, but in a more complicated setting. The complications arise from at least two factors:

(a) The presence of units in the imaginary quadratic field k, which always cause ambiguity in the set of generators of an ideal.

(b) The possible dependence of the GL_2-extension K with the maximal abelian extension k_{ab} of k. Usually these two fields intersect in the field generated over the rationals by all roots of unity, i.e. $\mathbf{Q}_{ab}$ (see Theorem 3.1), but it may happen that the intersection is bigger, by a finite extension which is an important invariant of the situation, and must be taken into account. So must the intersection of the GL_2-extension and the Hilbert class field H of k.

It is therefore natural to work with the composite extension Kk_{ab}, discussed in §3, giving rise to the probabilistic correspondence.

Practically, for the probabilistic model the above factor means that instead of parametrizing the probabilistic fiber at each prime by a single integer, we must now use pairs of integers. The first is a random variable for the trace of Frobenius in the GL_2-extension, and the second is a random variable for the trace of Frobenius in k. These will not be independent!

One can also consider the case of fixed trace from the quadratic field. For elliptic curves having complex multiplication, Mazur's "anomalous" primes (those with fixed trace $t_p = 1$, see [Ma]) happen to lie in certain quadratic progressions, for which Hardy and Littlewood had conjectured the asymptotic behavior. Both for later use, and also because it shows in a simple case how the arithmetic of the quadratic field affects the distribution, we recover the Hardy-Littlewood conjecture independently by making up a probabilistic model similar to the other cases. As an example, when $k = Q(i)$, the elements of trace 2 are those of the form $1 + ni$, and this special case gives the conjectured asymptotic behavior of those primes p which are of the form $n^2 + 1$. There is hardly any need to remind the reader that it is still unknown if in fact there exist infinitely many such primes.

Our main problem is to describe the constant $C(k, A)$.

We first show how one can define the constant $C(k, A)$ by taking a limit from finite levels, in a manner compatible with the Sato-Tate, Hecke, and Tchebotarev density properties. This is done in §4 and §5. We then show how the constant can be written as a quotient of integrals.

The denominator involves essentially the index of the Galois group $Gal(Kk_{ab}/H)$ in the "generic" Galois group, and the measure of this "generic" Galois group. It is convenient for our purposes to take additive Haar measures, so the measures of the multiplicative groups have to be computed.

The numerator involves the direct image of Haar measure into the space of conjugacy classes of integral Cartan elements corresponding to k. In §7, §8, §9 the measures and density functions involved are determined explicitly, and tabulated. Langlands suggested to us that what we were doing could be interpreted as computation of Harish transforms, cf. the comments at the end of §6. Theorem 7.1, Theorem 8.1, and the lemmas following Theorem 8.1 give a complete and systematic way of evaluating Harish transforms, and their integrals. The Harish transform in a neighborhood of the identity is computed in Sally-Shalika [S-Sh], but the context here is sufficiently different and the need for more systematic tables such that it was pointless to refer further to the literature.

Both the numerator and denominator admit an infinite product decomposition into local ℓ-factors. These are rational functions in $r = r(\ell) = 1/\ell$. If one writes the global constant $C(k, A)$ in terms of the discriminant, then these factors are of the form
$$1 + O(r^2),$$
and their product is absolutely convergent. If one writes the constant $C(k, A)$ in terms of the class number, then the ℓ-th factor is of the form
$$\left(1 - \left(\frac{k}{\ell}\right)\frac{1}{\ell}\right)^{-1}(1 + O(r^2))$$
and their product gives the value $L(1, \chi)$, times an absolutely convergent product. With our arguments, it is this second form which comes naturally. The constant is inversely proportional to $\sqrt{|D|}$.

In §10, §11 we consider the same special cases as in Part I, i.e. elliptic curves of Serre type, and $X_0(11)$, for which we obtain numerical values for the constant. This allows us to compare the predicted values with actual values for the asymptotic behavior of the primes in question, computed by machine. The fit is quite good on the whole, and is discussed in Part IV.

PART II

IMAGINARY QUADRATIC DISTRIBUTION

THE FIXED TRACE CASE

1.	Fixed traces from the quadratic field	77
2.	Computation of the constant for fixed trace	84

THE MODEL FOR THE MIXED CASE

3.	The mixed Galois representations	91
4.	The probabilistic model	104
5.	The asymptotic behavior	108
6.	The finite part of the constant as a quotient of integrals	112

COMPUTATIONS OF HARISH TRANSFORMS

7.	Haar measure under the trace-determinant map on Mat_2. General formalism.	123
8.	Relations with the trace-norm map on k	133
9.	Computation of C_ℓ for almost all ℓ	141
10.	The constant for Serre curves, $K \cap k_{ab} = Q_{ab}$	143
11.	The constant for $X_0(11)$	149

THE FIXED TRACE CASE

§1. Fixed traces from the quadratic field

In the probabilistic model, there is a random variable for each prime p, which will range over the integers. The probability function will again have a factor at infinity, and a factor at the finite primes.

We first describe the density function at infinity, which in the present instance is none other than the function of Hecke giving equidistribution of primes in sectors. We let

$$g''(\xi) = w \frac{1}{\pi} \frac{1}{\sqrt{1-\xi^2}}.$$

Then g'' is the distribution function of primes in k at infinity. The trace from k to Q will be abbreviated sometimes without subscript, e.g.

$$\text{Tr} = \text{Tr}_{k/Q}.$$

Because g'' blows up somewhat at the end points of the interval, it is convenient to define a truncation. We redefine $\xi(t)$ when $|t| < 2\sqrt{p}$ by letting

$$\xi(t) = \min\left\{\frac{t}{2\sqrt{p}},\ 1 - \frac{1}{2\sqrt{p}}\right\} \quad \text{if } t > 0$$

$$\xi(t) = \max\left\{\frac{t}{2\sqrt{p}},\ -1 + \frac{1}{2\sqrt{p}}\right\} \quad \text{if } t < 0.$$

We let P_k be the set of primes p which split completely in k,

$$p\mathfrak{o} = \mathfrak{p}\bar{\mathfrak{p}},$$

and such that the prime ideals $\mathfrak{p}, \bar{\mathfrak{p}}$ are principal, say $\mathfrak{p} = (\pi)$. The generator π is determined up to a unit in $\mathfrak{o}$.

Remark. By Tchebotarev, the density of primes which split completely in k is 1/2. Therefore the density of primes which have the above property is equal to 1/2h.

We let $P_k(x)$ be the set of primes $p \in P_k$ with $p \leq x$. For any interval J contained in $[-1,1]$, Hecke's theorem implies the density property:

$$\int_J g''(\xi) d\xi = \lim_{x \to \infty} \frac{1}{|P_k(x)|} \sum_{p \in P_k(x)} \#\{\pi \in \mathfrak{o}, \, \mathrm{Tr}\,\pi \in J, \, \pi\bar{\pi} = p\}.$$

In other words, $\xi(\mathrm{Tr}\,\pi)$ for prime elements π as above is distributed according to the function g'' on the interval $[-1,1]$. Note that such primes are counted without making any identification, e.g. $g(\xi(\mathrm{Tr}\,\pi)) = g(\xi(\mathrm{Tr}(-\pi)))$, but π and $-\pi$ are counted as distinct.

To deal with the presence of roots of unity in k, and to fix representatives of elements modulo such units, we can restrict our attention to smaller intervals. We say that a subinterval J of $[-1,1]$ is restricted if it is the projection of an arc in the upper half of the unit circle having length π/w. We let g''_J be the function which is equal to g'' on the interval J, and equal to 0 outside that interval.

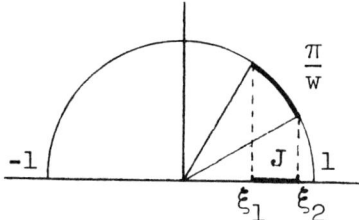

When J is restricted as above to be the projection of an arc with angle π/w, then we can also write the density relation in the form

$$\int_J g''(\xi) d\xi = \lim_{x \to \infty} \frac{\#\{p \in P_k(x), \, \mathrm{Tr}\,\pi \in J \text{ for some } \pi \text{ with } \pi\bar{\pi} = p\}}{|P_k(x)|}$$

We now pass to congruence conditions.

For a positive integer M we let $\mathfrak{o}(M) = \mathfrak{o}/M\mathfrak{o}$, and we let $\mathfrak{o}^*(M)$ be the image of $\mathfrak{o}^*$ in $\mathfrak{o}/M\mathfrak{o}$, actually in $\mathfrak{o}(M)^*$. If $M \geq 3$ then $\mathfrak{o}^* \to \mathfrak{o}^*(M)$ is injective, and consequently we also identify $\mathfrak{o}^*$ with the image $\mathfrak{o}^*(M)$. The factor group

$$\mathfrak{o}(M)^*/\mathfrak{o}^*$$

is going to play an important role in the sequel.

We let $P_{k,M}$ be the subset of primes of P_k which do not divide M, and $P_{k,M}(x)$ the subset of those primes $\leq x$. Let $\mathfrak{p}$ be a prime ideal above p. Given an integer s, we define

$$\tau_M(p,s) = \#\{\pi \text{ generating } \mathfrak{p} \text{ such that } \text{Tr}\,\pi \equiv s \bmod M\}.$$

We fix a positive integer $M \geq 3$. We define the function $F_M(s)$ by the formula

$$\frac{w}{M} F_M(s) = \lim_{x \to \infty} \frac{1}{|P_{k,M}(x)|} \sum_{p \in P_{k,M}(x)} \tau_M(p,s).$$

Except for taking into account the multiplicity of elements in a given class mod roots of unity $\mathfrak{o}^*$, the function F_M essentially counts the density relative to P_k of those primes p such that $\text{Tr}(\pi_p) \bmod M$ is a given congruence class mod M. In §2 we shall give a simple expression for $F_M(s)$ by using the Tchebotarev density theorem. For the moment, we pursue the probabilistic considerations.

From the definitions we immediately obtain

(1) $$\sum_{s \bmod M} \frac{1}{M} F_M(s) = 1.$$

We shall see below that

$$\lim_M F_M(s) = F(s)$$

exists.

We have fixed the interval J (corresponding to a sector of width π/w), and M. For each prime p we let the fiber of the probabilistic model at p be Z. We let the measure $\mu_{p,J}$ be represented with respect to counting measure by a function

$$f_{M,J}(s,p) \geq 0,$$

which we assume of the form

PR 1. $$f_{M,J}(s,p) = c_p g''_J(\xi(s,p)) F_M(s),$$

where c_p is a constant such that

(2) $$\sum_s f_{M,J}(s,p) = 1.$$

By considering the value

$$1 = \int_{-1}^{1} g''_J(\xi'')d\xi'',$$

and approximating Riemann sums

$$\frac{1}{2\sqrt{p}} \sum_s g''_J(\xi(s,p)), \qquad \text{for } p \to \infty,$$

it follows trivially, as in the first part, that we have the asymptotic relation

$$c_p \sim \frac{1}{2\sqrt{p}}.$$

For a fixed congruence class we have

$$\Pr\{y_p \equiv s_0 \bmod M\} = f_M(s_0, p) = c_p g''(\xi(s_0, p)) F_M(s_0).$$

Since

$$\xi(s_0, p) = s_0/2\sqrt{p},$$

it follows that

$$\lim_p g''(\xi(s_0, p)) = g''(0) = w/\pi.$$

Originally we had taken the density of p relative to P_k. We want the density relative to all primes, so we define the constant

$$\boxed{C(s_0, k) = \frac{1}{2h} \frac{w}{\pi} F(s_0)}$$

where h is the class number. The conjecture is that the number of primes in $P_k(x)$ with a fixed trace

$$\text{Tr}(\pi_p) = s_0$$

is given by the asymptotic formula

$$\boxed{N_{s_0,k}(x) \sim C(s_0, k)\pi_{\frac{1}{2}}(x),}$$

where
$$\pi_{\frac{1}{2}}(x) = \sum_{p \leq x} \frac{1}{2\sqrt{p}} \sim \frac{\sqrt{x}}{\log x}.$$

Example 1. Take $s_0 = 2$ and $k = Q(i)$. Then the elements with trace 2 in o are of the form
$$1 + ni$$
with some integer n, and therefore $N_{2,Q(i)}(x)$ is the number of primes $p \leq x$ such that p is of the form
$$p = n^2 + 1.$$

We shall determine the value of $F(2)$, thus recovering the Hardy-Littlewood conjecture concerning the asymptotics of such primes. From the tables of the next section, we find that the constant in this case is
$$C(2, Q(i)) = \frac{4}{\pi} \prod_{(k/\ell)=1} \frac{\ell(\ell-2)}{(\ell-1)^2} \prod_{(k/\ell)=-1} \frac{\ell^2}{\ell^2-1}.$$

On the other hand, let $L(1,\chi)$ be the value of the L-series of k with the non-trivial character. Then
$$L(1,\chi) = \frac{\pi}{4} = \prod_{(k/\ell)=1} \frac{1}{1-1/\ell} \prod_{(k/\ell)=-1} \frac{1}{1+1/\ell}.$$

(The products are written formally by abuse of notation. They should be combined and ordered by increasing ℓ, otherwise they don't converge.) From this we find
$$C(2, Q(i)) = \prod_{\ell \text{ odd}} \left(1 - \left(\frac{-1}{\ell}\right)\frac{1}{\ell-1}\right)$$

which is the constant of Hardy-Littlewood.

The 5,000th prime is 48,6811, and direct computation gives $\pi_{\frac{1}{2}}(48,611) \approx$ 26.4. Putting this in our asymptotic formula (and taking enough terms in the product for $C(2, Q(i))$ to make the relative error less than .01) we get 36.2 as an estimate for $N_{2,Q(i)}(48,611)$. It is trivial with the computer (and not hard by hand) to count the numbers of the form $n^2 + 1 \leq 48,611$ that are prime, and there turn out to be 37 of them. Similar calculations and counts give the following table.

$$N_{s,k}(48,611), \quad k = Q(i)$$

s	estimated	actual	s	estimated	actual
2	36.2	37	6	24.2	28
4	36.2	38	12	24.2	26
8	36.2	33			
16	36.2	33			
32	36.2	38			

There is good agreement between estimates and actual counts.

Example 2. There are prime elements of trace 1 in the quadratic field with discriminant D if and only if $D = 1-4m$, with m odd. It is then easy to check that the rational prime p has a complex factor with trace 1 if and only if it lies in the quadratic progression

$$m - Dx - Dx^2 .$$

For $m = 1, 3, 5, 9, 11$ one gets $D = -3, -11, -19, -35, -43$. The class number h is 2 for $D = -35$, and 1 for the others. The number of units w is 6 for $D = -3$, and 2 for the others. Estimated and actual counts are given in the table below.

$$N_{1,k}(48,611), \quad k = Q(\sqrt{D})$$

D	estimated	actual
−3	51.3	47
−11	8.9	10
−19	12.1	13
−35	6.0	7
−43	13.5	13

Agreement is again quite good. Since the occurrence of trace 1 has special interest [Ma], we include a table of the primes occurring in these five cases.

-3	-11	-19	-35	-43
7	3	5	79	11
19	223	43	709	97
37	619	233	6379	269
61	1213	1069	13309	1301
127	5569	1373	28429	3881
271	13093	2969	34729	6719
331	18043	3463	41659	9041
397	26953	8783		10331
547	30319	9619		14717
631	45763	14369		23747
919		21323		27961
1657		28163		30197
1801		34319		42667
1951				
2269				
2437				
2791				
3169				
3571				
4219				
4447				
5167				
5419				
6211				
7057				
7351				
8269				
9241				
10267				
11719				
12097				
13267				
13669				
16651				
19441				
19927				
22447				
23497				
24571				
25117				
26227				
27361				
33391				
35317				
42841				
45757				
47251				

Primes ≤ 48,611 with a principal factor of trace 1 in $Q(\sqrt{D})$, for $D = -3, -11, -19, -35, -43$.

§2. Computation of the constant for fixed trace

We use class field theory over k, and the notation established here concerning the maximal abelian extension of k will also be used later when we mix the situation with the GL_2 extension.

We let k_{ab} be the maximal abelian extension of k. Its Galois group is isomorphic to the finite part of the group of idele classes of k. We shall be interested in the Galois group $\mathcal{G}$ of k_{ab} over the Hilbert class field H. Let U denote the unit ideles of k, from which we project out the component at infinity. Thus

$$U = \prod_\ell o_\ell^*, \qquad U_\ell = o_\ell^*, \qquad \text{and } \bar{U} = U/o^*.$$

Then we have the class field isomorphism

$$\mathcal{G} = \text{Gal}(k_{ab}/H) \approx Uk^*/k^* \approx (U/(U \cap k^*) = U/o^*,$$

where o^* is the group of units in o, i.e. the roots of unity. In this manner we get a representation

$$\rho'' : \text{Gal}(k_{ab}/H) \longrightarrow \left(\prod_\ell o_\ell^*\right)/o^* = \bar{U}.$$

We let $o(M) = o/Mo$, and we let $o^*(M)$ be the image of o^* in o/Mo, actually in $o(M)^*$. We have a natural homomorphism

$$\bar{U} = U/o^* \longrightarrow o(M)^*/o^*(M).$$

If $M \geq 3$, then the map $o^* \to o^*(M)$ is injective (no two roots of unity are congruent mod M). We let

$$\rho''_{(M)} : \text{Gal}(k_{ab}/H) = \mathcal{G} \longrightarrow o(M)^*/o^*(M) = \bar{U}(M).$$

be the composition of ρ'' with the canonical homomorphism mod M. This reduction mod M of the representation has a kernel, whose fixed field is denoted by $k_{ab}(M)$, and we let $\mathcal{G}(M)$ denote its Galois group over H, so that we obtain an isomorphism

(for $M \geq 3$),
$$\rho''_{(M)} : \mathfrak{A}(M) \longrightarrow o(M)^*/o^* .$$

As usual, the prime ideals of k not dividing M (say above $\mathfrak{p} \in P_k$) are embedded in the idele classes, whence in U/o^* by mapping a prime element $\pi_{\mathfrak{p}}$ on the idele having $\pi_{\mathfrak{p}}^{-1}$ at the $\mathfrak{p}$-component, and 1 at all other components. Then
$$\rho''_{(M)}(\sigma_{\mathfrak{p}}) \equiv \pi_{\mathfrak{p}} \pmod{M},$$
because the Artin symbol of that idele has the same effect as the idele having $\pi_{\mathfrak{p}}$ at all components except for component 1 at $\mathfrak{p}$.

If X is a subset of $o(M)$, and s an integer, we let X_s be the subset of X consisting of those elements u such that
$$\text{Tr } u \equiv s \pmod{M} .$$

Theorem 2.1. *If* $M \geq 3$, *then*
$$\frac{1}{M} F_M(s) = \frac{|o(M)^*_s|}{|o(M)^*|} .$$

Proof. Let G be the Galois group of $k_{ab}(M)$ over Q. We define a function $\lambda_M(\sigma, s)$ for $\sigma \in G$ to be 0 unless σ lies in $\mathfrak{A}(M)$, that is σ leaves H fixed, and in that case,

$\lambda_M(\sigma, s)$ = number of elements in $\rho''_{(M)}(\sigma)$ (which is a coset of o^*) having a trace $\equiv$ s mod M .

Then λ_M is a function of conjugacy classes in G. We may view our previous function $\tau_M(p,s)$ as defined for any prime p, but equal to 0 except when p splits completely in k, and its factors are principal in k, and p does not divide M, in which case it has the value assigned to it previously. We let P be the set of all primes, and P(x) the subset of those primes $\leq$ x. Then

$$\frac{w}{M} F_M(s) = \lim_{x \to \infty} \frac{1}{|P_k(x)|} \sum_{p \in P_k(x)} \tau_M(p, s)$$

$$= \lim_{x \to \infty} \frac{2h}{|P(x)|} \sum_{p \in P(x)} \lambda_M(\sigma_p, s)$$

$$= \frac{2h}{|G|} \sum_{\sigma \in G} \lambda_M(\sigma, s) \qquad \text{(by Tchebotarev)}$$

$$= \frac{1}{|\mathcal{C}(M)|} |o(M)^*_s|$$

$$= \frac{w}{|o(M)^*|} |o(M)^*_s| \ .$$

This proves the theorem.

From Theorem 2.1 we can compute F_M by elementary congruence arguments, carried out in the next lemmas. The first one gives the multiplicativity over relatively prime factors. The second shows that the value of F_{ℓ^n} stabilizes at level 1 for odd primes, and at level 2 if $\ell = 2$. The third lemma gives an invariance under multiplicative translations, which reduces computations to just a few cases: 0 or 1 for odd ℓ and 0, 1, 2, 3 for even ℓ.

Lemma 1. *Suppose that* $M = M_1 M_2$ *is a product of two relatively prime factors. Then*
$$F_M(s) = F_{M_1}(s) F_{M_2}(s) \ .$$

Proof. Obvious.

Lemma 2. *Let ℓ be prime, $n \geq 1$ if ℓ is odd and $n \geq 2$ if $\ell = 2$. Let s, t be integers such that*

$$s \equiv t \bmod \ell \qquad \text{or} \qquad s \equiv t \bmod 4$$

according as ℓ is odd or even. Then

$$|o(\ell^n)^*_s| = |o(\ell^n)^*_t|$$

and

$$F_{\ell^n}(s) = F_{\ell^n}(t) = F_\ell(s) \qquad \text{if } \ell \text{ is odd}$$

$$F_{2^n}(s) = F_{2^n}(t) = F_4(s) \qquad \text{if } \ell = 2 \ .$$

Proof. The map
$$u \mapsto u + \frac{t-s}{2}$$
gives a bijection of $o(\ell^n)_s^*$ with $o(\ell^n)_t^*$ so the first assertion is obvious. The second follows from the first, and the fact that under the reduction map
$$o(\ell^m)^* \longrightarrow o(\ell^n)^*$$
there are $\ell^{2(m-n)}$ elements in each fiber, while for the reduction
$$Z(\ell^m) \longrightarrow Z(\ell^n)$$
there are ℓ^{m-n} elements in each fiber.

Lemma 3. *If* $a \in Z(M)^*$ *then* $F_M(as) = F_M(s)$.

Proof. The map $u \mapsto au$ is a bijection of $o(M)_s^*$ onto $o(M)_{as}^*$.

Putting these lemmas together, we obtain:

Theorem 2.2. *The constant* $C(s_0, k)$ *is given by*
$$C(s_0, k) = \frac{1}{2h} \frac{w}{\pi} F(s_0),$$
where
$$F(s_0) = F_4(s_0) \prod_{\ell \neq 2} F_\ell(s_0).$$

We must then tabulate $F_4(s_0)$ and $F_\ell(s_0)$.

Theorem 2.3. *Let* ℓ *be odd. The values of* F_ℓ *are given by the following table, where* $r = r(\ell) = 1/\ell$.

	$F_\ell(0)$	$F_\ell(1)$
$\left(\frac{k}{\ell}\right) = 1$	$\dfrac{1}{1-r}$	$\dfrac{1-2r}{(1-r)^2}$
$\left(\frac{k}{\ell}\right) = -1$	$\dfrac{1}{1+r}$	$\dfrac{1}{1-r^2}$
$\left(\frac{k}{\ell}\right) = 0$	$\dfrac{1}{1-r}$	$\dfrac{1}{1-r}$

Proof. In the first case, $\mathfrak{o}(\ell)^* \approx F_\ell^* \times F_\ell^*$. In the second case,

$$\mathfrak{o}(\ell)^* \approx F_{\ell^2}^*.$$

In the third case, if λ is a prime element above ℓ, then representatives for $\mathfrak{o}(\ell)$ are given by $a + b\lambda$ with $a, b \in \mathbf{Z}/\ell\mathbf{Z}$. It is then a routine matter in each case to determine the cardinality of $\mathfrak{o}(\ell)^*$, and then $F_\ell(0)$ and $F_\ell(1)$.

The case when $\ell = 2$ is slightly more complicated, especially when 2 ramifies in k, and two subcases arise. We say that the ramification is of first kind if there exists $\lambda \in \mathfrak{o}$ such that $\mathrm{Tr}(\lambda) \equiv 2 \pmod 4$, and λ is a local prime element lying above 2. The λ-adic expansion of an element of $\mathfrak{o}$ is then of type

$$a + b\lambda + c\lambda^2 + d\lambda^3 \pmod{4\mathfrak{o}},$$

where $a, b, c, d = 0, 1$. On the other hand, we say that the ramification is of second kind if $\mathrm{Tr}(\lambda) \equiv 0 \pmod 4$ for all local prime elements λ lying above 2. The table of values for F_4 is then as follows.

	$F_4(0)$	$F_4(1)$	$F_4(2)$	$F_4(3)$
$\left(\frac{k}{2}\right) = 1$	2	0	2	0
$\left(\frac{k}{2}\right) = -1$	2/3	4/3	2/3	4/3
$\left(\frac{k}{2}\right) = 0$ First kind	2	0	2	0
$\left(\frac{k}{2}\right) = 0$ Second kind	0	0	4	0

We leave the computations as an exercise.

THE MODEL FOR THE MIXED CASE

§3. The mixed Galois representation

Let

$$W = \prod_\ell GL_2(Z_\ell).$$

We let K be an infinite Galois extension of Q, let $G = \mathrm{Gal}(K/k)$, and let

$$\rho' : G \longrightarrow \prod GL_2(Z_\ell) = W$$

be a GL_2-representation (open embedding) as in Part I. We write ρ' instead of ρ because we shall use ρ'' for a representation related to k, and then we put the two together. As in Part I, we assume that there is a positive integer Δ such that the ℓ-adic representation ρ'_ℓ is unramified at p if $p \nmid \ell\Delta$. The Frobenius element σ'_p then has a characteristic polynomial

$$X^2 - t_p X + p,$$

assumed independent of ℓ, and with integer trace t_p. If ρ' is the representation associated with an elliptic curve, then Δ is the discriminant of the curve. Finally, since we do not explicitly assume that ρ' comes from an elliptic curve, we do assume the Riemann hypothesis that the roots of the characteristic polynomial have absolute value $\sqrt{p}$.

If L is a set of primes, we denote by G_L the projection of G under the representation

$$\rho'_L : G \longrightarrow \prod_{\ell \in L} GL_2(Z_\ell).$$

We may write

$$\rho'_L = \prod_{\ell \in L} \rho'_\ell.$$

We let K_L be the fixed field of the kernel of ρ'_L, so that

$$G_L = \mathrm{Gal}(K_L/Q).$$

If M is a positive integer, and L is the set of primes dividing M, then we also write

$$G_M = G_L \quad \text{and} \quad K_M = K_L.$$

On the other hand, if L is set of primes complementary to the primes dividing M, then we use the notation

$$G_L = G_{[M]}.$$

In the preceding section, we had described some facts concerning the class field theory above k. We now mix k_{ab} and the GL_2-extension K. The relevant lattice of fields is illustrated.

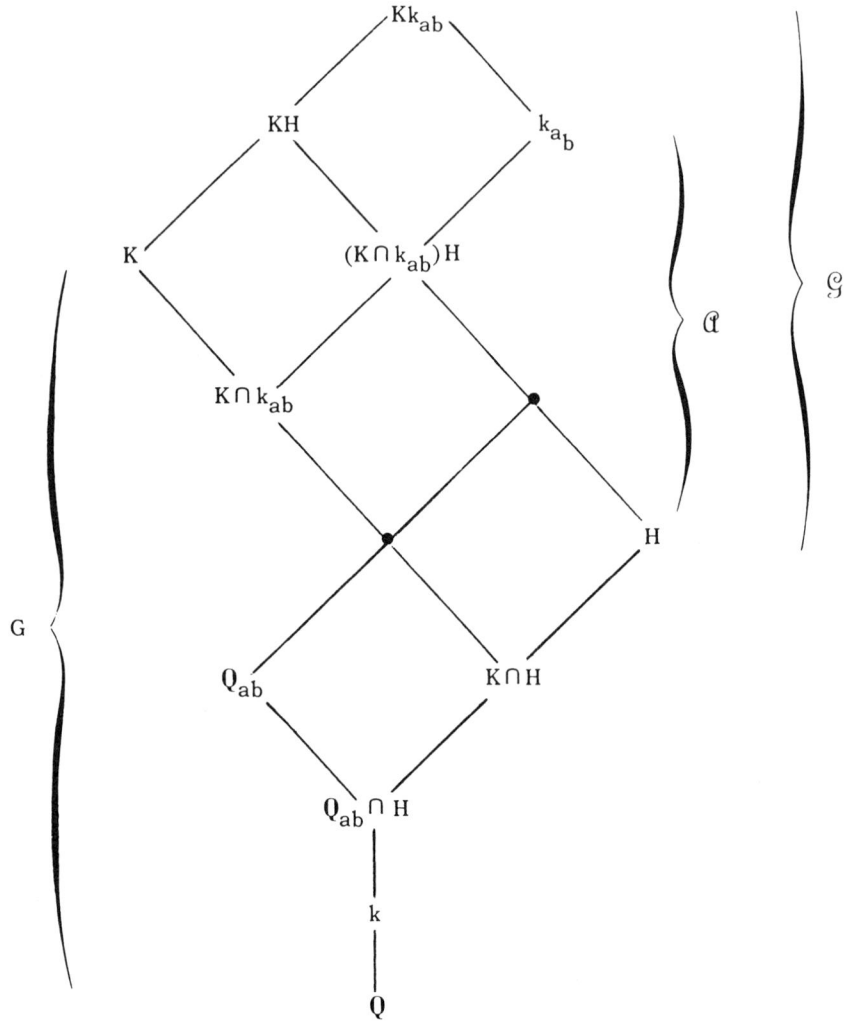

We shall consider the compositum $K = Kk_{ab}$, and let

$$\mathcal{G} = \text{Gal}(Kk_{ab}/H).$$

Then we have an embedding $\rho = (\rho', \rho'')$,

$$\rho : \mathcal{G} \longrightarrow G \times \mathfrak{A} \longrightarrow \prod GL_2(Z_\ell) \times \left(\prod \mathfrak{o}_\ell^*\right)/\mathfrak{o}^* = W \times \bar{U}.$$

The image of ρ is contained in the subgroup of the product consisting of those elements (σ', σ'') which have the same effect on the cyclotomic field Q_{ab}, which is contained in both K and k_{ab}. If $\sigma \in \mathcal{G}$, then σ' is its restriction to K and σ'' is its restriction to k_{ab}. We denote by a subscript N under the cross sign the fibering with respect to the effect on the roots of unity. We may therefore say that $\mathcal{G}$ is contained in

$$G \times_N \mathfrak{A}.$$

In terms of the matrix representations ρ' and ρ'', this means that the image of ρ is contained in the group of elements (a, u) such that

$$\det a = Nu,$$

where

$$\det : \prod GL_2(Z_\ell) \longrightarrow \prod Z_\ell^*$$

$$N : \prod \mathfrak{o}_\ell^* \longrightarrow \prod Z_\ell^*$$

are the determinant and norm maps respectively. The group of such matrices is denoted by the fibering notation

$$W \times_N \bar{U} = \prod GL_2(Z_\ell) \times_N \left(\prod \mathfrak{o}_\ell^*\right)/\mathfrak{o}^*.$$

We let $S\mathfrak{o}_\ell^*$ be the subgroup of $\mathfrak{o}_\ell^*$ consisting of those elements with norm 1 (the special linear group). Similarly, SG is the subgroup of G consisting of elements with determinant 1, and $S\mathcal{G}$ is the subgroup of $\mathcal{G}$ which fixes Q_{ab}. We have a pair of exact sequences:

$$\begin{array}{ccccccccc}
1 & \longrightarrow & SW \times S\bar{U} & \longrightarrow & W \times_N \bar{U} & \longrightarrow & \prod Z_\ell^* & \longrightarrow & 1 \\
& & \uparrow & & \uparrow & & \uparrow & & \\
1 & \longrightarrow & S\mathcal{G} & \longrightarrow & \mathcal{G} & \longrightarrow & \det \mathcal{G} & \longrightarrow & 1
\end{array}$$

and the bottom sequence maps injectively into the top one. The image of $S\mathcal{G}$ in

$$SW \times S\bar{U} = \prod SL_2(Z_\ell) \times \left(\prod So_\ell^*\right)/o^*$$

is of finite index, and similarly the image of $\mathcal{G}/S\mathcal{G}$ in $\prod Z_\ell^*$ is of finite index. Hence the image of $\mathcal{G}$ in the fiber product is of finite index, it is closed and hence open.

When dealing with these fiber products, it is convenient to get rid of the roots of unity, and to lift the Galois group as follows. We have a w to 1 covering

$$\prod GL_2(Z_\ell) \times_N \prod o_\ell^* \longrightarrow \prod GL_2(Z_\ell) \times_N \bar{U} .$$

We let $\tilde{\mathcal{G}}$ be the inverse image of $\mathcal{G}$ in this covering, so we have a commutative diagram:

$$\begin{array}{ccc} \tilde{\mathcal{G}} & \longrightarrow & \prod GL_2(Z_\ell) \times_N \prod o_\ell^* \\ \downarrow & & \downarrow \\ \mathcal{G} & \longrightarrow & \prod GL_2(Z_\ell) \times_N \left(\prod o_\ell^*\right)/o^* . \end{array}$$

Similarly, we can define $\tilde{\mathcal{Q}}$ and we can identify

$$\tilde{\mathcal{Q}} = U .$$

The rest of this section is devoted to two separate topics. First an analysis of the intersection $K \cap k_{ab}$. The reader may skip this, since it plays no role in the general determination of the desired constant. It is of course important to have for the determination in special cases.

Second, we discuss the reduction of the Galois group to finite levels. The reader should glance briefly at the definitions and then read into §4 immediately, referring to the formal development only if he needs it especially.

The intersection $K \cap k_{ab}$

This intersection will play a crucial role in the determination of the coincidence relations between Frobenius elements in G and Frobenius elements in k_{ab}.

We let $G_k = \text{Gal}(K/k)$. Then

$$[K \cap k_{ab} : Q_{ab}] = (G' : G'_k),$$

where the prime superscript indicates the closure of the commutator subgroup, cf. Part I. It is easy to see that the index on the right is finite, and it will follow from stronger lemmas to be proved below, concerning commutator subgroups of open subgroups of $GL_2(Z_\ell)$. By convention, when we speak of a commutator subgroup, we shall always mean the closure of the subgroup generated by commutators. We let $M_\ell = \text{Mat}_2(Z_\ell)$.

We begin with a lemma concerning the "generic" case, just to see what happens almost always.

Lemma 1. *Let q be a prime* ≥ 5. *Then*

$$GL_2(Z_q)' = SL_2(Z_q) = SL_2(Z_q)'.$$

Proof. Clearly, $GL_2(Z_q)'$ is a subgroup of $SL_2(Z_q)$, and is closed. For $q \geq 5$ it is standard finite group theory that

$$SL_2(F_q)' = SL_2(F_q).$$

Hence $SL_2(Z_q)'$ is a closed subgroup which reduces modulo q to $SL_2(F_q)$. The refinement lemma of Serre shows that $SL_2(Z_q)'$ must be equal to $SL_2(Z_q)$ and proves what we want.

If V is any open subgroup of $SL_2(Z_q)$ for any prime q, then V' is obviously open. This already makes it obvious that

$$(G' : G'_k)$$

is finite.

Theorem 3.1. *Let* q *be a prime* ≥ 5 *which divides* D. *Assume that*

$$G = GL_2(Z_q) \times G_L,$$

where L *is the complement of* $\{q\}$. *Then*

$$K \cap k_{ab} = Q_{ab}.$$

Proof. We have $G_q = GL_2(Z_q)$, and G_k is of index 2 in G. The group G_q has a unique subgroup of index 2, the subgroup E_q consisting of all elements σ such that

$$\det \sigma \in Z_q^{*2}.$$

Hence $E_q \times \{e_L\}$ is contained in G_k. Note that E_q contains $SL_2(Z_q)$. But

$$SL_2(Z_q)' = SL_2(Z_q)$$

by Lemma 1. Consequently $E_q' \supset SL_2(Z_q)$, and we find that

$$G_k' \supset SL_2(Z_q) \times \{e_L\}.$$

On the other hand, k cannot be contained in K_L because K_L is disjoint from K_q over the rationals, so that if $k \subset K_L$ then k would be contained in a cyclotomic field disjoint from the q^n-th roots of unity, whence would be unramified at q, which contradicts our hypothesis. If pr_L denotes the projection on the L-th factor, we obtain

$$pr_L G_k = G_L,$$

and therefore

$$pr_L G_k' = G_L',$$

Since $G_k' \supset SL_2(Z_q) \times \{e_L\}$, we conclude that

$$G_k' \supset SL_2(Z_q) \times G_L' = G',$$

thereby proving that $G_k' = G'$, and also proving the theorem.

We recall that

$$\mathcal{G} = \text{Gal}(Kk_{ab}/H) \qquad \text{and} \qquad \mathcal{Q} = \text{Gal}(k_{ab}/H).$$

We let
$$G \times_N \mathfrak{A}$$
be the set of pairs $(\sigma', \sigma'') \in G \times \mathfrak{A}$ such that σ' and σ'' have the same effect on Q_{ab}.

Theorem 3.2. *Assume that* $K \cap k_{ab} = Q_{ab}$. *Then*
$$\mathcal{G} = G \times_N \mathfrak{A}.$$

Proof. The situation is illustrated by the following diagram,

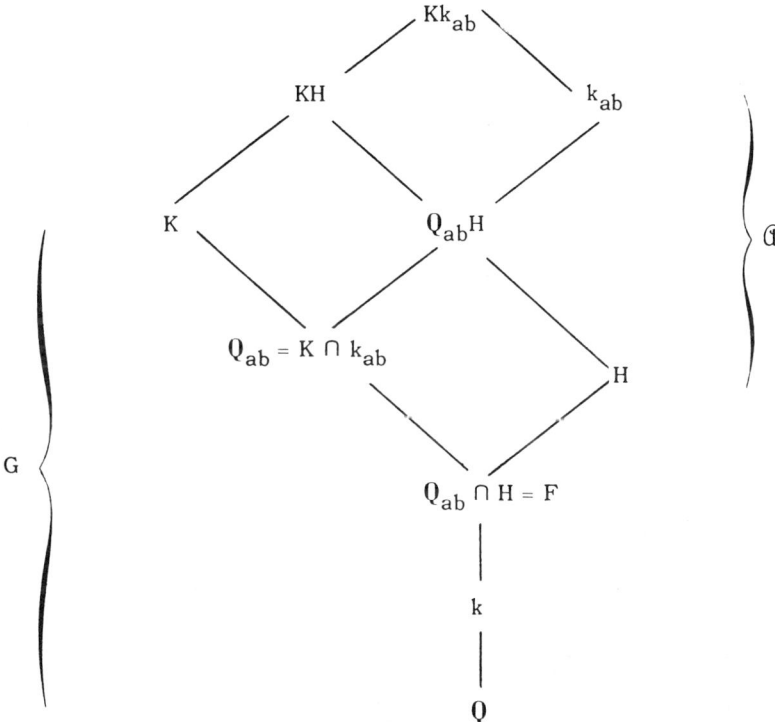

We let $F = Q_{ab} \cap H$. Given $\sigma' \in G$ and $\sigma'' \in \mathfrak{A}$ having the same effect on Q_{ab}, we have to show there exists $\sigma \in \mathcal{G}$ which restricts to σ' on K and σ'' on k_{ab}. Observe first that σ' is trivial on F. From the diagram, we can lift σ' to an element of Gal(KH/H), since H is disjoint from K over F. The element σ'' then coincides with σ' on $Q_{ab}H$. The two extensions KH and k_{ab} of

$Q_{ab}H$ are disjoint over $Q_{ab}H$. We can therefore find an automorphism σ of Kk_{ab} which gives σ'' on KH and σ' on k_{ab}, as desired.

In some applications, it is convenient to have a refinement of the above statement, taking into account only part of the Galois group. Let L be a set of primes. We may put L as an index in the preceding theorem in the following manner. We have the group G_L, which is the projection of G in the partial product

$$\prod_{\ell \in L} GL_2(Z_\ell),$$

and G_L is the Galois group of the field denoted by K_L. Similarly, we let

$$\mathcal{Q}_L = \left(\prod_{\ell \in L} o_\ell^*\right)/o^*,$$

so that $\mathcal{Q}_L$ is the Galois group of an extension of H which we denote by k_L^{ab}. We let

$$\mathcal{G}_L = \text{Gal}(K_L k_L^{ab}/H)$$

be the group of the mixed extension. We have the cyclotomic field Q_L^{ab} whose Galois group is

$$\prod_{\ell \in L} Z_\ell^*.$$

Theorem 3.2L. *Assume that* $K_L \cap k_L^{ab} = Q_L^{ab}$. *Then*

$$\mathcal{G}_L = G_L \times_N \mathcal{Q}_L.$$

Proof. The same as before, mutatis mutandis.

Under the representations ρ' and ρ'', the preceding theorem shows that

$$\tilde{\mathcal{G}} = \rho'(G) \times_N \prod_\ell o_\ell^*,$$

or omitting the ρ' for simplicity,

$$\tilde{\mathcal{G}} = G \times_N \prod_\ell o_\ell^*.$$

Example. Suppose that G is Serre's subgroup, cf. §10, below, or Part I,

$$G = S_{2q} \times \prod_{\ell \neq 2,q} GL_2(Z_\ell).$$

Let us define

$$\tilde{S}_{2q} = S_{2q} \times_N o_{2q}^*.$$

Theorem 3.3. *Assume that G is Serre's subgroup as above. Assume also that* $K \cap k_{ab} = Q_{ab}$. *Then*

$$\tilde{\mathcal{G}} = \tilde{S}_{2q} \times \prod_{\ell \neq 2,q} [GL_2(Z_\ell) \times_N o_\ell^*].$$

Proof. Special case of the previous theorem.

We shall now state the results to be proved in Part III concerning the intersection $K \cap k_{ab}$.

Theorem 3.4. *If* $D = -8, -24, -15, -20, -40, -55, -88$ *then* $K \cap k_{ab} = Q_{ab}$ *for all of our five curves*

$$A, B, C, D, X_0(11).$$

Proof. The case -8 is proved in Part III, Theorem 3.2, the case -24 in Theorem 3.3. The other cases are proved in Theorems 3.4, 3.5, 3.6.

Next we give a table of intersections, also giving the reference to the theorem in Part III which proves the assertion.

	$K \cap k_{ab}$
-3	$Q_{ab}(\Delta^{\frac{1}{3}})$, all curves except D, by 4.2
-4	$Q_{ab}(\Delta^{\frac{1}{4}})$, all curves, by 5.2
-43	$Q_{ab}(B_2, \Delta^{\frac{1}{4}})$ for curve B, by 5.1
-11	$Q_{ab}(X_0(11)_2, \Delta^{\frac{1}{4}})$ for $X_0(11)$, by 5.1
-3	$Q_{ab}(D_2, \Delta^{\frac{1}{3}}, \Delta^{\frac{1}{4}})$ for curve D, by 5.1

Reduction mod M

Let M be a positive integer. For each ℓ we let $W_{\ell,M}$ consist of those matrices $a \in GL_2(\mathbb{Z}_\ell)$ such that

$$a \equiv I \pmod{M}.$$

If $\ell \nmid M$ then this condition is empty. If $\ell \mid M$, then it means the usual congruence mod M $\text{Mat}_2(\mathbb{Z}_\ell)$.

Similarly, we let $U_{\ell,M}$ be the subgroup of elements u of $\mathfrak{o}_\ell^*$ such that $u \equiv 1 \pmod{M\mathfrak{o}_\ell}$.

We let

$$W_M = \prod_\ell W_{\ell,M} \quad \text{and} \quad U_M = \prod_\ell U_{\ell,M}.$$

If $M \geq 3$ then $U_M \cap \mathfrak{o}^* = \{1\}$. In this case we may view U_M as a subgroup of

$$\left(\prod \mathfrak{o}_\ell^*\right)/\mathfrak{o}^*.$$

Since $\rho(\mathcal{G})$ is open in the fiber product

$$\prod GL_2(\mathbb{Z}_\ell) \times_N \left(\prod \mathfrak{o}_\ell^*\right)/\mathfrak{o}^*,$$

there exists an integer M_0 having the following property.

If $M_0 \mid M$, then $\rho(\mathcal{G})$ contains the fiber product

$$W_M \times_N \overline{U}_M.$$

Such an integer M_0 will be said to stabilize ρ, and we let

$$\mathcal{G}_M = \rho^{-1}(W_M \times_N \overline{U}_M).$$

We often identify $\mathcal{G}_M$ with its matrix representation

$$\rho(\mathcal{G}_M) = W_M \times_N \overline{U}_M.$$

If we don't, then we write $\sigma = \sigma(a, u)$ to mean $(a, u) = \rho(\sigma)$. We let $\mathcal{G}(M) = \mathcal{G}/\mathcal{G}_M$.

Similarly, we let G_M be the subgroup of G corresponding to W_M, i.e. such that $\rho'(G_M) = W_M$, and $\mathcal{A}_M$ be the subgroup of $\mathcal{A}$ such that $\rho''(\mathcal{A}_M) = U_M$. We let

$$G(M) = G/G_M \quad \text{and} \quad \mathcal{A}(M) = \mathcal{A}/\mathcal{A}_M .$$

We let $K(M)$ and $k_{ab}(M)$ be the fixed fields of G_M and $\mathcal{A}_M$ respectively. We then get:

The fixed field of G_M is $K(M) k_{ab}(M)$.

We already had

$$\rho'_{(M)} : \mathrm{Gal}(K/\mathbb{Q}) = G \longrightarrow GL_2(M) = W(M) .$$

We therefore obtain

$$\rho_{(M)} : \mathcal{G} \longrightarrow W(M) \times_N \overline{U}(M) .$$

We let

$$\mathcal{K} = K k_{ab} \quad \text{and} \quad \mathcal{K}(M) = K(M) k_{ab}(M) ,$$

so that

$$\mathcal{G}(M) = \mathrm{Gal}(\mathcal{K}(M)/H) .$$

The lattice of fields and groups may be drawn as follows.

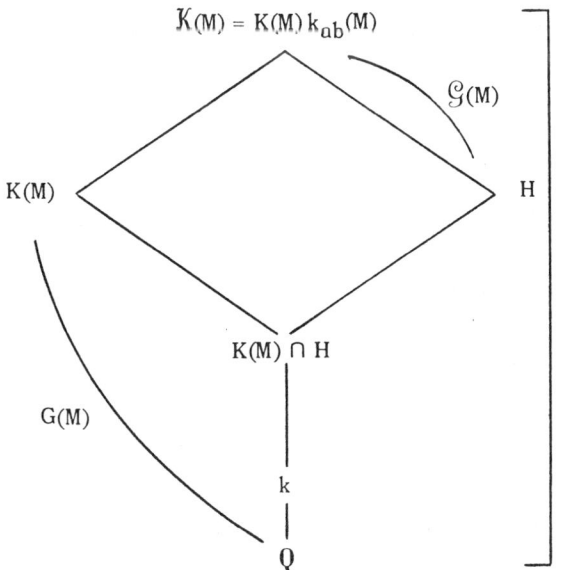

We let r_M denote reduction mod M. If M_0 stabilizes ρ, and $M_0|M$, then we have a commutative diagram:

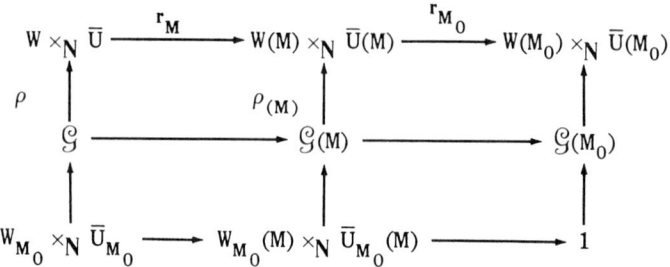

When we write the N as an index to the cross sign in a fiber product mod M, say in
$$W_{M_0}(M) \times_N \bar{U}_{M_0}(M) ,$$

the fibering is of course to be interpreted mod M. Elements of this fibered product are pairs $(\bar{a}, \bar{u})$ such that

$$\det \bar{a} \equiv N\bar{u} \pmod{M} .$$

Remark. *Given such a pair $(\bar{a}, \bar{u})$ with $\det \bar{a} = N\bar{u}$ mod M, there exists $(a, u) \in W_{M_0} \times_N \bar{U}_{M_0}$ such that*
$$r_M(a, u) = (\bar{a}, \bar{u}) .$$

This is obvious by solving a linear equation whose leading coefficient is a unit. In particular, the bottom row in our diagram is exact.

We can also express the stabilizing condition in a form analogous to that which we used previously for the simpler supersingular case.

If M_0 stabilizes ρ, and $M_0|M$, then $\mathcal{G}(M)$ is the inverse image of $\mathcal{G}(M_0)$ in
$$W(M) \times_N \bar{U}(M) ,$$

under the reduction map r_{M_0}. Similarly, $\mathcal{G}$ is the inverse image of $\mathcal{G}(M_0)$ in
$$W \times_N \bar{U} .$$

In the above statement, we identified $\mathcal{G}$ with its image in the corresponding matrix groups. We let $\widetilde{\mathcal{G}}_M$ be the projection on the factors with $\ell|M$ as usual.

We say that M splits $\rho = (\rho', \rho'')$ if $\tilde{\mathcal{G}}$ has an expression

$$\tilde{\mathcal{G}} = \tilde{\mathcal{G}}_M \times \prod_{\ell \nmid M} [\mathrm{GL}_2(\mathbb{Z}_\ell) \times_N \mathfrak{o}_\ell^*].$$

The above discussion shows that there always exists an integer M which splits and stabilizes ρ.

§4. The probabilistic model

We fix a positive integer Δ such that, for all M, if $p \nmid M\Delta$, then p is unramified in K(M) (that is, $\rho_{(M)}$ is unramified at p). If $p \nmid M\Delta D$, then p is unramified in $\tilde{K}(M) = K(M) k_{ab}(M)$. When ρ is the representation associated with an elliptic curve, then Δ is the discriminant.

Let M be a positive integer ≥ 3 and let t, s be integers. Let $\mathfrak{p}$ be a prime ideal above p in k. We define:

$$\tau_M(\mathfrak{p},t,s) = \begin{cases} 0 & \text{unless } p \nmid M\Delta D,\ \operatorname{tr}\rho'_{(M)}(\sigma'_p) \equiv t \bmod M, \\ & \text{p splits completely and its factors are principal in k.} \\ & \text{otherwise, number of generators } \pi \text{ of } \mathfrak{p} \\ & \text{such that } \operatorname{Tr} \pi \equiv s \bmod M. \end{cases}$$

We note that $\tau_M(\mathfrak{p},t,s) = 0$ unless $\operatorname{tr}\rho'_{(M)}(\sigma'_p) \equiv t \pmod M$, and also there is an element in $\rho''_{(M)}(\sigma''_p)$ whose trace is $\equiv s \pmod M$. The function τ_M counts the multiplicity of such elements in that situation. We define $F_M(t,s)$ by the condition:

$$\frac{w}{M^2} F_M(t,s) = \lim_{x \to \infty} \frac{1}{|P_k(x)|} \sum_{\mathfrak{p} \in P_k(x)} \tau_M(\mathfrak{p},t,s).$$

Thus, roughly speaking, the limit on the right is the density relative to P_k of those primes $\mathfrak{p} \in P_k$ such that $t \equiv \operatorname{tr}\rho'_{(M)}(\sigma'_p)$, and such that $s \equiv \operatorname{Tr}\rho''_{(M)}(\sigma''_p)$, except that this last congruence has to be explained accurately by means of the function $\tau_M(\mathfrak{p},t,s)$. Having defined F_M in this manner, we then have

(1) $$\sum_{(t,s) \bmod M} \frac{1}{M^2} F_M(t,s) = 1.$$

We remind the reader that $\mathfrak{G}(M) = \text{Gal}(K(M) k_{ab}(M)/\mathbb{Q})$ but

$$\mathcal{G}(M) = \text{Gal}(K(M) k_{ab}(M)/H).$$

Thus elements of $\mathcal{G}(M)$ leave H fixed, and if $\sigma \in \mathcal{G}(M)$ then $\sigma'' \in \mathfrak{A}(M)$. Cf. the diagram of §3.

We define a function $\lambda_M(\sigma, t, s)$ for any pair of integers t, s and $\sigma \in \text{Gal}(\tilde{K}(M)/\mathbb{Q}) = \mathfrak{G}(M)$ by:

$$\lambda_M(\sigma, t, s) = \begin{cases} 0 & \text{unless } \sigma \in \mathcal{G}(M) \text{ and } \text{tr}\, \rho'_{(M)}(\sigma') \equiv t \bmod M; \\ & \text{otherwise, number of elements in } \rho''_{(M)}(\sigma'') \\ & \text{whose trace is } \equiv s \bmod M. \end{cases}$$

We view $\rho''_{(M)}(\sigma'')$ as a coset $\pi \mathfrak{o}^*$ in $\mathfrak{o}(M)^*/\mathfrak{o}^*$. The definition is designed to transfer our counting problem to the Galois group, because if $p \nmid M\Delta D$, then

$$\lambda_M(\sigma_p, t, s) = \tau_M(p, t, s).$$

It is also useful to deal with a lifting of the Galois group which gets rid of $\mathfrak{o}^*$. We have a w to 1 covering

$$G(M) \times \mathfrak{o}(M)^* \longrightarrow G(M) \times \mathfrak{o}(M)^*/\mathfrak{o}^*$$

and we let $\tilde{\mathcal{G}}(M)$ be the inverse image of $\mathcal{G}(M)$ in this covering, so we have a commutative diagram:

$$\begin{array}{ccc} \tilde{\mathcal{G}}(M) & \longrightarrow & G(M) \times \mathfrak{o}(M)^* \\ \downarrow & & \downarrow \\ \mathcal{G}(M) & \longrightarrow & G(M) \times \mathfrak{o}(M)^*/\mathfrak{o}^* . \end{array}$$

We let $\tilde{\mathcal{G}}(M)_{s,t}$ = subset of elements σ in $\tilde{\mathcal{G}}(M)$ such that

$$\text{tr}\, \sigma' \equiv t \quad \text{and} \quad \text{Tr}\, \sigma'' \equiv s \bmod M.$$

Theorem 4.1. *If $M \geq 3$ then*

$$\frac{1}{M^2} F_M(t,s) = \frac{|\tilde{\mathcal{G}}(M)_{t,s}|}{|\tilde{\mathcal{G}}(M)|}.$$

397

Proof. The index of $\mathcal{G}(M)$ in $\mathfrak{S}(M)$ is precisely 2h, where h is the class number ($= (H:k)$). Thus from the definition, we obtain

$$\frac{w}{M^2} F_M(t,s) = \lim_{x \to \infty} \frac{2h}{|P(x)|} \sum_{p \in P(x)} \lambda_M(\sigma_p, t, s)$$

$$= \frac{2h}{|\mathfrak{S}(M)|} \sum_{\sigma \in \mathfrak{S}(M)} \lambda_M(\sigma, t, s) \quad \text{(by Tchebotarev)}$$

$$= \frac{1}{|\mathcal{G}(M)|} \sum_{\sigma \in \mathcal{G}(M)} \lambda_M(\sigma, t, s)$$

$$= w \frac{|\mathcal{G}(M)_{t,s}|}{|\mathcal{G}(M)|},$$

thereby proving the theorem.

We have fixed the interval J (corresponding to a sector of width π/w) and a positive integer M which splits ρ''. For each prime p, the fiber of the probabilistic model at p is now $Z \times Z$. The measure $\mu_{p,J}$ is represented with respect to counting measure by a function

$$f_{M,J}(t, s, p) \geq 0 ,$$

which we assume of the form

PR 3. $\quad f_{M,J}(t, s, p) = c_p g(\xi(t, p)) g''_J(\xi(s, p)) F_M(t, s) ,$

where c_p is a constant such that

(2) $$\sum_{(t,s)} f_{M,J}(t, s, p) = 1 .$$

We can now prove successively that $c_p \sim 1/4p$, and that the above assumption implies the Tchebotarev, Sato-Tate, and Hecke distribution properties. Thus again, we have picked a simple probabilistic model compatible with these properties. Note that **PR 3** amounts to a condition of independence for the behaviors at infinity (in the GL_2 extension and the imaginary quadratic field) from each other, and from the behavior under congruence conditions.

We give the proof that $c_p \sim 1/4p$ in full detail to show the reader how the somewhat more complicated system under which we now work, nevertheless parallels the previous one quite closely.

We start with the integral

$$1 = \mu_{g''}(J) = \int_{-1}^{1}\int_{-1}^{1} g(\xi')g''_J(\xi'')\,d\xi'\,d\xi''.$$

We have a product decomposition of the double sum,

$$\frac{1}{4p}\sum_{(t,s)} g(\xi(t,p))g''_J(\xi(s,p)) = \left(\frac{1}{2\sqrt{p}}\sum_t g(\xi(t,p))\right)\left(\frac{1}{2\sqrt{p}}\sum_s g''_J(\xi(s,p))\right).$$

The first factor is an approximating Riemann sum for the constant 1. The second factor is a lower sum, which approaches 1 as $p \to \infty$. Therefore the double sum on the left approaches 1 also. The double sum, summed only for pairs $(t,s) \equiv (t_0, s_0) \bmod M$ approaches

$$\frac{1}{M^2}.$$

Multiplying with $F_M(t_0, s_0)$ we obtain the limit

$$(3) \quad \lim_p \frac{1}{4p} \sum_{(t,s) \equiv (t_0,s_0) \bmod M} g(\xi(t,p))g''_J(\xi(s,p))F_M(t_0,s_0) = \frac{1}{M^2} F_M(t_0,s_0).$$

Summing over all congruence classes $(t_0, s_0) \bmod M$ yields the value 1. Comparing with the definition of $f_{M,J}(t,s,p)$ and the normalization of c_p concludes the proof that

$$c_p \sim 1/4p.$$

We have reached an analogous point to that reached in the supersingular case, where we can have lemmas corresponding to Lemma 1 and Lemma 2, to show compatibility of our probability model with Tchebotarev, Sato-Tate and Hecke for random sequences, using the law of large numbers. The routine is the same, and will be omitted, since it is not used in the sequel.

§5. The asymptotic behavior

We finally come to counting what we want, which amounts to computing the measure of the diagonal D in each fiber over each prime. We have

$$\mu_{p,J}(D) = \sum_t f_{M,J}(t,t,p)$$

$$= \sum_t c_p\, g(\xi(t,p))\, g''_J(\xi(t,p))\, F_M(t,t)$$

$$= \sum_{t_0 \bmod M} F_M(t_0, t_0) \sum_{t \equiv t_0} c_p\, g(\xi(t,p))\, g''_J(\xi(t,p))$$

The sum over $t \equiv t_0$ on the right is a Riemann sum, which is asymptotic to

$$\frac{1}{2\sqrt{p}} \frac{1}{M} \int_{-1}^{1} g(\xi)\, g''_J(\xi)\, d\xi\ .$$

We abbreviate

$$\boxed{\int_{-1}^{1} g(\xi)\, g''_J(\xi)\, d\xi = C_J^\infty(k,\rho)\ .}$$

It is the part at infinity for our constant, relative to the interval J. We then find

$$\mu_{p,J}(D) \sim C_J^\infty(k,\rho) \sum_{t_0 \bmod M} \frac{1}{M} F_M(t_0, t_0) \cdot \frac{1}{2\sqrt{p}}\ .$$

It follows from §7, Lemma 1 and Theorem 9.1 that the limit over suitable M exists:

$$\boxed{\lim_M \sum_{t_0 \bmod M} \frac{1}{M} F_M(t_0, t_0) = C^{\text{fin}}(k,\rho)\ ,}$$

defining the finite part of the desired constant. We then define

$$C_J(k,\rho) = \frac{1}{2h} C_J^\infty(k,\rho) C^{fin}(k,\rho).$$

We have reintroduced a factor $1/2h$ because originally, we determined F_M by the density of primes relative to P_k, but we want at the very end to have the density relative to the whole set of primes. We then define the total constant, summing over w disjoint intervals J covering $[-1, 1]$:

$$C(k,\rho) = \sum_J C_J(k,\rho) = \frac{1}{2h} C^\infty(k,\rho) C^{fin}(k,\rho),$$

where

$$C^\infty(k,\rho) = \int_{-1}^{1} \frac{w g(\xi)}{\pi \sqrt{1-\xi^2}} \, d\xi.$$

The conjecture is that

$$N_{k,\rho}(x) \sim C(k,\rho) \pi_{\frac{1}{2}}(x).$$

Remark. When the Sato-Tate function is given by $\phi(\theta) = \frac{2}{\pi} \sin^2\theta$, or in other words

$$g(\xi) = \frac{2}{\pi} \sqrt{1-\xi^2},$$

then we get a rather convenient unexpected cancellation, and the infinity part of the constant is given by

$$C_J^\infty(k,\rho) = \frac{2w}{\pi^2} \text{ (length of } J).$$

Thus

$$C^\infty(k,\rho) = 4w/\pi^2$$

and we find

$$C(k,\rho) = \frac{1}{2h} \frac{4w}{\pi^2} C^{fin}(k,\rho).$$

It will be proved in Theorem 6.3 that the finite part of the constant has a product decomposition

$$C^{fin}(k,\rho) = C_M \prod_{\ell \nmid M} C_\ell$$

for a suitable integer M. Furthermore, the non-special factors C_ℓ will be computed and tabulated in §9, showing that if

$$L(1,\chi)_\ell = \left(1 - \left(\frac{k}{\ell}\right)\frac{1}{\ell}\right)^{-1}$$

is the ℓ-component of the product expression for the non-trivial series of k at 1, then C_ℓ differs from this ℓ-component by a factor $O(1 + 1/\ell^2)$. Consequently it is useful to make a further transformation of the expression for the constant, and to give a name to these factors, which give rise to an absolutely convergent product.

For $\ell \nmid M$ we let

$$C'_\ell = \begin{cases} 1 & \text{if } \left(\frac{k}{\ell}\right) = 1 \\ 1 + \dfrac{2r^2}{(1-r^2)(1+r)} & \text{if } \left(\frac{k}{\ell}\right) = -1 \\ \dfrac{1}{1-r^2} & \text{if } \left(\frac{k}{\ell}\right) = 0 \,. \end{cases}$$

Recall that $r = 1/\ell$. Then Theorem 9.1 will show that

$$C_\ell = L(1,\chi)_\ell C'_\ell \,.$$

We define C'_M by the relation

$$C_M = C'_M \prod_{\ell \mid M} L(1,\chi)_\ell \,.$$

Then

$$C^{\text{fin}}(k,\rho) = L(1,\chi) C'_M \prod_{\ell \nmid M} C'_\ell \,.$$

For the quadratic field k we have the formula

$$L(1,\chi) = \frac{2\pi h}{w\sqrt{|D|}} \,.$$

Putting all this together yields:

Theorem 5.1. *We have the product decomposition*

$$C(k,\rho) = \frac{4}{\pi\sqrt{|D|}} C'_M \prod_{\ell \nmid M} C'_\ell \,.$$

As mentioned previously, the infinite product in this expression is absolutely convergent. The value $1/\sqrt{|D|}$ is of course known. Hence the computation of the constant is easily reduced to the exceptional factor C'_M, which depends on the part of the Galois group which cannot be easily split or decomposed, and has to be studied in special cases separately, depending on their idiosyncrasies. This factor C'_M is a quotient of C_M by a finite product of local L-series terms $L(1,\chi)_\ell$ for $\ell | M$. We shall see that C_M is determined by a singular measure on the M-component of a Galois group. Cf. Theorem 6.3.

§6. The finite part of the constant as a quotient of integrals

The finite part of the constant is expressed as a limit over M. We describe how it can be interpreted as an integral, first at finite level.

Let $\mathbf{M}$ denote Mat_2, so that $\mathbf{M}(M) = \mathrm{Mat}_2(M) = \mathrm{Mat}_2(\mathbf{Z}/M\mathbf{Z})$. We let

$$T' = (\mathrm{tr}, \det)$$

be the map sending a matrix to its trace and determinant. At finite level, we should write

$$T'_{(M)} : \mathbf{M}(M) \longrightarrow \mathbf{Z}(M)^2 ,$$

but we omit the subscript (M) and write simply T when M is fixed. Similarly, we have the trace-norm map

$$T'' = (\mathrm{Tr}, N)$$

on elements of $\mathfrak{o}$, whence

$$T''_{(M)} : \mathfrak{o}(M) \longrightarrow \mathbf{Z}(M)^2 .$$

Haar measure on any compact group, and especially on a finite group, will be assumed to be normalized to give the group measure 1, unless otherwise specified. Here we have in mind $\mathbf{M}(M)$, $\mathbf{Z}(M)$, and $\mathfrak{o}(M)$.

Similarly, we have the M-adic corresponding notions, namely:

$$\mathbf{M}_\ell = \mathrm{Mat}_2(\mathbf{Z}_\ell), \qquad \mathbf{M}_M = \mathrm{Mat}_2(\mathbf{Z}_M) = \prod_{\ell \mid M} \mathbf{M}_\ell$$

$$T' = T'_M : \mathbf{M}_M \longrightarrow \mathbf{Z}_M^2$$

on the GL_2 side, and

$$T'' = T''_M : \mathfrak{o}_M \longrightarrow \mathbf{Z}_M^2$$

on the quadratic field side, where

$$\mathfrak{o}_M = \prod_{\ell \mid M} \mathfrak{o}_\ell .$$

Let S be an open subset of M_M. Then $S(M) \subset M(M)$ is the reduction of S mod M.

We let μ denote Haar measure, with a subscript to indicate the corresponding group. If S is an open subset of the group, we let μ_S denote the restriction of Haar measure to S, and 0 on the complement of S. It will be proved in the next sections that the map T' has a continuous density function with respect to Haar measures. It is essentially clear from the computations of §2 that the similar map T'' on the quadratic field side also has a continuous density function. These two functions are denoted by h' and h'' respectively, although we sometimes write h instead of h'. By definition, we have the expression for the direct image of Haar measure M_M and o_M respectively; with $X \subset M_M$ and $Y \subset o_M$:

$$T'_* d\mu_X = h'_X d\mu_{Z^2_M} \quad \text{and} \quad T''_* d\mu_Y = h''_Y d\mu_{Z^2_M} .$$

We shall deal with subsets S of the fibered product

$$S \subset M_M \times_N o_M .$$

Such a subset is said to **decompose at level** M if it is a finite disjoint union of fibered products

$$S = \bigcup S_i = \bigcup X_i \times_N Y_i , \qquad S_i = X_i \times_N Y_i ,$$

where X_i is open in M_M, Y_i is open in o_M, and X_i, Y_i are stable at level M (in other words, are the inverse images of their reductions mod M).

For such a subset S, we define

$S(M)_{t,t}$ = set of elements $(g, a) \in S(M)$ such that tr g = Tr $a \equiv t$.

The equality on the right is to be viewed in $Z(M)$, i.e. as a congruence mod M. We define

$$C_{(M)}(S) = M \sum_{t \bmod M} \frac{|S(M)_{t,t}|}{|S(M)|} .$$

It is convenient to normalize the numerator and denominator of this expression by multiplying with M^5. Thus we let

$$M^5 \text{Num}_{(M)}(S) = M \, \# \, \{(g,a) \in S(M), \text{ tr } g = \text{Tr } a\} .$$

405

The denominator is just

$$M^5 \text{Den}_{(M)}(S) = |S(M)|.$$

We shall write the numerator and denominator as integrals, using the density functions $h'_{X(M)}$ and $h''_{Y(M)}$, thereby giving an expression for $C_{(M)}(S)$ as a quotient of integrals,

$$C_{(M)}(S) = \text{Num}_{(M)}(S)/\text{Den}_{(M)}(S).$$

Theorem 6.1. *We have:*

$$\text{Num}_{(M)}(S) = \sum_i \iint_{Z(M)^2} h'_{X_i(M)}(t,u) \, h''_{Y_i(M)}(t,u) \, du \, dt$$

$$\text{Den}_{(M)}(S) = \sum_i \iiint_{Z(M)^3} h'_{X_i(M)}(t',u) \, h''_{Y_i(M)}(t'',u) \, du \, dt' \, dt''$$

Proof. Without loss of generality, we may assume that

$$S = X \times_N Y.$$

For simplicity of notation, we write S, X, Y instead of S(M), X(M) and Y(M) respectively, so we work entirely under reduction mod M. Then the numerator is given by

$$M^5 \text{Num}_{(M)}(S) = M \,\#\{(g,a) \in X \times Y, \text{ tr } g = \text{tr } a, \det g = Na\}$$

$$= M \sum_{(t,u)} \#\{(g,a) \in X \times Y, \text{ tr } g = \text{Tr } a = t, \det g = Na = u\}$$

$$= M \sum_{(t,u)} \#\{(g \in X, \text{ tr } g = t, \det g = u\} \, \#\{a \in Y, \text{ Tr } a = t, Na = u\}$$

$$= M \sum_{(t,u)} \frac{M^4}{M^2} h'_X(t,u) \, \frac{M^2}{M^2} h''_Y(t,u)$$

$$= M M^2 \frac{M^4}{M^2} \frac{M^2}{M^2} \iint_{Z(M)^2} h'_X(t,u) \, h''_Y(t,u) \, d\mu_{Z(M)^2}(t,u)$$

$$= M^5 \iint_{Z(M)^2} h'_X(t,u) \, h''_Y(t,u) \, du \, dt$$

406

thereby proving the desired expression for the numerator. As for the denominator, we have

$$|S(M)| = \#\{(g,a) \in X \times Y, \det g = Na\}$$

$$= \sum_{(t',t'',u)} \#\{g \in X, \det g = u, \operatorname{tr} g = t'\} \#\{a \in Y, Na = u, \operatorname{Tr} a = t''\}$$

$$= \sum_{(t,t,u)} \frac{M^4}{M^2} h'_X(t', u) \frac{M^2}{M^2} h''_Y(t'', u)$$

$$= M^5 \iiint_{Z(M)^3} h'_X(t', u) h''_Y(t'', u) \, du \, dt' \, dt''$$

as was to be shown.

We may then pass to the limit. We define

$$C_M(S) = \lim_{n \to \infty} C_{(M^n)}(S).$$

It is also convenient to use the abbreviations

$$\operatorname{Num}_M(X \times_N Y) = \operatorname{Num}_M(X, Y) = \iint_{Z_M^2} h'_X(t, u) h''_Y(t, u) \, dt \, du$$

$$\operatorname{Den}_M(X \times_N Y) = \operatorname{Den}_M(X, Y) = \iiint_{Z_M^3} h'_X(t', u) h''_Y(t'', u) \, du \, dt' \, dt''.$$

Suppose that S is decomposed as a disjoint union of fibered products

$$S = \bigcup_i X_i \times_N Y_i.$$

Then we define

$$\operatorname{Num}_M(S) = \sum_i \operatorname{Num}_M(X_i, Y_i) \quad \text{and} \quad \operatorname{Den}_M(S) = \sum_i \operatorname{Den}_M(X_i, Y_i).$$

At the end of the next section, we shall prove:

Theorem 6.2. *Assume that the subset* S *of* $M_M \times_N o_M$ *is stable and decomposed at level* M, *as a disjoint union*

$$S = \bigcup X_i \times_N Y_i .$$

Then

$$C_M(S) = \text{Num}_M(S)/\text{Den}_M(S) .$$

In the applications, the set S will be the lifting $\tilde{\mathcal{G}}$ of the Galois group $\mathcal{G}$ in the product

$$\prod [GL_2(Z_\ell) \times_N o_\ell^*] .$$

We know from §3 that there always exists some integer $M \geq 3$ which splits and stabilizes $\rho = (\rho', \rho'')$. It is therefore also convenient to introduce the notation

$$\text{Num}_M = \text{Num}_M(GL_2(Z_M), o_M^*) ,$$

giving what we call the **generic numerator**, and similarly for the denominator.

In Theorem 4.1, we had found the expression

$$\frac{1}{M} F_M(t, t) = M \frac{|\tilde{\mathcal{G}}(M)_{t,t}|}{|\tilde{\mathcal{G}}(M)|} .$$

The sum of this expression over $t \bmod M$ is the M-th approximation to the finite part of the constant $c^{\text{fin}}(k,\rho)$, and is precisely equal to $C_{(M)}(S)$, where $S = \tilde{\mathcal{G}}$. In the next sections we shall discuss the (obvious) multiplicativity, and specific values of the integrals giving the limit value of the constant ℓ-adically. In the light of these results, we can then state:

Theorem 6.3. *Let* $M \geq 3$ *split and stabilize* ρ, *and assume that* $\tilde{\mathcal{G}}$ *is decomposed at level* M. *Then the finite part of the constant* $c^{\text{fin}}(k,\rho)$ *is equal to a product*

$$c^{\text{fin}}(k,\rho) = C_M(\tilde{\mathcal{G}}) \prod_{\ell \nmid M} C_\ell(\tilde{\mathcal{G}}_\ell) ,$$

where for $\ell \nmid M$,

$$\tilde{\mathcal{G}}_\ell = GL_2(Z_\ell) \times_N o_\ell^* ,$$

and $C_\ell = \text{Num}_\ell/\text{Den}_\ell$.

We observe that for almost all ℓ the set G_ℓ consists of only one piece, which is the fiber product of the full groups of invertible elements on both sides.

Furthermore, the computation of the denominators need only be done in the "generic" case, because all cases can be reduced to this one without further computation, by means of the next theorem.

Theorem 6.4. *Let*

$$\text{Den}_M = \text{Den}_M(GL_2(Z_M) \times_N o_M^*).$$

Let e *be the index of* $\tilde{\mathcal{G}}_M$ *in* $GL_2(Z_M) \times_N o_M^*$. *Then*

$$\text{Den}_M(\tilde{\mathcal{G}}_M) = e^{-1} \text{Den}_M.$$

Proof. This is actually obvious by taking the limit from finite levels, without integration, namely

$$\text{Den}_M(\tilde{\mathcal{G}}_M) = \lim_n [M^{-5n}|\tilde{\mathcal{G}}(M^n)|].$$

The index has a simple expression in most cases.

Theorem 6.5. *If* $\tilde{\mathcal{G}}_M = G_M \times_N \tilde{\mathcal{Q}}_M$, *then*

$$(GL_2(Z_M) \times_N \tilde{\mathcal{Q}}_M : \tilde{\mathcal{G}}_M) = (GL_2(Z_M) : G_M).$$

Proof. This is an immediate consequence of the standard isomorphism theorem for groups, and the fact that G_M, $GL_2(Z_M)$ have the same image under the norm map N.

The numerator will have to be computed piece by piece. In any given piece

$$X \times_N Y$$

the integral for the numerator can be written unsymmetrically by using the definition of the direct image of Haar measure,

$$\iint_{Z_M^2} h'_X(t,u) h''_Y(t,u)\, dt\, du = \int_Y h'_X(\text{Tr } z, Nz)\, dz,$$

$$= \int_X h''_Y(\text{tr } \sigma, \det \sigma)\, d\sigma,$$

where dz is Haar measure on o_M and $d\sigma$ is Haar measure on M_M. We call this the **seesaw principle**. *For computations, only the first integral, taken over* Y, *will play a role.* We then write h instead of h' to simplify the notation. From the positive values of h'_X near the identity, it will be obvious that the numerator is not equal to 0, and hence that the constant is not 0. Since the integral for the numerator has an obvious independent interest from the context in which it occurs here, we suggest the general notation

$$\boxed{H(X, Y) = \int_Y h_X(\text{Tr } z, Nz) \, dz \ .}$$

Fix a prime ℓ, and consider the case $M = \ell$. The map

$$T' : \text{Mat}_2(Q_\ell) \longrightarrow Q_\ell^2$$

is such that two non-scalar matrices have the same image under this map if and only if they are conjugate. Thus T' parametrizes conjugacy classes over $GL_2(Q_\ell)$. We define the **rational Cartan subset** $C(k_\ell)$ determined by k to be the set of semisimple matrices $\sigma \in GL_2(Q_\ell)$ such that

$$(\text{tr } \sigma, \det \sigma) = (\text{Tr } a, Na)$$

for some $a \in k_\ell^*$. If we omitted the condition of semisimplicity, we would obtain a larger set, containing certain unipotent elements, but which differs from the Cartan set by a set of measure 0.

The Cartan subset defined above is the rational one, and is invariant under conjugation by $GL_2(Q_\ell)$. In our problem, we deal with its integral points, namely the set

$$C = C_\ell = C(k_\ell) \cap GL_2(Z_\ell) \, ,$$

which will be called the **Cartan subset** of $GL_2(Z_\ell)$ determined by k, or simply the Cartan subset.

In case

$$\tilde{G}_\ell = GL_2(Z_\ell) \times_N o_\ell^*$$

we can write the numerator in the form

$$\boxed{\text{Num}_\ell(\tilde{G}_\ell) = \int_{C_\ell} h''(\text{tr } \sigma, \det \sigma) \, d\sigma \ .}$$

Following a suggestion of Langlands, we show that the function h'_M is the Harish transform with respect to the Cartan subgroup of the characteristic function of the set $\mathcal{C}_\ell$. As this is not used in the sequel, and is included only for the convenience of the reader who wants to connect with the literature on representation theory, we do not bother to normalize Haar measure on coset spaces carefully, and the following equalities between integrals are meant to hold only up to such normalizing constant factors.

For the rest of this section, we change notation to conform a little more to the formalism of Lie groups. So we use G to denote $GL_2(Q_\ell)$. We let B be the Cartan subgroup of G corresponding to k. The Harish transform is defined to be

$$H^B \psi(b) = |D(b)|^{\frac{1}{2}} \int_{B \backslash G} \psi(g^{-1} bg) \, d\dot{g} ,$$

where $d\dot{g}$ is Haar measure on $B \backslash G$, normalized so that

$$dg = d\dot{g} \, db .$$

An integral formula computing differentials shows that for any function ψ,

$$\int_{BG} \psi(\sigma) \, d\sigma = \int_B H^B \psi(b) |D(b)|^{\frac{1}{2}} \, db .$$

Warning: We use $D(b)$ to denote the discriminant of b, which is the square of the difference of eigenvalues, to fit the notation of §8.

If f is a function invariant under conjugation, then

$$H^B(f\psi) = f H^B(\psi) .$$

Replace ψ by $f\psi$, and then let ψ be the characteristic function of $\mathcal{C}_\ell$. Then we obtain

$$\int_{\mathcal{C}} f(\sigma) \, d\sigma = \int_B f(b) H^B \psi(b) |D(b)|^{\frac{1}{2}} \, db .$$

In §8, following Theorem 8.1, we shall see that

$$|D(z)|^{\frac{1}{2}} \, dz = dt \, ds .$$

Hence we obtain

$$\int_{\mathcal{C}} f_*(\operatorname{tr} \sigma, \det \sigma) d\sigma = \iint f_*(t,s)(H^B\psi)_*(t,s) dt ds .$$

The lower star indicates the same function with respect to the change of variables. By definition of the direct image of Haar measure, the left hand side is also equal to

$$\iint_{T'(\mathcal{C})} f_*(t,s) h'(t,s) dt ds .$$

This proves that

$$H^B \psi = h' ,$$

as we wanted.

The Harish transform of certain functions on SL_2 in the ℓ-adic case has been computed before, see Sally and Shalika [Sa-Sh 2], and also [Sa-Sh 1], [Sa-Sh 3]. In a sense, the next two sections possibly perform equivalent computations, but it would not have helped to refer to the literature at this point. Furthermore, we need more complete and systematic values for the density function h'_X, with small sets X, than would in any case be available, and we need them in a form which makes the connection with the quadratic field involved easily apparent.

The Harish transform in the higher dimensional case is not yet completely cleared up, cf. Harish-Chandra [H-C].

COMPUTATIONS OF HARISH TRANSFORMS

§7. Haar measure under the trace-determinant map on Mat_2. General formalism.

Let M denote Mat_2. We abbreviate

$$M_\ell = \text{Mat}_2(Z_\ell) \qquad \text{and} \qquad M_M = \prod_{\ell \mid M} M_\ell.$$

This section deals only with matrices, so we let

$$T = (\text{tr}, \det)$$

be the map sending a matrix to its trace and determinant. Then

$$T : M_M \longrightarrow Z_M^2.$$

Let R be an open subset of M_M. Then $R(M) \subset M(M)$ is the reduction of R mod M.

Given $(t,s) \in Z_M^2$ we let $R_{t,s}$ = subset of elements $\sigma \in R$ such that

$$\text{tr } \sigma = t \quad \text{and} \quad \det \sigma = s.$$

Haar measure on any compact group will be assumed to be normalized to give the group measure 1, unless otherwise specified. Let μ be Haar measure on M_M. We let μ_R be the restriction of μ to R, and 0 outside R. We shall see that the direct image $T_*\mu$ is represented by a continuous function. We define: h_R = density function of $T_*\mu_R$ with respect to Haar measure on Z_M^2. (We write h_R instead of h'_R in this section.)

Lemma 1. *Suppose that* $M = M_1 M_2$ *where* $(M_1, M_2) = 1$, *and that* $R = R_1 \times R_2$ *is a direct product of open sets in* M_{M_1} *and* M_{M_2} *respectively. Then*

$$h_R = h_{R_1} \otimes h_{R_2}.$$

Proof. Obvious.

Scalar matrices will play a special role in determining the density function. For any scalar a, we let

$$\psi_a : (t,s) \longmapsto (t-2a, a^2 - at + s).$$

This mapping ψ_a describes how the trace and determinant change under translation by a scalar matrix, and has ψ_{-a} for its inverse. We note that ψ_a preserves additive Haar measure on Z_ℓ^2.

Lemma 2. (i) *Let* $\tau \in \prod_{\ell | M} GL_2(Z_\ell)$. *Then*

$$h_{\tau R \tau^{-1}} = h_R.$$

(ii) *Let* $a \in \prod_{\ell | M} Z_\ell$. *Then*

$$h_{aI+R}(t,s) = h_R(t - 2a, a^2 - at + s) = h_R(\psi_a(t,s)).$$

Proof. Again obvious.

The prime power case

We now assume that $M = M_\ell$, where ℓ is a prime. We shall compute h_R locally, whence globally. We shall see that h depends on how close an element is to being a scalar matrix. Thus it is natural to consider a filtration of matrices according to their distance from scalar matrices.

We first compute h_R when R is the inverse image under reduction mod ℓ^n of some non-scalar matrix. By translation and dilation we can get h_R when R is the set of matrices congruent to a scalar modulo a high power of ℓ. The values are given in Lemmas 5 and 7.

In the next section, we then combine these computations with the quadratic field correspondence, i.e. give the values for

$$h_R(\text{Tr } z, Nz)$$

when $z \in \mathfrak{o}_\ell^*$.

All the way through, the number $1/\ell$ occurs to various powers, so it is convenient to use a special notation for it, and we put

$$r = r(\ell) = 1/\ell.$$

We denote reduction mod ℓ by a bar:

$$\sigma \longmapsto \bar{\sigma}, \qquad t \longmapsto \bar{t}.$$

Theorem 7.1. *Let $\sigma \in M$ be such that $\bar{\sigma}$ is not scalar. Let $R = \sigma + \ell^n M$, for some integer $n \geq 1$. Then*

$$h_R(t,s) = \begin{cases} r^{2n} & \text{if } (t,s) \in T(R), \text{ or equivalently,} \\ & \text{if } (t,s) \equiv T(\sigma) \bmod \ell^n M \\ 0 & \text{otherwise}. \end{cases}$$

Proof. After a conjugation, using Lemma 2, we may assume that

$$\bar{\sigma} = \begin{pmatrix} \bar{a} & \bar{b} \\ \bar{c} & \bar{d} \end{pmatrix} \qquad \text{with} \qquad \bar{b} \neq 0.$$

We select

$$\sigma_0 \in R, \qquad \sigma_0 = \begin{pmatrix} a & b \\ c & d \end{pmatrix}$$

where b is a unit. Let $(t_0, s_0) = T(\sigma_0)$. Define the coordinate maps

$$f : Z_\ell^4 \longrightarrow R \qquad \text{by} \qquad f(w,x,y,z) = \sigma_0 + \ell^n \begin{pmatrix} w & x \\ y & z \end{pmatrix}$$

and

$$g : Z_\ell^2 \longrightarrow Z_\ell^2 \qquad \text{by} \qquad g(u,v) = (t_0 + \ell^n u, s_0 + \ell^n v).$$

Then

$$T = gT_0 f^{-1} \qquad \text{where} \qquad T_0 : Z_\ell^4 \longrightarrow Z_\ell^2$$

is given by

$$T_0(w,x,y,z) = (w+z, az+wd-cx-by+\ell^n(wz-xy))$$
$$= (w+z, wd-cx+(a+\ell^n w)z-(b+\ell^n x)y).$$

We have the commutative diagram:

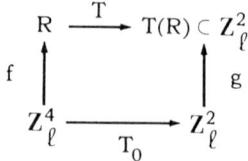

Define
$$\phi : Z_\ell^4 \longrightarrow Z_\ell^4$$
by
$$\phi(w, x, y, z) = (w, x, w + z, az + wd - cx - by + \ell^n(wz - xy))$$
$$= (w, x, T_0(w, x, y, z)) .$$

Because $b + \ell^n x$ is a unit, we can write a formula for ϕ^{-1} by solving for z and y, so ϕ is a bijection. Its Jacobian satisfies

$$|\mathrm{Jac}_\phi| = |b + \ell^n x| = 1 ,$$

where the absolute value is ℓ-adic. Consequently ϕ is measure preserving. Then

$$T_0 = \mathrm{pr}_{3,4} \circ \phi .$$

But f, g have constant Jacobian determinant, which is an obvious power of ℓ, and yields the value asserted in the theorem.

Let S_0 be the set of matrices $\sigma \in M$ such that $\bar{\sigma}$ is not scalar. Then S_0 is a union of sets $\sigma_i + \ell M$, of type considered in the previous theorem with $n = 1$. It is standard linear algebra that two non-scalar matrices over a field have the same characteristic polynomial if and only if they are conjugate by an element of GL_2 over that field. Over the prime field, one can count these matrices. Let $n(t, s)$ be the number of matrices σ in $M(\ell)$ such that

$$T(\bar{\sigma}) = (\bar{t}, \bar{s}) .$$

Then $n(t, s)$ is the index of the centralizer of $\bar{\sigma}$ in $GL_2(F_\ell)$. The centralizer is easily determined according to three types for σ (diagonal, unipotent, non-split Cartan), and $n(t, s)$ depends on the behavior of the polynomial

$$X^2 - \bar{t}X + \bar{s} \quad\quad \text{over} \quad\quad F_\ell .$$

One finds:

Lemma 3.
$$n(t,s) = \begin{cases} \ell^2 + \ell & \text{if two roots in } F_\ell \\ \ell^2 - 1 & \text{if one root in } F_\ell \\ \ell^2 - \ell & \text{if zero roots in } F_\ell. \end{cases}$$

Theorem 7.2.
$$h_{S_0}(t,s) = \begin{cases} 1+r & \text{if } X^2 - \bar{t}X + \bar{s} \text{ has two roots in } F_\ell \\ 1-r^2 & \text{if one root} \\ 1-r & \text{if zero roots}. \end{cases}$$

Proof. We sum the constant value found in Theorem 7.1 over all the non-scalar matrices of $M(\ell)$, and see that the desired answer is the product of ℓ^{-2} and $n(t,s)$, found above.

The function h_{S_0} depends only on the discriminant

$$\Delta(t,s) = t^2 - 4s,$$

and in fact on the discriminant reduced mod ℓ. We shall write

$$\boxed{h_{S_0}(t,s) = h_0(\Delta(t,s)).}$$

We observe that this discriminant is invariant under "translations," precisely,

$$\Delta(t,s) = \Delta(\psi_a(t,s)).$$

We can then give the density function for the translation of certain sets by scalar matrices by using this discriminant.

The next lemma shows what happens under dilation.

Lemma 4. Let R be an open subset of M. Then

$$h_{\ell^j R}(t,s) = \begin{cases} \ell^{-j} h_R(\ell^{-j}t, \ell^{-2j}s) & \text{if } \ell^j | t \text{ and } \ell^{2j} | s \\ 0 & \text{otherwise}. \end{cases}$$

Proof. We can factor T on R as in the following diagram:

$$\begin{array}{ccc} R & \xrightarrow{\ell^j} & \ell^j R \\ T \downarrow & & \downarrow T \\ Z_\ell \times Z_\ell & \xleftarrow{} & \ell^j Z_\ell \times \ell^{2j} Z_\ell \end{array}$$

where the bottom map is (ℓ^{-j}, ℓ^{-2j}). The absolute value of the Jacobian of the top map is ℓ^{-4j}, and the absolute value of the Jacobian of the bottom map is ℓ^{3j}, so the lemma is clear.

Writing the expression of Lemma 4 in terms of the discriminant, and making a translation by a scalar matrix aI, we obtain:

Lemma 5. *Let* $X = aI + \ell^j S_0$, *for some integer* $j \geq 0$. *Then*

$$h_X(t,s) = \begin{cases} \ell^{-j} h_0(\ell^{-2j} \Delta(t,s)) & \text{if } (\ell^j, \ell^{2j}) \text{ divides } \psi_a(t,s) \\ 0 & \text{otherwise}. \end{cases}$$

The value h_X found in Lemma 5 is the pivotal value, which gives the density function locally.

We define S_j = set of matrices $\sigma \in M$ such that

$$\sigma \equiv \text{scalar matrix} \mod \ell^j \text{ but not } \mod \ell^{j+1}.$$

Then we can write S_j as a disjoint union

$$S_j = \bigcup_{a \bmod \ell^j} aI + \ell^j S_0.$$

Lemma 6. *If* $(t,s) \in T(S_j)$, *then there exists a unique* $a \mod \ell^j$ *such that*

$$(t,s) \in T(aI + \ell^j S_0).$$

Proof. The assertion is obvious if ℓ is odd, and in this case we need only the trace to characterize a. In fact, we get

$$a = t/2.$$

In general, suppose we have $\sigma = aI + \ell^j \sigma_0$ and $\sigma' = a'I + \ell^j \sigma'_0$ such that

$$(t, s) = T(\sigma) = T(\sigma').$$

Then
$$\sigma - aI \equiv 0 \bmod \ell^j \qquad \text{and} \qquad \sigma' - aI = (a' - a)I + \ell^j \sigma'_0.$$

Then
$$\begin{aligned}
\det(\sigma - aI) &= \det \sigma - a \operatorname{tr} \sigma + a^2 \\
&= \det(\sigma' - aI) \\
&\equiv (a' - a)^2 + (a' - a)\ell^j \operatorname{tr} \sigma'_0 \pmod{\ell^{2j}} \\
&\equiv 0 \pmod{\ell^{2j}}
\end{aligned}$$

because $\sigma - aI \equiv 0 \bmod \ell^j$. Therefore

$$(a' - a)^2 \equiv (a' - a)\ell^j t'_0 \pmod{\ell^{2j}}$$

where $t'_0 = \operatorname{tr} \sigma'_0$. In any case we have $a' - a \equiv 0 \bmod \ell^{j-1}$. If this congruence cannot be improved to ℓ^j then either t'_0 is divisible by ℓ and we conclude that $(a' - a)^2 \equiv 0 \pmod{\ell^{2j}}$, whence the desired congruence between a' and a, or t'_0 is not divisible by ℓ, and the right hand side of the congruence cannot be a square mod ℓ^{2j}, whence a contradiction which proves the lemma.

From Lemma 6 and Lemma 5 we conclude at once:

Lemma 7.
$$h_{S_j}(t, s) = \begin{cases} \ell^{-j} h_0(\ell^{-2j} \Delta(t, s)) & \text{if } (t, s) \in T(S_j) \\ 0 & \text{otherwise.} \end{cases}$$

Of course there is no uniqueness of the index j such that $(t, s) \in T(S_j)$, and we have:

Lemma 8. If $(t, s) \in T(S_m)$ and $j < m$ then $(t, s) \in T(S_j)$.

Proof. Say $(t, s) = T(\sigma)$ where $\sigma = aI + \ell^m \sigma_0$, and write

$$\ell^{m-j} \sigma_0 = \begin{pmatrix} a & b \\ c & d \end{pmatrix}.$$

If $b \neq 0$, and $b = \ell^i u$ where u is a unit, replace $\ell^{m-j}\sigma_0$ by

$$\begin{pmatrix} a & u \\ \ell^i c & d \end{pmatrix}.$$

If $b = 0$, replace c by 1. This proves the lemma.

We note that in Lemma 7, the values for h_{S_j} approach 0 as $j \to \infty$. It is therefore a reasonable convention to let

$$S_\infty = \text{set of scalar matrices}.$$

Since S_∞ has measure 0, it follows that

$$h_{S_\infty}(t,s) = 0.$$

Observe that M is the disjoint union

$$M = \bigcup_{j=0}^{\infty} S_j,$$

and so we get

$$h_M = \sum_{j=0}^{\infty} h_{S_j}.$$

By Lemma 7 and Theorem 7.2 we know that $h_{S_j} \leq 2r^j$, and therefore the series converges uniformly, defining h_M as a continuous function. It is also clear that h_M is locally constant at any point (t,s) such that $\Delta(t,s) \neq 0$, because in this case, (t,s) lies in $T(S_j)$ for only a finite number of j. On the other hand, if $\Delta(t,s) = 0$, it is easy to see that $(t,s) \in T(S_j)$ for all j, so h_M may not be locally constant at (t,s).

The preceding computations give us the values of h_X for the calculations we want to make in specific cases. They are also sufficient to prove further lemmas concerning this density function, describing how it behaves, and necessary to prove the limit theorem 6.2. However, the reader who wishes to omit the proof of Theorem 6.2 may omit the rest of this section without impeding the understanding of the rest of the paper.

Let us say that a subset of M is elementary if it is of the form

$$\sigma + \ell^n M$$

for some integer $n \geq 1$. If R is such an elementary set, and does not contain a scalar matrix, then σ is not a scalar. We may then write σ in the form

$$\sigma = aI + \ell^j \sigma_0 + \ell^n M,$$

where σ_0 is not scalar mod ℓ, and it follows that $j < n$.

On the other hand, if R contains a scalar, then R is of the form

$$R = aI + \ell^n M.$$

If X is a subset of M, we denote by $X(\ell^n)$ its reduction mod ℓ^n. We recall that X is called stable of level m if it is the inverse image of its reduction mod ℓ^m. (As we deal with the prime power case, we use the exponent to index the stability level.) We say that X is stable if it is stable at some level. A set is stable if and only if it is a finite union of elementary sets.

Theorem 7.3. *If X is a stable set in M then h_X is continuous, and locally constant at any point (t,s) which does not lie in $T(X \cap S_\infty)$. In particular, if X contains no scalar matrix, then h_X is locally constant.*

Proof. We already know the theorem when $X = M$. An elementary set is obtained from M by translations and dilations, to which we can apply Theorem 7.1 and Lemmas 2, 4, which show the theorem to be true for elementary sets. Since a stable set is a finite union of elementary sets, the theorem follows at once.

Theorem 7.4. *For any stable set X,*

$$\lim_{n \to \infty} h_{X(\ell^n)}(t,s) = h_X(t,s).$$

Proof. Let X be stable at level m and let $n \geq m$. Let

$$E_n = (t,s) + \ell^n Z_\ell^2.$$

Then

$$\ell^{-2n} h_{X(\ell^n)}(t,s) = \int_{E_n(\ell^n)} h_{X(\ell^n)}(t',s') dt' ds'$$

$$= \mu_n(X(\ell^n) \cap T_n^{-1}(E_n(\ell^n))),$$

where μ_n is Haar measure on $M(\ell^n)$, and T_n is now indexed by n for clarity, being the trace-determinant map on $M(\ell^n)$. Since X is assumed stable at level n, it follows that

$$X \cap T^{-1}(E_n)$$

is also stable at level n, and hence our last expression is equal to

$$\mu(X \cap T^{-1}(E_n)) = \int_{E_n} h_X(t', s')\, dt'\, ds',$$

where μ is Haar measure on M. Thus we have shown that

$$h_{X(\ell^n)}(t, s) = \frac{1}{\mu_n(E_n)} \int_{E_n} h_X(t', s')\, dt'\, ds'.$$

Taking the limit as $n \to \infty$ we see that the right hand side approaches $h_X(t, s)$, as was to be shown.

Theorem 7.5. *Let* Y *be a stable set in* $\mathfrak{o}_\ell$, *and* X *stable in* M. *Then*

$$\lim_{n \to \infty} \int_{Y(\ell^n)} h_{X(\ell^n)}(\text{Tr } z, Nz)\, dz = \int_Y h_X(\text{Tr } z, Nz)\, dz.$$

Proof. The expression

$$h_{X(\ell^n)}(\text{Tr } z, Nz)$$

defines a function on $\mathfrak{o}(\ell^n)$. Suppose that Y is stable at level n. Then

$$\int_{Y(\ell^n)} h_{X(\ell^n)}(\text{Tr } z, Nz)\, dz = \int_Y h_{X(\ell^n)}(\text{Tr } z, Nz)\, dz.$$

Theorem 7.4 and the bounded convergence theorem conclude the proof.

This is the result which we needed to prove Theorem 6.2, expressing the basic constant as a limit from finite levels. Note that the handling of the denominator is trivial, by using Theorem 6.4, and the fact that the generic Galois group is stable at level 1.

§8. Relations with the trace-norm map on k

We have $\mathfrak{o}_\ell = \mathbf{Z}_\ell \otimes \mathfrak{o}$. We write $\mathfrak{o}_\ell = [1, \eta]$ to mean that 1 and η form a basis for $\mathfrak{o}_\ell$ over $\mathbf{Z}_\ell$. We can find such a basis with $\eta \in \mathfrak{o}$. We let

$$T'' = (\mathrm{Tr}, N)$$

be the trace-norm map from k to $\mathbf{Q}$. Then T'' extends uniquely to $\mathfrak{o}_\ell$, giving rise to a map still denoted by T'', namely

$$T'' : \mathfrak{o}_\ell \longrightarrow \mathbf{Z}_\ell^2 .$$

An element $z \in \mathfrak{o}_\ell$ is then a root of its characteristic equation

$$X^2 - (\mathrm{Tr}\, z) X + Nz = 0 .$$

The automorphism of k extends by continuity to k_ℓ, and induces an automorphism of $\mathfrak{o}_\ell$. We can define the discriminant

$$D(z) = D_\ell(z) = (z - \bar{z})^2 .$$

Then

$$D(z) = \Delta(\mathrm{Tr}\, z, Nz) = (\mathrm{Tr}\, z)^2 - 4Nz .$$

Let $L_z = [1, z]$ be the sublattice (over $\mathbf{Z}_\ell$) generated by 1 and z. Then its discriminant is well defined modulo the square of a unit in $\mathbf{Z}_\ell$, and we have

$$\mathrm{Discr}(L_z) = D(z)(\mathrm{unit})^2 .$$

There is an integer $j \geq 0$ such that

$$D(z) = \ell^{2j} (\mathrm{unit})^2 D ,$$

where the unit is of course an ℓ-adic unit, i.e. an element of $\mathbf{Z}_\ell^*$, and $D = D(\mathfrak{o})$ is the discriminant of k. We define

$$v(z) = j$$

to be this integer.

Lemma 1. Let $o_\ell = [1, \eta]$ and $z = x + y\eta$, where $x, y \in Z_\ell$. The following conditions are equivalent:

(i) $v(z) = j$.
(ii) $y = \ell^j u$ for some unit $u \in Z_\ell^*$.
(iii) $z \equiv a \pmod{\ell^j o_\ell}$ for some $a \in Z_\ell$, but z is not congruent to any element of Z_ℓ mod ℓ^{j+1}.

Proof. This is a simple exercise, which will be left to the reader.

In classical terminology, we could call ℓ^j the **conductor** of z, or the conductor of L_z. Then $v(z)$ is the order of the conductor.

Observe that if we coordinatize z by (x, y) such that

$$z = x + y\eta, \qquad x, y \in Z_\ell$$

and put $r = 1/\ell$, then

$$|y| = r^{v(z)}.$$

This (ℓ-adic) absolute value is independent of the choice of basis $o_\ell = [1, \eta]$. With such coordinates, we have

$$\boxed{dz = dx\, dy\, .}$$

Lemma 2. Let $z \in o_\ell$. Then

$$(\text{Tr } z, Nz) \in T'(S_j) \quad \text{if and only if} \quad j \leqq v(z).$$

Proof. Let

$$(\text{Tr } z, Nz) = (\text{tr } \sigma, \det \sigma)$$

with some matrix $\sigma = aI + \ell^j \sigma_0$, and σ_0 not scalar mod ℓ. Let

$$w = \ell^{-j}(z - a).$$

Then $(\text{Tr } w, Nw) = (\text{tr } \sigma_0, \det \sigma_0)$. Hence $w \in o_\ell$, and

$$z = a + \ell^j w,$$

so that $v(z) \geq j$ as desired.

Conversely, let $v = v(z)$. We can write

$$z = a + \ell^v w, \qquad \text{with} \qquad a \in Z_\ell, \; w \in o_\ell$$

and $w \in Z_\ell$. Under the regular representation on $o_\ell = [1, \eta]$, we obtain representing matrices such that

$$\sigma_z = aI + \ell^v \sigma_w,$$

which show that $(\mathrm{Tr}\, z, Nz) \in T'(S_m)$ for $m \geq v$. The first part of the proof shows that $m = v$. If $j \leq v$ we can use Lemma 8 of the preceding section to conclude the proof.

From Lemma 7 of the preceding section, we can write h_{S_j} in the form

$$h_{S_j}(\mathrm{Tr}\, z, Nz) = \begin{cases} \ell^{-j} h_0(\ell^{-2j} D(z)) & \text{if } j \leq v(z) \\ 0 & \text{otherwise.} \end{cases}$$

Hence we get h_{S_j} in a form convenient for computations:

Lemma 3. Let $r = r(\ell) = 1/\ell$.

$$h_{S_j}(\mathrm{Tr}\, z, Nz) = \begin{cases} r^j(1 - r^2) & \text{if } j < v(z) \\ 0 & \text{if } j > v(z) \end{cases}$$

and for $j = v(z) = v$:

$$h_{S_v}(\mathrm{Tr}\, z, Nz) = \begin{cases} r^v(1+r) & \text{if } \left(\frac{k}{\ell}\right) = 1 \\ r^v(1-r^2) & \text{if } \left(\frac{k}{\ell}\right) = 0 \\ r^v(1-r) & \text{if } \left(\frac{k}{\ell}\right) = -1. \end{cases}$$

Taking into account Lemma 8 of the preceding section, and summing a geometric series gives the value of h_M for the full set of matrices.

Let $M = M_\ell$ and $v = v(z)$. Then

$$h_M(\mathrm{Tr}\, z, Nz) = (1+r)(1-r^v) + h_{S_v}(\mathrm{Tr}\, z, Nz).$$

It is sometimes convenient to combine the terms involving $r^{v(z)}$ when we integrate. Thus we define the function

$$\psi_\ell(k) = \begin{cases} 0 & \text{if } \left(\frac{k}{\ell}\right) = 1 \\ r(1+r) & \text{if } \left(\frac{k}{\ell}\right) = 0 \\ 2r & \text{if } \left(\frac{k}{\ell}\right) = -1 \,. \end{cases}$$

and get:

Theorem 8.1.

$$\boxed{h_M(\text{Tr } z, Nz) = 1 + r - \psi_\ell(k)\, r^{v(z)} \,.}$$

Remark. Although we do not need it for the sequel, it may be convenient for the reader to see also the formula for h'', which reads:

$$h''(\text{Tr } z, Nz) = 2|D|^{-1}\, r^{-v(z)} \,.$$

The proof is easy. The Jacobian determinant of T'' is immediately computed to be

$$4yN\eta - y(\text{Tr } \eta)^2 = -yD \,.$$

Its absolute value is therefore $|D|\,|y|$. Since T'' is two-to-one, the formula follows at once. As $v(z) \to \infty$, note that

$$h(\text{Tr } z, Nz) = h'(\text{Tr } z, Nz)$$

remains bounded, and converges, while $h''(\text{Tr } z, Nz)$ tends to infinity. The way we choose our coordinates, and write the integral for the numerator $\text{Num}_\ell(X, Y)$, the density functions will always appear in their bounded form, i.e. $h'(\text{Tr } z, Nz)\, dz$.

We have the corresponding lemmas giving the behavior under dilation and translation, easier to handle when dealing with the quadratic field than with matrices.

Lemma 4. *Let R be an open set in M. Then*

$$h_{\ell^j R}(\text{Tr } z, Nz) = \begin{cases} \ell^{-j} h_R(\text{Tr}(z/\ell^j), N(z/\ell^j)) & \text{if } z \in \ell^j o_\ell \\ 0 & \text{otherwise} \,. \end{cases}$$

If Y is open in o_ℓ, then

$$\int_Y h_{\ell^j R}(\mathrm{Tr}\, z, Nz)\, dz = r^{3j} \int_{\ell^{-j} Y \cap o_\ell} h_R(\mathrm{Tr}\, z, Nz)\, dz.$$

Proof. The formula for $h_{\ell^j R}$ is obvious from Lemma 4 of the preceding section. We then make a change of variables by a dilation, which gives the value for the integral.

Lemma 5. *Let R be an open set in M. Let $a \in Z_\ell$. Then*

$$h_{aI+R}(\mathrm{Tr}\, z, Nz) = h_R(\mathrm{Tr}(z-a), N(z-a)).$$

If Y is open in o_ℓ, then

$$\int_Y h_{aI+R}(\mathrm{Tr}\, z, Nz)\, dz = \int_{Y-a} h_R(\mathrm{Tr}\, z, Nz)\, dz.$$

Proof. The formula for h_{aI+R} is obvious from Lemma 2 of the preceding section. We then make a translation of variable to get the other form for the integral.

Using the two lemmas, one can always reduce an integral of the type we shall consider to Theorem 7.1 or Theorem 8.1.

Indeed, the sets X in M which we consider are finite unions of sets of type

$$X = aI + \ell^j \sigma + \ell^n M, \quad \text{or} \quad X = aI + \ell^n M,$$

where σ is not scalar mod ℓ, and $j < n$. Using Lemmas 4 and 5, we see that

$$\int_Y h_X(\mathrm{Tr}\, z, Nz)\, dz = \int_{Y'} h_{X'}(\mathrm{Tr}\, z, Nz)\, dz \quad (\text{times } r^{3j} \text{ or } r^{3n}),$$

where Y' is a suitable transform of Y, and X' has the form

$$X' = \sigma + \ell^{n-j} M \quad \text{or} \quad X' = M.$$

We carry out a simple case as an example.

Lemma 6. *Let* $a \in Z_\ell$. *Then*

$$\int_{o_\ell} h_{aI+\ell^n M}(\text{Tr } z, Nz)\, dz = r^{3n}\left[(1+r) - \psi_\ell(k)\,\frac{1}{1+r}\right].$$

Proof.

(1) $$\int_{o_\ell} h_{aI+\ell^n M}(\text{Tr } z, Nz)\, dz = \int_{o_\ell} h_{\ell^n M}(\text{Tr } z, Nz)\, dz$$

(2) $$= \int_{\ell^n o_\ell} h_{\ell^n M}(\text{Tr } z, Nz)\, dz$$

(3) $$= r^{3n} \int_{o_\ell} h_M(\text{Tr } z, Nz)\, dz$$

(4) $$= r^{3n}\left[(1+r) - \psi_\ell(k)\,\frac{1}{1+r}\right]$$

as shown in the next lemma.

Lemma 7. *We have*

$$\int_{o_\ell} r^{v(z)}\, dz = \frac{1}{1+r}$$

and

$$\int_{o_\ell} h_M(\text{Tr } z, Nz)\, dz = (1+r) - \psi_\ell(k)\,\frac{1}{1+r}.$$

Proof. The first integral is equal to

$$\iint |y|\, dx\, dy$$

430

taken over $Z_\ell \times Z_\ell$. This makes it obvious. The second is obtained by substituting in Theorem 8.1.

In the next theorems we give explicitly some integrals which are used to compute the constants.

Theorem 8.2. *We have the following integral table.* ($r = 1/\ell$.)

	$\int_{o_\ell^*} dz = \mu(o_\ell^*)$	$\int_{o_\ell^*} r^{v(z)} dz$
$\left(\frac{k}{\ell}\right) = 1$	$(1-r)^2$	$\dfrac{(1-r)(1-r-r^2)}{1+r}$
$\left(\frac{k}{\ell}\right) = 0$	$1-r$	$\dfrac{1-r}{1+r}$
$\left(\frac{k}{\ell}\right) = -1$	$1-r^2$	$\dfrac{1-r^3}{1+r}$

Proof. We have

$$\mu(o_\ell^*) = \frac{|o(\ell)^*|}{|o(\ell)|},$$

which is computed explicitly and easily in the three cases. This gives the first column. For the second column, the integral over the units is expressed as the difference of the integral over o_ℓ and the integral over the non-units. One must characterize the units in each case.

If $\left(\frac{k}{\ell}\right) = 1$, then the non-units are of the form

$$\ell Z_\ell \times Z_\ell \quad \cup \quad Z_\ell^* \times \ell Z_\ell,$$

and the union is disjoint. The integral is then trivially computed.

If $\left(\frac{k}{\ell}\right) = -1$, then the non-units are the elements such that ℓ divides both x and y, which again makes the integral obvious.

Suppose that $\left(\frac{k}{\ell}\right) = 0$, so ℓ ramifies in k. We can pick an ℓ-adic basis $[1,\lambda]$ for o_ℓ over Z_ℓ such that λ is not a unit, e.g. a local parameter at a prime above ℓ. Then the units $x + y\lambda$ with $x, y \in Z_\ell$ are precisely those elements for which x is a unit. From this the table entry for the ramified case follows at once.

The table of Theorem 8.2 suffices for situations of general type. It is inadequate for the computations of Part III, where we study specific instances needing a slight generalization. The reader is advised to skip the next theorem until he needs it.

Theorem 8.3. *Let* $a \in o_\ell$. *Then*

$$\int_{a + \ell o} r^{v(z)} dz = \begin{cases} \dfrac{r^3}{1+r} & \text{if } \bar{a} \in Z(\ell) \\ r^2 = \mu(a + \ell o) & \text{otherwise .} \end{cases}$$

Proof. Note that $x + y\eta$ lies in $Z(\ell)$ if and only if y is not a unit. In this case, the desired integral is equal to

$$\iint_{\substack{|y| < 1 \\ x \equiv a}} |y|\, dx\, dy .$$

The single integral over $x \equiv a$ splits off to give the value r. For the integral over y, we write $y = \ell y'$, $dy = rdy'$, and the desired value drops out. On the other hand, if y is a unit, then we have to compute $\mu(a + \ell o)$, which is trivial.

§9. Computation of C_ℓ for almost all ℓ

We have seen that for almost all ℓ, the ℓ-adic part of the desired constant is given by a ratio of integrals

$$C_\ell = \text{Num}_\ell/\text{Den}_\ell ,$$

where the numerator is

$$\text{Num}_\ell = \int_{o_\ell^*} h'_M(\text{Tr } z, Nz)\, dz$$

and the denominator need not be handled by means of its integral expression, but from the elementary definition

$$\text{Den}_\ell = \lim_{n \to \infty} \ell^{-5n} |\text{GL}_2(\ell^n) \times_N o(\ell^n)^*| .$$

Theorem 9.1. *Let* $r = r(\ell) = 1/\ell$. *We have:*

$$\text{Den}_\ell = (1-r^2)\mu(o_\ell^*) .$$

The constant C_ℓ *and numerator* Num_ℓ *are given in the following table.*

	C_ℓ	Num_ℓ	Den_ℓ
$\left(\frac{k}{\ell}\right) = 1$	$\dfrac{1}{1-r}$	$(1+r)(1-r)^2$	$(1+r)(1-r)^3$
$\left(\frac{k}{\ell}\right) = 0$	$\dfrac{1}{1-r^2}$	$1-r$	$(1+r)(1-r)^2$
$\left(\frac{k}{\ell}\right) = -1$	$\dfrac{1}{1+r}\left[1 + \dfrac{2r^2}{(1+r)(1-r^2)}\right]$	$\dfrac{1-r}{1+r}[1+r+r^2-r^3]$	$(1+r)^2(1-r)^2$

Proof. We start with the denominator, and in this case, there is no need for the integral expression, the matter is totally elementary. We see immediately that the expression for the denominator is stable at level 1, i.e. independent of n for $n \geq 1$. Consequently

$$\text{Den}_\ell = \ell^{-5} |GL_2(Z(\ell)) \times_N o(\ell)^*|.$$

From the exact sequence

$$0 \longrightarrow GL_2(Z(\ell)) \times_N o(\ell)^* \longrightarrow GL_2(Z(\ell)) \times o(\ell)^* \longrightarrow Z(\ell)^* \longrightarrow 0$$

we get

$$\text{Den}_\ell = \frac{|GL_2(Z(\ell)) \times o(\ell)^*|}{(\ell-1)\ell^5}.$$

But

$$\mu(o_\ell^*) = \frac{|o(\ell)^*|}{|o(\ell)|} \quad \text{and} \quad |GL_2(Z(\ell))| = \ell(\ell-1)(\ell^2-1).$$

The desired value for the denominator drops out.

Next we deal with the numerator. We have

$$\text{Num}_\ell = \int_{o_\ell^*} h_M(\text{Tr } z, Nz) dz.$$

The appropriate expression for h_M was given in Theorem 8.1, from which we see that the integral for the numerator is a sum of integrals which have been computed in Theorem 8.2. This gives the asserted value for the numerator.

§10. The constant for Serre curves, $K \cap k_{ab} = Q_{ab}$

In Part I we had already discussed the special case of Serre curves. They have a special Galois group described as follows. Let q be an odd prime. We let:

$$E_2 = \{\sigma \in GL_2(\mathbb{Z}_2), \ \bar{\sigma} \text{ is an even permutation in } GL_2(2)\}$$

$$E_q = \{\sigma \in GL_2(\mathbb{Z}_q), \ \det \bar{\sigma} \text{ is a square in } F_q^*\}.$$

The notation $\bar{\sigma}$ denotes the reduction of σ mod 2 and mod q respectively. We note that E_2 and E_q are of index 2 in their respective GL_2, and we let O_2 and O_q be their respective cosets, which we call the odd cosets, as distinguished from E_2 and E_q which are called the even cosets.

We let
$$S_{2q} = (E_2 \times E_q) \cup (O_2 \times O_q).$$

Serre's subgroup is by definition

$$G = S_{2q} \times \prod_{\ell \neq 2, q} GL_2(\mathbb{Z}_\ell).$$

In this section, we want to compute the value $C_{2q}(\mathcal{S})$ as given in Theorem 6.2, where $\mathcal{S}$ is the set

$$\mathcal{S}_{2q} = \mathcal{S} = S_{2q} \times_N o_{2q}^*,$$

and $o_{2q}^* = o_2^* \times o_q^*$. The relevance of this computation for the applications was shown in Theorem 3.3.

The constant is a quotient,

$$C_{2q}(\mathcal{S}) = \text{Num}_{2q}(\mathcal{S})/\text{Den}_{2q}(\mathcal{S}).$$

The denominator is easily taken care of, by reduction to the "generic" case, handled in the preceding section.

Theorem 10.1. $\text{Den}_{2q}(\mathcal{S}) = \frac{1}{2} \text{Den}_{2q}(GL_2(\mathbb{Z}_{2q}) \times_N o_{2q}^*)$
$$= \frac{1}{2} \text{Den}_2 \cdot \text{Den}_q,$$

where Den_2 and Den_q are the denominators computed in Theorem 9.1.

Proof. The subgroup $\mathcal{S}$ is of index 2 in the full fiber product

$$GL_2(Z_{2q}) \times_N o_{2q}^* .$$

We can then apply Theorem 6.4.

Next we deal with the numerator,

$$\text{Num}_{2q}(\mathcal{S}) = \int h_{S_{2q}}(\text{Tr } z, Nz) dz ,$$

and the integral is taken over o_{2q}^2. Since S_{2q} is a disjoint union of

$$E_{2q} = E_2 \times E_q \quad \text{and} \quad O_{2q} = O_2 \times O_q ,$$

we have

$$h_{S_{2q}} = h_{E_{2q}} + h_{O_{2q}} ,$$

and the integral splits as a sum of two integrals. Each of these is a product, because the variables separate, for instance

$$h_{E_2 \times E_q}(\text{Tr } z, Nz) = h_{E_2}(\text{Tr } z_2, Nz_2) h_{E_q}(\text{Tr } z_q, Nz_q) .$$

This yields:

(1) $\quad \text{Num}_{2q}(\mathcal{S}) = \text{Num}_2(E_2 \times_N o_2^*) \text{Num}_q(E_q \times_N o_q^*)$
$\quad\quad\quad\quad\quad\quad\quad + \text{Num}_2(O_2 \times_N o_2^*) \text{Num}_q(O_q \times_N o_q^*) .$

We use abbreviations,

$$\text{Num}_\ell^+ = \text{Num}_\ell(E_\ell \times_N o_\ell^*) = \int_{o_\ell^*} h_{E_\ell}(\text{Tr } z, Nz) dz$$

$$\text{Num}_\ell^- = \text{Num}_\ell(O_\ell \times_N o_\ell^*) = \int_{o_\ell^*} h_{O_\ell}(\text{Tr } z, Nz) dz .$$

We also have the generic numerator,

$$\text{Num}_\ell = \text{Num}_\ell^+ + \text{Num}_\ell^- = \text{Num}_\ell(GL_2(Z_\ell) \times_N o_\ell^*) ,$$

which was computed in §9 for any prime ℓ.

With this notation, we have:

Theorem 10.2. $\text{Num}_{2q}(\mathcal{S}) = \text{Num}_2^+ \text{Num}_q^+ + \text{Num}_2^- \text{Num}_q^-$.

Furthermore, we can write

$$\text{Num}_\ell^+ = \text{Num}_\ell - \text{Num}_\ell^-,$$

so that $\text{Num}_{2q}(\mathcal{S})$ can be computed in terms of Num_ℓ, which we already know, and Num_ℓ^- (with $\ell = 2, q$) which we shall determine and tabulate. In the table, $o_{2,\text{even}}^*$ denotes the units in o_2 with even trace.

	$\mu(o_{2,\text{even}}^*)$	Num_2^-	Num_2^+
$\left(\frac{k}{2}\right) = 1$	1/4	3/16	3/16
$\left(\frac{k}{2}\right) = 0$	$\frac{1}{2}$	3/8	1/8
$\left(\frac{k}{2}\right) = -1$	1/4	3/16	17/48

The table for q follows. As usual, $r = r(q) = 1/q$.

	Num_q^-	Num_q^+
$\left(\frac{k}{q}\right) = 1$	$\frac{1}{2}(1-r)^2(1+r)$	$\frac{1}{2}(1-r)^2(1+r)$
$\left(\frac{k}{q}\right) = 0$	0	$1-r$
$\left(\frac{k}{q}\right) = -1$	$\frac{1}{2}(1-r)^2(1+r)$	$\frac{1}{2}\frac{1-r}{1+r}[1+r+3r^2-r^3]$

Computations at 2

The odd elements in $GL_2(F_2)$ are represented by the matrices

$$\begin{pmatrix} 0 & 1 \\ 1 & 0 \end{pmatrix}, \quad \begin{pmatrix} 1 & 1 \\ 0 & 1 \end{pmatrix}, \quad \begin{pmatrix} 1 & 0 \\ 1 & 1 \end{pmatrix}.$$

We see that $O_2(2)$ consists of the non-scalar matrices with even trace. Hence

$$h_{O_2}(t,u) = \begin{cases} 0 & \text{if } t \text{ is odd or } u \text{ is not a unit} \\ h_{S_0}(t,u) & \text{if } t \text{ is even and } u \text{ is a unit}. \end{cases}$$

Therefore by Theorem 7.2 we find with $r = r(2) = 1/2$:

$$h_{O_2}(t,u) = \begin{cases} 0 & \text{if } t \text{ is odd or } u \text{ is not a unit} \\ 1-r^2 & \text{if } t \text{ is even and } u \text{ is a unit}. \end{cases}$$

Hence

(2) $$\text{Num}_2^- = \int_{O_2} h_{O_2}(\text{Tr } z, Nz)\,dz = (1-r^2)\mu(o_{2,\text{even}}^*).$$

The values of $\mu(o_{2,\text{even}}^*)$ depend on $\left(\frac{k}{2}\right)$.

If 2 splits completely in k, then

$$o_2^* \approx Z_2^* \times Z_2^*$$

and every unit has even trace. Hence the measure of the units is $1/4$. The rest of the top line in the table is then obtained from the value for Num_2^- in (2), and the generic numerator of Theorem 9.1.

If 2 ramifies, every unit has even trace, and the units form the single additive coset of the non-units. Hence the measure of the units is $1/2$. The other table entries follow trivially.

We leave the third case to the reader.

Computations at q

By the definition of the odd elements at q, we find

$$h_{o_q}(t,u) = \begin{cases} h_{s_0}(t,u) & \text{if } \left(\frac{u}{q}\right) = -1 \\ 0 & \text{otherwise .} \end{cases}$$

If q ramifies in k, then the norm of every unit is a square, and hence we get 0 in the corresponding table entry.

The other two cases will be seen to give the same value. If $\left(\frac{u}{q}\right) = -1$ then

$$X^2 - \bar{t}X + \bar{u} = 0$$

has either zero or two roots in F_q. We have

$$\Delta(\text{Tr } z, Nz) = y^2 D \qquad \text{if} \qquad z = x + y\eta .$$

If Nz is a non-square unit then $\Delta(\text{Tr } z, Nz)$ is a unit because

$$(\text{Tr } z)^2 - 4Nz \not\equiv 0 \pmod{q} .$$

Hence y is a unit, and therefore

$$\left(\frac{\Delta}{q}\right) = \left(\frac{D}{q}\right) = \left(\frac{k}{q}\right) .$$

From Theorem 7.2 we therefore obtain the value:

Lemma 1.
$$h_{o_q}(\text{Tr } z, Nz) = \begin{cases} 0 & \text{unless } Nz \text{ is a non-square unit} \\ & \text{and otherwise:} \\ 1 + r & \text{if } \left(\frac{k}{q}\right) = 1 \\ 1 - r & \text{if } \left(\frac{k}{q}\right) = -1 . \end{cases}$$

Let U_q^- be the subset of o_q^* consisting of those elements whose norm is not a square in Z_q^*. We find:

(3) $$\text{Num}_q^- = \iint_{o_q} h_{o_q}(\text{Tr } z, Nz) dz = \begin{cases} (1+r)\mu(U_q^-) & \text{if } \left(\frac{k}{q}\right) = 1 \\ (1-r)\mu(U_q^-) & \text{if } \left(\frac{k}{q}\right) = -1 . \end{cases}$$

The norm map

$$N : o_q^* \longrightarrow Z_q^*$$

is surjective. Hence

(4) $$\mu(U_q^-) = \frac{1}{2}\mu(o_q^*),$$

where

(5) $$\mu(o_q^*) = \begin{cases} (1-r)^2 & \text{if } \left(\frac{k}{q}\right) = 1 \\ 1-r^2 & \text{if } \left(\frac{k}{q}\right) = -1. \end{cases}$$

In view of (3), this gives the first column in the table for Num_q^-. The second column is obtained as before, by subtracting from Num_q found in Theorem 9.1.

§11. The constant for $X_0(11)$

The Galois group of division points of $X_0(11)$ was determined in Part I, §8. We recall the result, and derive implications for the correspondence with imaginary quadratic fields.

We have
$$G_\ell = GL_2(Z_\ell) \quad \text{for} \quad \ell \neq 5.$$

We have
$$G \subset G_2 \times G_5 \times G_{11} \times \prod_{\ell \neq 2,5,11} GL_2(Z_\ell),$$

and
$$G = G_{110} \times \prod_{\ell \neq 2,5,11} GL_2(Z_\ell).$$

All we have to worry about is $G_{110} \subset G_2 \times G_5 \times G_{11}$.

We denote by v_{10} the multiplicative group $Z(11)^*$. We let v_2 and v_5 be its components of orders 2 and 5 respectively.

We have a homomorphism
$$\phi_2 : G_2 \longrightarrow \{\pm 1\}$$

corresponding to the even and odd elements. We denote
$$G_{2,1} = E_2 \quad \text{and} \quad G_{2,-1} = O_2,$$

the inverse image of 1 and -1 respectively by ϕ_2 in G_2.

In Part I, §8 we had described precisely the correspondence between G_5 and G_{22} (actually factoring through G_{11}). There is a homomorphism
$$\phi_5 : G_5 \longrightarrow Z(11)^*/\pm 1 \approx v_5$$

such that, in our previous notation,
$$\phi_5(\sigma) = (\pm 2)^{\psi(\sigma)},$$

where $\psi(\sigma) = a \bmod 5$ if
$$\sigma = \begin{pmatrix} 1+5a & 5b \\ 5c & u \end{pmatrix}.$$

We let $G_{5,\zeta} = \phi_5^{-1}(\zeta)$, for $\zeta \in v_5$.

Finally, we have the natural homomorphism
$$\det{}_{(11)} = \phi_{11} : G_{11} \longrightarrow Z(11)^* \approx v_{10} = v_2 \times v_5$$

If $\zeta \in Z(11)^*$ we let $G_{11,\zeta} = \phi_{11}^{-1}(\zeta)$.

Then Theorem 8.3 of Part I can be formulated by saying that

$$\boxed{G_{110} = \bigcup_\zeta G_{11,\zeta} \times G_{2,\zeta(2)} \times G_{5,\zeta(5)}}$$

where the union is taken for $\zeta \in v_{10}$. We have used the notation
$$\tilde\zeta(2) \quad \text{and} \quad \tilde\zeta(5)$$

to denote the canonical image of ζ in v_2 and v_5 respectively. Observe that the products are automatically fibered over $Q(\sqrt{\Delta})$, which is contained in the field of 11-th roots of unity.

The next theorem gives us the denominator of the constant, by the proxies of Theorems 6.4 and 9.1.

Theorem 11.1. *For* $X_0(11)$ *we have*
$$(GL_2(Z_{110}) : G_{110}) = 1{,}200 \ .$$

Proof. We note that G_{110} has index 10 in $G_2 \times G_5 \times G_{11}$. Furthermore, G_5 has index 120 in $GL_2(Z_5)$. The theorem follows at once.

We then come to the computation of the numerator.

We first treat the case when $K \cap k_{ab} = Q_{ab}$. Then all but the 110-factors split off and are generic, so what we have to compute is
$$\text{Num}(\tilde{\mathcal{G}}_{110}) = \text{Num}_{110}(\tilde{\mathcal{G}}) \ ,$$

and for this we shall use a decomposition into fiber products to get:

Theorem 11.2. Let $\mathcal{S}_{2\cdot 11}$ be Serre's subgroup. Assume that $K \cap k_{ab} = Q_{ab}$. Then
$$\text{Num}_{110}(\tilde{\mathcal{G}}) = \frac{1}{5} \text{Num}_5(\tilde{\mathcal{G}}_5) \text{Num}_{2\cdot 11}(\mathcal{S}_{2\cdot 11}).$$

Proof. We need a lemma.

Lemma. *The values*
$$\text{Num}_5(G_{5,\zeta(5)}, o_5^*)$$
are independent of ζ, *and equal to* $\frac{1}{5} \text{Num}_5(G_5, o_5^*)$.

Proof. Let X_a be the set of matrices
$$\begin{pmatrix} 1+5a & 5b \\ 5c & u \end{pmatrix}$$
with a, b, c $\in Z_5$ and u $\in Z_5^*$. Then
$$X_a + 5dI = X_{a+d}.$$
Therefore
$$\int_{o_5^*} h_{X_{a+d}}(\text{Tr } z, Nz) dz = \int_{o_5^*} h_{5dI+X_a}(\text{Tr } z, Nz) dz$$
$$= \int_{o_5^*} h_{X_a}(\text{Tr } z, Nz) dz,$$
as was to be shown.

Returning to the theorem, we have
$$\text{Num}_{110}(\tilde{\mathcal{G}}) = \sum_\zeta \text{Num}_{11}(G_{11,\zeta}, o_{11}^*) \text{Num}_2(G_{2,\zeta(2)}, o_2^*) \text{Num}_5(G_{5,\zeta(5)}, o_5^*).$$

By the lemma, we can replace $\text{Num}_5(G_{5,\zeta(5)}, o_5^*)$ by its constant value. The sum then gives precisely the numerator for the Serre group, as was to be shown.

Theorem 10.2 gives the numerator for the Serre group, so we are reduced to computing $\text{Num}_5(\tilde{\mathcal{G}}_5)$. This is merely a matter of putting together all the techniques which we already know, but we give the details.

Theorem 11.3. *Let* $r = r(5) = 1/5$.

$$\text{Num}_5(G_5, o_5^*) = \begin{cases} 7r^4 + r^3 & \text{if } \left(\frac{k}{5}\right) = 1 \\ r^3 & \text{if } \left(\frac{k}{5}\right) = 0 \\ r^3(1+r) - \frac{2r^4}{1+r} & \text{if } \left(\frac{k}{5}\right) = -1 \end{cases}$$

Proof. Let $d = 1, 2, 3, 4$. Let

$$\sigma(d) = \begin{pmatrix} 1 & 0 \\ 0 & d \end{pmatrix}.$$

Let $R(d) = \sigma(d) + 5M_5$. Then we have a disjoint union

$$G_5 = \bigcup_{d=1}^{4} R(d).$$

Define

$$f(d) = \int_{o_5} h_{R(d)}(\text{Tr } z, Nz) dz.$$

Then

$$\text{Num}_5(\tilde{G}_5) = \sum_{d=1}^{4} f(d).$$

Let $T_{d+1,d} = \{z \in o_5, \text{Tr } z \equiv d+1 \text{ and } Nz \equiv d \text{ mod } 5\}$. It is immediately verified that:

If $\left(\frac{k}{5}\right) \neq 1$ and $d \neq 1$ then $T_{d+1,d}$ is empty.

If $\left(\frac{k}{5}\right) = 1$ and $d \neq 1$ then taking into account the isomorphism

$$o_5 \approx Z_5 \times Z_5,$$

the set $T_{d+1,d}$ consists of the pairs (u, w) such that mod 5,

$w \equiv 1$ and $u \equiv d$ or $w \equiv d$ and $u \equiv 1$.

From this one finds from Theorem 7.1:

(3) $$\sum_{d=2}^{4} f(d) = \begin{cases} 6r^4 & \text{if } \left(\frac{k}{5}\right) = 1 \\ 0 & \text{if } \left(\frac{k}{5}\right) = 0 \\ 0 & \text{if } \left(\frac{k}{5}\right) = -1 \end{cases}$$

There remains to evaluate $f(1)$. We have:

$$f(1) = \int_{o_5} h_{I+5M}(\text{Tr } z, Nz)\, dz ,$$

which was evaluated in formula (5) of §8. A final addition yields the values stated in the theorem.

Remark. In Theorem 11.3, with $r = 1/5$, we see that the first value, in case

$$\left(\frac{k}{5}\right) = 1$$

is substantially bigger (about twice as big) as the value in the two other cases. This means that when $k = Q(\sqrt{D})$ and

$$D \equiv 1, 4 \pmod{5}$$

one expects the frequency of occurrences of k to be correspondingly bigger. Indeed, in the tables, both the actual and predicted values for the occurrences of such k are big compared to the other cases, when $D \equiv 0, 2, 3 \pmod{5}$.

In Part III we shall determine the constant for the Serre group even when $K \cap k_{ab}$ is larger than Q_{ab} for a number of cases. We can apply this to $X_0(11)$ because the same type of argument as before will give the same factor $\frac{1}{5} \text{Num}_5(\tilde{\mathcal{G}})_5$. It is convenient to put this result here since it follows the same pattern that we just encountered.

Theorem 11.4. *Let* $k = Q(\sqrt{D})$ *and assume that* $5 \nmid D$. *Then*

$$\text{Num}_{330}(\tilde{\mathcal{G}}_{330}) = \frac{1}{5} \text{Num}_5(G_5, o_5^*) \, \text{Num}_{66}(\tilde{\mathcal{G}}_{66}) .$$

Proof. We assume that the reader is acquainted with the fibering terminology of Part III and the commutator manipulations of Part III, §3. Let L be a set of primes not containing 5 or 11. Then

$$G_{55L} = G_5 \times_{\phi_5} G_{11L}$$

$$G_{k,55L} = G_5 \times_{\phi_5} G_{k,11L}$$

whence

$$G'_{55L} = G'_5 \times G'_{11L}$$

$$G'_{k,55L} = G'_5 \times G'_{k,11L} \; .$$

In each of the cases considered, we have a field F such that F is abelian over k, with Galois group B, and

$$K \cap k_{ab} = Q_{ab}F \; .$$

Then $G_{330} = G_{30q}$ (with $q = 11$) is defined by a fibering

$$\phi = (\phi', \phi'') \; ,$$

$$\phi' : G_{30q} \longrightarrow \text{Gal}(F/k) \quad \text{and} \quad \phi'' : \prod_{\ell \mid 30q} \mathfrak{o}^*_\ell \longrightarrow \text{Gal}(F/k) \; ,$$

so that

$$\tilde{\mathcal{G}}_{30q} = G_{k,30q} \times_{N,\phi} \prod_{\ell \mid 30q} \mathfrak{o}^*_\ell$$

$$= (G_5 \times_{\phi_5} G_{k,6q}) \times_{N,\varphi} (\mathfrak{o}^*_5 \times \mathfrak{o}^*_{6q}) \; ,$$

and

$$G_{30q} = \bigcup_\zeta G_{5,\zeta(5)} \times G_{11,\zeta} \times G_{2,\zeta(2)} \times G_3 \; .$$

This gives

$$G_{k,30q} = \bigcup_{\zeta,\beta} [G_{5,\zeta(5)} \times (G_{11,\zeta} \times G_{2,\zeta(2)} \times G_3)_\beta] \; ,$$

where β ranges over B, and the index β follows our usual notation, denoting inverse image under ϕ. Therefore, we get

$$\bigcup_{\zeta,\beta} (G_{5,\zeta(5)} \times_N \mathfrak{o}^*_5) \times [(G_{11,\zeta} \times G_{2,\zeta(2)} \times G_3)_\beta \times_N (\mathfrak{o}^*_{11} \times \mathfrak{o}^*_2 \times \mathfrak{o}^*_3)_\beta] \; .$$

The use of the lemma concludes the proof.

PART III

SPECIAL COMPUTATIONS

In the second part, we had worked out a general formula for the constant giving the conjectured asymptotic behavior of Frobenius automorphisms in a GL_2-extension of the rationals K, with a representation of its Galois group

$$\rho' : G \longrightarrow \prod GL_2(Z_\ell) .$$

For all but a finite number of imaginary quadratic fields k, we had seen that

$$K \cap k_{ab} = Q_{ab} ,$$

and the constant was worked out explicitly in these cases.

We now come to the study of the exceptional quadratic fields such that $K \cap k_{ab} \neq Q_{ab}$. This requires special techniques of local class field theory, and is designed for individual curves which will be Serre curves, and the Shimura curve $X_0(11)$. The quadratic fields $Q(\sqrt{-1})$, $Q(\sqrt{-3})$, $Q(\sqrt{\Delta})$, $Q(\sqrt{-\Delta})$ play a special role.

The complications arise in these special cases because the exceptionally large intersection $K \cap k_{ab}$ forces us to evaluate the integrals giving the numerator of the constant by decomposing the domain of integration over fairly small sets, determined by the dependence relations of Galois and class field theory on this intersection.

For discriminants whose absolute value is < 100 we work out all cases except one, for our five curves, determining $K \cap k_{ab}$ and the corresponding constant. The single case we have not worked out, for the curve with $\Delta = -2^6 3^5$, $k = Q(\sqrt{-3})$, would have required additional complications. On the other hand, it seemed that it would be somewhat repetitive, without much additional insight arising from it, and was not worth the effort.

For those which we include, we already have to make an analysis of the manner in which the non-abelian operation of the matrices in the GL_2-extension corresponds to the abelian operation of the k-ideles. The most interesting case is that of §12. Applied to $X_0(11)$, it gives the theoretical explanation (conjecturally, of course) for the unusually large occurrence of the quadratic field

$k = Q(\sqrt{-11})$, namely 88 times, which represents a confluence of several forces, including the fact that $-11 \equiv -1 \pmod 5$, cf. Theorem 11.3 of Part II, as well as the fact that $K \cap k_{ab} \neq Q_{ab}$. Actually, the predicted value is a little high, see the comments on numerical results in Part IV.

An analogous situation arises for the curve with discriminant -43, and the field $k = Q(\sqrt{-43})$ has a high frequency. In this case, the predicted value is perfectly in line with the actual count.

In all three cases $k = Q(\sqrt{-3})$, $k = Q(i)$, $k = Q(\sqrt{\Delta})$, certain Galois symmetries give rise to simplifications in the computation of the desired integral for the numerator of the constant. The results are given in Theorems 7.3, 7.4, 7.5 for $Q(\sqrt{-3})$, Theorem 9.1 for $Q(i)$, and Theorem 12.1 for $Q(\sqrt{\Delta})$. It turns out that for $Q(i)$, the answer is the same as for the Serre fibering of Part II, §10. For $Q(\sqrt{-3})$, it is the same in half the cases, and close to it in the other half. However, for $Q(\sqrt{\Delta})$, there are substantial differences.

In the first sections, which we call general lemmas, we give a detailed series of lemmas on $GL_2(Z_2)$ and $GL_2(Z_3)$, especially concerning their subgroups of index 2, and the commutator subgroups, whose fixed fields are precisely the intersection $K \cap k_{ab}$. In §3 we determine those cases when $K \cap k_{ab} = Q_{ab}$ which were not covered in Part II, because they required a somewhat finer argument than that contained in Part II, Theorem 3.1. We then go through systematically the fields $k = Q(\sqrt{-3})$, $k = Q(\sqrt{-1})$ and $k = Q(\sqrt{\Delta})$.

PART III

SPECIAL COMPUTATIONS

GENERAL LEMMAS

1.	Lemmas on commutator subgroups	163
2.	$G_2 = GL_2(\mathbf{Z}_2)$	165
3.	Cases when $K \cap k_{ab} = \mathbf{Q}_{ab}$	174
4.	$K \cap k_{ab}$ when $k = \mathbf{Q}(\sqrt{-3})$ and $GL_2(\mathbf{Z}_3)$ splits	181
5.	$K \cap k_{ab}$ in other cases	185

$$k = \mathbf{Q}(\sqrt{-3})$$

6.	The action of $\mathcal{C}$ on $k(\Delta^{1/3})$	191
7.	The constant for Serre fiberings, $k = \mathbf{Q}(\sqrt{-3})$, $M = 2q$, q odd prime $\neq 3$, $\Delta = \pm q^n$	195
8.	Computation of integrals	201

$$k = \mathbf{Q}(i)$$

9.	The constant for Serre fiberings, q odd $\neq 3$	209

$$k = \mathbf{Q}(\sqrt{\Delta})$$

10.	The action of $\mathcal{C}$ on $k(A_2, \Delta^{1/4})$ when $k = \mathbf{Q}(\sqrt{\Delta})$	215
11.	The action of matrices on $k(A_4)$	218
12.	Computation of integrals and the constant	221

GENERAL LEMMAS

§1. Lemmas on commutator subgroups

Lemma 1. *Let q be a prime number. Let $r \geq 1$. Let*

$$W_{q,r} = W_r = I + q^r M_q .$$

Then

$$W'_r = W_{2r} \cap SL_2(Z_q) .$$

Proof. We write a commutator from $I + q^r M_q$,

$$(I+q^r X)(I+q^r Y)(I-q^r X+q^{2r}X^2)(I-q^r Y+q^{2r}Y^2) \quad (\text{mod } q^{2r+1})$$

$$= I + q^{2r}(XY - YX) \quad (\text{mod } q^{2r+1}) .$$

It is easy to see that the vector space over F_q generated by the matrices of the form

$$XY - YX$$

has dimension 3, and consists of the matrices with trace 0. Hence W'_r is the unique closed subgroup of W_{2r} having this space as associated vector space at level $2r$. Furthermore, W'_r is contained in $SL_2(Z_q)$, and the associated vector space of

$$W_{2r} \cap SL_2(Z_q)$$

at level $2r$ has dimension ≤ 3. This proves the desired equality.

Lemma 2. *Let q be a prime ≥ 3. Let r denote reduction mod q. Let*

$$V = r^{-1}(SL_q(q)) = (I+qM_q)SL_2(Z_q) .$$

Then

$$V/V' \approx SL_2(F_q)/SL_2(F_q)' ,$$

and

$$V' = r^{-1}(SL_2(F_q)') \cap SL_2(Z_q) .$$

Proof. There are commutators from V of the form

$$\sigma(I+qX)\sigma^{-1}(I-qX) = I + q(X - \sigma X \sigma^{-1}) \pmod{q^2},$$

and $\sigma X \sigma^{-1}$ depends only on σ mod q. It is easy to get three linearly independent matrices mod q, out of the expression

$$X - \sigma X \sigma^{-1},$$

where X has trace 0 mod q. This shows that the associated vector space in M_q/qM_q to the closed subgroup V′ consists of the matrices of trace 0. On the other hand, the closed subgroup of $I + qM_q$ belonging to this space is

$$(I+qM_q) \cap SL_2(Z_q).$$

The lemma follows at once.

§2. $G_2 = GL_2(\mathbb{Z}_2)$

In this section we gather together mostly group theoretic facts about $GL_2(\mathbb{Z}_2)$, which we denote by G_2.

We are interested first in the (closed) subgroups of index 2. They correspond to characters of order 2. If need be, any group of order 2 is identified with $\{\pm 1\}$.

Suppose that the character factors through the determinant homomorphism, and hence amounts to a homomorphism of

$$\mathbb{Z}_2^*/\mathbb{Z}_2^{*2} \approx Z(8)^*.$$

Note that 1, 3, 5, 7 mod 8 represent the elements of $Z(8)^*$. We have three possible characters, denoted by χ_2, χ_{-2}, χ_i such that

$$\text{Ker } \chi_2 = \{1, 7\}, \qquad \text{Ker } \chi_{-2} = \{1, 3\}, \qquad \text{Ker } \chi_i = \{1, 5\}.$$

If G_2 is the Galois group of an extension of $\mathbb{Q}$, inducing the determinant character on the roots of unity, then the indices 2, -2, i indicate the quadratic field fixed by the kernel of the character, namely

$$\mathbb{Q}(\sqrt{2}), \qquad \mathbb{Q}(\sqrt{-2}), \qquad \mathbb{Q}(i)$$

respectively.

On the other hand we have the sign homomorphism

$$\varepsilon : GL_2(\mathbb{Z}_2) \longrightarrow GL(2) \approx S_3 \longrightarrow \{\pm 1\},$$

where S_3 is the permutation group. We then obtain four subgroups

$$E_2 = \text{Ker } \varepsilon, \qquad \text{Ker } \varepsilon \chi_2, \qquad \text{Ker } \varepsilon \chi_{-2}, \qquad \text{Ker } \varepsilon \chi_i.$$

Lemma 1. *The above subgroups of index* 2 *are the only ones. The subgroups*

$$\text{Ker } \chi_2, \qquad \text{Ker } \chi_{-2}, \qquad \text{Ker } \varepsilon \chi_2, \qquad \text{Ker } \varepsilon \chi_{-2}$$

do not contain $I + 4M_2$. *The other three subgroups contain* $I + 4M_2$.

Proof. The first statement will follow from the determination of $GL_2(\mathbb{Z}_2)'$ in Theorem 2.1. The characterization of the kernels in terms of $I + 4M_2$ is immediate and left to the reader.

Theorem 2.1. *The homomorphism* $\varepsilon \times \det$ *is the maximal abelianizing homomorphism of* $GL_2(\mathbb{Z}_2)$. *In other words,*

$$GL_2(\mathbb{Z}_2)' = E_2 \cap SL_2(\mathbb{Z}_2).$$

Proof. Since E_2 has index 2 in G_2 it suffices to prove that

$$(SL_2(\mathbb{Z}_2) : G_2') = 2.$$

For the rest of this section, as we deal only at the prime 2, we abbreviate:

$$E_2 = E, \qquad G_2 = G, \qquad M_2 = M.$$

To do what we want, we now see that it suffices to do it mod 4.

Lemma 2.
(i) $\qquad\qquad\qquad I + 2M \subset E.$

(ii) $\qquad\qquad G' \supset (I+2M)' = (I+4M) \cap SL_2(\mathbb{Z}_2)$

(iii) $\qquad\qquad\qquad (I+4M) \cap G' \subset E'$

(iv) *Reduction mod 4 gives an isomorphism*
$$SL_2(\mathbb{Z}_2)/G' \approx SL_2(4)/G(4)'.$$

Proof. That $I + 2M \subset E$ is obvious. Lemma 2 of §1, applied to the prime 2, yields (ii). Hence the inverse image of $G(4)'$ in $SL_2(\mathbb{Z}_2)$ is G'. The rest of the lemma then follows obviously.

Lemma 3. *Let* $2M(4)_0$ *be the set of matrices in* $2M(4)$ *with trace* 0. *Then* $I + 2M(4)_0$ *is contained in* $G(4)'$.

Proof. Let

$$\gamma = \begin{pmatrix} -1 & -1 \\ 1 & 0 \end{pmatrix} \quad \text{and} \quad \beta = \begin{pmatrix} 0 & 1 \\ 1 & 0 \end{pmatrix}.$$

Forming commutators of elements of type $I + 2A$ with γ and β yields elements of the form
$$I + 2(A - \gamma A \gamma^{-1}) \quad \text{and} \quad I + 2(A - \beta A \beta^{-1}) \mod 4.$$

It is immediately seen that the associated vector space of such elements in $2M/4M$ has dimension 3, in other words, is the space of matrices with trace 0, as desired.

Returning to the theorem, via Lemma 2, we consider the sequence of subgroups
$$G(4) \supset SL_2(4) \supset G(4)'.$$

It is clear that $SL_2(4)$ has index 2 in $G(4)$. Note that $G(4)'$ has $G(2)'$ as factor group, and that $G(2)'$ is the commutator group of $G(2) \approx S_3$, and has order 3. The group
$$I + 2M(4)_0$$
has order 8, and is contained in the kernel of the reduction mod 2:
$$G(4)' \longrightarrow G(2)',$$
so that $G(4)'$ has order at least $3 \cdot 8 = 24$. But $G(4)$ has order $4 \cdot 24$. Hence
$$(G(4) : G(4)') \leq 4.$$

Since $G(4)' \neq SL_2(4)$ (because of the existence of ε), it follows that the index is exactly 4, and also that
$$(SL_2(4) : G(4)') = 2,$$
thereby proving the theorem.

We can also formulate Lemma 3 in a 2-adic way.

Theorem 2.2. $GL_2(\mathbb{Z}_2)' \cap (I + 2M_2) = SL_2(\mathbb{Z}_2) \cap (I + 2M_2)$.

Proof. We have seen that this is true mod 4, and the result follows by refinement because of Lemma 2, (ii) (after level 4, the group has the right associated vector space mod higher powers of 2).

The next theorem determines V' when V is any one of the subgroups of $GL_2(\mathbb{Z}_2)$ not containing $I + 4M$.

Theorem 2.3. *Let* V *be any one of the subgroups of* $GL_2(Z_2)$ *of index* 2, *which does not contain* $I + 4M$, *i.e.*

$$\text{Ker } \chi_2, \quad \text{Ker } \chi_{-2}, \quad \text{Ker } \varepsilon\chi_2, \quad \text{Ker } \varepsilon\chi_{-2}.$$

Then
$$V' = GL_2(Z_2)'.$$

Proof. By hypothesis,
$$V(I + 4M_2) = GL_2(Z_2).$$

Hence we obtain an isomorphism
$$V/[V \cap (I + 4M_2)] \approx GL_2(Z_2)/(I + 4M_2).$$

On the other hand, if Y is a normal subgroup of X, then we have the formula
$$(X/Y)' = X'/(X' \cap Y).$$

Consequently we obtain an isomorphism
$$V'/[V' \cap (I + 4M_2)] \approx GL_2(Z_2)'/[GL_2(Z_2)' \cap (I + 4M_2)].$$

Since $GL_2(Z_2)' \subset SL_2(Z_2)$, it suffices to prove:

Lemma 5. *Let* V *be any one of the subgroups of* $GL_2(Z_2)$ *which does not contain* $I + 4M$. *Then*
$$V' \cap (I + 4M_2) = SL_2(Z_2) \cap (I + 4M_2).$$

Proof. Note that
$$V' \cap (I + 4M_2) \supset [V \cap (I + 2M_2)]'.$$

For matrices A, B the commutator of $I + 2A$ and $I + 2B$ is
$$I + 4(AB - BA) \mod 8.$$

We can pick matrices A, B such that $I + 2A$ and $I + 2B$ lie in
$$V \cap (I + 2M_2)$$
and such that their commutators give enough elements to show that the vector subspace of $4M_2/8M_2$ associated with the closed subgroup

$$[V \cap (I+2M_2)]'$$

is the space of matrices with trace 0. Since this subspace determines the associated group uniquely, the lemma will be proved. There remains to exhibit the matrices A and B.

For χ_{-2} and $\varepsilon\chi_{-2}$ we use for instance:

$$A = \begin{pmatrix} 0 & 1 \\ 0 & 0 \end{pmatrix} \quad \text{and} \quad B = \begin{pmatrix} 0 & 0 \\ 1 & 0 \end{pmatrix}, \quad AB - BA = \begin{pmatrix} 1 & 0 \\ 0 & -1 \end{pmatrix}$$

$$A = \begin{pmatrix} 1 & 0 \\ 0 & 0 \end{pmatrix} \quad B = \begin{pmatrix} 0 & 1 \\ 0 & 0 \end{pmatrix}, \quad AB - BA = \begin{pmatrix} 0 & 1 \\ 1 & 0 \end{pmatrix}$$

$$A = \begin{pmatrix} 1 & 0 \\ 0 & 0 \end{pmatrix} \quad B = \begin{pmatrix} 0 & 0 \\ 1 & 0 \end{pmatrix}, \quad AB - BA = \begin{pmatrix} 0 & 0 \\ 1 & 0 \end{pmatrix}$$

For χ_2 and $\varepsilon\chi_2$ we take the same A, B in the first line, but use $-A$ instead of A in the second and third line. This concludes the proof.

Next we handle Ker ε = E and Ker χ_i. We shall not treat Ker $\varepsilon\chi_i$, which is a little more complicated, and is not needed in the range of computations which we worked out. Whereas we shall see that we can treat the first two groups mod 4, it would require going to congruences mod 8 to treat Ker $\varepsilon\chi_i$ in an analogous way.

Let

$$SW_{2,2} = (I+4M) \cap SL_2(Z_2).$$

Then if V = E = Ker ε, or Ker χ_i, we shall prove that $V' \supset SW_{2,2}$, and we shall obtain a diagram where the horizontal arrows induce an isomorphism of factor groups at each level.

$$\begin{array}{ccc} SL_2(Z_2) & \longrightarrow & SL_2(4) \\ | & & | \\ G' & \longrightarrow & G'(4) \\ | & & | \\ V' & \longrightarrow & V'(4) \\ | & & | \\ SW_{2,2} & \longrightarrow & 1 \end{array}$$

Each group on the left is the inverse image of the corresponding group on the right.

The even and odd matrices mod 2 play a pervasive role. We remind the reader of what they look like:

$$O(2): \begin{pmatrix} 0 & 1 \\ 1 & 0 \end{pmatrix}, \quad \begin{pmatrix} 1 & 1 \\ 0 & 1 \end{pmatrix}, \quad \begin{pmatrix} 1 & 0 \\ 1 & 1 \end{pmatrix}$$

$$E(2): \begin{pmatrix} 1 & 0 \\ 0 & 1 \end{pmatrix}, \quad \begin{pmatrix} 0 & 1 \\ 1 & 1 \end{pmatrix}, \quad \begin{pmatrix} 1 & 1 \\ 1 & 0 \end{pmatrix}.$$

We note that $O(2)$ and the zero matrix form an additive group, and also $E(2)$ and the zero matrix form an additive group. Therefore

$$I + 2O(2) \cup I \pmod 4 \quad \text{and} \quad I + 2E(2) \cup I \pmod 4$$

form multiplicative subgroups of $GL_2(4) = G(4)$.

Theorem 2.4.
(i) *The group* E *is generated by*

$$I + 2M \quad \text{and} \quad \gamma = \begin{pmatrix} -1 & -1 \\ 1 & 0 \end{pmatrix}.$$

(ii) $E' \supset SW_{2,2}$.

(iii) $E'(4) = (I + 2O(2)) \cup I$.

(iv) *Reduction* mod 4 *gives isomorphisms*

$$G'/E' \approx G(4)'/E(4)' \quad \text{and} \quad E'/SW_{2,2} \approx E'(4).$$

Proof. Since E contains $I + 2M$, and γ has order 3 (mod 2), the first assertion is obvious. The second repeats part of Lemma 2, (ii). The third is obtained by computing commutators in a trivial way. Then (iv) follows from Lemma 2 (iii), as was to be shown.

Theorem 2.5. *Let* $V = \mathrm{Ker}\, \chi_i$. *Then:*
(i) $V = SL_2(\mathbb{Z}_2)(I + 4M)$.

(ii) $V' \supset SW_{2,2}$.

(iii) $V'(4) = SL_2(4)'$ *has order* 12. *It is generated by*

$$I + 2O(2)$$

and by any element of order 3 in G(4). It contains all the elements of order 3 in G(4).

(iv) *Reduction mod 4 gives an isomorphism*
$$V'/SW_{2,2} \approx V'(4).$$

Proof. From $I + 4M \subset V$ and Lemma 1 of §1 we conclude that
$$(I + 8M) \cap SL_2(\mathbb{Z}_2) \subset V'.$$

Statement (ii) now amounts to showing that the associated vector space of V' in $4M/8M$ has dimension 3, i.e. consists of the elements with trace 0. For this we have to find enough commutators.

Pick first
$$A = \begin{pmatrix} 0 & 1 \\ 0 & 0 \end{pmatrix} \quad \text{and} \quad B = \begin{pmatrix} 0 & 0 \\ 1 & 0 \end{pmatrix}.$$

Then

(1) $$[I + 2A, I + 2B] \equiv I + 4 \begin{pmatrix} 1 & 0 \\ 0 & 1 \end{pmatrix} \pmod{8}.$$

We then form commutators
$$[I + 2A, \sigma] \equiv I + 2(A - \sigma A \sigma^{-1}) \pmod{4}$$

with two cases:

(2) $\quad A = \begin{pmatrix} 1 & 1 \\ 0 & 1 \end{pmatrix} \quad \text{and} \quad \sigma = \begin{pmatrix} 0 & 1 \\ -1 & 0 \end{pmatrix}, \quad [I + 2A, \sigma] \equiv I + 2 \begin{pmatrix} 0 & 1 \\ -1 & 0 \end{pmatrix}$

(3) $\quad A = \begin{pmatrix} 0 & 1 \\ 0 & 0 \end{pmatrix} \quad \text{and} \quad \sigma = \begin{pmatrix} -1 & 1 \\ -1 & 0 \end{pmatrix}, \quad [I + 2A, \sigma] \equiv I + 2 \begin{pmatrix} 1 & 1 \\ 1 & -1 \end{pmatrix}$

Squaring the commutators in the second and third case yields
$$I + 4 \begin{pmatrix} -1 & 1 \\ -1 & -1 \end{pmatrix} \quad \text{and} \quad I + 4 \begin{pmatrix} 2 & 0 \\ 1 & 0 \end{pmatrix}, \quad \pmod{8}.$$

This gives us the 3-dimensional space of trace 0 in $4M/8M$.

We also see that $V'(4)$ contains $I + 2O(2)$. Pick any element of order 3, say γ, and form its commutator with

$$\begin{pmatrix} 0 & 1 \\ -1 & 0 \end{pmatrix}.$$

We find an element of order 3, which lies in $V'(4)$. Hence $V'(4)$ has order at least 12. Let H be the subgroup generated by

$$I + 2O(2)$$

and by an element of order 3 in $V'(4)$. Then one sees that $-I \notin H$, and H is of index 2 in $V(4)/\pm I$, whence normal, and also normal in $V(4)$ since $-I$ lies in the center. The factor group is abelian, and hence $H \supset V'(4)$. Therefore $H = V'(4)$. It now follows that $V'(4)$ has order 12, and $V'(4) = SL_2(4)'$, so that $V'(4)$ is normal in $G(4)$. The rest of (iii) follows at once, because the 3-Sylow groups are conjugate, and (iv) is clear, thereby proving the theorem.

The purely group theoretic arguments which precede are used in the context when $G(4) = GL_2(4)$ is the Galois group of the 4-division points of an elliptic curve A, i.e.

$$G(4) = \text{Gal}(Q(A_4)/Q).$$

It is a classical fact that $\Delta^{1/4}$ is contained in the field of 4-division points. The corresponding lattice of fields for Theorems 2.4 and 2.5 can then be drawn as on the figure. The small numbers near each line indicate the degree of the corresponding field extension.

The field $Q(x(A_4))$ is the field of x-coordinates of the 4-division points – we could call it the **modular subfield** of 4-division points.

Since for the moment we only need the existence of $\Delta^{1/4}$ in the field of 4-division points, we don't go any further into the matter. In §11, we shall prove this fact, along with more precise information on how the matrices and idele class groups operate on $Q(A_4)$.

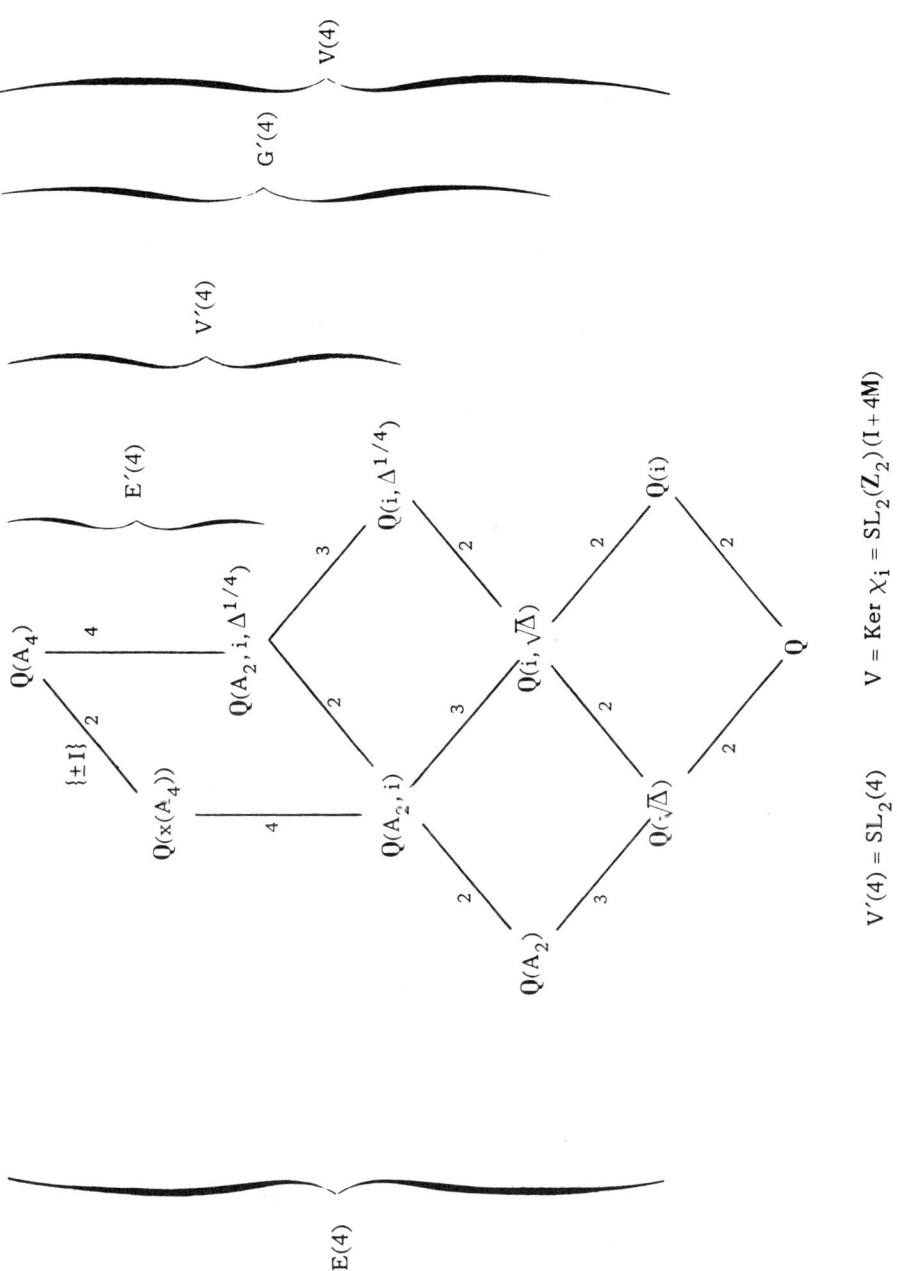

§3. Cases when $K \cap k_{ab} = Q_{ab}$

In Part II, §3, we gave a table of those cases when $K \cap k_{ab}$ is equal to Q_{ab}. We now want to prove the results justifying the table. We begin with lemmas of general group theory.

Let G_1, G_2 be groups, and let B be an abelian group. Let

$$\phi_1 : G_1 \longrightarrow B \quad \text{and} \quad \phi_2 : G_2 \longrightarrow B$$

be surjective homomorphisms. We define the fiber product over ϕ,

$$G = G_1 \times_{(\phi_1, \phi_2)} G_2 = G_1 \times_\phi G_2$$

to be the set of pairs (σ_1, σ_2) such that $\phi_1(\sigma_1) = \phi_2(\sigma_2)$. We say that the fibering ϕ **dissolves under commutators** if

$$G' = G'_1 \times G'_2 \, .$$

Lemma 1. *Let* $G = G_1 \times_\phi G_2$ *be a fiber product as above. Assume that* $\phi_2^{-1}(B)$ *contains a subset of elements which map onto* B *under* ϕ_2 *and which commute with each other. Then*

$$G' = G'_1 \times G'_2 \, ,$$

i.e. the fibering dissolves under commutators.

Proof. We have $\text{pr}_1 G' = G'_1$ and $\text{pr}_2 G' = G'_2$ because the first and second projections are surjective. It suffices therefore to prove that G' contains $G'_1 \times \{e_2\}$. Given $\sigma_1, \tau_1 \in G_1$, we can find $\sigma_2, \tau_2 \in G_2$ which commute with each other, and such that

$$\phi_1(\sigma_1) = \phi_2(\sigma_2), \qquad \phi_1(\tau_1) = \phi_2(\tau_2) \, .$$

The commutator of (σ_1, σ_2) and (τ_1, τ_2) gives what we want.

Remarks. The proof shows that the lemma holds under the weaker hypothesis that given a pair of elements $\sigma_1, \tau_1 \in G_1$ we can find σ_2, τ_2 in G_2 commuting with each other, and having the same images under ϕ_1, ϕ_2 respectively.

The condition under which the lemma holds is obviously satisfied if the group B is cyclic.

We apply the lemma when G is the Galois group of our usual GL_2-extension. We need more systematic notation for a set of primes and its complement. If M is a positive integer, we let [M] denote the set of primes not dividing M. Then

$$G \subset G_M \times G_{[M]}.$$

Let M, L be disjoint sets of primes. Suppose that $Q(\sqrt{\Delta})$ is contained in the fields K_M and K_L, whose Galois groups are G_M and G_L respectively. We write

$$G_M \times_\Delta G_L$$

for the fibered product with respect to the homomorphisms

$$\phi_1 : G_M \longrightarrow \mathrm{Gal}(Q(\sqrt{\Delta})/Q) \quad \text{and} \quad \phi_2 : G_L \longrightarrow \mathrm{Gal}(Q(\sqrt{\Delta})/Q).$$

We call this **Serre's fibering**.

As we have already seen in dealing with Serre subgroups, we always have

$$\sqrt{\Delta} \in K_2.$$

We denote by Δ_0 the discriminant of the field $Q(\sqrt{\Delta})$. We shall be specifically interested in the case when

$$\Delta_0 = \pm q$$

where q is an odd prime, since all the five curves which we consider have this property. In this case, only q can ramify in $Q(\sqrt{\Delta})$, and consequently we also have

$$Q(\sqrt{\Delta}) \subset K_q.$$

Let k be a quadratic imaginary extension of Q as usual. As before, $G_k = \mathrm{Gal}(K/k)$. We first give a result concerning G'/G'_k.

Theorem 3.1. *Let q be an odd prime and let $k = Q(\sqrt{\Delta})$. Assume:*
(i) $\Delta_0 = -q$

(ii) $G = GL_2(\mathbb{Z}_2) \times_\Delta G_{[2]}$.

(iii) $G_{[2]} = GL_2(\mathbb{Z}_q) \times_\phi G_{[2q]}$ is a cyclic fibering.

Then
$$G'/G'_k = GL_2(\mathbb{Z}_2)'/E'_2 \times G'_q/E'_q .$$

If $q \geq 5$, then the term G'_q/E'_q is 1, and can be omitted.

Proof. First we have by Lemma 1,

(1) $$G' = G'_2 \times G'_q \times G'_{[2q]} .$$

Next, we have

(2) $$G_k = E_2 \times G_{[2],k} ,$$

because the left hand side and right hand side of this expression both have index 4 in $G_2 \times G_{[2]}$, and the right hand side is obviously contained in the left hand side. Since q is the only ramified prime in k, we also have

(3) $$G_{[2],k} = E_q \times_\phi G_{[2q]} ,$$

for some cyclic fibration ϕ, which is in fact of order 2. Taking commutators and using Lemma 1, we find

(4) $$G'_k = E'_2 \times E'_q \times G'_{[2q]} .$$

The theorem follows at once, because $SL_2(\mathbb{Z}_q) \subset E_q$, and for $q \geq 5$, we know from Part II, Lemma 1 of §3 that $SL_2(\mathbb{Z}_q)' = SL_2(\mathbb{Z}_q)$.

Remark. The hypotheses of Theorem 3.1 apply to a Serre curve, with Galois group
$$G = S_{2q} \times G_{[2q]} .$$

According to Part I, §8, they also apply to $X_0(11)$.

Theorem 3.2. Let $k = \mathbb{Q}(\sqrt{-8})$. Assume that
$$G = GL_2(\mathbb{Z}_2) \times_\phi G_{[2]}$$
is a fibering where $\phi = (\varepsilon, \phi_{[2]})$. Then
$$G' = G'_k, \qquad \text{that is} \qquad K \cap k_{ab} = \mathbb{Q}_{ab} .$$

Proof. We first note that $k \subset K_2$, and so
$$G_k = G_{k,2} \times_\phi G_{[2]}.$$
Hence by Lemma 1,
$$G'_k = G'_{k,2} \times G'_{[2]}$$
and also
$$G' = GL_2(\mathbf{Z}_2)' \times G'_{[2]}.$$
The theorem follows from Theorem 2.3, with $V = G_{k,2}$.

Remark. In the preceding theorem, and also the two subsequent ones, the essential factors are G_2 and G_q, with a possible further fibering between G_q and $G_{[2q]}$. As we have already encountered in studying $X_0(11)$, it is useful to consider only partial products G_L instead of $G_{[2q]}$, where L is a set of primes not containing 2 or q. All these theorems apply mutas mutandis to this case. Stating them with the added L into the notation gets heavy, and it seemed preferable to make the remark instead.

Any discriminant of a quadratic field can be factored uniquely into a product of discriminants each of which has only one prime factor. If D_1, D_2 are discriminants, and also $D_1 D_2$, then we let
$$\chi_{D_1 D_2} = \chi_{D_1} \chi_{D_2}$$
be the character associated with the corresponding quadratic field. They factor through the determinant homomorphism, and are defined on $\mathbf{Z}^*_{D_1 D_2}$.

Theorem 3.3. *Let* $k = Q(\sqrt{-24})$. *Assume that* $\Delta_0 = \pm q$ *where* q *is an odd prime. Assume furthermore:*

(i) $G = GL_2(\mathbf{Z}_2) \times_\Delta G_{[2]}$

(ii) G_3 *splits in* $G_{[2]}$.

Then $G' = G'_k$.

Proof. We distinguish two cases.

Case 1. $q \neq 3$.

We can write -24 as a product of discriminants,
$$-24 = 8(-3).$$
Then
$$G_k = GL_2(\mathbb{Z}_2) \times_\phi G_{[2]},$$
where $\phi = (\phi_1, \phi_2)$ is a homomorphism into $Z(2) \times Z(2)$, and
$$\phi_1 = \varepsilon \times \chi_2, \qquad \phi_2 = \chi_{q*} \times \chi_{-3}.$$
As usual, $q^* = \pm q$, taking the sign which makes $q^* \equiv 1 \pmod 4$. Since $q \neq 3$, the characters χ_{q*} and χ_{-3} factor through distinct factors of $G_{[2]}$, because we assumed that G_3 splits. Hence ϕ_2 satisfies the condition of Lemma 1, and ϕ dissolves under commutators. So does the fibering for G itself, and the theorem follows.

Case 2. $q = 3$.

Then $\Delta_0 = -3$ since Δ_0 is a discriminant. We can write
$$G = (GL_2(\mathbb{Z}_2) \times_\Delta G_3) \times G_{[6]}$$
and
$$G_k = (G_{k,2} \times_\Delta G_3) \times G_{[6]},$$
where
$$G_{k,2} = \operatorname{Ker} \varepsilon \chi_2.$$
Then Lemma 1 gives
$$G' = GL_2(\mathbb{Z}_2)' \times G'_3 \times G'_{[6]}$$
$$G'_k = G'_{k,2} \times G'_3 \times G'_{[6]}.$$
Hence
$$G'/G'_k = GL_2(\mathbb{Z}_2)'/G'_{k,2}.$$
We let $V = G_{k,2}$, and apply Theorem 2.3 to conclude the proof.

Theorem 3.4. Let $k = \mathbb{Q}(\sqrt{-8q})$ where q is an odd prime, and $D = -8q$. Assume that $\Delta_0 = \pm q$. Assume further that
$$G = GL_2(\mathbb{Z}_2) \times_\Delta G_q \times_\phi G_{[2q]}$$

where ϕ is a fibering which dissolves under commutators, between G_q and $G_{[2q]}$. Then
$$G' = G'_k.$$

Proof. We have
$$G_k = (G_{k,2} \times_\Delta G_q) \times_\phi G_{[2q]}$$
and
$$G_{k,2} = \text{Ker } \varepsilon \chi_{D/\Delta_0}.$$
Then
$$G' = GL_2(\mathbb{Z}_2)' \times G'_q \times G'_{[2q]}$$
and
$$G'_k = G'_{k,2} \times G'_q \times G'_{[2q]}$$
by the associativity of the fiberings, because the fibering over Δ is only between $GL_2(\mathbb{Z}_2)$ and G_q under the present hypotheses. We put $V = G_{k,2}$ and use Theorem 2.3 to conclude the proof.

Theorem 3.5. *Let* $k = Q(\sqrt{D})$ *and suppose that* $5|D$. *Assume that*
$$G = G_5 \times_\phi G_{[5]},$$
and that ϕ *is cyclic of order* 5. *Then*
$$G' = G'_k.$$

Proof. Since D is a discriminant, we cannot have $D = -5$, and hence some other prime divides D. Therefore
$$G_k = G_5 \times_{\phi,D} G_{[5]},$$
and the fibering for G_k is cyclic of order 10, so dissolves under commutators. The theorem follows at once.

The assumptions of Theorems 3.4 and 3.5 are of course designed for application to $X_0(11)$.

Again consider the case which will be applied to $X_0(11)$, namely assume that

$$G = G_5 \times_\phi G_{[5]},$$

and assume that ϕ dissolves under commutators.

Let $k = Q(\sqrt{D})$, and suppose that $5 \nmid D$. Then

$$G_k = G_5 \times_\phi G_{[5],k},$$

and consequently

$$G'_k = G'_{[5]}/G'_{[5],k}.$$

This gets rid of the factor at 5, and as we already mentioned in a remark, we conclude:

Theorem 3.6. *Let G be the Galois group of division points of $X_0(11)$. If k is any one of the fields with discriminant*

$$D = -8, -15, -20, -24, -40, -55, -88,$$

then $G' = G'_k$, that is,

$$K \cap k_{ab} = Q_{ab}.$$

Proof. We know from Part I, Theorem 8.3, that

$$G_{[5]} = S_{22} \times \prod_{\ell \neq 2,5,11} GL_2(Z_\ell)$$

so the Serre curve results apply.

§4. $K \cap k_{ab}$ when $k = Q(\sqrt{-3})$ and $GL_2(Z_3)$ splits

Throughout this section, we let $k = Q(\sqrt{-3})$.

We suppose that the representation

$$\rho' : G \longrightarrow \prod GL_2(Z_\ell)$$

is that associated with an elliptic curve A with discriminant Δ. It is a classical fact that $\Delta^{1/3}$ is contained in the field $Q(A_3)$ of division points of order 3. Indeed, one can see this from the fact that $j^{1/3}$ is a modular function of level 3 (see for instance [L 1], Chapter 18, §5, Theorem 8), and

$$j^{1/3} = g_2/\Delta^{1/3} .$$

The field generated by the x-coordinates (Weber-coordinate) of the points of order 3 is the same as the field generated by the values of the modular functions of level 3, cf. [L 1], Chapter 9, §3. Since $k = Q(\sqrt{-3})$ we see that

$$k(\Delta^{1/3})$$

is abelian, and in fact cyclic over k.

If the Galois group of $Q(A_3)$ over Q is all of $GL_2(3)$, then it follows that Δ is not a rational cube, and the above cyclic extension has precise degree 3. The subgroup of $GL_2(3)$ leaving k fixed is precisely $SL_2(3)$.

For future reference, we recall some facts concerning $GL_2(3)$. It operates as a permutation group of the subspaces of dimension 1 of F_3^2, and with this operation we have an isomorphism

$$GL_2(3)/\pm 1 \approx S_4 .$$

This induces an isomorphism

$$SL_2(3)/\pm 1 \approx A_4$$

with the alternating group, which has 12 elements. It is then easy to verify that the factor commutator group is given by

$$A_4/A_4' = Z(3) .$$

In fact, one can display explicitly the commutator group $SL_2(3)'$ and its cosets in $SL_2(3)$ as follows:

$$C_0 = SL_2(3)' = \left\{ \pm I, \quad \pm \begin{pmatrix} 0 & 1 \\ -1 & 0 \end{pmatrix}, \quad \pm \begin{pmatrix} 1 & 1 \\ 1 & -1 \end{pmatrix}, \quad \pm \begin{pmatrix} -1 & 1 \\ 1 & 1 \end{pmatrix} \right\}$$

$$C_1 = \left\{ \pm \begin{pmatrix} 1 & 1 \\ 0 & 1 \end{pmatrix}, \quad \pm \begin{pmatrix} 0 & 1 \\ -1 & -1 \end{pmatrix}, \quad \pm \begin{pmatrix} 1 & -1 \\ 1 & 0 \end{pmatrix}, \quad \pm \begin{pmatrix} 1 & 0 \\ -1 & 1 \end{pmatrix} \right\}$$

$$C_2 = \left\{ \pm \begin{pmatrix} 1 & -1 \\ 0 & 1 \end{pmatrix}, \quad \pm \begin{pmatrix} 0 & 1 \\ -1 & 1 \end{pmatrix}, \quad \pm \begin{pmatrix} 1 & 0 \\ 1 & 1 \end{pmatrix}, \quad \pm \begin{pmatrix} -1 & -1 \\ 1 & 0 \end{pmatrix} \right\}.$$

As a matter of notation, we let

$$E_3 = SL_2(\mathbb{Z}_3)(I + 3M_3).$$

Then

$$SL_2(3) = E_3(3).$$

The elements of the three cosets above can be characterized as follows, in $SL_2(3)$.

$$SL_2'(3) = E_3'(3) = \begin{cases} I \text{ and } -I \\ \text{all elements with trace } 0 \mod 3 \end{cases}$$

$$\text{The two cosets in } SL_2(3) = \begin{cases} \text{elements } \neq I \text{ or } -I \\ \text{all non scalar } \mod 3 \\ \text{all elements with trace } \neq 0 \mod 3. \end{cases}$$

Lemma 1. *Let* r *be reduction* mod 3. *Then*

$$E_3' = r^{-1}(SL_2(3)') \cap SL_2(\mathbb{Z}_3)$$

and we have an isomorphism

$$SL_2(\mathbb{Z}_3)/E_3' \approx SL_2(3)/SL_2(3)'.$$

In particular,

$$(SL_2(\mathbb{Z}_3) : E_3') = 3.$$

Proof. The first part of the lemma is independent of the prime 3 and was proved as Lemma 2 of §1. The second part follows from the preceding remarks on $GL_2(Z_3)$.

The corresponding field diagram on the field of 3-division points can be drawn as follows.

$$E_3(3) = SL_2(3) \left\{ \begin{array}{c} k(A_3) \\ | \\ k(\Delta^{1/3}) \\ | \\ k \end{array} \right\} E'_3(3)$$

Note that $GL'_2(3) = SL_2(3)$ and $GL'_2(Z_3) = SL_2(Z_3)$.

The preceding purely group theoretic lemma can be applied in the concrete situation of division points. In our notation, we let $G_k = \text{Gal}(K/k)$, and

$$G_{k,3} = \text{Gal}(K_3/k) = \text{Gal}(Q(A^{(3)})/k) \,.$$

Theorem 4.1.

(i) If $G_3 = GL_2(Z_3)$ then $G_{k,3} = E_3$.

(ii) If $G_{k,3} = E_3$ then
$$K_3 \cap k_{ab} = Q_{ab,3}(\Delta^{1/3}) \,.$$

Proof. The first assertion is merely a lifting 3-adically of the statement already made that $SL_2(3)$ is the subgroup of $GL_2(3)$ leaving k fixed. The second statement is due to the fact that $\Delta^{1/3}$ generates an abelian extension of degree 3 over the cyclotomic field Q_{ab}, and the fact that

$$(SL_2(Z_3) : E'_3) = 3$$

in the preceding lemma.

It was convenient to visualize the theorem locally at 3, but in the applications, we use it in situations when 3 splits.

Theorem 4.2. *Let L be a set of primes not containing 3.*

(i) *If*
$$G_{3L} = GL_2(\mathbb{Z}_3) \times G_L$$
then
$$G_{k,3L} = E_3 \times G_L.$$

(ii) *If $G_{3L} = G_3 \times G_L$ and $G_{k,3L} = E_3 \times G_L$, then $G'_{k,3L} = E'_3 \times G'_L$, and*
$$K_{3L} \cap k_{ab,3L} = Q_{ab,3L}(\Delta^{1/3}).$$

Proof. The first assertion follows from Theorem 4.1, since k is already contained in K_3. The second follows from the lemma, and the fact that
$$(G'_{3L} : G'_{k,3L}) = (G'_3 : E'_3) = 3.$$

The lattice of fields illustrating the preceding discussion can be drawn as follows.

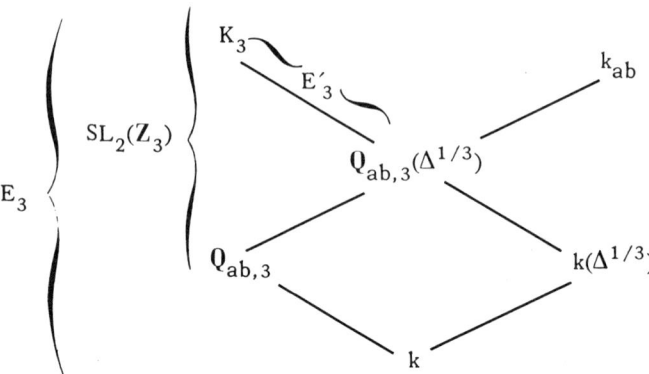

§5. $K \cap k_{ab}$ in other cases

Theorem 5.1. *Let* $k = Q(\sqrt{\Delta})$, $\Delta_0 = -q$ *where q is an odd prime. Assume that*

$$G_2 = GL_2(Z_2) \quad \text{and} \quad G = G_2 \times_\Delta G_{[2]}.$$

Then $K \cap k_{ab} = Q_{ab}F$, *where F is an abelian extension of k, and*

$$[F : k] = [Q_{ab}F : Q_{ab}].$$

Furthermore:

(i) *If* $q \neq 3$, *then* $F = k(\Delta^{1/4}, A_2)$ *and* $[F : k] = 6$.

(ii) *If* $q = 3$, *then* $F = k(A_2, \Delta^{1/4}, \Delta^{1/3})$ *and* $[F : k] = 18$.

Proof. The field diagram is as follows.

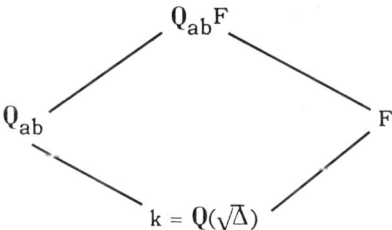

We note that k is contained in both K_2 and K_q. We have trivially $G_{2,k} = E_2$ and hence

$$G_k = E_2 \times G_{[2],k}.$$

By Theorem 3.1 and Lemma 4 (iii) of §2, we find

$$G'/G'_k = G'_2/G'_{k,2} = G'_2/E'_2 \approx G(4)'/E(4)',$$

and this last factor group has order 6. Since we have exhibited a cyclic extension of degree 6, namely F over k, whose intersection with Q_{ab} is obviously k, the theorem follows when $q \neq 3$.

Suppose that $q = 3$. Then we have to take into account the extra cyclic group of order 3 appearing in Theorem 3.1. Note that only 3 ramifies in $k(\Delta^{1/3})$. On the other hand, 2 must ramify in $Q(A_2)$ and 2 is unramified in $k = Q(\sqrt{-3})$. Hence 2 must ramify in $k(A_2)$. Hence $\Delta^{1/3}$ does not lie in $k(A_2)$, and it follows that

$$[k(A_2, \Delta^{1/4}, \Delta^{1/3}) : k] = 18 .$$

This combined with Theorem 3.1 immediately implies the case $q = 3$ of our theorem.

Theorem 5.2. *Let* $k = Q(i)$. *Assume that* $k \neq Q(\sqrt{\Delta})$. *Assume also that*

$$G = GL_2(Z_2) \times_\Delta G_{[2]}$$

is Serre's fibering. Then:

(i) *We have*

$$K \cap k_{ab} = Q_{ab} F \qquad where \qquad F = k(\Delta^{1/4}) .$$

Proof. We have

$$G_k = (\text{Ker } \chi_i) \times_\Delta G_{[2]} ,$$

whence by Lemma 1 of §3,

$$G'_k = (\text{Ker } \chi_i)' \times G'_{[2]}$$

$$G' = G'_2 \times G'_{[2]} .$$

Thus we have an inclusion of groups

$$SL_2(Z_2) \supset G'_2 \supset G'_k \supset (I+4M) \cap SL_2(Z_2) ,$$

with successive indices 2, 2, 12 as one sees from Theorem 2.5. This proves (i).

Since $\Delta \nmid -4Q^4$, we get $[F:k] = 4$, and $\Delta^{1/2}$ is contained in Q_{ab}. We know that $\Delta^{1/4}$ is contained in the field of 4-th division points, so the assertion (ii) is immediate from (i).

The lattice of fields is shown in the diagram.

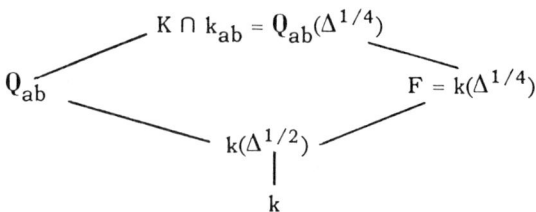

Remark. The hypothesis in both theorems that

$$G = GL_2(\mathbb{Z}_2) \times_\Delta G_{[2]}$$

is of course satisfied for Serre curves and for $X_0(11)$, whose Galois group was determined in Part I, §8.

$k = \mathbf{Q}(\sqrt{-3})$

§6. The action of $\mathcal{G}$ on $k(\Delta^{1/3})$

Eventually we want to mix the GL_2-extension with k_{ab}, and for this we have to make some properties of class field theory over k explicit on the cubic extension $k(\Delta^{1/3})$. We assume that $\Delta = \pm q^n$, where q is a prime > 3, and n is not divisible by 3. We let

$$k = \mathbb{Q}(\sqrt{-3}), \qquad F = k(\Delta^{1/3}) = k(q^{1/3}).$$

We let

$$\phi'' : \mathcal{G} \longrightarrow \mu_3 \approx \text{Gal}(F/k)$$

be the restriction homomorphism, composed with the identification of $\text{Gal}(F/k)$ with the group of cube roots of unity, through Kummer theory. We have to determine ϕ'' as explicitly as possible.

Case 1. $q \equiv \pm 1 \pmod{9}$

In this case, F/k is ramified only at q, because $\Delta^{1/3}$ has one conjugate in $\mathbb{Z}_3$, i.e. $X^3 - \Delta$ has one root mod 9, so has one root mod 27 (because a cube mod 9 is also a cube mod 27), and we can refine such a root to a root of $X^3 - \Delta$ in $\mathbb{Z}_3$.

It follows that ϕ'' factors through

$$\mathcal{G} \longrightarrow \mathcal{G}_q = \mathfrak{o}_q^*/\mathfrak{o}^*.$$

In addition, since $q \neq 3$, it must also be that ϕ'' factors through

$$\mathcal{G} \longrightarrow \mathfrak{o}(q)^*/\mathfrak{o}^*.$$

Case 1a. $q \equiv -1 \pmod{9}$

In this case, we must have $\left(\dfrac{k}{q}\right) = -1$. Then

$$\mathfrak{o}(q)^* = \mathbb{F}_{q^2}^*$$

and we can factorize ϕ'',

$$\mathfrak{o}_q^* \longrightarrow \mathfrak{o}(q)^* \longrightarrow \mu_3.$$

The kernel of ϕ'' consists of the elements which are third powers, i.e.

$$\mathfrak{o}_q^{*3}.$$

Representatives for the cosets are given by $1, \omega, \omega^2$.

Case 1b. $q \equiv 1 \pmod 9$

In this case, we must have $\left(\frac{k}{q}\right) = 1$, and we have a factorization

$$q = \pi \bar\pi \quad \text{in } \mathfrak{o}.$$

The extension F/k ramifies only over π and $\bar\pi$, and ϕ'' factors through

$$\mathfrak{o}(q)^* = \mathfrak{o}(\pi)^* \times \mathfrak{o}(\bar\pi)^* \approx F_q^* \times F_q^*.$$

Let $\theta, \bar\theta$ be such that

$$\theta^3 = \pi \quad \text{and} \quad \bar\theta^3 = \bar\pi.$$

For each extension $k(\theta)$ and $k(\bar\theta)$ we have the corresponding class field theory map

$$\phi''_\pi : \mathfrak{o}(\pi)^* \longrightarrow \text{Gal}(k(\theta)/k) \quad \text{and} \quad \phi''_{\bar\pi} : \mathfrak{o}(\bar\pi)^* \longrightarrow \text{Gal}(k(\bar\theta)/k).$$

Each of these Galois groups can be identified with μ_3, and

$$\phi''_q = \phi''_\pi \otimes \phi''_{\bar\pi}.$$

For $a \in \mathfrak{o}(\pi)^* \approx Z(q)^*$, let $\phi''_\pi(a) = \zeta$. Then

$$\phi''_q(a^{-1}) = \phi''_\pi(a)\phi''_{\bar\pi}(a^{-1}) = \zeta\zeta^{-1} = 1.$$

Hence if $\beta \in \mathfrak{o}(q)^*$ and $N\beta = 1$ then $\phi''_q(\beta) = 1$. Consequently, ϕ''_q factors through the norm map,

$$\phi''_q : \mathfrak{o}_q^* \longrightarrow \mathfrak{o}(q)^* \longrightarrow Z(q)^* \longrightarrow \mu_3.$$

Since $Z(q)^*$ is cyclic, it has a unique subgroup of index 3, which consists of the cubes. The kernel

$$\mathfrak{o}_{q,1}^{*'}$$

consists of the inverse image of these cubes under the norm map, composed with reduction mod q. Equivalently, it is the inverse image of the cubes in Z_q^* under the norm map

$$\mathfrak{o}_q^* \xrightarrow{N} Z_q^*.$$

Case 2. $q \not\equiv \pm 1 \pmod 9$

In this case, both 3 and q are ramified in $F = k(q^{1/3})$, and hence the class field theory map ϕ'' factors through

$$\phi''_{3q} : (\mathfrak{o}_3^* \times \mathfrak{o}_q^*)/\mathfrak{o}^* \longrightarrow \mu_3 .$$

Furthermore,

$$\phi''_{3q} = \phi''_3 \otimes \phi''_q .$$

The kernel of ϕ''_{3q} must then be determined through local class field theory, as the norm group locally.

Lemma 1. *Let* $\lambda = \sqrt{-3}$. *Then*

$$(1 + 3Z_3)(1 + 3\lambda Z_3) = 1 + 3\mathfrak{o}_3 .$$

Proof. The set $1 + 3\lambda Z_3$ is not a group, but it is easily verified that the left hand side is a group, and it is also easily verified recursively that every element of $1 + 3\mathfrak{o}_3$ can be written as an element of the left hand side. We leave the details to the reader.

Lemma 2. $\qquad (1 + \lambda \mathfrak{o}_3)^3 = 1 + 3\lambda \mathfrak{o}_3 .$

Proof. The left hand side is clearly contained in the right hand side. Conversely, suppose given an element of the form

$$1 + 3\lambda a, \qquad a \in \mathfrak{o}_3 .$$

Then $1 + \lambda a$ is a cube root mod 9. Furthermore, given an element of the form

$$1 + 9b ,$$

then $1 + 3b$ is a cube root mod 27. From here on, any standard refinement procedure takes hold to show that a cube root can be extracted, as desired.

Theorem 6.1. *Let* $q \not\equiv \pm 1 \pmod 9$. *Then*

$$\mathrm{Ker}\, \phi''_3 = \pm(1 + 3\mathfrak{o}_3) .$$

Proof. We note that q, and hence q^2 is a local norm, and

$$q^2 \not\equiv 1 \pmod 9 .$$

Consequently the closed subgroup generated by q^2 is $1 + 3\mathbf{Z}_3$, and is contained in the kernel of ϕ''_3. This kernel also contains all cubes, and therefore by Lemmas 1, 2 it contains $\pm(1+3\mathfrak{o}_3)$. Since $\pm(1+3\mathfrak{o}_3)$ has index 3 in $\mathfrak{o}_3^*$, the kernel is what we have stated.

Remark. The cosets of the kernel in Theorem 6.1 are the sets

$$\pm(\zeta + 3\mathfrak{o}_3)$$

where ζ ranges over the cube roots of unity. We shall denote them by

$$\mathfrak{o}_{3,\zeta}^* .$$

Observe that this is what we denote also by

$$r^{-1}(\zeta) ,$$

where r is reduction mod 3.

§7. The constant for Serre fiberings
$$k = Q(\sqrt{-3}), \quad M = 2q, \quad q \text{ odd prime} \neq 3, \quad \Delta = \pm q^n$$

We assume the conditions stated in the title of the section, and also that

$$G_{6q} = S_{2q} \times GL_2(Z_3),$$

where $S_{2q} = G_{2q}$ is Serre's subgroup. We determine the constant in Theorems 7.3, 7.4, 7.5. We first have to determine $\tilde{\mathcal{G}}_{6q}$, and decompose it as a union of products so that we can apply Part II, Theorem 6.1. We let

$$F = k(\Delta^{1/3}) = k(q^{1/3})$$

as before, and Δ has the stated value $\pm q^n$. From Theorem 4.2 we know that

$$K_{6q} \cap k_{6q}^{ab} = Q_{6q}^{ab}(\Delta^{1/3}).$$

We note that $\tilde{\mathcal{G}}_{6q}$ is of index 3 in

$$\tilde{\mathcal{G}}_{2q} \times (GL_2(Z_3) \times_N \mathcal{C}_3).$$

The mixing is due to the two maps ϕ' and ϕ'',

$$G \longrightarrow GL_2(3) \xrightarrow{\phi'} \mu_3 \approx \text{Gal}(F/k)$$

$$\mathcal{C} \longrightarrow \mathcal{C}_{3q} \xrightarrow{\phi''} \mu_3.$$

Consequently

(1) $$\mathcal{G}_{6q} = G_{6q} \times_{N, \phi' = \phi''} \mathcal{C}_{6q} = G_{k,6q} \times_{N, \phi' = \phi''} \mathcal{C}_{6q}.$$

To determine this fiber product, we write a decomposition for $G_{k,6q}$ and $\mathcal{C}_{6q}$.

We abbreviate as before

$$E_3 = SL_2(Z_3)(I + 3M_3).$$

Then

(2) $$G_{k,6q} = [(E_2 \times E_q) \cup (O_2 \times O_q)] \times E_3.$$

For each $\zeta \in \mu_3$ we let X_ζ be the set of elements $x \in X$ such that $\phi(x) = \zeta$, and $\phi = \phi'$ or $\phi = \phi''$ depending on the context. Thus for instance

$$E_{3,\zeta} = \{\sigma' \in E_3, \phi'(\sigma') = \zeta\}$$

$$\mathcal{C}_{q,\zeta} = \{\sigma'' \in \mathcal{C}_q, \phi''(\sigma'') = \zeta\}$$

and so forth. From (1) and (2) we get the decomposition

(3) $$\mathcal{G}_{6q} = \bigcup_{\zeta \in \mu_3} (E_2 \times E_{3,\zeta} \times E_q) \times_N (\mathcal{C}_2 \times (\mathcal{C}_3 \times \mathcal{C}_q)_\zeta)$$

$$\bigcup_{\zeta \in \mu_3} (O_2 \times E_{3,\zeta} \times O_q) \times_N (\mathcal{C}_2 \times (\mathcal{C}_3 \times \mathcal{C}_q)_\zeta).$$

We lift this to $\tilde{\mathcal{G}}$, to obtain

(4) $$\boxed{\begin{aligned}\tilde{\mathcal{G}}_{6q} &= \bigcup_\zeta [(E_2 \times E_{3,\zeta} \times E_q) \times_N (o_2^* \times (o_3^* \times o_q^*)_\zeta)]\\ &\quad \bigcup_\zeta [(O_2 \times E_{3,\zeta} \times O_q) \times_N (o_2^* \times (o_3^* \times o_q^*)_\zeta)].\end{aligned}}$$

The set $(o_3^* \times o_q^*)_\zeta$ is the inverse image of ζ under the map

$$\phi''_{3q} = \phi''_3 \otimes \phi''_q : o_3^* \times o_q^* \longrightarrow \mu_3.$$

We observe that the 2-factor in (4) splits off. Consequently we obtain:

(5) $$\text{Num}_{6q}(\tilde{\mathcal{G}}) = \text{Num}_2^+ \sum_\zeta \iint_{(o_3^* \times o_q^*)_\zeta} h_{E_{3,\zeta}}(\text{Tr } z_3, Nz_3) h_{E_q}(\text{Tr } z_q, Nz_q) dz_3 dz_q$$

$$+ \text{Num}_2^- \sum_\zeta \iint_{(o_3^* \times o_q^*)_\zeta} h_{E_{3,\zeta}}(\text{Tr } z_3, Nz_3) h_{O_q}(\text{Tr } z_q, Nz_q) dz_3 dz_q.$$

To give the value of $\text{Num}_{6q}(\tilde{\mathcal{G}})$ we have to distinguish cases depending on the congruence properties of q.

If $q \equiv \pm 1 \pmod 9$ then

$$(o_3^* \times o_q^*)_\zeta = o_3^* \times o_{q,\zeta}^*,$$

and consequently the 3-adic integral splits off. We obtain:

Theorem 7.1. *Let $\tilde{\mathcal{G}}_{6q}$ be the group associated with a Serre curve, and let k, 2q, Δ be as in the title of the section. Assume in addition that $q \equiv \pm 1 \pmod 9$. Then*

$$\text{Num}_{6q}(\tilde{\mathcal{G}}) = \text{Num}_2^+ \sum_\zeta \text{Num}_3(E_{3,\zeta}, o_3^*) \text{Num}_q(E_q, o_{q,\zeta}^*)$$

$$+ \text{Num}_2^- \sum_\zeta \text{Num}_3(E_{3,\zeta}, o_3^*) \text{Num}_q(O_q, o_{q,\zeta}^*).$$

On the other hand, if $q \not\equiv \pm 1 \pmod 9$, then

$$\text{Num}_{6q}(\tilde{\mathcal{G}}) = \text{Num}_2^+ \sum_\zeta \sum_{\zeta_3 \zeta_q = \zeta} \text{Num}_3(E_{3,\zeta}, o_{3,\zeta_3}^*) \text{Num}_q(E_q, o_{q,\zeta_q}^*)$$

$$+ \text{Num}_2^- \sum_\zeta \sum_{\zeta_3 \zeta_q = \zeta} \text{Num}_3(E_{3,\zeta}, o_{3,\zeta_3}^*) \text{Num}_q(O_q, o_{q,\zeta_q}^*).$$

The next section is devoted to evaluating the terms occurring in these sums.

In the following section, we shall establish a table of values for the numerator at the prime 3. We denote by ω, ω^2 the roots of unity in μ_3 which are $\neq 1$. The table is as follows.

		\multicolumn{3}{c}{o_3^*}		
		1	ω	ω^2
E_3	1	2/27	0	0
	ω	8/81	8/81	8/81
	ω^2	8/81	8/81	8/81

The table gives $\text{Num}_3(E_{3,\zeta}, o^*_{\zeta'})$. We note that these values are all equal to each other if $\zeta \neq 1$, and equal to 8/81. Thus we find for instance

$$\text{Num}_3(E_{3,\zeta}, o^*_3) = 8/27 \quad \text{if} \quad \zeta \neq 1 .$$

$$\text{Num}_3(E_{3,1}, o^*_3) = 2/27 .$$

We shall also find:

Theorem 7.2. *For all values of ζ,*

$$\text{Num}_q(O_q, o^*_{q,\zeta}) = \frac{1}{6}(1+r)(1-r)^2$$

is independent of ζ, where $r = 1/q$.

This allows us to make simplifications in the sums of Theorem 7.1. Start with the case $q \equiv \pm 1 \pmod 9$. We separate the first sum over ζ for $\zeta = 1$ and $\zeta = \omega, \omega^2$. In the second sum, we can replace the q-terms by their constant value, and then the sum over ζ for the terms $\text{Num}_3(E_{3,\zeta}, o^*_3)$ give $\text{Num}_3(E_3, o^*_3)$. We get:

$$\text{Num}_{6q}(\tilde{\mathcal{G}}_{6q}) = \text{Num}_2^+ \left[\frac{2}{27} \text{Num}_q(E_q, o^*_{q,1}) + \frac{16}{27} \text{Num}_q(E_q, o^*_{q,\omega}) \right]$$
$$+ \frac{1}{3} \text{Num}_2^- \text{Num}_3^+ \text{Num}_q^- .$$

Since we are in the ramified case, we know from the tables in Part II, §9 and §10 that

$$\text{Num}_3^+ = \text{Num}_3 = \frac{2}{3} .$$

Consequently the above expression can be rewritten in a more generic form as follows.

Theorem 7.3. *Suppose that $q \equiv \pm 1 \pmod 9$. Then*

$$\text{Num}_{6q}(\tilde{\mathcal{G}}_{6q}) = \frac{1}{3} \text{Num}_3 \left[\text{Num}_2^+ \left(\frac{1}{3} \text{Num}_q(E_q, o^*_{q,1}) + \frac{8}{3} \text{Num}_q(E_q, o^*_{q,\omega}) \right) + \text{Num}_2^- \text{Num}_q^- \right] .$$

Consider the other case $q \not\equiv \pm 1 \pmod 9$. Again in the first sum we take $\zeta = 1$ and $\zeta = \omega, \omega^2$ separately. We see that

$$\sum_{\zeta = \omega, \omega^2} \sum_{\zeta_3 \zeta_q = \zeta} \text{Num}_3(E_{3,\zeta}, o^*_{3,\zeta_3}) \text{Num}_q(E_q, o^*_{q,\zeta_q})$$

is equal to

$$\sum_{\zeta=\omega,\omega^2} \sum_{\zeta_3\zeta_q=\zeta} \frac{8}{81} \text{Num}_q(E_q, o_{q,\zeta_q}^*) = \frac{2\cdot 8}{81} \text{Num}_q^+$$

because the sum over ζ_3 becomes trivial.

On the other hand, we can use Theorem 7.2 for the second sum, and then the terms summed over ζ_3 add up to $\text{Num}_3(E_{3,\zeta}, o_3^*)$. We can then sum over ζ. We find:

$$\text{Num}_{6q}(\tilde{\mathcal{G}}) = \text{Num}_2^+\left[\frac{2}{27}\text{Num}_q(E_q, o_{q,1}^*) + \frac{16}{81}\text{Num}_q^+\right]$$
$$+ \frac{1}{3}\text{Num}_2^- \text{Num}_3^+ \text{Num}_q^- .$$

Factoring out in the same manner as in the preceding case, we obtain:

Theorem 7.4. *Assume that* $q \not\equiv \pm 1 \pmod 9$. *Then*

$$\text{Num}_{6q}(\tilde{\mathcal{G}}_{6q}) = \frac{1}{3}\text{Num}_3\left[\text{Num}_2^+\left(\frac{1}{3}\text{Num}_q(E_q, o_{q,1}^*) + \frac{8}{9}\text{Num}_q^+\right) + \text{Num}_2^- \text{Num}_q^-\right].$$

Remark. The expressions of Theorem 7.4 and 7.3 are identical, except for the terms

$$\frac{8}{9}\text{Num}_q^+ \quad \text{and} \quad \frac{8}{3}\text{Num}_q(E_q, o_{q,\omega}^*) .$$

However, these two terms are close together for large values of q.

We still have to give the values for the numerators involving E_q.

Theorem 7.5. *Suppose* $\left(\frac{k}{q}\right) = 1$. *Then for all* ζ,

$$\text{Num}_q(E_q, o_{q,\zeta}^*) = \frac{1}{6}(1+r)(1-r)^2 = \frac{1}{3}\text{Num}_q^+ .$$

Hence in this case,

$$\text{Num}_{6q}(\tilde{\mathcal{G}}_{6q}) = \frac{1}{3}\text{Num}_3 \text{Num}_{2q}(\mathcal{S}_{2q}) .$$

where $\mathcal{S}_{2q}$ *is Serre's group of Part II, §10.*

Proof. Obvious from Theorems 7.3, 7.4, and Part II, Theorem 10.2.

Corollary. If $\left(\frac{k}{q}\right) = 1$, then

$$\text{Den}_{6q}(\tilde{\mathcal{G}}_{6q}) = \frac{1}{3} \text{Den}_3 \, \text{Den}_{2q}(\mathcal{S}_{2q})$$

and

$$C_{6q}(\tilde{\mathcal{G}}_{6q}) = C_3 \, C_{2q}(\mathcal{S}_{2q}) .$$

Proof. The value for the denominator follows from Theorem 6.4 of Part II, because $\tilde{\mathcal{G}}_{6q}$ is obtained from a fibering of degree 3 over the Serre group. The value for the constant itself is then obvious.

Theorem 7.6. *Suppose* $\left(\frac{k}{q}\right) = -1$. *Then*

$$\text{Num}_q(E_q, \mathfrak{o}_{q,\zeta}^*) = \frac{1}{6}(1+r)(1-r)^2 - 2r \int_{\mathfrak{o}_{q,\zeta}^*} r^{v(z)} \, dz .$$

Proof. Given in the next section, formula (4).

The values for the integral are given in Lemmas 1 and 2 at the end of the next section.

§8. Computation of integrals

We compute the appropriate integrals to justify the values given in the preceding section.

Computations at 3

We have $k = \mathbb{Q}(\sqrt{-3})$. Let $\lambda = 1 - \omega$. We note that

$$\mathfrak{o}^*(3) = \{\pm 1, \pm\omega, \pm\omega^2\} = \{\pm 1, \pm(1+\lambda), \pm(1-\lambda)\}.$$

On the other hand, we had already seen in §4:

$$E_{3,1}(3) = \begin{cases} I \text{ and } -I \\ 6 \text{ matrices with trace } 0 \mod 3 \end{cases}$$

and for $\zeta \neq 1$:

$$E_{3,\zeta}(3) = \begin{cases} 8 \text{ matrices which are non-scalar } \mod 3 \\ 4 \text{ have trace } 1 \mod 3 \\ 4 \text{ have trace } -1 \mod 3. \end{cases}$$

Observe that all elements of $\mathfrak{o}^*(3)$ have trace $\not\equiv 0 \mod 3$. This gives rise to several orthogonalities in the integrals we have to evaluate.

From the description of $E_{3,\zeta}$ for $\zeta \neq 1$, and Part II, Theorem 7.1 we find:

$$h_{E_{3,\zeta}}(t,s) = \begin{cases} 4r^2 & \text{if } t \equiv 0 \text{ and } s \equiv 1 \mod 3 \\ 0 & \text{otherwise}. \end{cases}$$

Therefore for $\omega \neq 1$, we get

$$\mathrm{Num}_3(E_{3,\omega}, \mathfrak{o}^*_{3,\zeta}) = \tfrac{4}{9}\mu(\mathfrak{o}^*_{3,\zeta}) = 8/81,$$

and the same value with ω replaced by ω^2.

Since the traces of $\mathfrak{o}^*_3$ don't match the traces of matrices with trace 0 in $E_{3,1}$, we get

$$\text{Num}_3(E_{3,1}, o^*_{3,1}) = \int_{o^*_{3,1}} [h_{I+3M} + h_{-I+3M}(\text{Tr } z, Nz)] dz .$$

The integral of h_{I+3M} over $o^*_{3,1}$ is the same as its integral over $1 + 3o_3$. The integral of h_{-I+3M} over $o^*_{3,1}$ is the same as its integral over $-1 + 3o_3$. Hence their sum is equal to

$$2 \int_{3o_3} h_{3M}(\text{Tr } z, Nz) dz = 2/27$$

by Part II, §8, Lemma 6.

Next, let $\lambda = 1 - \omega$, so λ is the prime in o. We get:

$$\text{Num}_3(E_{3,1}, o^*_{3,\omega}) = \int_{\pm(1+\lambda)+3o_3} h_{I+3M} + \int_{\pm(1+\lambda)+3o_3} h_{-I+3M}$$

$$= \int_{1+\lambda+3o_3} h_{I+3M} + \int_{-(1+\lambda)+3o_3} h_{-I+3M}$$

$$= \int_{\lambda+3o_3} h_{3M} + \int_{-\lambda+3o_3} h_{3M}$$

$$= 0 .$$

(See Lemmas 4, 5 of Part II, §8.)

This concludes the proof of the evaluation of the entries of the table of 3-values.

Computations at q

We first deal with the terms involving O_q because they come out more simply and uniformly. Note that all elements of O_q are non-scalar mod q because the scalars have square determinant mod q.

Since q is an odd prime $\neq 3$, it follows that q is unramified in k.

In Part II, §10, Lemma 1, we had found:

$$h_{O_q}(\text{Tr } z, Nz) = \begin{cases} 0 & \text{unless } Nz \text{ is a non-square unit and otherwise:} \\ 1+r & \text{if } \left(\frac{k}{q}\right) = 1 \\ 1-r & \text{if } \left(\frac{k}{q}\right) = -1 \,. \end{cases}$$

On the other hand, for any ζ, under the norm map,

$$N : o_{q,\zeta}^* \longrightarrow Z_q^*$$

the inverse images N^{-1}(squares) and N^{-1}(non squares) differ by a multiplicative translation. Indeed, there exists an element $b \in o_q^*$ such that Nb is not a square (because q is unramified). Hence Nb^3 is not a square, and $b^3 \in o_{q,1}^*$. Multiplication by b^3 on $o_{q,\zeta}^*$ permutes the subsets whose norms are squares and non-squares respectively.

In particular, the measures of the sets of elements in $o_{q,\zeta}^*$ whose norms are squares or non-squares respectively, are equal, and in fact equal to

$$\tfrac{1}{2}\,\mu(o_{q,\zeta}^*) \,.$$

We now find:

$$\int_{o_{q,\zeta}^*} h_{O_q}(\text{Tr } z, Nz)\,dz = \tfrac{1}{2}\,\mu(o_{q,\zeta}^*) \cdot \begin{cases} 1+r & \text{if } \left(\frac{k}{q}\right) = 1 \\ 1-r & \text{if } \left(\frac{k}{q}\right) = -1 \end{cases}$$

(1) $$= \tfrac{1}{6}(1-r)^2(1+r)$$

as stated in Theorem 7.2.

We shift to E_q and M. In case $\left(\frac{k}{q}\right) = 1$, the factor $\psi_q(k) = 0$ in Theorem 8.2 of Part II. Hence we get

(2) $$\int_{o_{q,\zeta}^*} h_M(\text{Tr } z, Nz)\,dz = (1+r)\mu(o_{q,\zeta}^*) = \tfrac{1}{3}(1+r)(1-r)^2 \,.$$

Consequently by subtraction,

(3) $$\int_{o_{q,\zeta}^*} h_{E_q}(\text{Tr } z, Nz)\,dz = \tfrac{1}{6}(1-r)^2(1+r) \,.$$

For the rest of this section, we assume

$$\left(\frac{k}{q}\right) = -1.$$

In this case, $\psi_q(k) = 2r$. Using Theorem 8.1 of Part II, we find

(4) $$\int_{o_{q,\zeta}^*} h_M(\operatorname{Tr} z, Nz)\,dz = \tfrac{1}{3}(1+r)(1-r)^2 - 2r \int_{o_{q,\zeta}^*} r^{v(z)}\,dz.$$

Subtracting (1) gives the integral with M replaced by E_q. By Part II, Theorem 8.3, we get:

$$\int_{o_{q,\zeta}^*} r^{v(z)}\,dz = \frac{r^3}{1+r}\,|Z(q)^* \cap o(q)_\zeta^*|$$
$$+ r^2\,|o(q)_\zeta^* - Z(q)^*|.$$

We deal first with $\zeta = 1$. Then $o_{q,1}^*$ consists of the cubes in o_q^*. Since $\left(\frac{k}{q}\right) = -1$, all elements of $Z(q)^*$ are cubes, and $Z(q)^*$ is contained in $o(q)_1^*$. Hence

$$|o(q)_1^* \cap Z(q)^*| = q-1$$

$$|o(q)_1^* - Z(q)^*| = \frac{q^2-1}{3} - (q-1).$$

This gives

Lemma 1. *Assume* $\left(\frac{k}{q}\right) = -1$. *Then*

$$\int_{o_{q,1}^*} r^{v(z)}\,dz = \frac{r^2(1-r)}{1+r} + \frac{1-r^2}{3} - r + r^2.$$
$$= \frac{(1-r)(1-r+r^2)}{3(1+r)}.$$

Lemma 2. *Assume* $\left(\frac{k}{q}\right) = -1$. *If* $\zeta \neq 1$ *then*

$$\int_{o_{q,\zeta}^*} r^{v(z)}\,dz = \tfrac{1}{3}(1-r^2).$$

Proof. In this case, $\mathfrak{o}(q)^*_\zeta \cap Z(q)^*$ is empty, so the integral is equal to

$$r^2 |\mathfrak{o}(q)^*_\zeta| = \mu(\mathfrak{o}^*_{q,\zeta}) = \frac{1}{3} \mu(\mathfrak{o}^*_q),$$

which gives the desired value.

$k = \mathbf{Q}(i)$

§9. The constant for Serre fiberings, q odd ≠ 3

Throughout this section we assume that $k = \mathbb{Q}(i)$, and that

$$\Delta = -qc^4, \qquad \Delta_0 = -q,$$

where $c \in \mathbb{Z}$ and q is an odd prime. We let

$$F = k(\Delta^{1/4}),$$

so that by Theorem 5.2, F is cyclic over k of degree 4. We let B be the Galois group, $B = \text{Gal}(F/k)$, identified by Kummer theory with the group $\{\pm 1, \pm i\}$. We assume that

$$G_{2q} = GL_2(\mathbb{Z}_2) \times_\Delta GL_2(\mathbb{Z}_q).$$

The purpose of this section is to determine the constant

$$\text{Num}_{2q}(\tilde{\mathcal{G}}) = \text{Num}_{2q}(\tilde{\mathcal{G}}_{2q}).$$

This amounts to finding a decomposition of $\tilde{\mathcal{G}}_{2q}$, and computing integrals. We note that $k(\sqrt{\Delta})$ is unramified over k at 2, so the inertia group at 2 is the subgroup $\{\pm 1\}$ of B. From local class field theory, we have two local maps

$$\phi''_2 : \mathfrak{o}_2^*/N_2 \longrightarrow B \quad \text{and} \quad \phi''_q : \mathfrak{o}_q^*/N_q \longrightarrow B.$$

The image of ϕ''_2 is $\{\pm 1\}$. We put elements ζ of B as indices to indicate the inverse image of ζ under the maps ϕ''_2, ϕ''_q and the combined map

$$\phi'' = \phi''_2 \otimes \phi''_q : \mathfrak{o}_{2q}^* \longrightarrow B.$$

Then by definition,

(1) $$\mathfrak{o}_{2q,\zeta}^* = (\mathfrak{o}_{2,1}^* \times \mathfrak{o}_{q,\zeta}^*) \cup (\mathfrak{o}_{2,-1}^* \times \mathfrak{o}_{q,-\zeta}^*)$$

and

$$\tilde{\mathfrak{a}}_{2q} = \bigcup_\zeta \mathfrak{o}_{2q,\zeta}^*.$$

Furthermore,
$$o^*_{q,-\zeta} = - o^*_{q,\zeta}.$$

So much for the class field theory side, we don't need to know any more about the maps ϕ''. On the matrix side, we let

$$G_{k,2} = V = \operatorname{Ker} \chi_i \quad \text{in} \quad GL_2(\mathbb{Z}_2).$$

We have $G_2 = GL_2(\mathbb{Z}_2)$, $G_q = GL_2(\mathbb{Z}_q)$ and

$$G_{k,2q} = V \times_\Delta G_q.$$

We have homomorphisms

$$\phi'_2 : G_{k,2} \longrightarrow B \quad \text{and} \quad \phi'_q : G_q \longrightarrow B,$$

and by definition, for $\zeta \in B$, the inverse image under $\phi'_{2q} = \phi'_2 \otimes \phi'_q$ is

$$G_{k,2q,\zeta} = V_\zeta \times G_{q,\zeta^2}.$$

Then

(2) $$G_{k,2q} = \bigcup_\zeta (V_\zeta \times G_{q,\zeta^2}).$$

Putting (1) and (2) together, we obtain the decomposition

(3) $$\tilde{\mathcal{G}}_{2q} = \bigcup_\zeta \left[\left(V_\zeta \times_N o^*_{2,1} \right) \times \left(G_{q,\zeta^2} \times_N o^*_{q,\zeta} \right) \right]$$
$$\bigcup_\zeta \left[\left(V_\zeta \times_N o^*_{2,-1} \right) \times \left(G_{q,\zeta^2} \times_N o^*_{q,-\zeta} \right) \right].$$

By definition of even and odd elements (for an odd prime q, the elements in G_q whose determinant is a square or not), we see that

$$G_{q,1} = E_q \quad \text{and} \quad G_{q,-1} = O_q.$$

We also have $-E_q = E_q$ and $-O_q = O_q$. Making the change of variables $z \mapsto -z$ in the integral, we see that if $X_q = E_q$ or O_q,

502

$$\text{Num}_q(X_q, o_{q,\zeta}^*) = \text{Num}_q(X_q, o_{q,-\zeta}^*).$$

Since $o_{2,1}^* \cup o_{2,-1}^* = o_2^*$, we get the formula

(4) $$\text{Num}_{2q}(\tilde{\mathcal{G}}_{2q}) = \sum_{\zeta=1,-1} \text{Num}_2(V_\zeta, o_2^*)\text{Num}_q(E_q, o_{q,1}^*)$$

$$+ \sum_{\zeta=i,-i} \text{Num}_2(V_\zeta, o_2^*)\text{Num}_q(O_q, o_{q,i}^*).$$

Observe also the further symmetry

$$\text{Num}_2(V_\zeta, o_2^*) = \text{Num}_2(V_{-\zeta}, o_2^*)$$

which comes from the change of variables $z \mapsto -z$ and

$$-V_\zeta = V_{-\zeta}.$$

Theorem 9.1. Let $\mathcal{S}_{2q}$ be Serre's group. Then

$$\text{Num}_{2q}(\tilde{\mathcal{G}}_{2q}) = \frac{1}{2}\text{Num}_{2q}(\mathcal{S}_{2q}).$$

Proof. Serre's group was defined in Part II, §10, as

$$\mathcal{S}_{2q} = (E_2 \times_N o_2^*) \times (E_q \times_N o_q^*) \cup (O_2 \times_N o_2^*) \times (O_q \times_N o_q^*).$$

Note that

$$V \cap E_2 = V_1 \cup V_{-1} \quad \text{and} \quad V \cap O_2 = V_i \cup V_{-i}.$$

In $E_q \times_N o_q^*$ we may replace o_q^* by $o_{q,1}^* \cup o_{q,-1}^*$ without changing the value, because $E_q \times_N o_{q,i}^* = E_q \times_N o_{q,-i}^* = 0$. Similarly, in $O_q \times_N o_q^*$ we may replace o_q^* by $o_{q,i}^* \cup o_{q,-i}^*$. Then we obtain

$$\text{Num}_{2q}(\mathcal{S}_{2q}) = \sum_{\zeta=1,-1}\sum_{\zeta'=1,-1} \text{Num}_2(V_\zeta, o_2^*)\text{Num}_q(E_q, o_{q,\zeta'}^*)$$

$$+ \sum_{\zeta=i,-i}\sum_{\zeta'=i,-i} \text{Num}_2(V_\zeta, o_2^*)\text{Num}_q(O_q, o_{q,\zeta'}^*).$$

Each term is invariant under $\zeta \mapsto -\zeta$ or $\zeta' \mapsto -\zeta'$. All the terms are equal within each double sum, and $\text{Num}_{2q}(\tilde{\mathcal{G}}_{2q})$ consists of those terms for which $\zeta = \zeta'$. This proves the theorem.

Theorem 9.2. *We have*

$$\text{Den}_{2q}(\tilde{\mathcal{G}}_{2q}) = \frac{1}{2} \text{Den}_{2q}(\mathcal{S}_{2q})$$

and hence for the constant, the same value as in Part II, §10,

$$C_{2q}(\tilde{\mathcal{G}}_{2q}) = C_{2q}(\mathcal{S}_{2q}).$$

Proof. The group $\tilde{\mathcal{G}}_{2q}$ is obtained by a fibration which is cyclic of degree 2 over the field of roots of unity. The first assertion concerning the denominator follows from Part II, Theorem 6.4. The second is then obvious by the preceding theorem.

$$k = \mathbf{Q}(\sqrt{\Delta})$$

§10. The action of $\mathcal{G}$ on $k(A_2, \Delta^{1/4})$ when $k = Q(\sqrt{\Delta})$

Throughout this section, we assume that $\Delta = \Delta_0 c^4$, $c \in Z$, *and* $\Delta_0 = -q$, *where q is an odd prime,*

$$-q \equiv 5 \pmod{8}.$$

[In the applications, $q = 11$ or 43.] We let $k = Q(\sqrt{\Delta})$.

We are interested in the class field theory map

$$\phi_{2q} : \mathfrak{o}^*_{2q} = \mathfrak{o}^*_2 \times \mathfrak{o}^*_q \longrightarrow \text{Gal}(k(A_2, \Delta^{1/4})/k).$$

We begin with the kernel of the map

$$\phi_q : \mathfrak{o}^*_q \longrightarrow \text{Gal}(k(A_2, \Delta^{1/4})/k).$$

Theorem 10.1. *Let k_q be the completion of k at q. Then the norm group in $\mathfrak{o}^*_q$ from $k_q(\Delta^{1/4})$ is the group of squares, and -1 is not a square, so not a norm.*

Proof. Since q ramifies in k, we see that

$$\mathfrak{o}(q)^* \approx Z(q)^* \times Z(q),$$

and $\mathfrak{o}(q)^*$ is cyclic of order $q(q-1)$, so has a unique subgroup of index 2, which consists of the squares. The same is therefore true of $\mathfrak{o}^*_q$, as desired. It is obvious that -1 is not a square.

In the cases of interest to us for the completion of our tables, we have a special fact.

Theorem 10.2. *Let the elliptic curve be either the curve*

$$y^2 + y = x^3 + x^2,$$

or $X_0(11)$. Then $q = 43$ in the first case, and 11 in the second. In both

cases, we have
$$k_q(A_2) = k_q,$$
i.e. the prime above q in k splits completely in $k(A_2)$.

Proof. A translation in y puts the curve in the form $Y^2 = g(x)$, where
$$g(x) = x^3 + x^2 + 1/4 \quad \text{and} \quad g(x) = x^3 - x^2 - 10x - 20 + 1/4$$
respectively. In both cases, the derivative $g'(x)$ has no multiple roots mod q, and q ramifies in k. Hence there is a factorization
$$g(x) \equiv g_1(x) g_2(x)^2 \pmod{q}$$
where $g_1(x)$, $g_2(x)$ are linear. Since q ramifies in k, the existence of a prime of degree 1 implies the assertion of the theorem.

In particular, for the special case of Theorem 10.2 we find that
$$k_q(A_2, \Delta^{1/4}) = k_q(\Delta^{1/4}).$$

Next we work at the prime 2. We do not need to determine the kernel of the local map
$$\phi_2 : o_2^* \longrightarrow \text{Gal}(k(A_2, \Delta^{1/4})/k)$$
explicitly. The congruence $-q \equiv 5 \pmod 8$ implies that 2 remains prime in k.

Theorem 10.3. Let $F = k(A_2, \beta)$ where $\beta^4 = -q$. Then
$$[Fk_2 : k_2] = 6,$$
and Fk_2 is totally ramified over k_2. The norm group N_2 in o_2^* has index 6. Its cosets are represented by
$$\pm 1, \quad \pm \omega, \quad \pm \omega^2,$$
where $\omega^3 = 1$ and $\omega \neq 1$.

Proof. Since 2 remains prime in k and ramifies in $Q(A_2)$, it follows that 2 ramifies in $k(A_2)$ over k. Hence $k_2(A_2)$ has degree 3 over k_2 and is totally ramified of order 3. On the other hand, if we put

$$\lambda = \beta - 1,$$

then it is easy to see that λ satisfies an equation

$$\lambda^2 + 2\lambda - 2\eta = 0$$

where η is defined by

$$\sqrt{-q} = 1 + 2\eta.$$

Then η is a root of the equation

$$\eta^2 + \eta + m = 0, \qquad \text{where} \qquad -q = 1 - 4m,$$

m is odd, and so η is a unit at 2, and the equation for λ is an Eisenstein equation, which shows that F is also ramified of order 2, whence totally ramified at 2.

All the elements of $1 + 2\mathfrak{o}_2$ are cubes, and hence the cube roots of unity represent $\mathfrak{o}_2^*$ modulo cubes.

Finally, we have to see that -1 is not a local norm at 2. Let

$$(-1, 1, 1, \cdots)$$

be the idele which has component -1 at 2 and component 1 at all other primes of k. Then its Artin symbol is the same as the idele

$$(1, -1, -1, -1, \cdots).$$

The only ramified primes in F are 2 and q. We have already seen in Theorem 11.1 that -1 is not a local norm at q. Hence the Artin symbol of this idele is not trivial. This proves that -1 is not a local norm at 2, and concludes the proof of the theorem.

Remark. It can be shown that the norm group N_2 is generated by $1 + 4\mathfrak{o}_2$ and $-1 + 2\eta$. We do not need this in the sequel.

§11. The action of matrices on $k(A_4)$

We start with an elliptic curve over a field, defined by the equation

$$y^2 = f(x) = x^3 + c_4 x + c_6.$$

We let e_1, e_2, e_3 be the roots of $f(x)$. We let

$$\delta = 4(e_1 - e_2)(e_2 - e_3)(e_3 - e_1), \qquad \delta^2 = \Delta.$$

If (x_i, y_i) $(i = 1, 2, 3)$ are points on the curve, then we have the addition formula

$$x_1 + x_2 + x_3 = \left(\frac{y_1 - y_2}{x_1 - x_2}\right)^2$$

whenever

$$(x_1, y_1) + (x_2, y_2) + (x_3, y_3) = 0,$$

the addition being addition on the curve, and 0 the origin, i.e. the point at infinity. We then also have

$$\frac{y_1 - y_2}{x_1 - x_2} = \frac{y_2 - y_3}{x_2 - x_3} = \frac{y_3 - y_1}{x_3 - x_1}.$$

The points of order 2 are given in terms of coordinates by

$$(e_i, 0), \quad i = 1, 2, 3.$$

The Weierstrass form yields

$$e_1 + e_2 + e_3 = 0.$$

We need the coordinates for points of order 4 which divide these three points of order 2. If $P = (x, y)$ and $2P = (e_i, 0)$, then the x-coordinate of P is of the form

$$x = e_i \pm u_i \quad \text{where} \quad u_i^2 = 3e_i^2 + \sum_{j \neq j'} e_j e_{j'}.$$

and the four points P such that $2P = (e_i, 0)$ are of the form

$$(e_i + u_i, \pm v_i) \quad \text{where} \quad v_i^2 = u_i^2(3e_i + 2u_i)$$

$$(e_i - u_i, \pm v'_i) \quad \text{where} \quad v'^2_i = u_i^2(3e_i - 2u_i).$$

We fix a choice of u_1, v_1, u_2, v_2 and label

$$P_1 = (e_1 + u_1, v_1), \qquad P_2 = (e_2 + u_2, v_2).$$

We then define u_3, v_3 by the formula

$$P_3 = (e_3 + u_3, v_3) = -P_1 - P_2$$

so that

$$P_1 + P_2 + P_3 = 0.$$

Finally, we define v'_i to be those elements such that

$$(e_i + u_i, v_i) + (e_{i+1}, 0) + (e_i - u_i, v'_i) = 0.$$

We observe in passing that

$$16(u_1 u_2 u_3)^2 = -\delta^2,$$

so that we can express $\sqrt{-1}$ explicitly in the field of 4-division points, but we won't need this. We define

(1) $$w_i = \frac{v_i - v'_i}{(e_i + u_i) - (e_i - u_i)} = \frac{v_i - v'_i}{2u_i}.$$

We define

$$W = w_1 w_2 w_3.$$

The addition formula implies that

(2) $$w_i^2 = e_i - e_{i-1},$$

because

$$w_i^2 = e_i + u_i + e_{i+1} + (e_i - u_i) = 2e_i + e_{i+1} = e_i - e_{i-1}.$$

Therefore

(3) $$4W^2 = -\delta,$$

and we have obtained a root $\Delta^{1/4}$ in the field of 4-division points explicitly.

We let
$$\tau = \begin{pmatrix} -1 & 2 \\ 2 & 1 \end{pmatrix} \quad \text{and} \quad \gamma = \begin{pmatrix} -1 & -1 \\ 1 & 0 \end{pmatrix}.$$

Theorem 11.1. *Let* $F = k(A_2, \beta)$, $B = \text{Gal}(F/k)$.

(i) *For one of the choices of* $\sqrt{-q}$ *defining* η *and* β, *the matrix* τ *changes* β *to* $-\beta$. *For the other choice, the matrix* τ *leaves* β *fixed.*

(ii) *The effect of* γ *on* F *generates the cyclic subgroup of order* 3 *in* B.

Proof. Since the determinant of τ is not $\equiv 1 \pmod 4$, it follows that τ changes i to $-i$. Hence if τ leaves some fourth root of Δ fixed, it has to have a non-trivial effect on the other fourth root obtained by multiplication with i. This proves (i). As for (ii), since γ has period 3 and acts non-trivially on A_2, the assertion is obvious.

From now on, we assume that β *is chosen so that* $\tau\beta = \beta$. *Then*

$$E_1(4) = E'(4) \cup E'(4)\tau.$$

<pre>
                        ⎧   Q(A₄)
                        ⎪ |
                E'(4) ⎨ | 4
                        ⎪ |
                        ⎩ Q(A₂, β, i)
      E₁(4) |
                            | 2
                            |
                       F = Q(A₂, β)
                        2 ╱   ╲ 3   E(4)
        B   k(A₂)   k(β)
                    3 ╲   ╱ 2
                       k = Q(√Δ)
</pre>

§12. Computation of integrals and the constant

Throughout this section, we work under the same conditions as §10. We assume that $\Delta = \Delta_0 c^4$, $c \in \mathbb{Z}$, *and* $\Delta_0 = -q$, *where* q *is an odd prime,*

$$-q \equiv 5 \pmod 8.$$

We let $k = \mathbb{Q}(\sqrt{\Delta})$, *and* $\beta^4 = -q$. *We let* $F = k(A_2, \beta)$. *We let* $B = \mathrm{Gal}(F/k)$. *We assume that* $G_{2q} = S_{2q}$ *is Serre's fibering.*

We have the field diagram:

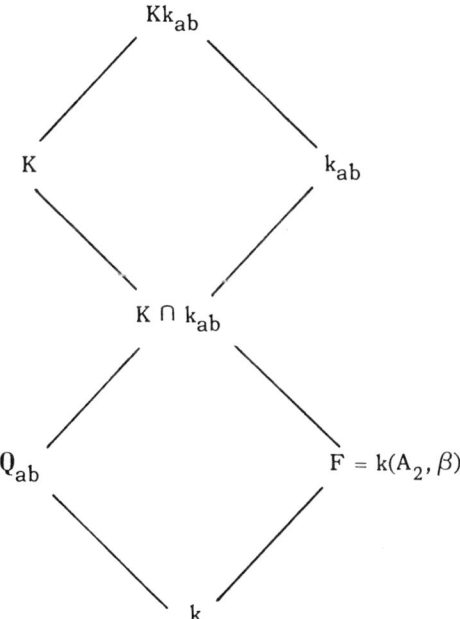

We know from Theorem 5.1 that $Q_{ab} F = K \cap k_{ab}$. We have two maps

$$\phi'_{2q} : S_{2q} \longrightarrow B$$

$$\phi''_{2q} : \tilde{\mathbb{Q}}_{2q} \approx \mathfrak{o}^*_{2q} \longrightarrow B$$

giving the Galois action on F for the GL_2-extension and for k_{ab}. We put

$$E_{2q} = E_2 \times E_q$$

as usual, where E_2 is the group of even elements, and E_q is the unique subgroup of index 2 in $GL_2(Z_q)$. Then

$$G_{k,2q} = E_{2q}.$$

We find

$$\tilde{\mathcal{G}}_{2q} = \{(\sigma_{2q}, a_{2q}) \in E_{2q} \times_N o^*_{2q}, \ \phi'_{2q}(\sigma_{2q}) = \phi''_{2q}(a_{2q})\}$$

or in other words,

(1)
$$\tilde{\mathcal{G}}_{2q} = \bigcup_{\zeta \in B} E_{2q,\zeta} \times_N o^*_{2q,\zeta}$$

where we denote elements of B by ζ, and index by ζ the elements of E_{2q} and o^*_{2q} which lie in the inverse image of ζ by the corresponding map ϕ' or ϕ''.

We sometimes write -1 also for the unique element of B which has period 2. This is unambiguous, since B is cyclic. Thus if $\zeta \in B$, then $-\zeta$ is equal to ζ times this unique element.

Since $F \subset k(A_4)$, and $k(A_4) \cap K_q = k$ (because G_{2q} is the Serre fibering) we see that E_q acts trivially on F, whence

$$E_{2q,\zeta} = E_{2,\zeta} \times E_q.$$

Since -1 is not a square in o^*_q, we see that

$$o^*_{q,-1} = \text{non-squares in } o^*_q.$$

We obtain:

$$o^*_{2q,\zeta} = (o^*_{2,\zeta} \times o^*_{q,1}) \cup (o^*_{2,-\zeta} \times o^*_{q,-1}),$$

and therefore we obtain the decomposition:

(2)
$$\boxed{\begin{array}{c} E_{2q,\zeta} \times_N o^*_{2q,\zeta} = \\ (E_{2,\zeta} \times_N o^*_{2,\zeta}) \times (E_q \times_N o^*_{q,1}) \\ \cup \ (E_{2,\zeta} \times_N o^*_{2,-\zeta}) \times (E_q \times_N o^*_{q,-1}). \end{array}}$$

Remark. We have

$$-E_{q,\zeta} = E_{q,-\zeta} \quad \text{and} \quad o^*_{2,-\zeta} = -o^*_{2,\zeta}.$$

The first equality results from the fact that $-I$ induces -1 on F, and the second likewise, in view of the fact that -1 is not a local norm at 2 (Theorem 10.3).

Lemma 1. $\text{Num}_q(E_q, o^*_{q,1}) = \text{Num}_q(E_q, o^*_{q,-1}) = \frac{1}{2} \text{Num}_q^+$.

Proof. We have

$$h_{E_q}(t,s) = h_{E_q}(-t,s)$$

because the map $\sigma \mapsto -\sigma$ reverses traces, does not change the determinant, and is measure preserving. Therefore

$$\int_{o^*_{q,1}} h_{E_q}(\text{Tr } z, Nz)\, dz = \int_{o^*_{q,1}} h_{E_q}(-\text{Tr } z, Nz)\, dz$$

$$= \int_{o^*_{q,-1}} h_{E_q}(\text{Tr } z, Nz)\, dz.$$

This comes from $-o^*_{q,1} = o^*_{q,-1}$, because -1 is not a square in $Z(q)^*$, and hence represents the coset of non-squares in the residue class field of o_q, which is the same as $Z(q)$. Since $o^*_{q,1} \cup o^*_{q,-1} = o^*_q$, the lemma follows from the definition of Num_q^+.

Let us define

$$Y_\zeta = o^*_{2,\zeta} \cup o^*_{2,-\zeta}, \qquad X_\zeta = E_{2,\zeta} \cup E_{2,-\zeta}.$$

The decomposition (2) and Lemma 1 give:

(3) $\qquad \text{Num}_{2q}(E_{2q,\zeta}, o^*_{2q,\zeta}) = \frac{1}{2} \text{Num}_q^+ \text{Num}_2(E_{2,\zeta}, Y_\zeta),$

and (1) shows that

$$\text{Num}_{2q}(\widetilde{\mathcal{G}}_{2q}) = \sum_\zeta \text{Num}_{2q}(E_{2q,\zeta}, o^*_{2q,\zeta})$$

$$= \frac{1}{2}\text{Num}^+_q \sum_\zeta \text{Num}_2(E_{2,\zeta}, Y_\zeta)$$

$$= \frac{1}{2}\text{Num}^+_q \sum_{\zeta=1,\omega,\omega^2} \text{Num}_2(X_\zeta, Y_\zeta)$$

We must therefore compute each term of this last sum. Observe that each such term is invariant under $\zeta \mapsto -\zeta$.

By Part II, Theorem 10.2 we know that $\text{Num}^+_q = 1 - 1/q$, because we are in the ramified case. The appropriate integrals will be computed to give:

Theorem 12.1. *Let the assumptions be those stated at the beginning of the section. Then*
$$\text{Num}_{2q}(\widetilde{\mathcal{G}}_{2q}) = \frac{11}{48} \frac{1}{2} (1-1/q)$$

Proof. We use Theorem 2.4. As in Theorem 11.1, let

$$\tau = \begin{pmatrix} -1 & 2 \\ 2 & 1 \end{pmatrix} \quad \text{and} \quad \gamma = \begin{pmatrix} -1 & -1 \\ 1 & 0 \end{pmatrix}$$

Since we work at the prime 2, we omit the subscript 2, and write E_ζ instead of $E_{2,\zeta}$. By our assumption on τ, we get

$$E'(4) = (I + 2O(2)) \cup I$$

$$E_1(4) = E'(4) \cup E'(4)\tau .$$

The elements of $O(2)$ have been tabulated (cf. Theorem 2.4), and all have trace $\equiv 0 \pmod 2$. The group E_1 is by definition the inverse image of $E_1(4)$ under reduction mod 4. We then get

$$E = X_1 \cup \gamma X_1 \cup \gamma^2 X_1 .$$

We have to tabulate the trace and determinant properties (mod 4) for the three sets X_1, γX_1, $\gamma^2 X_1$. We note that

$$X_1(4) = \pm E_1(4) \cup \pm E_1(4)\tau .$$

The table follows, giving the trace and determinant mod 4.

$$
\begin{array}{lll}
 & \text{trace} & \text{det} \\
X_1(4) = \begin{cases} \pm I, \pm I + 2O(2) \\ \pm\tau, \pm\tau + 2O(2) \end{cases} & \begin{array}{c} 2 \\ 0 \end{array} & \begin{array}{c} 1 \\ -1 \end{array} \\
\\
\gamma X_1(4) = \begin{cases} \gamma, \gamma + 2O(2) \\ -\gamma, -\gamma + 2O(2) \\ \gamma\tau, \gamma\tau + 2O(2) \\ -\gamma\tau, -\gamma\tau + 2O(2) \end{cases} & \begin{array}{c} -1 \\ 1 \\ 1 \\ -1 \end{array} & \begin{array}{c} 1 \\ 1 \\ -1 \\ -1 \end{array} \\
\\
\gamma^2 X_1(4) = \begin{cases} \gamma^2, \gamma^2 + 2O(2) \\ -\gamma^2, -\gamma^2 + 2O(2) \\ \gamma^2\tau, \gamma^2\tau + 2O(2) \\ -\gamma^2\tau, -\gamma^2\tau + 2O(2) \end{cases} & \begin{array}{c} -1 \\ 1 \\ -1 \\ 1 \end{array} & \begin{array}{c} 1 \\ 1 \\ -1 \\ -1 \end{array}
\end{array}
$$

We observe that Y_1 has index 3 in o_2^*, and consists of the cubes, whence

$$Y_1 = 1 + 2o_2 .$$

The cosets are represented by $1, \omega, \omega^2$. We have

$$X_\zeta = \gamma X_1 \quad \text{or} \quad \gamma^2 X_1, \qquad \text{if} \qquad \zeta = \omega \quad \text{or} \quad \omega^2,$$

and it is not necessary for us to know which.

We now compute the desired terms for $\zeta = \omega$ or ω^2.

From the table, we see that the elements of γX_1 and $\gamma^2 X_1$ are non scalar mod 2. For each one of these sets, every combination of 1 and -1 occurs, the same number of times, namely 4 times. From Part II, Theorem 7.1 we find:

$$h_{\gamma X_1}(\text{Tr } z, Nz) = h_{\gamma^2 X_1}(\text{Tr } z, Nz) = \begin{cases} \frac{1}{4} & \text{if Tr } z \text{ is odd,} \\ & Nz \text{ is a unit} \\ 0 & \text{otherwise.} \end{cases}$$

517

All elements of Y_ω or Y_{ω^2} have odd trace, and their norm is a unit. Consequently for $\zeta = \omega$ or ω^2 we get

(4) $$\text{Num}_2(X_\zeta, Y_\zeta) = \frac{1}{4}\mu(Y_\zeta) = \frac{1}{12}\mu(o_2^*) = \frac{1}{16}.$$

Finally, we compute the term with $\zeta = 1$.

All elements of Y_1 have even trace, and the elements of o_2^* which are not in Y_1 have odd trace. All elements of X_1 have even trace. Hence

$$\text{Num}_2(X_1, Y_1) = \text{Num}_2(X_1, o_2^*) = \text{Num}_2(X_1, o_2^*).$$

This leaves us with four integrals to compute. By Part II, §8, Lemma 6,

(5) $$\int_{o_2} h_{I+4M}(\text{Tr } z, Nz)\,dz = 5/384.$$

This integral will be counted 2 times.

Next, let $\sigma \in O_2$. We use constantly Part II, §8, Lemmas 4, 5, 6 without further reference. We obtain

(6) $$\int_{o_2} h_{I+2\sigma+4M}(\text{Tr } z, Nz)\,dz = \frac{1}{8}\int_{1+2o_2} h_{\sigma+2M}$$
$$= \frac{r^2}{8}\mu(1+2o_2)$$
$$= \frac{1}{128}.$$

This integral will be counted 6 times.

Next we write $\tau = I + 2\gamma \pmod 4$. We use Part II, Theorem 7.1. We get:

(7) $$\int_{o_2} h_{\tau+4M}(\text{Tr } z, Nz)\,dz = \frac{1}{8}\int_{o_2} h_{\gamma+2M}(\text{Tr } z, Nz)\,dz$$
$$= \frac{1}{8}\frac{1}{4}[\mu(\omega+2o_2) + \mu(\omega^2+2o_2)]$$
$$= 1/64.$$

This integral will be counted 2 times.

Next and last, we get

(8)
$$\int_{\mathfrak{o}_2} h_{\tau+2\sigma+4M}(\text{Tr } z, Nz)\,dz = \frac{1}{8} \int_{\mathfrak{o}_2} h_{\gamma+\sigma+2M}(\text{Tr } z, Nz)\,dz$$
$$= 0.$$

because $\gamma + \sigma$ has odd trace and even determinant, which is incompatible with any element of $\mathfrak{o}_2$.

Taking the sum of the last four integrals appropriately weighted yields

(9)
$$\text{Num}_2(X_1, Y_1) = \frac{2\cdot 5}{384} + \frac{6}{128} + \frac{2}{64} = \frac{5}{3} \cdot \frac{1}{16}.$$

Hence by (4) and (9) we obtain

$$\sum_{\zeta=1,\omega,\omega^2} \text{Num}_2(X_\zeta, Y_\zeta) = \frac{2}{16} + \frac{5}{3} \cdot \frac{1}{16} = \frac{11}{48}.$$

This concludes the proof.

PART IV

NUMERICAL RESULTS

PART IV

NUMERICAL RESULTS

SUPERSINGULAR AND FIXED TRACE DISTRIBUTION

1. General discussion of results	235
2. Tables	
Table I : Fixed trace distributions	239
Table II : Supersingular primes	240
Table III: Primes with $t_p = 1$	241
Table IV: Traces of Frobenius	242

IMAGINARY QUADRATIC DISTRIBUTION

3. General discussion of results	249
4. Tables	
Table V : Imaginary quadratic distributions	253
Table VI : Primes associated with fields of small discriminant, for curves A and B	258
Table VII: Distribution of primes associated with small discriminants	260

EXTENDED RESULTS FOR $X_0(11)$

5. Discussion and description of tables	265
Table VIII: Supersingular primes	267
Table IX : Imaginary quadratic distribution	268
Table X : Distribution of primes for fields with small discriminants	269
Remarks on the Computations	271
Bibliography	273

SUPERSINGULAR AND FIXED TRACE DISTRIBUTION

§1. General discussion of results

In this final part we present the results of numerical calculations for five curves. Four of them, which are arbitrarily labeled A, B, C, D appear in Serre [S 2] as examples 5.5.6, 5.5.7, 5.5.8 and 5.9.2. They have the property that the group G is of index 2 in $\prod GL_2(\mathbb{Z}_\ell)$. See §5 and §7 of Part I. The fifth is the modular curve $X_0(11)$ whose Galois group is determined in §8, and which also appears in [S 2] as example 5.5.2.

The following table gives the equation, conductor N, discriminant Δ, and j-invariant for each curve.

		N	Δ	j
A	$y^2 + y = x^3 - x$	37	37	$2^{12}3^3/37$
B	$y^2 + y = x^3 + x^2$	43	-43	$-2^{12}/43$
C	$y^2 + xy + y = x^3 - x^2$	53	53	$-3^3 5^3/53$
D	$y^2 = x^3 + 6x - 2$	$2^6 3^3$	$-2^6 3^5$	$2^9 3$
$X_0(11)$	$y^2 + y = x^3 - x^2 - 10x - 20$	11	-11^5	$-2^{12}31^3/11^5$

We used the machine to compute the values of the Frobenius traces t_p for each of these curves for the first 5,000 primes, from $p_1 = 2$ to $p_{5,000} = 48,611$. More extensive calculations for the curve $X_0(11)$ are reported in §5. We also calculated the constant $C(t, A)$ for $-7 \leq t \leq 7$ given by the formulas developed in §5 and §8 of Part I.

In table I, the first column for each curve gives the number of primes among the first 5,000 with the given trace. The second column gives the value predicted by our conjecture, i.e.

$$E_t = E_t(48,611) = C(t, A)\pi_{1/2}(48,611) = C(t, A)\, 26.434 ,$$

where the value of $\pi_{1/2}(48,611)$ is obtained by direct calculation. The number in the third column indicates roughly how well the data fit the conjecture for each particular case. Its meaning is explained further below. In statistical terms, the overall fit is quite good. The standard χ^2-test involves calculating the sum of

$$\frac{(\text{Predicted value} - \text{Actual value})^2}{\text{Predicted value}}$$

over all the cases. For the data in table I, we get the following results (omitting the traces $\equiv 1 \mod 5$ for $X_0(11)$ where the predicted value is 0).

Curve	χ^2-value	Probability
A	8.4	.91
B	14.8	.46
C	16.8	.32
D	10.8	.75
$X_0(11)$	5.6	.93

The column headed "probability" gives the chance that a random sample using the theoretical frequencies would give a χ^2-value greater than that observed, i.e. would give no better fit. (The values are from standard statistical tables, using 15 degrees of freedom for A, B, C, and D where there are 15 possibilities, and 12 degrees of freedom for $X_0(11)$.) By this criterion there is very good fit for A and $X_0(11)$, and reasonably good fit in all cases. The single prime that has $t_p = 1$ for $X_0(11)$ is the prime 5, which (unlike all other primes) is congruent to 0 modulo 5.

The third column for each curve in table I is calculated as follows. For a truly random model, the probability of getting a count of k when the theoretical frequency is m is well approximated by the formula for the Poisson distribution

$$p_k = e^{-m} m^k / k!$$

The value given in the table is $\sum p_i$, with the sum taken over all i such that $p_i \leq p_k$. Thus the value in the table is the chance that a random experiment based on the given theoretical frequency would not give a more likely result. Note that the value is 1 (printed as .99) if m comes anywhere between k and k + 1, since then p_k is the maximum of the p_i. No significance should be given to the precise value printed, but it does give a reasonable indication of how well the data agree with the prediction.

There are some rather low values, at traces −5 and 1 for B, and traces −6 and 4 for C, where the probability is about 1 in 30. It should be noted, however, that having as many as four events, each with probability about 1/30, occurring in a total of 72 "experiments," is not unlikely. Somewhat more disturbing is that in 8 out of 9 cases with trace ±4, the predicted value is noticeably less than the actual value. We don't see any easy way of pushing computations so much farther as would make it clear if this phenomenon persists or disappears.

In table II we list for each curve the supersingular primes $\leq 48,611$ (that is, those among the first 5,000 primes), and in table III, the primes with $t_p = 1$, which are of special interest in the light of Mazur's work [Ma]. Table IV shows the values of t_p for the first 200 primes.

Table I

	A			B			C		
t	N_t	E_t	Pr	N_t	E_t	Pr	N_t	E_t	Pr
-7	7	10.3	.43	10	10.4	.99	14	10.3	.27
-6	22	24.9	.69	25	24.8	.92	14	24.8	.03
-5	13	10.3	.35	4	10.4	.04	13	10.3	.35
-4	26	20.7	.23	26	20.7	.23	25	20.7	.32
-3	14	12.4	.57	10	12.4	.67	12	12.4	.99
-2	18	20.7	.66	18	20.7	.66	21	20.7	.91
-1	13	10.3	.53	13	10.4	.35	14	10.3	.27
0	21	27.3	.25	23	28.0	.39	28	27.4	.85
1	10	10.3	.99	18	10.4	.03	6	10.3	.21
2	21	20.7	.91	20	20.7	.99	20	20.7	.99
3	11	12.4	.89	12	12.4	.99	11	12.4	.89
4	25	20.7	.32	24	20.7	.44	11	20.7	.03
5	10	10.3	.99	11	10.4	.76	7	10.3	.43
6	19	24.9	.27	27	24.8	.62	24	24.8	.99
7	9	10.3	.88	13	10.4	.35	9	10.3	.88

	D			$X_o(11)$		
t	N_t	E_t	Pr	N_t	E_t	Pr
-7	9	12.4	.40	13	13.2	.99
-6	28	29.0	.99	35	31.2	.47
-5	10	13.1	.49	10	12.5	.57
-4	21	18.6	.56	0	0.0	--
-3	7	8.3	.86	21	15.8	.21
-2	19	18.6	.91	23	26.0	.62
-1	9	12.4	.40	12	13.2	.89
0	33	32.3	.86	33	34.7	.87
1	12	12.4	.99	1	0.0	--
2	24	18.6	.20	24	26.0	.77
3	5	8.3	.38	14	15.8	.80
4	25	18.6	.16	33	26.0	.17
5	10	13.1	.49	12	12.5	.99
6	24	29.0	.40	0	0.0	--
7	16	12.4	.32	12	13.2	.89

Fixed trace distributions

Table II

A	B	C	D	$X_O(11)$
17	7	5	101	19
19	37	11	251	29
257	1109	239	1637	199
311	1361	751	2383	569
577	1531	1201	2411	809
2161	2069	1459	2441	1289
2243	2281	1663	3631	1439
3511	2543	2089	4241	2539
7951	3011	2111	4889	3319
10399	5351	2153	8081	3559
11261	6569	2411	10979	3919
11579	7211	3001	11399	5519
15331	9923	4283	11471	9419
21031	12289	5443	12503	9539
21991	16339	5867	13577	9929
23369	19819	6113	15053	11279
23567	21191	8999	15131	11549
30319	29741	9127	15887	13229
31481	31531	9227	16619	14489
38167	31547	10079	17417	17239
40127	32411	10667	19913	18149
	38711	12071	22271	18959
	41077	15269	23057	19319
		29221	23561	22279
		32467	23981	24359
		33619	25163	27529
		45613	27799	28789
		46817	28439	32999
			30839	33029
			33767	36559
			35591	42899
			41399	45259
			46307	46213

Supersingular primes ≤ 48,611

Table III

A	B	C	D	$X_o(11)$
53	103	71	7	5
127	127	97	97	
443	541	1063	1447	
599	1429	3061	1663	
3989	1657	4993	6277	
23269	2087	5903	6397	
29131	3733		7039	
30529	8641		8821	
44179	8753		13147	
47237	10903		13417	
	18313		15919	
	21467		19333	
	34739			
	34897			
	41453			
	44579			
	46261			
	47857			

Primes $\leq 48{,}611$ with $t_p = 1$

Table IV Traces of Frobenius

p	A	B	C	D	$X_o(11)$
2	-2	-2	-1	*	-2
3	-3	-2	-3	*	-1
5	-2	-4	0	2	1
7	-1	0	-4	1	-2
11	-5	3	0	2	*
13	-2	-5	-3	-1	4
17	0	-3	-3	-6	-2
19	0	-2	-5	5	0
23	2	-1	7	6	-1
29	6	-6	-7	8	0
31	-4	-1	4	8	7
37	*	0	5	5	3
41	-9	5	6	-8	-8
43	2	*	-2	4	-6
47	-9	4	-2	-10	8
53	1	-5	*	4	-6
59	8	-12	-2	-14	5
61	-8	2	-8	-3	12
67	8	-3	-12	13	-7
71	9	2	1	-4	-3
73	-1	2	-4	9	4
79	4	-8	-1	11	-10
83	-15	15	-1	12	-6
89	4	-4	-14	2	15
97	4	7	1	1	-7
101	3	-9	-2	0	2
103	18	1	-1	-3	-16
107	-12	-12	6	-2	18
109	-16	7	16	10	10
113	-18	-20	15	10	9
127	1	1	13	20	8
131	-12	8	-2	-16	-18
137	-6	6	12	6	-7
139	4	19	-20	-5	10
149	-5	12	-5	-12	-10
151	16	-20	-3	-9	2
157	23	-10	-4	-2	-7
163	-18	14	-6	11	4
167	-12	-9	21	14	-12
173	9	6	10	-24	-6
179	18	20	11	12	-15
181	5	10	-2	19	7
191	-4	-16	-21	-6	17
193	-26	3	-16	-21	4
197	3	2	-18	2	-2
199	2	14	4	-19	0
211	-13	2	-2	-1	12
223	-17	-28	-14	16	19
227	-16	-4	6	-12	18
229	7	-15	21	-26	15

* indicates bad reduction

Table IV Traces of Frobenius (continued)

p	A	B	C	D	$X_0(11)$
233	6	6	-8	24	24
239	-6	16	0	12	-30
241	14	-12	-11	13	-8
251	-2	-23	20	0	-23
257	0	-24	-28	4	-2
263	19	-18	-28	-12	14
269	-6	-25	9	6	10
271	-31	23	-14	-21	-28
277	12	-32	-8	2	-2
281	12	19	17	-24	-18
283	4	21	-9	-20	4
293	-2	-26	26	-18	24
307	-17	-7	-16	12	8
311	0	15	16	-30	12
313	22	22	-6	-17	-1
317	22	9	3	12	13
331	-2	-26	28	-29	7
337	-25	-3	10	-31	-22
347	-10	28	-30	4	28
349	6	14	24	5	30
353	8	-31	-18	-4	-21
359	-15	19	9	-14	-20
367	8	-32	22	-15	-17
373	-19	32	-10	-11	-26
379	15	11	11	-19	-5
383	20	32	16	-16	-1
389	4	6	12	18	-15
397	-5	-6	18	14	-2
401	18	5	-32	-28	2
409	20	-24	19	13	-30
419	7	-28	-12	26	20
421	-24	-10	14	19	22
431	-30	-21	-36	30	-18
433	9	-12	-7	-2	-11
439	28	17	-34	-8	40
443	1	-4	-33	-36	-11
449	36	30	-11	-34	35
457	18	-18	38	2	-12
461	30	30	-9	30	12
463	-22	4	-23	-19	-11
467	-2	6	-28	2	-27
479	14	21	3	12	20
487	-24	36	20	-11	23
491	-28	-6	27	14	-8
499	12	-8	-23	24	20
503	16	6	16	6	-26
509	-31	-15	2	-6	15
521	-33	14	-45	-42	-3
523	-22	12	-42	-15	-16
541	20	1	22	-37	-8

535

Table IV   Traces of Frobenius (continued)

p	A	B	C	D	$X_o(11)$
547	8	-29	-38	15	8
557	-18	-3	14	14	-2
563	-30	37	-1	24	4
569	-24	7	24	10	0
571	7	-14	-4	-19	-28
577	0	-20	10	3	33
587	-32	2	28	6	28
593	-5	-16	25	36	44
599	1	-1	-30	8	40
601	-22	-4	22	-26	2
607	-32	-4	-8	41	-22
613	15	-18	-16	9	-16
617	17	-21	-42	-42	18
619	-1	36	16	9	-25
631	-28	-6	7	-1	7
641	-1	12	46	-4	-33
643	14	-36	-8	12	29
647	-8	-12	38	24	-7
653	-24	-14	-38	-44	-41
659	-15	-19	12	-36	10
661	-28	31	-1	-23	37
673	27	14	-10	23	14
677	-11	34	40	-14	-42
683	18	-9	2	-24	-16
691	-20	-40	41	-4	17
701	-12	-2	12	30	2
709	40	-1	6	23	-25
719	39	8	24	-52	15
727	16	16	-30	48	3
733	7	32	14	34	-36
739	-9	-10	15	20	50
743	21	-24	-26	-42	4
751	25	-6	0	-25	-23
757	-50	28	-43	-33	-22
761	-35	20	-28	42	12
769	26	-42	-2	-17	20
773	-9	-4	40	-36	-6
787	-5	4	-4	-3	-32
797	52	-42	16	-24	53
809	2	26	-16	20	0
811	47	-14	42	20	-38
821	-47	49	-24	2	22
823	-16	-1	26	43	39
827	22	-36	47	2	-52
829	-4	44	-30	-41	25
839	44	-40	-30	-4	-5
853	26	-29	-56	-15	14
857	-48	-10	42	-28	8
859	-20	-32	-58	19	-15
863	-24	-6	-30	38	24

Table IV   Traces of Frobenius (continued)

p	A	B	C	D	$X_o(11)$
877	50	41	-2	17	-12
881	-14	37	16	6	-43
883	48	31	-36	31	4
887	25	-22	-7	12	-22
907	52	47	46	17	-12
911	26	-22	-32	-48	12
919	-58	-49	15	48	10
929	18	-6	38	-12	-30
937	37	32	-55	-25	8
941	-10	33	10	22	42
947	12	-33	-12	6	-27
953	61	22	-42	-6	34
967	-14	37	52	27	-32
971	-8	-13	42	-34	47
977	28	34	-20	-44	-27
983	9	-24	46	42	39
991	-18	2	34	-7	-8
997	-42	4	-13	54	38
1009	-47	-18	16	-11	-10
1013	36	-22	39	-26	39
1019	46	-30	15	60	-10
1021	-62	-14	-52	54	22
1031	-4	46	-22	22	32
1033	3	13	-54	-59	-16
1039	-59	34	44	-23	5
1049	-4	6	3	20	-55
1051	-16	-60	-30	4	2
1061	-62	-40	-31	54	-13
1063	7	12	1	31	44
1069	-30	32	-23	-46	-20
1087	12	20	-25	59	8
1091	30	40	-13	20	-58
1093	-36	-46	-14	-22	-51
1097	36	12	-1	-2	-42
1103	-8	54	-26	24	-51
1109	-35	0	-11	28	-30
1117	33	4	-53	-45	48
1123	-22	60	-52	-41	24
1129	50	-26	-5	41	50
1151	-25	-30	10	-18	2
1153	18	21	6	37	-31
1163	-36	32	-12	44	34
1171	-22	-52	11	-53	-3
1181	57	-40	55	22	-18
1187	-33	-48	3	-26	-12
1193	-11	32	-54	-18	-21
1201	44	-55	0	-5	2
1213	-12	-59	31	47	-41
1217	-41	-30	48	-14	-42
1223	30	-6	-12	-4	14

IMAGINARY QUADRATIC DISTRIBUTION

§3. General discussion of results

For each of the curves already described and each of the first 5,000 primes, we used the machine to find the square-free part of $4p - t_p^2$, and thus determine the quadratic field k, (excluding supersingular primes and those with bad reduction).

Table V gives data for each curve on the number of primes found for the 31 fields of discriminant D with $|D| < 100$. Two sets of data are given for each curve. The first refers to primes among the first 5,000, and the second to primes p such that
$$p_{72} = 359 < p \leq p_{5,000} = 48,611.$$

Our conjecture deals with asymptotic behavior, and our heuristic arguments involve the assumption that p is large compared to $|D|$. It is clearly very unlikely for large values of D to occur with small p, but small values of p contribute heavily to the factor
$$\pi_{1/2}(x) = \sum_{p \leq x} \frac{1}{2\sqrt{p}}.$$

We therefore thought it appropriate to look at the result of eliminating some of the small primes from consideration. Choosing to discard exactly 72 primes was arbitrary, but convenient because we had already subdivided the data in that way (see the description of Table VII below).

The table has a page for each curve. The first column gives the field discriminant, the next three pertain to the first 5,000 primes, and the last three to the primes with the first 72 primes omitted. In each group of 3 columns, the first gives the actual number of primes found, the second gives the expected number E_D calculated from the formulas of Parts II and III, and the third gives the probability that a random experiment would not give a more likely result. (See the discussion of these probabilities given in §1.)

As in the fixed trace case, there is generally good agreement between actual and predicted counts. There is a very conspicuous discrepancy at $D = -11$ for the curve $X_0(11)$, when all primes are considered, but the discrepancy becomes

541

much less significant when the low primes are discarded. There are several other cases where the probability falls as low as .01, but it is statistically reasonable for a few such low probabilities to come up among 154 cases. (See §5 for further discussion of the curve $X_0(11)$.)

A χ^2 computation using the data on only the first 15 discriminants (to avoid excessively small frequencies) gives the following values. (See the discussion in §1 for an explanation. There are 15 degrees of freedom, except for curve D, which has 14 because the predicted value for the field $Q(\sqrt{-3})$ was not calculated.)

	5,000 primes		4,928 primes	
Curve	χ^2	Probability	χ^2	Probability
A	17.8	.27	16.9	.32
B	7.8	.93	8.0	.92
C	15.6	.41	15.3	.44
D	7.1	.93	7.5	.91
$X_0(11)$	26.7	.03	14.6	.47

By this criterion, the fit is reasonably good except for $X_0(11)$ when all 5,000 primes are considered. This high value for χ^2 can be attributed to the bad fit at $D = -11$, and when the small primes are discarded, the result is quite acceptable.

For most entries in the table, we have $K \cap k_{ab} = Q_{ab}$, and the predicted values are found from the formulas of Part II. It was shown in Part III that the same formulas apply for $k = Q(i)$, and if $q \equiv 1 \pmod{3}$, for $k = Q(\sqrt{-3})$, even though

$$K \cap k_{ab} \neq Q_{ab}$$

in these cases. For the curve C and $X_0(11)$, with $k = Q(\sqrt{-3})$, we have

$$q \equiv -1 \pmod{3},$$

so the formulas of Part III apply. The value obtained, however, differs by less than one per cent from the value one gets simply applying the formulas of Part II.

The situation is very different for the curve B, with $D = -43$, and the curve $X_0(11)$, with $D = -11$, where $k = Q(\sqrt{\Delta})$. Taking the intersection $K \cap k_{ab}$ fully into account increases the prediction by a factor of almost 2. For $D = -43$

there is a sharp difference between the actual values for the curve B and the other Serre curves, perfectly in line with the prediction. For $D = -11$ and $X_0(11)$, the situation is further complicated by the factor at 5, see Theorem 11.3 of Part II.

Except for $k = Q(\sqrt{-43})$, one notices that the predicted values for A, B, C are identical to the first decimal place. This is not surprising, because the tables of §10 show that the constants differ by a factor of the form $1 + O(1/q)$, and the values of q for the three curves are 37, 43, 53. The value of q for curve D is 3, so slight differences appear.

In Table VI we give a complete list of all the primes corresponding to the indicated discriminants for the curve A and B. Visually, the table looks like a parabola with kinks, as it should. The other cases we have computed look roughly the same, so we did not feel it was worthwhile to include them.

In Table VII we present a more detailed breakdown of how the primes belonging to different fields are distributed. If one divides the primes up to $p_{5,000} = 48,611$ into 12 segments

$$X_i = \{p, p_{i-1} < p \leq p_i\}$$

so as to make the sums

$$\sum_{p \in X_i} \frac{1}{2\sqrt{p}}$$

as nearly equal as possible, the division points come out as follows.

Segment	i	p_i
1	16	53
2	72	359
3	187	1117
4	371	2539
5	632	4673
6	976	7691
7	1407	11719
8	1930	16691
9	2548	22807
10	3263	30169
11	4080	38713
12	5000	48611

According to our conjecture, one would have the primes associated with a given field equally distributed over the 12 segments. Of course, since the number predicted for an individual segment is small, considerable statistical fluctuation is to be expected.

In the table, the first column gives the field discriminant, and the second gives the total count for each field. The last 12 columns give the counts $N_D(X_i)$ for the individual segments. The bottom line gives the total counts for each segment. The total count for the first segment is noticeably low, which is not surprising since it contains only 16 primes altogether. Otherwise, the distribution across segments appears reasonably uniform.

We discarded the first two segments to get the reduced data used for the second group of columns in Table V. Since the sums

$$\sum_{p \in X_i} \frac{1}{2\sqrt{p}}$$

are almost equal for $i = 1, 2, \cdots, 12$, the theoretical frequencies for the reduced set of primes are very nearly $10/12$ of the corresponding frequencies for the full set.

Table V

D	N_D	E_D	Pr	N_D	E_D	Pr
-3	40	34.9	.40	30	29.1	.85
-4	24	28.5	.45	21	23.7	.68
-7	18	17.0	.81	18	14.2	.29
-8	21	18.0	.48	18	15.0	.44
-11	17	15.8	.71	16	13.2	.41
-15	13	10.9	.54	12	9.1	.32
-19	13	13.6	.99	12	11.3	.77
-20	19	10.9	.02	15	9.1	.06
-23	6	8.1	.60	6	6.7	.99
-24	5	10.7	.09	4	8.9	.13
-31	4	7.5	.27	3	6.3	.31
-35	5	8.9	.24	5	7.4	.46
-39	8	6.5	.55	8	5.4	.27
-40	7	8.9	.74	6	7.4	.85
-43	14	9.9	.20	13	8.2	.11
-47	2	5.5	.19	2	4.6	.34
-51	10	8.0	.48	9	6.7	.33
-52	10	8.0	.48	10	6.7	.18
-55	2	5.8	.14	2	4.8	.26
-56	8	6.3	.43	7	5.3	.38
-59	7	6.6	.84	7	5.5	.52
-67	10	8.1	.48	9	6.7	.33
-68	7	5.9	.54	7	4.9	.36
-71	2	4.3	.46	1	3.6	.28
-79	3	4.7	.64	3	4.0	.99
-83	3	5.8	.30	2	4.9	.26
-84	4	5.7	.67	4	4.8	.99
-87	1	4.4	.15	0	3.7	.06
-88	5	6.4	.84	2	5.4	.19
-91	7	6.4	.69	7	5.3	.39
-95	4	3.8	.80	3	3.2	.99

Curve A

Table V

D	N_D	E_D	Pr	N_D	E_D	Pr
-3	34	34.9	.99	29	29.1	.99
-4	26	28.5	.71	23	23.7	.99
-7	18	17.0	.81	16	14.2	.59
-8	14	18.0	.41	9	15.0	.15
-11	12	15.8	.45	12	13.2	.89
-15	8	10.9	.45	7	9.1	.62
-19	9	13.6	.28	8	11.3	.45
-20	10	10.9	.99	9	9.1	.99
-23	5	8.1	.38	4	6.7	.44
-24	12	10.7	.64	11	8.9	.40
-31	4	7.5	.27	3	6.3	.31
-35	8	8.9	.99	7	7.4	.99
-39	7	6.5	.70	6	5.4	.67
-40	9	8.9	.87	9	7.4	.58
-43	23	25.1	.76	20	20.9	.99
-47	1	5.5	.05	1	4.6	.10
-51	13	8.0	.11	12	6.7	.05
-52	8	8.0	.99	7	6.7	.85
-55	6	5.8	.83	6	4.8	.49
-56	9	6.3	.31	8	5.3	.27
-59	6	6.6	.99	5	5.5	.99
-67	10	8.1	.48	10	6.7	.24
-68	1	5.9	.04	1	4.9	.11
-71	4	4.3	.99	3	3.6	.99
-79	7	4.7	.25	6	4.0	.30
-83	4	5.8	.68	4	4.9	.99
-84	1	5.7	.05	0	4.8	.02
-87	3	4.4	.81	3	3.7	.99
-88	2	6.4	.11	2	5.4	.19
-91	6	6.4	.99	5	5.3	.99
-95	4	3.8	.80	4	3.2	.57

Curve B

Table V

D	N_D	E_D	Pr	N_D	E_D	Pr
-3	37	34.9	.67	33	29.1	.46
-4	16	28.5	.01	15	23.7	.08
-7	13	17.0	.40	12	14.2	.69
-8	13	18.0	.29	11	15.0	.37
-11	15	15.8	.99	15	13.2	.58
-15	12	10.9	.65	11	9.1	.50
-19	13	13.6	.99	11	11.3	.99
-20	8	10.9	.45	5	9.1	.24
-23	5	8.1	.38	5	6.7	.70
-24	14	10.7	.28	14	8.9	.09
-31	7	7.5	.99	5	6.3	.84
-35	6	8.9	.40	5	7.4	.46
-39	8	6.5	.55	7	5.4	.51
-40	4	8.9	.13	3	7.4	.14
-43	12	9.9	.43	9	8.2	.73
-47	9	5.5	.13	9	4.6	.05
-51	8	8.0	.99	7	6.7	.85
-52	3	8.0	.08	3	6.7	.24
-55	3	5.8	.40	3	4.8	.64
-56	7	6.3	.69	7	5.3	.38
-59	10	6.6	.17	9	5.5	.13
-67	10	8.1	.48	8	6.7	.56
-68	8	5.9	.40	6	4.9	.65
-71	2	4.3	.46	2	3.6	.60
-79	5	4.7	.82	5	4.0	.61
-83	4	5.8	.68	4	4.9	.99
-84	6	5.7	.83	6	4.8	.49
-87	5	4.4	.64	5	3.7	.43
-88	7	6.4	.69	7	5.4	.39
-91	10	6.4	.16	10	5.3	.05
-95	1	3.8	.20	1	3.2	.39

Curve C

Table V

D	N_D	E_D	Pr	N_D	E_D	Pr
-3	46	--	--	43	--	--
-4	27	26.3	.84	23	21.9	.75
-7	14	17.0	.54	13	14.2	.89
-8	20	18.0	.64	18	15.0	.44
-11	14	15.8	.80	13	13.2	.99
-15	11	10.9	.88	9	9.1	.99
-19	13	14.2	.89	13	11.9	.66
-20	11	10.9	.88	11	9.1	.50
-23	5	8.1	.38	5	6.7	.70
-24	4	5.3	.83	4	4.4	.99
-31	4	7.5	.27	4	6.3	.54
-35	9	8.9	.87	9	7.4	.58
-39	4	6.5	.43	2	5.4	.19
-40	9	8.2	.73	7	6.8	.85
-43	6	10.4	.21	4	8.6	.13
-47	7	5.5	.52	6	4.6	.48
-51	4	10.5	.04	2	8.8	.02
-52	4	7.4	.27	3	6.2	.31
-55	1	5.8	.04	1	4.8	.10
-56	1	6.3	.03	0	5.3	.01
-59	10	6.6	.17	8	5.5	.28
-67	7	8.5	.86	5	7.1	.57
-68	7	5.9	.54	7	4.9	.36
-71	6	4.3	.34	6	3.6	.18
-79	2	4.7	.35	2	4.0	.45
-83	4	5.8	.68	4	4.9	.99
-84	3	2.9	.77	2	2.4	.99
-87	7	4.4	.22	7	3.7	.11
-88	9	5.9	.21	6	4.9	.65
-91	5	6.7	.70	5	5.6	.99
-95	4	3.8	.80	4	3.2	.57

Curve D

Table V

D	N_D	E_D	Pr	N_D	E_D	Pr
-3	29	35.5	.31	26	29.6	.58
-4	67	84.9	.05	58	70.7	.14
-7	15	17.2	.72	14	14.3	.99
-8	25	18.3	.13	23	15.2	.05
-11	88	120.3	.0026	84	100.2	.11
-15	13	13.1	.99	13	10.9	.54
-19	32	40.8	.18	29	33.9	.44
-20	11	13.0	.68	11	10.9	.88
-23	5	8.2	.38	5	6.8	.70
-24	26	32.0	.33	22	26.6	.44
-31	19	22.5	.53	15	18.8	.49
-35	8	10.7	.54	8	8.9	.99
-39	15	19.4	.36	14	16.2	.71
-40	5	10.7	.09	5	8.9	.24
-43	8	10.0	.64	8	8.3	.99
-47	3	5.6	.39	3	4.6	.64
-51	29	24.1	.31	28	20.1	.09
-52	6	8.1	.60	6	6.8	.99
-55	3	6.9	.18	3	5.8	.40
-56	15	18.9	.49	14	15.7	.80
-59	16	19.8	.50	15	16.5	.81
-67	8	8.2	.99	8	6.9	.57
-68	3	6.0	.30	3	5.0	.50
-71	11	13.0	.68	11	10.8	.88
-79	10	14.2	.35	10	11.9	.77
-83	4	5.9	.68	4	4.9	.99
-84	17	17.2	.99	15	14.3	.79
-87	2	4.5	.34	2	3.7	.60
-88	0	3.3	.09	0	2.7	.12
-91	21	19.2	.65	20	16.0	.32
-95	5	4.6	.81	5	3.8	.44

Curve $X_o(11)$

Table VI A

Primes ≤ 48,611 associated with discriminants ≥ −43

for the curve $A : y^2 + y = x^3 - x$

−3	−4	−7	−8	−11	−15	−19	−20	−23	−24	−31	−35	−39	−40	−43
3	2	487	113	157	139	11	29	2393	193	283	6521	367	251	59
7	5	877	137	619	1021	19	61	2927	769	5701	7151	607	1019	509
13	317	1373	179	1433	1549	853	109	8821	5281	17977	12301	2131	3251	821
31	541	2251	563	2689	2659	1109	281	9929	11113	37309	18691	2311	3449	3911
79	797	2543	1889	2777	3019	2063	401	13417	44119		33679	3121	17041	5431
127	1061	3637	2897	2957	3391	4547	449	19553				6301	39769	6043
211	2437	3691	3539	3617	12049	5573	2801					27901	44249	8807
229	3089	4657	4139	4801	17491	5711	3221					28687		8839
241	3169	5303	4409	11719	21961	5869	4441							9001
313	5581	6469	4603	13183	26839	13687	5009							
439	8893	6911	6011	19429	29311	32491	5881							18289
733	11353	12517	6763	22639	32059	40189	8069							30931
1039	14629	12671	8089	27241	35491	43963	12101							31147
1231	15749	16349	13963	29297		45491	14549							35081
1753	17497	19763	17449	30187			18121							39209
2671	19889	20143	17939	45137			20921							
3733	28657	22691	20113	47777			22409							
4621	30113	33721	25321				26881							
4813	32233		34273				35281							
5119	36493		41771											
7549	36697		42569											
8161	39857													
8707	43321													
9403	47521													
14479														
14947														
15937														
16033														
16231														
21157														
21517														
31153														
32713														
33769														
34303														
38923														
39667														
41221														
41941														
44959														

	-3	-4	-7	-8	-11	-15	-19	-20	-23	-24	-31	-35	-39	-40	-43
	13	2	7	3	421	61	251	29	59	199	293	11	181	809	47
	127	5	79	19	1367	1129	6271	641	1871	3391	3187	709	673	4001	109
	223	197	239	73	1423	1549	8689	1669	11069	6361	22483	2689	1249	4409	359
	313	593	701	137	1973	1831	14251	2729	11801	13879	45589	6091	33589	6211	977
	349	1489	2671	331	10607	8329	15739	3001	16273	16231		10301	36919	9929	1559
	1021	2833	2731	1409	13619	16921	21221	3821		31687		10691	39199	24281	1741
	1657	3049	5261	1723	16831	34549	30841	8369		32719		19541	43627	24859	2003
	1801	3373	5791	2459	18461	35521	34261	20089		33601		44809		31771	3881
	2089	4933	6619	5153	19891		39251	28649		34033				33049	4357
	2851	5981	8831	6113	24533			31741		37567					4813
	3067	6397	12511	6833	26987					38791					6217
	3343	7349	16349	18059	43591					48337					9431
	3919	8093	17749	29483											11057
	6043	8761	19867	45827											17909
	6121	10061	20147												18587
	9067	12553	23143												18869
	9199	13933	29327												20743
	9391	15889	30319												30097
	14389	19037	38851												39079
	14983	23293													39929
	18013	24109													41849
	18919	25057													44893
	20479	28933													45077
	21169	43577													
	21751	43777													
	24889	47057													
	29803														
	32353														
	33409														
	39409														
	39631														
	41269														
	42649														
	47809														

Table VI B

Primes $\leq 48{,}611$ associated with discriminants ≥ -43

for the curve $B : y^2 + y = x^3 + x^2$.

Note the large frequency for $D = -43$.

Table VII

Curve A

D	N_D											
-3	40	4	6	3	2	3	3	3	5	2	0	4	5
-4	24	2	1	3	1	2	1	2	2	2	2	3	3
-7	18	0	0	2	2	4	3	0	3	3	0	1	0
-8	21	0	3	1	1	5	2	1	1	3	1	1	2
-11	17	0	1	1	1	4	1	1	1	2	2	1	2
-15	13	0	1	1	1	3	0	0	1	2	2	2	0
-19	13	1	0	2	1	1	3	0	1	0	0	1	3
-20	19	1	3	2	0	3	2	1	2	3	1	1	0
-23	6	0	0	0	1	1	0	2	1	1	0	0	0
-24	5	0	1	1	0	0	1	1	0	0	0	0	1
-31	4	0	1	0	0	0	1	0	0	1	0	1	0
-35	5	0	0	0	0	0	2	0	1	1	0	1	0
-39	8	0	0	2	2	1	1	0	0	0	2	0	0
-40	7	0	1	1	0	2	0	0	0	1	0	0	2
-43	14	0	1	2	0	1	2	3	0	1	0	3	1
	Totals	8	19	21	12	30	22	14	18	22	10	19	19

Curve B

D	N_D											
-3	34	1	4	1	3	4	2	3	2	5	2	2	5
-4	26	2	1	1	1	3	4	3	3	1	4	0	3
-7	18	0	2	1	0	2	3	1	2	3	2	1	1
-8	14	2	3	0	3	0	3	0	0	1	1	0	1
-11	12	0	0	1	3	0	0	1	1	3	2	0	1
-15	8	0	1	0	3	0	0	1	0	1	0	2	0
-19	9	0	1	0	0	0	1	1	2	1	0	2	1
-20	10	1	0	1	1	3	0	1	0	1	1	1	0
-23	5	0	1	0	1	0	0	1	2	0	0	0	0
-24	12	0	1	0	0	1	1	0	2	0	0	5	2
-31	4	0	1	0	0	1	0	0	0	1	0	0	1
-35	8	1	0	1	0	1	1	2	0	1	0	0	1
-39	7	0	1	1	1	0	0	0	0	0	0	2	2
-40	9	0	0	1	0	2	1	1	0	0	2	2	0
-43	23	1	2	1	3	2	2	2	0	4	1	0	5
	Totals	8	18	9	19	19	18	17	14	22	15	17	23

Distribution of primes associated with small discriminants

Table VII

D	N_D				Curve C							
-3	37	3	1	3	3	2	3	1	6	5	4	3	3
-4	16	0	1	0	2	1	1	3	2	2	1	3	0
-7	13	1	0	3	0	1	1	2	2	0	0	1	2
-8	13	1	1	3	1	1	1	1	1	3	0	0	0
-11	15	0	0	2	1	2	3	1	3	1	1	1	0
-15	12	0	1	0	2	1	2	0	2	1	2	0	1
-19	13	0	2	2	2	1	0	0	0	4	1	0	1
-20	8	0	3	0	0	1	2	1	0	0	0	1	0
-23	5	0	0	1	1	2	0	0	0	0	0	0	1
-24	14	0	0	2	1	0	1	4	1	0	1	2	2
-31	7	0	2	0	0	2	0	1	1	1	0	0	0
-35	6	0	1	0	1	0	0	1	1	0	1	0	1
-39	8	0	1	0	2	1	0	0	1	1	0	0	2
-40	4	0	1	0	1	1	0	0	0	0	0	1	0
-43	12	2	1	1	2	1	0	0	0	1	1	3	0
Totals		7	15	17	19	17	14	15	20	19	12	15	13

D	N_D				Curve D							
-3	46	1	2	6	6	3	4	4	4	5	3	4	4
-4	27	3	1	1	1	3	3	0	2	2	4	3	4
-7	14	0	1	1	0	3	0	3	1	3	1	0	1
-8	20	1	1	1	3	4	0	2	2	1	1	3	1
-11	14	0	1	2	1	1	0	2	0	5	2	0	0
-15	11	1	1	1	0	2	1	0	1	2	1	0	1
-19	13	0	0	1	1	2	1	1	2	1	2	0	2
-20	11	0	0	3	1	3	0	1	1	1	0	1	0
-23	5	0	0	0	0	0	2	0	1	1	0	1	0
-24	4	0	0	0	0	1	1	0	0	0	2	0	0
-31	4	0	0	0	1	1	0	0	0	0	0	0	2
-35	9	0	0	0	1	0	1	1	2	1	2	1	0
-39	4	1	1	0	0	0	0	0	1	0	1	0	0
-40	9	1	1	2	1	0	1	1	0	1	0	1	0
-43	6	0	2	1	0	1	0	0	0	0	1	1	0
Totals		8	11	19	16	24	14	15	17	23	20	15	15

Distribution of primes associated with small discriminants

Table VII

D	N_D				Curve $X_o(11)$							
-3	29	1	2	1	3	1	6	4	1	3	2	3	2
-4	67	4	5	4	10	4	7	5	7	7	5	5	4
-7	15	0	1	2	0	3	3	1	0	3	1	1	0
-8	25	0	2	1	0	1	1	2	2	2	3	3	8
-11	88	2	2	7	10	5	10	7	7	13	5	6	14
-15	13	0	0	1	1	2	1	1	2	1	2	0	2
-19	32	1	2	3	1	3	2	1	3	4	4	3	5
-20	11	0	0	1	3	1	0	2	1	1	1	1	0
-23	5	0	0	0	1	0	0	0	1	1	0	0	2
-24	26	1	3	2	2	2	2	1	3	2	4	2	2
-31	19	1	3	1	3	3	0	1	1	2	4	0	0
-35	8	0	0	1	1	1	1	1	0	1	0	0	2
-39	15	0	1	3	1	1	0	2	1	1	1	4	0
-40	5	0	0	1	0	1	0	0	1	0	2	0	0
-43	8	0	0	0	0	1	1	1	1	2	0	1	1
	Totals	10	21	28	36	29	34	29	31	43	34	29	42

Distribution of primes associated with small discriminants

EXTENDED RESULTS FOR $X_o(11)$

§5. Discussion and description of tables

As explained below in "remarks on the computations," extended calculations for the curve $X_0(11)$ could be done much more cheaply than for our other examples. We therefore decided to push the calculation further, especially to see how the agreement between conjecture and experiment would come out for this curve and the field $Q(\sqrt{-11})$. The computation was taken as far as

$$p_{189,521} = 2,590,717$$

for which

$$\pi_{\frac{1}{2}}(2,590,717) \approx 130.7 .$$

The stopping point was quite arbitrary. We wanted to get far enough to have $\pi_{\frac{1}{2}}$ at least four times the value 26.4 reached in the earlier calculation, and it turned out that we could go somewhat further and still stay within our budget.

Table VIII simply lists the supersingular primes and needs no further explanation.

As before, we divided the primes into twelve segments making nearly equal contributions to the sum $\pi_{\frac{1}{2}}(X)$. The division points came out as follows.

Segment	i	p_i
1	615	4523
2	3177	29221
3	8173	83773
4	15860	174299
5	26411	305021
6	39961	479377
7	56616	700949
8	76467	971483
9	99592	1293857
10	126058	1670129
11	155926	2101657
12	189251	2590717

Table IX is just like Table V except that the second group of columns (corresponding to dropping the first two of the twelve segments) now refers to primes p such that

$$p_{3177} = 29221 < p \leq p_{189251} = 2950717 \ .$$

Table X is like Table VII (referring of course to the segments just listed), except that with the larger frequencies there is some point in including all the discriminants with absolute value less than 100.

Table IX shows good agreement on the whole between the conjectured predictions and actual counts, again with the exception of $Q(\sqrt{-11})$ when the initial segments are not left out. We find it quite reasonable to attribute this discrepancy to slowness of the approach to asymptotic behavior.

There may be identifiable reasons for particularly slow asymptotic approach in this case. Our heuristic arguments involve estimating "probabilities" for coincidences modulo finite M, and then taking a limit as M becomes more and more divisible. The estimates for finite M are arrived at under the assumption that we have primes p with $\sqrt{p} \gg M$. Also we depend on the Tchebotarev density theorem, applied to a Galois extension of degree proportional to M^4, and presumably convergence here is slower when M is large. In many cases the factor we get with a small value of M (say 12) is not very far off the limiting value. For $X_0(11)$, however, the factor 5 makes a big difference and (at least for $D = -11$) the factor 11 is important too. Thus one has to take a fairly large M, and should not expect asymptotic behavior until one reaches very large primes.

It may also be reasonable to expect counts corresponding to very high predicted frequencies to be too low when one looks at an initial segment of primes, simply because there are not enough primes to go around. The relevant factor is the ratio of the number of primes $\leq X$ to $\pi_{\frac{1}{2}}(X)$, and of course it gets larger as X increases. Note that in Table IX the actual frequencies fall noticeably below the predictions whenever the latter are over 100, and that for all the predictions over 150, the probability measure of agreement improves when the initial segments of data are discarded.

We do not offer the above remarks as anything more than suggestive possibilities, but they do reinforce our opinion that the data are consistent with our conjecture.

Table VIII

Curve $X_o(11)$

19	66809	475229	1354649
29	67289	522839	1379659
199	79229	539339	1403239
569	88259	578789	1503149
809	110339	597869	1543319
1289	131479	613999	1562279
1439	150169	628909	1621679
2539	159209	654779	1720109
3319	168869	664109	1728119
3559	196919	666079	1739839
3919	199669	715189	1769539
5519	202109	716299	1792709
9419	209569	724079	1808039
9539	213949	733169	1847999
9929	220279	747449	1875499
11279	226789	749909	1896109
11549	228959	783359	1898759
13229	234869	833509	1981669
14489	243119	860779	2030069
17239	268969	913259	2032799
18149	272999	918839	2077919
18959	273929	920729	2099059
19319	281069	945059	2103389
22279	282889	983429	2122469
24359	289559	1018789	2124359
27529	294989	1024669	2125939
28789	324839	1025939	2146289
32999	325079	1077179	2238529
33029	339539	1088089	2268389
36559	360989	1104659	2405339
42899	372059	1125329	2432719
45259	384089	1147639	2543459
46219	395449	1207439	2547599
49529	402859	1239179	2556739
51169	410009	1269559	2572379
52999	418259	1286099	
55259	464939	1324949	

Supersingular primes ≤ 2,590,717

Table IX

D	N_D	E_D	Pr	N_D	E_D	Pr
-3	165	175.8	.42	141	146.5	.65
-4	376	419.8	.03	318	349.8	.09
-7	86	85.2	.93	72	71.0	.90
-8	88	90.3	.81	74	75.3	.88
-11	528	595.0	.006	461	495.8	.12
-15	63	64.9	.81	52	54.1	.84
-19	185	201.5	.24	162	168.0	.65
-20	56	64.4	.29	46	53.7	.34
-23	40	40.4	.99	37	33.7	.55
-24	134	158.2	.05	113	131.8	.10
-31	100	111.5	.28	81	92.9	.22
-35	54	52.8	.84	48	44.0	.55
-39	88	96.1	.41	77	80.1	.73
-40	48	52.9	.54	43	44.1	.94
-43	44	49.5	.48	37	41.3	.59
-47	27	27.5	.99	25	22.9	.60
-51	110	119.3	.40	85	99.4	.15
-52	41	40.2	.87	37	33.5	.54
-55	32	34.2	.80	29	28.5	.93
-56	96	93.3	.78	87	77.7	.29
-59	91	98.1	.47	77	81.8	.60
-67	39	40.7	.88	33	33.9	.99
-68	27	29.7	.71	24	24.8	.99
-71	61	64.2	.69	53	53.5	.99
-79	64	70.4	.45	57	58.7	.90
-83	28	29.3	.93	26	24.4	.69
-84	86	85.1	.93	74	70.9	.72
-87	28	22.2	.20	27	18.5	.06
-88	11	16.1	.26	11	13.4	.68
-91	99	95.1	.69	86	79.3	.45
-95	23	22.7	.92	20	18.9	.73

Curve $X_o(11)$

Table X

D	N_D				Curve $X_o(11)$							
-3	165	8	16	15	11	16	18	20	15	9	11	16	10
-4	376	26	32	30	24	29	36	36	35	27	32	35	34
-7	86	6	8	1	11	3	6	5	8	8	7	15	8
-8	88	4	10	17	6	8	5	5	9	8	6	3	7
-11	528	26	41	40	39	47	42	47	37	51	62	48	48
-15	63	4	7	6	2	7	7	5	8	4	4	5	4
-19	185	10	13	16	15	17	20	15	19	17	17	12	14
-20	56	5	5	1	7	1	6	5	8	6	1	4	7
-23	40	1	2	3	7	3	2	4	7	5	3	2	1
-24	134	10	11	11	6	12	15	7	9	12	19	15	7
-31	100	11	8	3	10	10	7	14	4	6	8	9	10
-35	54	3	3	6	6	7	3	6	5	5	2	2	6
-39	88	6	5	7	9	9	6	8	8	9	8	5	8
-40	48	2	3	2	8	5	3	7	1	3	3	5	6
-43	44	2	5	5	4	0	3	6	5	2	4	6	2
-47	27	0	2	1	0	2	0	5	4	5	4	3	1
-51	110	10	15	11	7	9	9	4	6	7	11	12	9
-52	41	2	2	3	4	3	3	2	7	1	4	4	6
-55	32	2	1	0	1	7	4	3	1	1	8	3	1
-56	96	5	4	10	5	12	8	8	4	10	9	12	9
-59	91	6	8	8	13	9	7	6	3	7	7	7	10
-67	39	4	2	4	1	4	4	0	3	6	2	4	5
-68	27	1	2	0	3	0	3	4	2	1	1	5	5
-71	61	3	5	7	7	5	5	8	6	4	1	4	6
-79	64	2	5	8	3	3	7	5	10	6	6	2	7
-83	28	1	1	2	5	1	2	5	2	1	1	4	3
-84	86	5	7	9	6	8	8	5	5	9	10	5	9
-87	28	1	0	2	2	2	2	4	5	2	3	4	1
-88	11	0	0	0	0	1	0	1	1	1	3	0	4
-91	99	3	10	13	10	7	7	5	7	6	12	8	11
-95	23	0	3	2	3	3	1	2	6	0	0	1	2
Totals		169	236	243	235	250	249	257	250	239	269	260	261

Distribution of primes associated with discriminants ≥ -95.

Remarks on the computations

We conclude with some comments on the machine computations. For $X_0(11)$, t_p was found as the coefficient of q^p in the q-expansion

$$q \prod_{n=1}^{\infty} (1-q^n)^2 (1-q^{11n})^2 .$$

Cf. Shimura [Sh] and Ligozat [Li], where such products are given also for other modular curves. Euler's formula gives

$$\prod_{n=1}^{\infty} (1-q^n)$$

as a series in which all coefficients are zero except for 1's and −1's occurring for n in certain quadratic progressions. Straightforward multiplication of the series is an easy computation if one wants only 50,000 or so terms.

For the other curves, the calculation was based on the formula

$$t_p = 1 + p - N_p ,$$

where N_p is the number of rational points on the curve over F_p. The equation of the curve can be rewritten if necessary to have the form

$$y^2 = f(x) = 4x^3 + a_2 x^2 + a_4 x + a_6 ,$$

without changing the value of N_p for any $p > 2$. The calculation is then done by first constructing a table of values

$$r(0), r(1), \cdots, r(p-1) ,$$

where

$$r(i) = 1 + \left(\frac{i}{p}\right) ,$$

i.e. the number of solutions of $y^2 = r(i) \bmod p$ [Sw − D 2]. Then

$$N_p - 1 = \sum_{i=1}^{p} r(f(i)).$$

It is worth noting that successive values of $f(i)$ (and the quadratic residues needed to construct the table $r(i)$), can be built up by successive addition from constant third (second) differences. By checking each number occurring in the calculation when it is formed, reduction mod p can always be carried out by a single addition or subtraction of p. This avoids the comparatively slow machine operation of integer division, and leads to significantly faster calculation than the more obvious method of calculating $f(x)$ by multiplications and additions, followed by reduction mod p. The computing time needed to calculate t_p by this method is clearly proportional to p, so the cost of finding t_p for all $p \leq x$ rises rapidly with x. We picked the 5,000th prime as a stopping point as a reasonable compromise between cost of computation and the need for a significant sample of data.

The equation for the points of order 5 on $X_0(11)$ was found using a program which calculates the polynomials for the n-th division points (x-coordinate), given in most books, e.g. in Weber, and sometimes referred to as $\psi_n(x)$. It was factored with the help of a program which uses a version of Berlekamp's algorithm for factoring polynomials modulo primes, cf. [B].

In the more extended calculations for $X_0(11)$, the relevant terms of the series could not all be kept in core storage at one time, so a new program was required to use auxiliary storage for intermediate results. By far the greatest part of the computer time was spent on the final multiplication

$$\prod (1-q^n) \times \prod (1-q^n)(1-q^{11n})^2 .$$

As noted above, the series for $\prod (1-q^n)$ has very few non-zero terms (which is what makes the calculation feasible at all). In the factor on the right, the density of non-zero coefficients turns out to be a little over one-half and (perhaps surprisingly) to be increasing steadily, though very slowly, as one goes to higher order terms.

BIBLIOGRAPHY

[B]   E. R. BERLEKAMP, Algebraic coding theory, McGraw Hill, N.Y., 1968.

[H-L]   G. H. HARDY and J. LITTLEWOOD, Partitio Numerorum, Acta Math. 44 (1923), pp. 1-70, especially p. 48.

[H-C]   HARISH-CHANDRA, Harmonic analysis on reductive p-adic groups, Proceedings of symposia in pure mathematics, AMS, Williamstown conference (1973), pp. 167-192.

[I]   Y. IHARA, Hecke polynomials as congruence zeta functions in elliptic modular case, Ann. of Math. 85 (1967), pp. 267-295.

[L 1]   S. LANG, Elliptic functions, Addison Wesley, Reading, 1973.

[L 2]   ———, Algebraic Number Theory, Addison Wesley, Reading, 1971

[Le]   D. H. LEHMER, Note on the distribution of Ramanujan's Tau function, Math. Comp. 24 (1970), pp. 741-743.

[Li]   G. LIGOZAT, Courbes modulaires de genre 1, to appear.

[Ma]   B. MAZUR, Rational points of abelian varieties with values in towers of number fields, Inv. Math. 18 (1972), pp. 183-266.

[Man]   J. MANIN, Periods of parabolic forms and p-adic Hecke series, Mat. Sbornik n.s. 92 (1973), 378-401, English translation, Math. USSR Sbornik 21 (1973), 371-393.

[R]   K. RIBET, On ℓ-adic representations attached to modular forms, to appear.

[Sa-Sh 1] SALLY and J. SHALIKA, Characters of the discrete series of representations of SL(2) over a local field, Proc. Nat. Acad. Sci. USA 61 (1968), pp. 1231-1237.

[Sa-Sh 2] ———, The Plancherel formula for SL(2) over a local field, Proc. Nat. Acad. Sci. USA 63 (1969), pp. 661-667.

[Sa-Sh 3] ———, The Fourier transform on SL_2 over a non-archimedean local field, to appear.

[S 1]   J. P. SERRE, Groupes de Lie ℓ-adiques attachés aux courbes élliptiques, Colloque Clermont-Ferrand, Les tendances géomètriques en algèbre et théorie des nombres, 1964, sections 3.4 and 4.3.

[S 2]   J. P. SERRE, Propriétés Galoisiennes des points d'ordre fini des courbes élliptiques, Inv. Math. 15 (1972), pp. 259-331.

[S 3]   _____, Abelian ℓ-adic representations and elliptic curves, Benjamin, 1968.

[Sh]   G. SHIMURA, A reciprocity law in non-solvable extensions, J. reine angew. Math. 221 (1966), pp. 209-220.

[SwD]   P. SWINNERTON-DYER, On ℓ-adic representations and congruences for coefficients of modular forms, Springer Lecture Notes 350 (Antwerp conference).

[SwD 2] _____, An application of computing to class field theory, in Algebraic Number Theory (Brighton Conference, edited by J. W. S. Cassels and A. Frölich) Thompson Book Co., Washington, 1967.

[T]   J. TATE, The Arithmetic of elliptic curves, Invent. Math. 23 (1974), pp. 179-206.

[Tu]   T. A. TUSKINA, A numerical experiment on the calculation of the Hasse invariant for certain curves, Izv. Akad. Nauk SSSR Ser. Mat. 29 (1965), 1203-1204, English translation, AMS Translations (Ser. 2) 66, (1968), 204-205.

[Y]   H. YOSHIDA, On an analogue of the Sato conjecture, Invent. Math. 19 (1973), pp. 261-277.

Units in the Modular Function Field.
IV. The Siegel Functions are Generators

Daniel Kubert* and Serge Lang*

Department of Mathematics, Cornell University, Ithaca, NY 14853, USA
Department of Mathematics, Yale University, New Haven, CT 06520, USA

In [KL II] we constructed a full set of units in the modular function field F_N of level N, by means of the Klein forms, which are modular forms of weight -1 and level $2N^2$, and whose $2N$-th power have level N. Their q-expansion coefficients lie in the field

$$Q_N = Q(e^{2\pi i/N})$$

of N-th roots of unity. Let $F = \bigcup F_N$ be the modular function field. Let QR denote the integral closure of $Q[j]$ in F. Then these units generate (modulo constants) a free abelian group in the group of units $(QR)^*$.

In this paper, we show that they generate the group of all units (modulo constants), and the proof yields an independent confirmation of the fact that they have the appropriate rank. In the light of [KL II] and [KL D], where the regulator of these units was expressed as a product of character sums, we obtain a proof that the generalized Bernoulli numbers $B_{2,\chi}$ are $\neq 0$ for even characters χ, a fact whose only known proof was via the direct study of Dirichlet L-series, cf. Iwasawa [Iw].

The method of proof consists in reducing multiplicative relations among the q-expansions of units to additive relations of an appropriately chosen Fourier coefficient (more or less, the leading coefficient after the constant term). In analysing a modular function such that some power can be expressed as a product of Siegel functions, we are led to taking roots of q-products. By a theorem of Shimura, we know that a suitably reduced form of the q-product has integral coefficients, and we see that this can happen only if the modular function itself is already expressible as a power product of Siegel functions.

The additive relations of divisibility occur among elements of the cyclotomic fields, and hence we devote a section to describing an appropriate basis for the integers of these fields, which allow us to recognize easily when the coefficients are integral.

---

* Supported by NSF grants

Roots of modular units are then analysed in the prime power case and the general case respectively, and the main results are stated as Theorems 1 and 2 of §1. More precise versions are given later in Theorems 3, 4 and 5.

Other authors have considered functions in the modular function field of what is usually denoted by $X_0(N)$, which have their divisors concentrated at cusps (or as we also say, at infinity), e.g. Newman [N] and Ogg [O]. These functions were obtained as quotients of the discriminant function $\Delta(N\tau)/\Delta(\tau)$, and cases are determined in these papers when power products or such functions are modular of level N. Ogg also determines the order of the cusp in the divisor class group for $X_0(N)$ when N is prime. More recently, see also Klimek [Kl], for the case of $X_1(N)$ with N prime.

§1. Statement of Results

We adopt the notation of [KL II], and we do not repeat the general definition of the Klein forms here, although when we need it later, we shall recall their q-expansions. We let η be the Dedekind eta function. If $a=(a_1,a_2)\in Q^2$ and $a\neq 0$, we have the klein form $\mathfrak{k}_a$. We let the **Siegel function** be

$$g_a = \mathfrak{k}_a \eta^2,$$

so that g_a has weight 0, and is therefore a function on the upper half plane. If a is changed by an element of Z^2, then g_a gets multiplied by a root of unity (cf. [KL II], formula **K 2**). We are only interested in modular functions up to constant factors in this paper. Hence for $a\in Q^2/Z^2$ we mean by g_a the class of functions modulo constants. By abuse of language, we still speak of g_a as a function. We say that such a function is **modular** if it lies in the modular function field F_N for some level N (strictly speaking, if some representative function and therefore any representative function lies in a modular function field). We shall investigate when a root of a power product of Siegel functions is modular.

To begin with, we note that the Siegel functions themselves are modular. It is clear from the explicit transformation law for a Klein form $\mathfrak{k}_a$ that it is modular [KL II], §1, formula **K 3**. It is well known for η^2, and can be proved in the same style. Indeed, up to a constant factor we have (e.g. from q-expansions)

$$\eta^{-6} = \mathfrak{k}_{(1/2,0)} \mathfrak{k}_{(1/2,1/2)} \mathfrak{k}_{(0,1/2)}$$
$$\eta^{-8} = \mathfrak{k}_{(1/3,0)} \mathfrak{k}_{(0,1/3)} \mathfrak{k}_{(1/3,1/3)} \mathfrak{k}_{(1/3,-1/3)},$$

and therefore $\eta^2 = \eta^8/\eta^6$ is modular whence g_a is modular.

For convenience of expression, it will be useful to adopt the notation g_a above only when $2a\neq 0$. If $2a=0$, let us put

$$h_a = \mathfrak{k}_a \eta^2.$$

We see from the distribution relation (cf. [KL III]) that

$$h_{(1/2,0)} = \lambda h^2_{(1/4,0)} h^2_{(1/4,1/2)},$$

where λ is a constant. Thus it is clear that $h_{(1/2,0)}^{1/2}$ is modular. Likewise, if $2a=0$, it is clear that $h_a^{1/2}$ is modular. In this light, if $2a=0$, we *define*

$$g_a = h_a^{1/2} = (\mathfrak{f}_a \eta^2)^{1/2},$$

so that also in this case g_a is modular.

The next two theorems give a more precise result than just describing the group of modular units. We let U be the group generated by the Siegel functions g_a for all $a \in Q^2/Z^2, a \neq 0$, and the constants $\neq 0$. Let F be the modular function field, equal to the union of all the modular fields F_N of all levels N. Let QR be the integral closure of $Q[j]$ in F. In view of the fullness of the group of Siegel units [KL II], [KL D], it follows that the factor group of modular units $(QR)^*$ modulo U is a torsion group. It turns out that they are equal except for 2-torsion. Furthermore, let U_N be the subgroup of U generated by the Siegel functions g_a such that $Na=0$, and the non-zero constants.

Theorem 1. *Let $N = p^r$ be a prime power. Then U_N is equal to its own division group in the group of modular units. In other words, if g is a modular function such that some positive power of g lies in U_N, then g also lies in U_N.*

Theorem 2. *Let N be an arbitrary integer >1. Let l be an odd integer. If g is a modular function such that g^l lies in U_N, then g lies in U_N.*

In the proof of Theorem 1 it will clearly be sufficient to assume that $g^l \in U_N$ for some prime l, and in the proof of Theorem 2 it will suffice to assume that l is an odd prime.

The rest of the paper gives the proofs of the above theorems. We begin by some lemmas giving an appropriate basis for the cyclotomic integers. The main part of the proof consists in reducing the study of multiplicative relations among the units to additive relations among such integers, by projecting on the first coefficient of the q-expansion. Using a theorem of Shimura (see Lemma 3.1) to the effect that the Fourier coefficients of modular forms are algebraic integers, we can formulate versions of Theorems 1 and 2 involving only formal power series arguments, and no other property of modular forms. This is done in Theorems 3 and 5.

§2. Cyclotomic Integers

Let $Q_N = Q(e^{2\pi i/N})$, and let $\mathfrak{o}_N$ be the algebraic integers in Q_N. It is standard that

$$\mathfrak{o}_N = Z[e^{2\pi i/N}].$$

We want an appropriate Z-basis for $\mathfrak{o}_N$.

Suppose first that $N = p^r$ is a prime power. Let μ_N as usual be the group of N-th roots of unity. Let S_j $(j=1, \ldots, N/p)$ be the cosets of μ_p in μ_N. For each coset, choose arbitrarily an element ζ_j in S_j.

Lemma 2.1. *The set*

$$\bigcup (S_j - \{\zeta_j\})$$

forms a **Z**-*basis for* $\mathfrak{o}_{p^r}$.

Proof. We note that the number of cosets is p^{r-1}, so that

$$\bigcup (S_j - \{\zeta_j\})$$

has $p^r - p^{r-1}$ elements, which is the right number. Since the elements of μ_N generate $\mathfrak{o}_N$ over **Z**, we need only show that ζ_j can be recovered. To do this, multiply the identity

$$\sum_{\zeta \in \mu_p} \zeta = 0$$

by ζ_j. We get

$$\zeta_j + \sum_{\zeta \neq 1} \zeta_j \zeta = 0.$$

But the sum is the same as that taken over all elements of $S_j - \{\zeta_j\}$, thus proving the lemma.

We next extend the lemma to arbitrary N. We let

$$N = \prod p^{n(p)}$$

be the prime factorization of N. For each element $\zeta \in \mu_N$ we let

$$\zeta = \prod \zeta_p$$

be the factorization into prime power roots of unity. For each prime p we form the set

$$S(p) = \bigcup (S_j^{(p)} - \zeta_j^{(p)})$$

as in the preceding lemma.

Lemma 2.2. *Let S be the set of N-th roots of unity ζ such that $\zeta_p \in S(p)$ for each $p|N$. Then S is a* **Z**-*basis for* $\mathfrak{o}_N$.

Proof. Again, the cardinality of S is the right one. Let ω be an arbitrary N-th root of unity. Write

$$\omega = \prod_{p \in S(p)} \omega_p \prod_{p \notin S(p)} \omega_p.$$

As shown in Lemma 2.1, we may write ω_p for $p \notin S(p)$ as an integral linear combination of elements in $S(p)$. We make this substitution and expand out the product by distributivity to prove the lemma.

Next we wish to determine when sums of the form

$$\alpha = \sum a_\zeta \zeta, \quad a_\zeta \in \mathbf{Z},$$

taken over all N-th roots of unity, are divisible by a prime number l. We begin with the prime power case.

Modular Function Field. IV.

Lemma 2.3. *Let $N = p^r$. A prime l divides $\sum a_\zeta \zeta$ if and only if, for each pair of elements ω, ξ in the same coset of μ_p, l divides $a_\omega - a_\xi$.*

Proof. Let S_j as before be the cosets of μ_p. Since

$$-\zeta_j = \sum_{S_j - \{\zeta_j\}} \zeta$$

we obtain

$$\sum a_\zeta \zeta = \sum_j \sum_{\zeta \in S_j - \{\zeta_j\}} (a_\zeta - a_{\zeta_j}) \zeta.$$

Since the union of the sets $S_j - \{\zeta_j\}$ forms a basis for $\mathfrak{o}_N$ over $\mathbf{Z}$, the lemma is obvious.

Next we deal with arbitrary N. We first describe some notation. Let ω, ξ be N-th roots of unity such that for each $p|N$, ω_p and ξ_p lie in the same coset of μ_p and $\omega_p \neq \xi_p$. Let

$$\Phi(\omega, \xi)$$

denote the set of N-th roots of unity φ such that $\varphi_p = \omega_p$ or $\varphi_p = \xi_p$. For $\varphi \in \Phi(\omega, \xi)$ we define

sign $\varphi = (-1)^k$, where k is the cardinality of the set of primes p such that $\varphi_p = \omega_p$.

Lemma 2.4. *Let l be a prime number. The following two conditions are equivalent:*
  (i) *l divides $\alpha = \sum a_\zeta \zeta$.*
  (ii) *For every pair of N-th roots of unity ω, ξ such that $\omega_p/\xi_p \in \mu_p$, and $\omega_p \neq \xi_p$ for all $p|N$, the prime l divides*

$$\sum_{\varphi \in \Phi(\omega, \xi)} (\text{sign}\, \varphi) a_\varphi.$$

Proof. Assume (i). We express $\sum a_\zeta \zeta$ in terms of the special basis of Lemma 2.2, making the choice of ξ_p as the element which has been excluded from its coset relative to μ_p. By definition of the basis S in Lemma 2.2, we conclude that $\omega \in S$, i.e. ω is a basis element. It follows at once that

$$\sum_{\varphi \in \Phi(\omega, \xi)} (\text{sign}\, \varphi) a_\varphi$$

is the coefficient of ω, in the expression of $\alpha = \sum a_\zeta \zeta$ in terms of the basis S. It is then clear that (i) implies (ii).

Conversely, suppose that for every choice of ω, ξ, condition (ii) is satisfied. Write $N = \prod p^{r(p)}$. We choose one element

$$\xi_{p,j(p)} \in S_{j_p}(p)$$

from each coset of μ_p in $\mu_{p^{r(p)}}$, and we form all possible elements

$$\xi = \prod_p \xi_{p, j(p)}.$$

Given $\omega \in S$, there will be precisely one element ξ above satisfying the hypothesis of condition (ii), and

$$\sum_{\varphi \in \Phi(\omega, \xi)} (\text{sign}\, \varphi) a_\varphi$$

is the coefficient of ω in the expression of $\sum a_\zeta \zeta$ in terms of the basis S. Thus each coefficient with respect to this basis is divisible by l, whence α is divisible by l, thereby proving the lemma.

§3. Remarks on q-Expansions

As mentioned already, we shall reduce multiplicative properties of modular units to additive properties of some of their Fourier coefficients. We recall a theorem of Shimura [Sh], Theorem 3.52, p. 85.

The space of cusp forms for $\Gamma(N)$ over $\mathbf{C}$ is generated by forms whose Fourier coefficients are rational integers.

As a consequence, we obtain:

Lemma 3.1. *Let f be a modular function with divisor concentrated at the cusps. If the q-expansion of f at some cusp has algebraic coefficients in a number field K, then these coefficients have bounded denominators, and are integral at all but a finite number of primes of K.*

Proof. A conjugate g of f will have the given q-expansion at infinity. Then $\Delta^n g$ is a cusp form for suitably large n. Let $\Delta^n g = h$. The Fourier coefficients of h are a finite linear combination over K of a $\mathbf{Q}$-basis for modular forms of weight $12n$. By Shimura's theorem, it is clear that h has bounded denominators, and $g = h/\Delta^n$. Since $q^{-1}\Delta$ is invertible as a power series over $\mathbf{Z}$, the lemma follows.

Let K be a field with a valuation, assumed non-archimedean. We can extend the valuation to the power series whose coefficients have bounded valuation, by the maximal value of the coefficients (Gauss lemma). In particular, if f is such a power series, n is a positive integer, and f^n has integral coefficients for the valuation, so does f. This will be applied to coefficients of modular forms.

We let $q_{\tau/N} = q^{1/N} = e^{2\pi i \tau/N}$. We shall consider the power series field

$$\mathbf{Q}_N((q^{1/N})).$$

If

$$f = \sum \alpha_n q_N^n, \qquad \alpha_n = \alpha_n(f),$$

and $\alpha_r q^{r/N}$ is the lowest term ($\alpha_r \neq 0$), then we define

$$f^* = f/(\alpha_r q^r).$$

We call f^* the **reduced power series**, or **reduced form**, of f. If f is a modular function, then of course f^* is usually not a modular function. It is a power series in $q^{1/N}$, whose lowest term is 1. We give examples of this by recalling the q-products for the Siegel functions.

Writing $q_a = q_\tau^{a_1} e^{2\pi i a_2}$, we have

$$g_a = q^{\frac{1}{2}B_2\langle a_1 \rangle}(1-q_a)\prod_{n=1}^{\infty}(1-q_\tau^n q_a)(1-q_\tau^n/q_a),$$

where $\langle x \rangle$ is the smallest real number ≥ 0 in the residue class of $x \bmod \mathbf{Z}$, and

$$B_2(X) = X^2 - X + \tfrac{1}{6}.$$

Then for (a_1, a_2) not of period 2 $\bmod \mathbf{Z}^2$, we have:

$$g_a^* = (1-q_a) \prod (1 - q_\tau^n q_a)(1 - q_\tau^n/q_a) \quad \text{if } \langle a_1 \rangle \neq 0,$$
$$g_a^* = \prod (1 - q_\tau^n q_a)(1 - q_\tau^n/q_a) \quad \text{if } \langle a_1 \rangle = 0.$$

On the other hand, if $2a = 0$, then

$$g_a^* = \prod_{n=1}^\infty (1 - q_\tau^n/q_a).$$

In writing the above products, it is understood that a_1, a_2 are chosen equal to their representatives $\langle a_1 \rangle, \langle a_2 \rangle$. Furthermore, since $g_a = g_{-a}$ (up to a constant factor), we shall always assume that a_1 has the property

$$0 \leq a_1 \leq N/2.$$

Thus we consider only representative elements a of $(\mathbf{Q}^2/\mathbf{Z}^2)/\pm 1$.

Let N be the least common multiple of the denominators of a_1, a_2. We call N the **primitive denominator** of a, and say that a has **primitive period** N. If we have merely $Na = 0$, then we call N a **denominator**, and say that N is a **period** of a.

Suppose that a has primitive period N. We shall be interested in the coefficient of $q^{1/N}$ in the q-expansion of g_a^*. Let α_1 be this coefficient, so that

$$g_a^* = 1 + \alpha_1 q^{1/N} + \dots .$$

Then $\alpha_1 \neq 0$ only in case $a_1 = 1/N$ [remember we just excluded the case when $a_1 = (N-1)/N$ for $N > 2$]. Furthermore,

$$\text{if } a = \left(\frac{1}{N}, a_2\right) \quad \text{then} \quad \alpha_1 = -e^{2\pi i a_2}.$$

This is obvious from the q-products, and is valid also for $N = 2$ according to the special definition of §1.

From the q-products for the Siegel functions, we see that the reduced power series g_a^* have coefficients in the ring of cyclotomic integers $\mathfrak{o}_N$. It follows at once that any element g of the group generated by the Siegel functions and the non-zero constants also has a reduced form g^* of the same type, i.e. with coefficients in $\mathfrak{o}_N$.

Furthermore, if g is a power series such that some power g^l lies in that group, then g^* also has this property, in view of the Gauss lemma for power series recalled above.

We shall be interested in power products of the functions g_a. Let

$$m: \mathbf{Q}^2/\mathbf{Z}^2 \to \mathbf{Z}$$

be a function such that $m(a) = 0$ for almost all a (all but a finite number). We call m **even** if $m(a) = m(-a)$, and **normalized** if $m(0) = 0$. Throughout the sequel, we assume that such functions m are even and normalized, unless otherwise specified.

We say that a denominator d **occurs** in m if there is some a such that $m(a) \neq 0$ and $da = 0$. We say that N is a **denominator for** m if $Na = 0$ for all a such that $m(a) \neq 0$. We say that a **occurs** in m if $m(a) \neq 0$.

We define
$$g(m) = \prod g_a^{m(a)},$$
where the product is taken over $a \in (Q^2/Z^2)/\pm 1$.

Then $g(m)$ is modular. Similarly we define the power series
$$g^*(m) = \prod g_a^{*m(a)},$$
whose leading term is 1. Since we can take roots of such power series formally by means of the binomial expansion, we may then also define the power series
$$g^*(m/n) = g^*(m)^{1/n} = \prod g_a^{*m(a)/n},$$
for any positive integer n.

Let
$$A_Z = \prod Z_p$$
be the integral adeles. For any $\sigma \in \mathrm{GL}_2(A_Z)$ we define σm by
$$(\sigma m)(a) = m(a\sigma^{-1}),$$
and
$$g^*(\sigma m) = \prod g_{a\sigma}^{*m(a)}.$$

If N is a denominator for m, then the effect of σ in this last formula is determined by the image of σ in $\mathrm{GL}_2(Z/NZ)$.

With this notation, Lemma 3.1, and the above remarks (Gauss lemma), we have:

Lemma 3.2. *Let g be modular. Let l be prime. If*
$$g^l = \prod g_a^{m(a)},$$
then the power series $g^*(\sigma m/l)$ has l-integral coefficients for every $\sigma \in \mathrm{GL}_2(A_Z)$.

Finally we recall the

Distribution relation. *Given $b \in (Q^2/Z^2)/\pm 1$, and a positive integer D, there is a constant λ such that*
$$\prod_{Da=b} g_a = \lambda g_b.$$

This is a special case of the Klein-Ramachandra-Robert relations, cf. [KL III], which can be proved even more easily, being essentially immediate from the q-expansions. In [KL III] the distribution relation is stated on Q^2/Z^2, but with our adjusted definition of g_a when $2a = 0$ the relation holds as stated above on
$$(Q^2/Z^2)/\pm 1.$$

§4. The Prime Power Case

The first lemma will be used both in this case and the general composite case, providing the beginning of induction arguments.

Lemma 4.1. *Let l be a prime number. Let g be modular, and suppose that*

$$g^l = \prod g_a^{m(a)}.$$

Assume that every a occurring in m has prime period. Then there exists a representation

$$g = \lambda \prod g_a^{m'(a)},$$

where λ is constant, and every denominator occurring in m' also occurs in m.

In the light of the remarks of § 3, it will suffice to prove the following version.

Lemma 4.2. *Let l be a prime number. Assume that the power series*

$$g^*(\sigma m/l) = \prod g_{a\sigma}^{*m(a)/l}$$

has l-integral coefficients for every $\sigma \in \mathrm{GL}_2(A_\mathbf{Z})$. Assume also that every a occurring in m has prime period. Then there exists a representation

$$\prod g_a^{m(a)} = \lambda \prod g_a^{m'(a)}$$

where λ is a constant, and:
  (i) *If a occurs in m' then l divides $m'(a)$.*
  (ii) *Every denominator occurring in m' also occurs in m.*

Proof. For each prime p define

$$g^*_{(p)}(m/l) = \prod_{pa=0} g_a^{*m(a)/l}.$$

Then $g^*_{(p)}(m/l)$ is a power series in $q^{1/p}$ with leading term 1. Furthermore, we have

$$g^*(m/l) = \prod_p g^*_{(p)}(m/l).$$

It follows that the coefficients of $q^{1/p}$ in $g^*(m/l)$ and $g^*_{(p)}(m/l)$ are equal. The same also applies to $g^*(\sigma m/l)$ and $g^*_{(p)}(\sigma m/l)$.

Let us call a, b **independent** if the cyclic groups (a) and (b) generated by a and b have only 0 as their intersection.

We now prove:

If $pa = pb = 0$ then l divides $m(a) - m(b)$.

If a, b are not independent we choose c such that $pc = 0$, and such that c, a are independent. Then

$$m(a) - m(b) = m(a) - m(c) + m(c) - m(b),$$

and if l does not divide $m(a) - m(b)$, then l does not divide $m(a) - m(c)$ or $m(c) - m(b)$. Thus without loss of generality, we may assume that a, b are independent.

We then find an automorphism σ such that

$$a\sigma = \left(\frac{1}{p}, 0\right) \quad \text{and} \quad b\sigma = \left(\frac{1}{p}, \frac{1}{p}\right).$$

The coefficient of $q^{1/p}$ in $g^*_{(p)}(\sigma m/l)$ is equal to the coefficient of $q^{1/p}$ in the product

$$\prod_c \left[(1-q_c) \prod_{n=1}^{\infty} (1-q^n q_c)(1-q^n/q_c)\right]^{m(c)/l},$$

where the product is taken over c satisfying

$$c = a\sigma + v(b\sigma - a\sigma), \quad v = 0, \ldots, p-1.$$

Therefore this coefficient is equal to

$$-\sum_c \frac{1}{l} m(c) \zeta_c, \quad \text{where} \quad \zeta_c = e^{2\pi i v/p}.$$

By hypothesis, this coefficient is l-integral, and by Lemma 2.3, we conclude that l divides $m(a) - m(b)$, as desired.

Let us fix a primitive element d such that $pd = 0$. Write

$$\prod_{pa=0} g_a^{m(a)} = \prod_{pa=0} g_a^{m(a)-m(d)} \prod_{pa=0} g_a^{m(d)}.$$

The distribution relation tells us that

$$\prod_{pa=0} g_a = \lambda \text{ is constant}.$$

Having proved that l divides $m(a) - m(d)$ also proves the lemma.

Note that in the case $p = 2$, the product above is just

$$\prod_c (1-q_c) \prod_{n=2}^{\infty} (1-q^n q_c),$$

by our convention concerning the definition of g_c in the case of level 2. Again

$$-\sum \frac{1}{l} m(c) \zeta_c$$

is the appropriate coefficient, and the same proof works.

We shall now prove Theorem 1, and formulate a formal version concerning bounded denominators, as for the previous lemmas.

Theorem 3. *Let l be a prime number. Assume that the power series*

$$g^*(\sigma m/l) = \prod g_{a\sigma}^{*m(a)/l}$$

has l-integral coefficients for every $\sigma \in \mathrm{GL}_2(A_\mathbf{Z})$, and that if a occurs in m, then a has prime power denominator. Then there exists an even function m' such that

$$\prod g_a^{m(a)} = \lambda \prod g_a^{m'(a)},$$

where λ is constant, and:

   (i) *If a occurs in m' then l divides $m'(a)$.*
   (ii) *Denominators for m' can be taken among prime power denominators for m.*

Modular Function Field. IV.

Proof. By induction on the largest prime power denominator. The prime case is taken care of by Lemma 4.2. Let p^r be the largest denominator, and assume the theorem for lower cases.

We prove first:

Let a, b have period p^r. If $p(a-b)=0$, then l divides $m(a)-m(b)$.

For suppose not. If $(a-b)=(a)$, i.e. $(a)=(b)$, we choose c such that $p(a-c)=0$, $(c) \neq (a)$ and c has period p^r. Then

$$m(a)-m(b) = m(a)-m(c)+m(c)-m(b),$$

and l does not divide one of the two terms on the right. Without loss of generality, we may therefore assume that $(a) \neq (b)$.

Choose t such that $p^{r-1}t = b-a$. Then a, t generate

$$\frac{1}{p^r} \mathbf{Z}^2/\mathbf{Z}^2.$$

By taking an appropriate conjugation σ, we may assume that

$$a = \left(\frac{1}{p^r}, 0\right) \quad \text{and} \quad t = \left(0, \frac{1}{p^r}\right).$$

We shall now see that in the expansion of $g^*(m/l)$, the coefficient of q^{1/p^r} is not l-integral, which will yield a contradiction.

Again we can restrict our attention to the primitive p^r-vectors a, and in fact to the set

$$a+vt, \quad \text{with} \quad 0 \leq v \leq p^r-1.$$

The coefficient of q^{1/p^r} is equal to

$$-\sum_v \frac{1}{l} m\left(\frac{1}{p^r}, \frac{v}{p^r}\right) e^{2\pi i v/p^r}.$$

We have $a = (1/p^r, 0)$ and $b = (1/p^r, 1/p)$. Lemma 2.3 shows that $l | m(a) - m(b)$, as desired.

Given a coset B of $\left(\frac{1}{p}\mathbf{Z}^2/\mathbf{Z}^2\right) / \pm 1$ in $\left(\frac{1}{p^r}\mathbf{Z}^2/\mathbf{Z}^2\right) / \pm 1$, the distribution relation shows that

$$\prod_{b \in B} g_b = \lambda g_v, \quad \text{where} \quad v = pb, \; b \in B.$$

Let $d \in B$. Then

$$\prod_{b \in B} g_b^{m(b)} = \prod_{b \in B} g_b^{m(b)-m(d)} \prod_{b \in B} g_b^{m(d)}$$

$$= \lambda^{m(d)} \prod_{b \in B} g_b^{m(b)-m(d)} g_v^{m(d)}.$$

However,

$$\prod_B \prod_{b\in B} g_b^{(m(b)-m(d))/l}$$

has l-integral Fourier coefficients because we have seen that l divides $m(b)-m(d)$. Since v has denominator p^{r-1}, we are done by induction.

§5. The Composite Case

Theorem 4. *Let l be a prime number $\neq 2$. Let g be modular, and suppose that*

$$g^l = \prod g_a^{m(a)}.$$

Let N be a period for every a occurring in m. Then there exists a representation

$$g = \lambda \prod g_a^{m'(a)},$$

where λ is constant, and N is also a period for every a occurring in m'.

As in the prime power case, we can reformulate Theorem 5.1 only in terms of power series, because of Shimura's theorem (Lemma 3.1).

Theorem 5. *Let l be a prime $\neq 2$. Assume that the power series*

$$g^*(\sigma m/l) = \prod g_{a\sigma}^{*m(a)/l}$$

has l-integral coefficients for every $\sigma \in \mathrm{GL}_2(A_{\mathbf{Z}})$. Let N be a period for every a occurring in m. Then there exists a representation

$$\prod g_a^{m(a)} = \lambda \prod g_a^{m'(a)},$$

where λ is constant and:
 (i) *If a occurs in m' then l divides $m'(a)$.*
 (ii) *If a occurs in m' then N is a period for a.*

Proof. Again by induction on the largest primitive denominator occurring in m. We let M be maximal such that there exists a with $m(a) \neq 0$ and a has primitive period M, with $M|N$.

It will be convenient to abbreviate

$$\frac{1}{n}\mathbf{Z}^2/\mathbf{Z}^2 = Z(n) \quad \text{and} \quad Z(n)^* = \text{subset of } Z(n) \text{ of elements}$$
with primitive denominator equal to n.

We shall carry out a development applying to the 2-dimensional situation of $Z(n)$, similar to that carried out for roots of unity in §2. We let M_0 be the product of all the primes dividing M, taken to the first power. If a, a' are in the same coset of $Z(M_0)$ and $a_p \neq a'_p$ for all $p|M$, then we define

$$\Phi(a,a') = \text{set of } x \in Z(M)^* \text{ such that } x_p = a_p \text{ or } a'_p \text{ for all } p|M.$$

Observe that the condition a, a' to be in the same coset of $Z(M_0)$ means that

$$p(a_p - a'_p) = 0 \quad \text{for all} \quad p|M.$$

Modular Function Field. IV.   235

Lemma 5.1. *Let a, a' be in the same coset of $Z(M_0)$, and assume $a_p \neq a'_p$ for all $p|M$. Let m be the function of Theorem 5. Then*

$$\sum_{x \in \Phi(a,a')} (\operatorname{sign} x) m(x) \equiv 0 \pmod{l}.$$

Proof. Suppose first that $(a_p - a'_p) \cap (a_p) = 0$ for all $p|M$. We can find d primitive such that $(d) \cap (a) = 0$, and $a - a' \in (d)$. We put $b = a + d$. Then b has primitive denominator M, and

$$a' = a + k(b - a)$$

with some integer k. For some σ we have

$$a\sigma = \left(\frac{1}{M}, 0\right) \quad \text{and} \quad b\sigma = \left(\frac{1}{M}, \frac{1}{M}\right).$$

Then the coefficient α of $q^{1/M}$ in

$$g^*(\sigma m/l) = \prod_{x \in Z(N)} g_{x\sigma}^{*m(x)/l} = \ldots + \alpha q^{1/M} + \ldots$$

is

$$\alpha = \sum_c \frac{1}{l} m(c) \check{\zeta}_c,$$

where $c = a + v(b - a)$, $0 \leq v < M$, since M was chosen to be the largest primitive denominator. By Lemma 2.4 we conclude that

$$\sum_{x \in \Phi(a,a')} (\operatorname{sign} x) m(x) \equiv 0 \pmod{l},$$

and the lemma is proved in this case.

We now wish to remove the condition that $(a_p - a'_p)$ and (a_p) are independent. This is done by induction. Given a, a' we induct on the cardinality of the number of primes such that $(a_p) = (a'_p)$. We assume the result for cardinality s, and consider the case $s + 1$. Let p_0 be a prime such that $(a_{p_0}) = (a'_{p_0})$. Choose d_{p_0} of the same primitive denominator as a_{p_0}, lying in the same coset of $Z(p_0)$, i.e.

$$d_{p_0} \equiv a_{p_0} \bmod Z(p_0),$$

and such that

$$(d_{p_0}) \neq (a_{p_0}) = (a'_{p_0}).$$

Let

$$d = \sum_{p \neq p_0} a_p + d_{p_0} \quad \text{and} \quad d' = \sum_{p \neq p_0} a'_p + d_{p_0}.$$

Then by induction, all congruences being mod l, we have

$$0 \equiv \sum_{x \in \Phi(a,d')} (\operatorname{sign} x) m(x)$$

$$\equiv \sum_{x_{p_0} = a_{p_0}} + \sum_{x_{p_0} = d_{p_0}} (\operatorname{sign} x) m(x)$$

and similarly for $x \in \Phi(d, a')$ instead of $x \in \Phi(a, d')$. Then

$$\sum_{\substack{x_{p_0} = a_{p_0} \\ x \in \Phi(a, d')}} \equiv - \sum_{\substack{x_{p_0} = d_{p_0} \\ x \in \Phi(a, d')}}$$

and

$$\sum_{\substack{y_{p_0} = a'_{p_0} \\ y \in \Phi(d, a')}} \equiv - \sum_{\substack{y_{p_0} = d_{p_0} \\ y \in \Phi(d, a')}}$$

The expressions on the right of these last two congruences are the negative of each other, as one sees at once from the definition of the sign. Hence

$$\sum_{\substack{x_{p_0} = a_{p_0} \\ x \in \Phi(a, d')}} \equiv - \sum_{\substack{y_{p_0} = a'_{p_0} \\ y \in \Phi(d, a')}}$$

The definitions now show that

$$\sum_{\substack{x_{p_0} = a_{p_0} \\ x \in \Phi(a, d')}} + \sum_{\substack{y_{p_0} = a'_{p_0} \\ y \in \Phi(d, a')}} = \sum_{x \in \Phi(a, a')} (\operatorname{sign} x) m(x),$$

and the lemma is proved.

For the next lemma, we define some notation. Let $p|M$. We write

$$Z(M)^*/Z(p)$$

for the set of congruence classes of elements of $Z(M)^*$ modulo $Z(p)$.

Lemma 5.2. *Let* $m: Z(M)^* \to A$ *be a function into an abelian group, not necessarily even. Then the following two conditions are equivalent.*

(1) *For all pairs* $a, a' \in Z(M)^*$ *satisfying* $a \equiv a' \bmod Z(M_0)$ *and* $a_p \neq a'_p$ *for all* $p|M$, *we have*

$$\sum_{\substack{x \in Z(M)^* \\ x \in \Phi(a, a')}} (\operatorname{sign} x) m(x) = 0.$$

(2) *For each* $p|M$, *there exists a function*

$$\psi_p: Z(M)^*/Z(p) \to A$$

such that if $a \in Z(M)^*$ *then*

$$m(a) = \sum_{p|M} \psi_p(r_p(a)),$$

where $r_p(a)$ *is the residue class of* $a \bmod Z(p)$.

Proof. We first prove that the prime decomposition of the function m as in condition (2) implies that m satisfies the linear relations of condition (1). We write

$$\sum_{x \in \Phi(a, a')} (\operatorname{sign} x) m(x) = \sum_{x \in \Phi(a, a')} \operatorname{sign} x \sum_{p|M} \psi_p(r_p(x))$$

$$= \sum_{p|M} \sum_{x \in \Phi(a, a')} (\operatorname{sign} x) \psi_p(r_p(x)).$$

Fix a prime p_0. To each $x \in \Phi(a, a')$ we associate x' such that

$$x_p = x'_p \quad \text{if} \quad p \neq p_0 \quad \text{and} \quad x_{p_0} \neq x'_{p_0}.$$

Then

$$x \equiv x' \bmod Z(p_0) \quad \text{and} \quad \text{sign}\, x' = -\text{sign}\, x.$$

Thus elements in the sum occur in pairs, giving contributions which cancel each other, and we have shown that (2) implies (1).

We shall now prove that (1) implies (2). We write M as a prime power product,

$$M = \prod p^{v_p}.$$

We denote by $\{s_{j_p}\}$ the elements of $Z(p^{v_p})^*/Z(p)$. Thus j_p is an index for the residue classes of $Z(p^{v_p})^*$ modulo $Z(p)$. We fix an arbitrary choice of an element $a_{j_p} \in S_{j_p}$ for each j_p. We define, relative to this choice,

$X(M)$ = set of elements $x \in Z(M)^*$ such that $x_p = a_{j_p}$ for at least one prime $p|M$, and $j_p = j_p(x)$ is the index for the residue class of x itself.

Lemma 5.3. *Let $m: Z(M)^* \to A$ be a function into an abelian group, not necessarily even, and satisfying condition (1). If m_1 is another such function such that*

$$m(x) = m_1(x) \quad \text{for all} \quad x \in X(M),$$

then $m = m_1$. In other words, m is determined by its values on $X(M)$.

Proof. Suppose $y \in X(M)$ so $y_p \neq a_{j_p(y)}$ for all $p|M$. Let

$$a(y) = \sum_{p|M} a_{j_p(y)}.$$

Then

$$\sum_{z \in \Phi(a(y), y)} (\text{sign}\, z) m(z) = 0.$$

But $\Phi(a(y), y) = [\Phi(a(y), y) \cap X] \cup \{y\}$. This shows that $m(y)$ is determined by values of m on $X(M)$, thereby proving the lemma.

In order to solve for the functions ψ_p of Lemma 5.2, we shall have to consider a filtration of $X(M)$ as follows. Let t be the number of prime factors of M. We let

$X_k(M)$ = set of $x \in X(M)$ such that the number of primes $p|M$ such that $x_p = a_{j_p(x)}$ is exactly equal to k.

For each prime $p_0|M$, we define a function

$$f_{p_0}: X(M)/Z(p_0) \to \{0, \ldots, t-1\}$$

by

$f_{p_0}(y) =$ number of primes $p|M$, $p \neq p_0$ such that $y_p = a_{j_p(y)}$.

We define a function

$$f: \bigcup_{p|M} X(M)/Z(p) \to \{0, \ldots, t-1\}$$

in the natural way, since the union is disjoint. For each prime $p|M$ we let

$[X(M)/Z(p)]_k$ = set of elements $S \in X(M)/Z(p)$ such that

$$f(y) = k \quad \text{for all} \quad y \in S.$$

Then we have a disjoint union

$$X(M)/Z(p) = \bigcup_{k=0}^{t-1} [X(M)/Z(p)]_k.$$

We define ψ_p by descending induction, starting with

$[X(M)/Z(p)]_{t-1}$.

We have to satisfy

$$m(x) = \sum_{p|M} \psi_p(r_p(x)), \quad \text{all} \quad x \in X_t(M).$$

We observe that each $r_p(x)$ occurs in only one of these equations. Consequently it is possible to solve for the ψ_p in this case. Inductively, we define ψ_p on

$$\bigcup_{k=i-1}^{t-1} [X(M)/Z(p)]_k$$

for all $p|M$ such that for all $x \in X_t(M) \cup \ldots \cup X_i(M)$ ew have

$$m(x) = \sum_{p|M} \psi_p(r_p(x)).$$

If $x \in X_{i-1}(M)$ then

$$r_p(x) \in [X(M)/Z(p)]_k \quad \text{with} \quad k = i-1 \quad \text{or} \quad i-2.$$

If $k = i-1$, the definition of ψ_p is done by induction. If $k = i-2$, we have the same phenomenon as in the first step, namely each $r_p(x)$ occurs in only one of the equations

$$m(y) = \sum_{p|M} \psi_p(r_p(y)), \quad \text{with} \quad y \in X_t(M) \cup \ldots \cup X_{i-1}(M)$$

and therefore we can solve for the functions ψ_p on $[X(M)/Z(p)]_{i-1}$.

Finally, we define for any $x \in Z(M)^*$,

$$m^*(x) = \sum_{p|M} \psi_p(r_p(x)).$$

Then m^* is a solution to the system of equations (1) on $Z(M)^*$ by the first part of the proof, i.e. (2), implies (1), and agrees with m on $X(M)$. Therefore $m^* = m$, and the proof of the lemma is complete.

For the application, we of course want to deal with an even function m, in which case we also want the functions ψ_p to be even. This is taken care of by the next lemma.

Lemma 5.4. *Let l be a prime $\neq 2$, and let*

$$m: Z(M)^* \to \mathbf{Z}/l\mathbf{Z}$$

be a function which is even (i.e. $m(a) = m(-a)$), satisfying condition (1) of Lemma 5.2. Then the functions ψ_p of (2) can be selected to be even for each $p|M$.

Modular Function Field. IV.

Proof. With the functions ψ_p of (2), we define
$$\psi_p^*(y) = \tfrac{1}{2}[\psi_p(y) + \psi_p(-y)].$$
This is an even function which again satisfies (2), as desired.

We come to the main part of the proof of Theorem 5.2. We write the product of Siegel functions in the form
$$\prod g_a^{m(a)} = \prod_{a \in Z(M)^*} g_a^{m(a)} \cdot \text{other factors}.$$
We now view $m(a)$ as lying in the abelian group $\mathbf{Z}/l\mathbf{Z}$, and solve for the functions ψ_p mod l, so that we have for $a \in Z(M)^*$,
$$m(a) \equiv \sum_{p \mid M} \psi_p(r_p(a)) \bmod l.$$
We lift the functions ψ_p to $\mathbf{Z}$ in any way. Then
$$\prod_{a \in Z(M)^*} g_a^{m(a)} = \prod_{a \in Z(M)^*} \prod_{p \mid M} g_a^{\psi_p(r_p(a))} f^l,$$
where f is modular. If $pa = pb$ then $r_p(a) = r_p(b)$, and therefore
$$\prod_{a \in Z(M)^*} g_a^{m(a)} = \prod_{p \mid M} \prod_{a \in Z(M)^*} g_a^{\psi_p(r_p(a))} f^l$$
breaks up into a product over equivalence classes of elements mod $Z(p)$. In each such equivalence class, the exponent is constant. Thus for each equivalence class $S \in Z(M)^*/Z(p)$ we have
$$\prod_{a \in S} g_a^{\psi_p(r_p(a))} = \left(\prod_{a \in S} g_a \right)^{\psi_p(S)}.$$
Set $t = pa$. Suppose first that $(M/p, p) \neq 1$. Then by the distribution relations,
$$\prod_{a \in S} g_a = \lambda g_t$$
for some constant λ. If $(M/p, p) \neq 1$, then there exists $t' \in Z(M/p)^*$ such that $pt' = t$. Then by the distribution relations we have
$$\prod_{a \in S} g_a = \lambda g_{t'}/g_t.$$
Hence in either case,
$$\prod g_a^{m(a)}$$
has been expressed as an l-th power times a product of Siegel functions of lower level, and the proof of the theorem is complete by induction.

§6. Multiplicative Dependence of Siegel Functions

Theorem 6. *Suppose that a product*
$$\prod g_a^{m(a)} = \lambda$$
is constant, so modular. Let M be the largest integer which occurs in m as a primitive denominator. If $a, a' \in Z(M)^*$ satisfy $a_p \equiv a'_p \bmod Z(p)$ and $a_p \neq a'_p$ for all $p \mid M$, then
$$\sum_{x \in \Phi(a, a')} (\mathrm{sign}\, x) m(x) = 0.$$

Proof. If this sum is $\neq 0$ for some a, a', then there exists a prime l such that the sum is $\not\equiv 0 \bmod l$. This contradicts Lemma 5.1.

Remark. Shimura's theorem used via Lemma 3.1 was used only to go from Theorem 4 to Theorem 5. As Theorem 6 depends only on Theorem 5 and Lemma 5.1, we are not using Shimura's theorem in the proof of Theorem 6. As we shall see later, this means that only formal power series considerations are used in the proof that $B_{2,\chi} \neq 0$, not properties of modular forms.

Theorem 7. *Let n be a positive integer. Let $e_n \in Z(p^n)^*$. Then the functions g_a, with $a \in Z(p^n)^*$ and $a \neq e_n$, are multiplicatively independent modulo constants.*

Proof. We give the proof by induction on n, and it is convenient to pick a sequence $e = (e_1, e_2, \ldots)$ such that $e_n \in Z(p^n)^*$ and $pe_{n+1} = e_n$. We suppose that there is some product which is constant,

$$\prod g_a^{m(a)} = \lambda,$$

where the product is taken for $a \in Z(p^n)^*$ and $a \neq e_n$. Consider first the case $n = 1$. Let $M = p$ in Theorem 6. Then

$$m(a') = m(a)$$

for any pair, $a, a' \in Z(p)^*$ such that $a \neq a'$. In particular, if $a = e_1$ we have $m(e_1) = 0$ by assumption, and therefore for each a we have $m(a) = 0$ and we are done.

Assume now the theorem for n, we prove it for $n+1$. We let $M = p^{n+1}$ in Theorem 6. If $a, a' \in Z(p^{n+1})^*$ and $a \equiv a' \bmod Z(p)$, then

$$m(a) = m(a').$$

In particular, if $a \equiv e_{n+1} \bmod Z(p)$, we must have $m(a) = 0$.

Let $\{S\}$ range over the equivalence classes $\bmod Z(p)$ in $Z(p^{n+1})^*/\pm 1$. Then

$$\prod g_a^{m(a)} = \prod_S \prod_{a \in S} g_a^{m(S)}$$

$$= \lambda_1 \prod_S g_{pS}^{m(S)}$$

by the distribution relation. Here we use the notation g_{pS} for g_{pa} with any $a \in S$. If $e_n = pS$ then $m(S) = 0$. Then

$$\prod_S g_{pS}^{m(S)} = \prod g_b^{m(b)},$$

where the product is taken over $b \in [Z(p^n)^* - \{e_n\}]/\pm 1$, and we are done by induction.

Theorem 7 proves the independence of the Siegel functions in the prime power case. In the composite case, one has to use [Ku 2], and we choose a set of Siegel functions as in that paper. Using Theorem 4 one may then prove that they are multiplicatively independent in a manner analogous to the above proof. The rank of these elements is the maximal possible one. However, in [KL II] and [KL D], we show that the maximality of the rank is equivalent to the non-vanishing $B_{2,\chi} \neq 0$ for even characters χ with appropriate conductor. In this manner, using simple facts about cyclotomic integers, and the q-expansions of the Siegel functions, we have proved the non-vanishing of these generalized Bernoulli numbers. As the prime

§7. Dependence of Δ

We conclude with remarks about Δ. When does Δ belong to the group generated by the Klein forms? The question in the most general case is closely linked to the possibility of taking modular square roots from the group generated by the Siegel functions. We give a complete answer to the question in the prime power case.

Let $\mathfrak{K}_N$ be the group generated by the Klein forms $\mathfrak{k}_a$, where a has period N, modulo constants.

Theorem 8. *Suppose $N = p^r$ is a prime power. Then Δ belongs to $\mathfrak{K}_N$ if and only if $(p^2 - 1)/2$ divides 12.*

Proof. If $(p^2-1)/2$ divides 12, then $p \leq 5$ and we can write Δ explicitly as a product of Klein forms as follows. We observe that for any prime p we have

$$\prod_{a \in Z(p)^*/\pm 1} \mathfrak{k}_a = \begin{cases} \lambda \Delta^{-(p^2-1)/24} & \text{if } p \neq 2 \\ \lambda \Delta^{-3/12} & \text{if } p = 2, \end{cases}$$

where λ is constant. Indeed, the weights of both sides are equal. The expression on the left is invariant under $SL_2(\mathbf{Z})$, and if $(p^2-1)/2$ divides 12, then an appropriate integral power yields Δ.

Conversely we use induction on r. Suppose

$$\prod \mathfrak{k}_a^{m(a)} = \lambda \Delta,$$

with some constant λ, then the product $\prod g_a^{m(a)}$ is constant. Note that $\sum m(a) = 12$. It will suffice therefore to prove:

Lemma 6.1. *Suppose that*

$$\prod g_a^{m(a)} = constant,$$

and that every a occurring in m has period N where $N = p^r$ is a prime power. Then $(p^2-1)/2$ divides $\sum m(a)$.

Proof. By induction on r. Take first $r=1$. By Theorem 6, for each pair $a \neq a' \in Z(p)$ we have

$$m(a) = m(a') = m.$$

Thus

$$\sum m(a) = \sum_{a \in Z(p)^*} m = m(p^2-1)/2,$$

and the lemma is proved in this case. Assume the lemma for some r, we shall prove it for $N = p^{r+1}$. By Theorem 6, if

$$a \equiv a' \pmod{Z(p)}$$

then $m(a) = m(a')$. Let $\{S\}$ be the elements of $Z(p^{r+1})^*/Z(p)$. Then

$$\prod_{a \in Z(p^{r+1})^*} g_a^{m(a)} = \prod_S \prod_{a \in S} g_a^{m(a)} = \prod_S g_{pS}^{m(S)}.$$

However

$$\sum_{a \in S} m(a) - m(S) = (p^2 - 1)m(S) \equiv 0 \bmod p^2 - 1.$$

We are therefore done by induction.

Remark. In general, for the composite case, it can be shown that if N has prime factors p_i, $(i = 1, ..., t)$ and

g.c.d. $(p_i^2 - 1)/2$ divides 12,

then Δ belongs to $\mathfrak{R}_N$. Conversely, if Δ belongs to $\mathfrak{R}_N$, the non-2-part of the above g.c.d. must divide 3. A complete statement in this case will involve determining exactly when the square root of a modular unit is a modular unit.

In conclusion, we note that for the prime power case, the unit group in the modular function field is completely determined: It is the union of the unit groups U_{p^r} for $r = 1, 2, \ldots$. One may also ask what is the unit group at level p^r for fixed r. The theorems of this paper together with the theorems of [Ku 1], §4, answer this question completely.

The next paper of the series will put the present result, [Ku 1], [KL II], [KL III], and [KL D] together to obtain on $X(p)$ and $X_1(p)$ (for p prime) the eigenspace decomposition with respect to the Cartan group of the p-primary part of the cuspidal divisor class group, and obtain theorems entirely analogous to the classical results of Herbrand, Kummer, and more recently Ribet. Furthermore, classical difficulties of number theory will have no analogue here, so the results are essentially complete, in terms of $B_{2,\chi}$ (instead of $B_{1,\chi}$ in the cyclotomic case).

References

[Kl]   Klimek, S.: Thesis, Berkeley, 1975
[Ku 1] Kubert, D.: Quadratic relations for generators of units in the modular function field. Math. Ann. **225**, 1—20 (1977)
[Ku 2] Kubert, D.: A system of free generators for the universal even ordinary distribution on Q^{2k}/Z^{2k}. Math. Ann. **224**, 21—31 (1976)
[KL I] Kubert, D., Lang, S.: Units in the modular function field. I. Diophantine applications. Math. Ann. **218**, 67—96 (1975)
[KL II] Kubert, D., Lang, S.: Units in the modular function field. II. A full set of units. Math. Ann. **218**, 175—189 (1975)
[KL III] Kubert, D., Lang, S.: Units in the modular function field. III. Distribution relations. Math. Ann. **218**, 273—285 (1975)
[KL D] Kubert, D., Lang, S.: Distributions on toroidal groups. Math. Zeitschrift **148**, 33—51 (1976)
[KL V] Kubert, D., Lang, S.: Units in the modular function field V. The p-primary component of the cuspidal divisor class group of the modular curve $X(p)$. To appear
[N]   Newman, M.: Construction and application of a class of modular functions. Proc. London Math. Soc. **3**, 334—350 (1957)
[O]   Ogg, A.: Rational points on certain elliptic modular curves. AMS conference, St. Louis, pp. 221—231, 1972
[Sh]   Shimura, G.: Introduction to the arithmetic theory of automorphic functions. Iwanami Shoten and Princeton University Press 1971

Received February 3, 1976

PRIMITIVE POINTS ON ELLIPTIC CURVES

BY S. LANG AND H. TROTTER[1]

Communicated by Olga Taussky Todd, September 23, 1976

A well-known conjecture of Artin predicts the density of primes for which a given rational number is a primitive root (cf. the introduction to his collected works). Our purpose here is to formulate an analogous conjecture on elliptic curves A, say defined over the rationals for concreteness. Let a be a rational point of infinite order. We ask for the density of those primes p such that the group $\bar{A}(F_p)$ of rational points mod p is cyclic, generated by the reduction $\bar{a}$ of a mod p. We shall use the Galois extensions $K_l = Q(A_l, l^{-1}a)$ analogous to the splitting fields of the equations $X^l - a = 0$ when a is in the multiplicative group. We may say that *a is primitive* for such primes. We let $\langle a \rangle$ be the cyclic group generated by a.

The affine group, equal to the extension of the translation group A_l by $GL_2(l)$, operates on $l^{-1}a$. For simplicity we fix an element $u_0 \in l^{-1}a$. Then we may represent an element σ in the affine group by a pair (γ, τ) with $\gamma \in GL_2(l)$ and a translation $\tau \in A_l$, such that

$$(\gamma, \tau)u = u_0 + \gamma(u - u_0) + \tau.$$

The Galois group $\text{Gal}(K_l/Q(A_l))$ can be identified with a group of translations, subgroup of A_l, and is equal to A_l for almost all l by a theorem of Bashmakov [Ba]. If $\sigma = (\gamma, \tau)$ as above, we have

$$\sigma u = u \quad \text{if and only if} \quad (\gamma - 1)(u_0 - u) = \tau.$$

Let Δ be the discriminant of the curve. We want to give a condition on the Frobenius element $\sigma_p = (\gamma_p, \tau_p)$ in G_l when $p \nmid \Delta l$ in order that the index of $\langle \bar{a} \rangle$ in $\bar{A}(F_p)$ is divisible by l. Note that l divides the order of $\bar{A}(F_p)$ if and only if γ_p has eigenvalue 1. Furthermore, $\bar{A}(F_p) = \text{Ker}(\gamma_p - 1)$.

If $\gamma_p = 1$ then the index of $\langle \bar{a} \rangle$ is divisible by l.

Suppose on the other hand that $\text{Ker}(\gamma_p - 1)$ is cyclic of order l. Then the index of $\langle \bar{a} \rangle$ is divisible by l if and only if there exists $b \in \bar{A}$ with $lb = \bar{a}$ and b is fixed by σ_p. Indeed, if $\bar{a}$ has period divisible by l, and the index is divisible by l, then $\bar{a}$ is divisible by l in $\bar{A}(F_p)$, otherwise $\bar{A}(F_p)$ would contain $Z(l)^2$. The converse is clear. If $\bar{a}$ has period not divisible by l then $lb = \bar{a}$ for some b in $\langle \bar{a} \rangle$, so the assertion is also clear in this case.

We see that *the index of $\langle \bar{a} \rangle$ is divisible by l if and only if σ_p lies in the*

AMS (MOS) subject classifications (1970). Primary 12A75, 14G25.
[1] Both authors supported by NSF grants.

Copyright © 1977, American Mathematical Society

set S'_l consisting of all elements (γ, τ) such that γ has eigenvalue 1, and either $\gamma = 1$, or $\text{Ker}(\gamma - 1)$ is cyclic and $\tau \in \text{Im}(\gamma - 1)$.

Let $S_l = G_l - S'_l$. For any set of primes L we let $S_L = \Pi_{l \in L} S_l$. We let K_L be the compositum of all fields K_l with $l \in L$, and $G_L = \text{Gal}(K_L/\mathbf{Q})$. We let $P_{L,S}(x)$ be the set of primes $p \leq x$ such that $p \nmid \Delta$ and the Frobenius element $(p, K_l/\mathbf{Q}) \in S_l$ for all $l \in L$ and $l \neq p$. If L is finite, we have by Tchebotarev,

$$\lim_{x \to \infty} \frac{|P_{L,S}(x)|}{\pi(x)} = \frac{|S_L \cap G_L|}{|G_L|} = \delta_L(S).$$

It is easy to see that the limit $\delta(S) = \text{Lim}_L \delta_L(S)$ exists (for L increasing to include all primes). The conjecture states that the limit is equal to the density of primes for which a is primitive.

There exists a finite set M of primes such that for any L containing M we have

$$\delta_L = \delta_M \prod_{l \in L - M} \delta_l, \quad \text{where} \quad \delta_l = 1 - \frac{|S'_l \cap G_l|}{|G_l|}.$$

This allows for effective computation of the conjectured density. One finds in all cases that $\delta_l = 1 + O(1/l^2)$, so that the product is absolutely convergent. When there is no complex multiplication, for instance, and G_l is the full affine group (which occurs for almost all l), then one finds

$$\delta_l = 1 - \frac{|S'_l|}{|G_l|} \quad \text{where} \quad \frac{|S'_l|}{|G_l|} = \frac{l^3 - l - 1}{l^2(l^3 - l^2 - l + 1)}.$$

We have computed numerical values for some "Serre curves" (cf. [LT]) whose Galois group of torsion points is of index 2 in the full product of all $GL_2(\mathbf{Z}_l)$, for instance the three curves

$$y^2 + y = x^3 - x, \quad y^2 + y = x^3 + x^2, \quad y^2 + xy + y = x^3 - x^2.$$

In each case the point of infinite order is the point $(0, 0)$, and the predicted density comes within three decimals of 0.440. Among the first 200 primes, the expected number is then 88, and the actual numbers are 91, 96, 91 respectively. For 180 primes (leaving out the first 20) the expected number is 79.2, and the actual numbers are 79, 84, 78 respectively, a good fit. One can of course handle in a similar way the density of primes p such that $\bar{A}(\mathbf{F}_p)$ is cyclic (forget about a).

We have also considered the more general problem dealing with a free subgroup Γ of rational points, rather than an infinite cyclic one. Let M be an integer > 1. We wish to characterize the possibility that the image under reduction

$$\Gamma \to \bar{A}(\mathbf{F}_p)/M\bar{A}(\mathbf{F}_p)$$

is surjective by a condition on the Frobenius element σ_p. For simplicity, fix a section $\lambda\colon \Gamma \to M^{-1}\Gamma$ such that $M\lambda a = a$ for all $a \in \Gamma$. This corresponds to choosing u_0 when Γ is infinite cyclic, and determines a homomorphism $\tau\colon \Gamma \to A_M$ such that

$$\sigma u = \gamma(u - \lambda u) + \lambda M u + \tau M u$$

for $u \in M^{-1}\Gamma$. We can identify an element σ of the affine group in this case with a pair (γ, τ). We see this time that $\sigma u = u$ if and only if $(\gamma - 1)(u - \lambda M u) = -\tau M u$. If we let T_p be the kernel of the map

$$\Gamma \to \bar{A}(\mathbf{F}_p)/M\bar{A}(\mathbf{F}_p),$$

then $T_p = \tau_p^{-1}((\gamma_p - 1)A_M)$, where $\sigma_p = (\gamma_p, \tau_p)$ is the Frobenius element. Note that we have an equality of indices

$$(\bar{A}(\mathbf{F}_p)\colon MA(\mathbf{F}_p)) = |(\bar{A}(\mathbf{F}_p) \cap \bar{A}_M)| = |\mathrm{Ker}(\gamma_p - 1)|.$$

In particular, take $M = l$ prime. We find:

The index of $\bar{\Gamma}$ in $\bar{A}(\mathbf{F}_p)$ is prime to l if and only if

$$\dim \tau(\Gamma) - \dim(\tau(\Gamma) \cap \mathrm{Im}(\gamma - 1)) = \dim \mathrm{Ker}(\gamma - 1).$$

We define the bad set S'_l to consist of those $\sigma = (\gamma, \tau)$ such that either $\mathrm{Ker}(\gamma - 1)$ is cyclic and $\tau(\Gamma) \subset \mathrm{Im}(\gamma - 1)$, or $\mathrm{Ker}(\gamma - 1) = A_l$ and rank $\tau(\Gamma) = 0$ or 1. Then we obtain:

The index of $\bar{\Gamma}$ in $\bar{A}(\mathbf{F}_p)$ is divisible by l if and only if the Frobenius element σ_p lies in S'_l (for $p \nmid \Delta l$).

The same type of limit as before yields the conjectured density of primes such that $\bar{\Gamma} = \bar{A}(\mathbf{F}_p)$, (say for which Γ is primitive).

Trying to prove the conjecture from the Riemann hypothesis in line with Hooley's work for the Artin case [H] met difficulties having to do with the larger degrees, behaving like $>><< l^6$ (or l^4 in the complex multiplication case) rather than l^2 in the Artin case. It also leads to the problem of proving the analogue of the Brun-Titchmarsh theorem, to given an upper bound for the number of primes $p \leq x$ such that the Frobenius at p operating on A_l has eigenvalue 1.

BIBLIOGRAPHY

[Ba] M. I. Bašmakov, *Un théorème de finitude sur la cohomologie des courbes elliptiques*, C. R. Acad. Sci. Paris Sér. A–B **270** (1970), A999–A1101. MR **42** #4548.

[Go] L. Goldstein, *Analogues of Artin's conjecture*, Trans. Amer. Math. Soc. **149** (1970), 431–442. MR **43** #4792.

[Ho] C. Hooley, *On Artin's conjecture*, J. Reine Angew. Math. **225** (1967), 209–220. MR **34** #7445.

[LT] S. Lang and H. Trotter, *Frobenius distributions in GL_2-extensions*, Lecture Notes in Math., vol. 504, Springer-Verlag, Berlin and New York, 1975.

[We] P. Weinberger, *A counterexample to an analogue of Artin's conjecture*, Proc. Amer. Math. Soc. **35** (1972), 49–52. MR **45** #8630.

DEPARTMENT OF MATHEMATICS, YALE UNIVERSITY, NEW HAVEN, CONNECTICUT 06520

DEPARTMENT OF MATHEMATICS, PRINCETON UNIVERSITY, PRINCETON, NEW JERSEY 08540

ISBN 0-387-98803-3